GROWTH, NUTRITION,

AND METABOLISM OF CELLS IN CULTURE

Volume I

CONTRIBUTORS

Vincent J. Cristofalo

Norman C. Dulak

Charles T. Gregg

William Kelley

Paul M. Kraemer

William F. McLimans

William J. Mellman

M. K. Patterson, Jr.

John Paul

Robert W. Pierson, Jr.

George H. Rothblat

Arthur A. Spector

Howard M. Temin

Charity Waymouth

GROWTH, NUTRITION, AND METABOLISM OF CELLS IN CULTURE

EDITED BY

George H. Rothblat and Vincent J. Cristofalo

WISTAR INSTITUTE OF ANATOMY AND BIOLOGY
PHILADELPHIA, PENNSYLVANIA

VOLUME I

 1972

ACADEMIC PRESS   New York and London

ACADEMIC PRESS, INC.
111 Fifth Avenue, New York, New York 10003

United Kingdom Edition published by
ACADEMIC PRESS, INC. (LONDON) LTD.
24/28 Oval Road, London NW1 7DD

LIBRARY OF CONGRESS CATALOG CARD NUMBER: 71-182625

PRINTED IN THE UNITED STATES OF AMERICA

CONTENTS

8. Fatty Acid, Glyceride, and Phospholipid Metabolism

Arthur A. Spector

9. Cellular Sterol Metabolism

George H. Rothblat

10. Human Diploid Cell Cultures: Their Usefulness in the Study of Genetic Variations in Metabolism

William J. Mellman and Vincent J. Cristofalo

Contents

LIST OF CONTRIBUTORS

Numbers in parentheses indicate the page on which the authors' contributions begin.

VINCENT J. CRISTOFALO, Wistar Institute of Anatomy and Biology, Philadelphia, Pennsylvania (327)

NORMAN C. DULAK, McArdle Laboratory, University of Wisconsin, Madison, Wisconsin (49)

CHARLES T. GREGG, Biomedical Research Group, Los Alamos Scientific Laboratory, University of California, Los Alamos, New Mexico (84)

WILLIAM N. KELLEY, Division of Rheumatic and Genetic Diseases, Duke University Medical Center, Durham, North Carolina (211)

PAUL M. KRAEMER, Biomedical Research Group, Los Alamos Scientific Laboratory, University of California, Los Alamos, New Mexico (371)

WILLIAM F. McLIMANS, Roswell Park Memorial Cancer Institute, New York State Department of Health, Buffalo, New York (137)

WILLIAM J. MELLMAN, Department of Pediatrics and Medical Genetics, University of Pennsylvania School of Medicine, Philadelphia, Pennsylvania (328)

M. K. PATTERSON, JR., Biomedical Division, The Samuel Roberts Noble Foundation, Inc., Ardmore, Oklahoma (171)

JOHN PAUL, Beatson Institute for Cancer Research, Royal Beatson Memorial Hospital, Glasgow, Scotland (1)

ROBERT W. PIERSON, JR., McArdle Laboratory, University of Wisconsin, Madison, Wisconsin (49)

GEORGE H. ROTHBLAT, Wistar Institute of Anatomy and Biology, Philadelphia, Pennsylvania (298)

ARTHUR A. SPECTOR, Department of Internal Medicine and Biochemistry, The Clinical Research Center, University of Iowa, Iowa City, Iowa (258)

HOWARD M. TEMIN, McArdle Laboratory, University of Wisconsin, Madison, Wisconsin (49)

CHARITY WAYMOUTH, The Jackson Laboratory, Bar Harbor, Maine (12)

PREFACE

The use of cell and tissue culture as a tool for the study of a wide variety of fundamental biological problems has grown rapidly over the last two decades. Early in this period one of the major uses of cells in culture was for virus research. These studies have yielded considerable information about the basic molecular biology of viruses, as well as of the mechanisms of viral pathogenesis, infectivity, and immunity.

Concurrently, it was recognized that cells in culture could be used as an experimental system for studying the cell under controlled environmental conditions, and a steady stream of research has been directed at describing and understanding details of the biochemistry and physiology of cells in culture and their relationship to the *in vitro* environment. As a result of this basic work, new areas of research are now emerging in which problems which were heretofore unapproachable can be studied with cell cultures. For example, the methodology for the cultivation of normal human cells has become almost routine so that now the nature of a wide variety of human diseases and inborn metabolic errors can be elucidated at the molecular level. Techniques for the cultivation of differentiated functional systems, such as muscle and nerve, are now available. The cultivation of cells from poikilothermic vertebrates, invertebrates, and plants has added another dimension to the use of cell cultures in biological research.

The potential of cell culture can be fully realized only if the vast and complex literature which has accumulated can be critically evaluated and summarized. Therein lies the scope and purpose of these volumes. We have attempted to bring together in this two-volume treatise a comprehensive series of reviews that summarize the current status of knowledge of the growth, nutrition, and metabolism of various types of cell cultures. The chapters are both detailed and comprehensive enough for the specialist and broad enough to provide a general background for the nonspecialist.

Volume I is comprised of contributions that describe the uptake, synthesis, and degradation of biologically important compounds, particularly the major components usually present in tissue culture medium. Volume II deals with specialized mammalian, plant, and invertebrate cell systems and techniques. In these chapters the culturing of specific classes of cells, including the establishment of their special nutritional requirements and metabolic features, are discussed.

We hope that these two volumes will meet the needs of investigators who routinely use cell culture techniques, as well as those of students and individuals in associated areas of cell and molecular biology. If they do, it is because of the efforts of the authors who contributed their chapters with care and enthusiasm. We wish to thank them all. In addition we wish to give our special thanks to Dr. H. Koprowski of the Wistar Institute whose interest and enthusiasm encouraged us to undertake this task.

GEORGE ROTHBLAT
VINCENT J. CRISTOFALO

CONTENTS OF VOLUME II

1

GENERAL INTRODUCTION

John Paul

In few areas of science is it as easy as in the field of tissue culture research to discern how the structure of present knowledge is based on the individual contributions of a large number of investigators. The goal has always been obvious to thinking biologists: the precise control of the environment in which cells from multicellular organisms grow so that meaningful quantitative experiments can be done. The questions raised by the behavior of these cells have such profound significance, relating, as they do, to the physical nature of all plants and animals, including man himself, that the field has attracted some of the best scientific minds of this century. Among the Nobel Prize winners who have done research with tissue culture at one time or another are Carrel, Warburg, Lippman, Enders, Crick, Watson, and Medawar. Some of these scientists stayed on in the field to make contributions to it, while others, emulating the man who is generally given credit for initiating modern tissue culture, made their contributions and then turned to other things.

This founder of modern tissue culture was Ross Grenville Harrison, whose paper "Observation on the Living Developing Nerve Fiber" appeared in the *Proceedings of the 23rd Meeting of the Society for Experimental Biology and Medicine,* reported in *The Anatomical Record* of June 1, 1907. In this paper he described how he took fragments of the medullary tube from frog embryos and implanted them in a lymph clot. He observed these cultures frequently and was able to show that, as nerve cells developed, they formed fibers which grew out into the lymph clot. This not only resolved a dispute about the origin of nerve fibers but it demonstrated, in a most dramatic way, that cells could survive and develop in tissue culture and could be used effectively to tackle biological questions.

Harrison's experiments were not the first attempts to maintain tissues *in vitro*. In the 1880s chick embryos had been successfully maintained in saline by Roux (1885), and Arnold (1887) had studied amphibian lymphocytes in culture. Hence, successful whole embryo culture and culture of hemopoietic cells preceded Harrison's experiments by some twenty years. His work is, however, rightly considered the point of departure for modern tissue culture because it created so much excitement and interest that the technique was taken up by many scientists immediately thereafter. From then until now contribution upon contribution has been steadily added to the impressive edifice which is the technology of modern cell and tissue culture.

Not very long after Harrison's experiments people began to think about carrying out biochemical studies with tissue in culture. Looking back, we can admire the courage and imagination of the early pioneers, but the magnitude of the problems of cellular biochemistry was not at all appreciated fifty years or so ago and quite naturally the questions asked were naive by today's standards. This implies no criticism of the early investigators; the questions we are asking today may well seem naive in another twenty or thirty years. The reason for remarking on it is simply that it enables us to recognize a rather clear watershed in studies on the nutrition and metabolism of cells in culture, which occurred in the early 1950s. In the early days of tissue culture the biochemical questions being asked were on the whole of a very general nature. Workers were concerned with trying to define the general nutritional requirements of animal cells and studying rather general parameters of metabolism, such as respiration. What distinguishes the modern era, starting in the mid-1950s, is the exploitation of cell and tissue culture techniques to answer specific problems, especially in the fields of virology and molecular and cellular biology. This transition was a result of different factors. First, there was the impact of the biochemical knowledge which had been accumulating with accelerating speed since the mid-1930s and which at last defined the true dimensions of the problems. Then there was a revolution in cell culture techniques, partly a result of improvements developed by cell culturists themselves and partly the result of a great invasion of the field by virologists. Finally, there was the emergence of the discipline of molecular biology: Cell cultures have now joined the T-even phages and *E. coli* in the molecular biologists' armamentarium.

The early development of cell culture nutrition is outlined in some detail by Dr. Waymouth (herself a pioneer of biochemical studies in tissue culture) in Chapter 2, Volume 1. The credit for foreseeing the importance of controlling the cellular environment and defining it in chemically precise terms should probably be given to Lewis and Lewis (1911a,b), who in the years before World War I carried out experiments on the use of quite simple media for culturing tissues. The subject really got under way in the mid-1930s when Vogelaar and Erlichman (1933) and Baker (1936) produced media which are recognizable as precursors of today's media. A major advance was the introduction by Fischer (1941) of the idea of using dialyzed plasma as a basal medium to identify the small molecular components needed to supplement it. The medium which he and his colleagues (Fischer *et al.*, 1948) published has considerable similarities to some of those now commonly used, as has the medium published in the following year by White (1949).

At that time it was fashionable to recognize "synthetic" and "analytical" approaches to defining cellular nutrients. The former, of course, implied the arbitrary inclusion in mixtures of substances which had been demonstrated to be of some metabolic importance from general biochemical research; the latter implied the demonstration in tissue culture medium of nutrient substances. The distinction was always highly artificial and is never made now. It may be remarked, however, that the "synthetic" approach on the whole proved more fruitful.

In the decade following the publication of Fischer's and White's media many synthetic mixtures were published; finally, in 1955, Eagle published the first version of his medium. This was based on studies similar to Fischer's and was designed to contain only those small molecular components which were necessary to maintain the growth of some common cell lines. It came at a time when virologists were using tissue cultures intensively and when sophisticated biochemical studies were beginning to be undertaken; moreover, it proved to be a satisfactory general medium for most purposes. Hence, although many different media have been developed since that time, the appearance of Eagle's medium represents the culmination of the era of developing media for general purposes. A more recent trend has been the development of media for special purposes, and this will be discussed later.

One important general question of cell nutrition has, however, still not been answered. From the early studies of Fischer and later of Eagle, it became apparent that most cells had a requirement not only for small molecular species, but also for certain macromolecules present in serum. Although a few cells, in special circumstances, can grow without serum, virtually all animal cells, particularly primary cells, are dependent on the presence of factors in serum for growth. The role and nature of these serum factors have therefore been studied intensively. They have variously been identified as α-glycoproteins, fetuin, macroglobulins, and serum-bound small molecules. Until relatively recently serum was thought to have a somewhat nonspecific effect in cell nutrition, but much new interest has been engendered by the observation that serum is in some way connected with the phenomenon of density control, i.e., the phenomenon which leads to a cessation of growth in cells when they have reached a certain density. Some aspects of this problem are discussed by Drs. Temin, Pierson, and Dulak in Chapter 3, Volume 1.

Even before reliable defined media for tissue culture cells were developed, general studies of the metabolism of tissue cultures were being conducted. Many of these had to do with carbohydrate metabolism. It was established quite early that glucose was the main source of energy for cells in culture (Lewis, 1922; Krontowski and Jazimirska-Krontowska, 1926; Krontowski and Bronstein, 1926; Warburg, 1930), and the quantitative requirements for oxygen were determined (Laser, 1933; Warburg and Kubowitz, 1927). In those early days, too, it was demonstrated that energy could be derived by deamination of amino acids (Holmes and Watchorn, 1927; Warburg and Kubowitz, 1927) and at about the same time the distinction between glycolytic and oxidative metabolism was made. Warburg (1930) reported that tumors almost invariably exhibited higher glycolysis than normal tissues. This claim was the subject of contention for many years and was satisfactorily resolved only when improved tissue culture technology made it possible to maintain cells in strictly controlled conditions. The reason for the difference is still not clear.

Two other general questions which excited interest in the early phases of cellular studies were whether respiration was essential for survival of vertebrate cells *in vitro,* and whether there was any special relationship of respiration to the cell cycle (in view of experiments which seemed to indicate that dependence on an intact electron transport system was absolute up to a certain stage of the cell cycle). It was rather clearly established that many cells in culture can go through a cell cycle in the presence of high concentrations of inhibitors of electron transport systems (Pomerat and Willmer, 1939) but the question of whether oxygen requirement is absolute in all conditions for prolonged maintenance of all eukaryotic cells is still open.

With the elucidation in detail of many metabolic pathways, most of these general questions assumed less importance and were replaced by more precise questions concerning the presence and amounts of different enzymes and inhibitors in specific kinds of cells. In Chapter 4, Volume 1, Dr. Gregg reviews our knowledge of energy metabolism in mammalian cells, and in Chapter 5, Volume 1, Dr. McLimans discusses the importance of the gaseous environment, while some special aspects of carbohydrate metabolism are also discussed in Chapters 10 and 11, Volume 1.

In the late 1940s increasing interest began to be taken in two specific fields of biochemical interest, nucleic acid and protein metabolism.

The earliest questions relating to protein metabolism arose in connection with studies on cell nutrition; they were concerned mainly with the importance of amino acids, serum proteins, and peptides in nutrition. Surprisingly, only in one of these areas have reasonably complete answers been obtained to date, in relation to the uptake of amino acids and their utilization in protein synthesis. This subject is reviewed by Dr. Patterson in Chapter 6, Volume 1.

Experiments on the behavior of nucleic acids in cells in culture were already carried out in the 1940s (Davidson, 1947; Davidson et al., 1949). With the enunciation of Watson and Crick's theories concerning the nonconservative nature of DNA replication, important questions arose about DNA in mammalian cells. Did it behave during replication like the DNA of viruses and bacteria? Experiments by Graham and Siminovitch (1957) and Thomson et al. (1956, 1958), in which the metabolic stability of DNA and RNA was studied, provided a positive answer to this question. These experiments also demonstrated the potential use of cell cultures in tackling problems of this kind, and initiated the long series of studies in many laboratories which has provided us with an extensive understanding of the kinetics of synthesis of DNA and RNA in eukaryotic cells. A particularly important aspect of this field, which has attracted much attention recently, is the detailed metabolism of purines and pyrimidines, since, beside its own intrinsic interest, this knowledge has been turned to very good effect in the isolation and identification of mutants of cell lines which can be used in somatic cell genetic studies. This field is reviewed by Dr. Kelley in Chapter 7, Volume 1.

The role of lipids in cellular metabolism has also been the subject of intermittent interest since the early observations that cells in culture often accumulated lipid droplets. Few of the early studies were very revealing, and it was only with the relatively recent acquisition of detailed knowledge of lipid and steroid metabolism that the kinds of studies outlined by Dr. Spector and Dr. Rothblat in Chapters 8 and 9, Volume 1, became possible.

Culturing animal tissues in defined media did not by any means solve all the problems of studying their metabolism. It soon emerged that the patterns of metabolism in isolated cells were very sensitive to changes in the immediate environment, especially changes in oxygen tension, pH, glucose levels, and so on. It has emerged that for many studies it is necessary to maintain a rather constant environment, for example, by continuous perfusion, as discussed by Dr. Kruse in Chapter 2, Volume 2.

Until quite recently most of the interest in cell culture metabolism was concentrated on investigating special questions in established cell lines which have few characteristics distinguishing one from another. A marked trend in the past few years, which may well mark the beginning of a new era, is the posing of similar questions about specific cell types. For this reason there has been much interest in culturing specific classes of cells, in establishing their special nutritional requirements, and in studying those metabolic features which characterize them.

The study of differentiated cells is not, of course, a new idea. Ross Harrison's very first experiment, and even those experiments which preceded Harrison's work, was aimed at culturing specific kinds of tissue, and a great deal of the literature up to 1950 is concerned with the culturing of special cell types, such as pigmented retinal cells, contracting cardiac muscle cells, nerve cells, and so on. However, interest in the metabolism of special tissues has tended to be swamped during the past twenty years by studies with cell cultures because of the obvious simplicity of working with them and the clear-cut results which can be obtained with them. Nevertheless, during this era (which I sometimes consider the "microbiological phase" of tissue culture), a good deal of work continued with primary cultures. Cytogeneticists, in particular, were interested in them, since established cells develop gross karyotypic abnormalities. Virologists also recognized that the spectrum of viral susceptibility differed between primary and established cell lines; tumor virologists, in particular, recognized that the two kinds of cell lines could behave in quite different ways and that primary lines could be transformed with tumor viruses. This phenomenon was originally demonstrated by Temin and Rubin (1958) and has generated an enormous amount of interest in primary lines. These primary lines were of a relatively undistinguished kind, most of them being fibroblasts. A number of workers did, however, appreciate the value of maintaining cells or cell lines with the characteristics of differentiated tissues. Sato's group (Sato and Buonassisi, 1964; Sato and Yasumura, 1966), interested in some of the special metabolism of some differentiated tissues such as adrenal cortex, pursued studies on growing cells of hormonal tissue in continuous culture. Some embryologists also maintained interest in culturing differentiated cells. For example, some studies by Konigsberg (1963), Cahn and Cahn (1966), and Coon (1966) showed that differentiated cells would remain differentiated throughout successive clonings. Partly because of demonstrations of this kind, partly because many of the general questions of cell culture have now been answered, and partly because techniques in general are so much better, there has been a recent renaissance of interest in the behavior of specific tissues in culture; some aspects of these studies are covered in Chapters 3–6, Volume 2.

The culture of early embryos, which was one of the very first challenges to tissue culturists, has seen a very great revival of interest. In particular, cultivation of mammalian embryos has received a great deal of attention in the past decade and the procedure has become so reliable that biochemical studies have been carried out in some detail. Dr. Brinster reviews some of this work in Chapter 7, Volume 2.

In most of Volumes 1 and 2 the discussion centers on studies with mammalian and avian cells; indeed by far the greatest volume of research in tissue culture has been done with them. However, there has always been a good deal of interest in culturing other materials. Cells from poikilothermic vertebrates and arthropods have turned out to be cultivable using the same principles as for cells of warm-blooded vertebrates. An increasing amount of work has been done with them recently; some features of these studies are the subjects of Chapters 8 and 9, Volume 2.

Plant tissue culture has grown up as a separate discipline from animal tissue culture, partly because the nutritional requirements of plants are so very different and partly because botanists and zoologists have always tended to go their separate ways. Within the past fifteen years the revelation that plant cells and animal cells have more in common than had previously been realized has tended to bring them together again. It is therefore appropriate that they should have some mention in these volumes otherwise devoted to animal tissue culture (Chapter 10, Volume 2).

Volumes 1 and 2 are intended to provide a review of the present state of studies on nutrition and metabolism of cells in culture in the year 1971. Where are these studies likely to lead us in the future? Insofar as nutrition is concerned, it is now rather clear that general requirements of continuous cell lines for low-molecular-weight substances have been defined and that only details of individual variations have to be added. Future developments are most likely to have to do with the elucidation of special requirements for special cells. For example, it is possible that cells from hormone-dependent tissues may require hormones for continued survival *in vitro*. Moreover, the role of serum factors in controlling growth may be tied up with the chalones and similar substances which are thought to regulate the growth of cells in animals; within the next decade our ideas about the nutrition of cells from special tissues may have to take such substances into account. Perhaps with increasing interest in cultivation of special tissues and more knowledge of substances of this kind, we shall see an improved ability to grow specialized cells from a wide variety of organs. This may have enormous implications in many fields; an obvious outcome will be the development of special metabolic studies on each of these tissues. For example, some of the problems of hormonal control of metabolism are likely to be worked out in detail in cell types from specific target organs.

One further development which one feels is now predictable, with the development of reliable means for cloning cells and selecting mutants, is the analysis of metabolic pathways by genetic means, as has been done in microorganisms. We already have examples of a number of mutant cell lines in which specific enzymes are missing. These can be highly informative in working out details of metabolic relationships, particularly of control mechanisms. Their increased use for this purpose seems highly likely.

In studies of metabolism in cell cultures the general movement seems to have been a progression from the general to the particular—from studies of general questions to studies of particular ones and from studies of general cellular metabolism to studies of the metabolism of specific cell types. These volumes illustrate these trends with, on the one hand, reviews of some of the new rather well-established general principles of cell metabolism, and, on the other, accounts of studies with special cell types. It seems rather likely that in the future the acquisition of this detailed information will lead to the enunciation of new sets of generalizations. These are often the greatest rewards of research, for which investigators should always be seeking.

REFERENCES

Arnold, J. (1887). Ueber Theilungsvorgänge an den Wanderzellen, ihre progressiven und regressiven Metamorphosen. *Arch. Mikrosk. Anat.* **30**, 205–310.

Baker, L. E. (1936). Artificial media for the cultivation of fibroblasts, epithelial cells and monocytes. *Science* **83**, 605–606.

Cahn, R. D., and Cahn, M. B. (1966). Heritability of cellular differentiation: Clonal growth and expression of differentiation in retinal pigment cells *in vitro*. *Proc. Nat. Acad. Sci. U.S.* **55**, 106–114.

Coon, H. G. (1966). Clonal stability and phenotypic expression of chick cartilage cells *in vitro*. *Proc. Nat. Acad. Sci. U.S.* **55**, 66–73.

Davidson, J. N. (1947). Some factors influencing the nucleic acid content of cells and tissues. *Cold Spring Harbor Symp. Quant. Biol.* **12**, 50–59.

Davidson, J. N., Leslie, I., and Waymouth, C. (1949). The nucleoprotein content of fibroblasts growing *in vitro*. 4. Changes in the ribonucleic acid phosphorus (RNAP) and deoxyribonucleic acid phosphorus (DNAP) content. *Biochem. J.* **44**, 5–17.

Eagle, H. (1955). The specific amino acid requirements of a mammalian cell (strain L) in tissue culture. *J. Biol. Chem.* **214**, 839–852.

Fischer, A. (1941). Die Bedeutung der Aminosäuren für die Gewebzellen *in vitro*. *Acta Physiol. Scand.* **2**, 143–188.

Fischer, A., Astrup, T., Ehrensvard, G., and Oehlenschlager, V. (1948). Growth of animal tisue cells in artificial media. *Proc. Soc. Exp. Biol. Med.* **67**, 40–46.

Graham, A. F., and Siminovitch, L. (1957). Conservation of RNA and DNA phosphorus in strain L (Earle) mouse cells. *Biochim. Biophys. Acta* **26**, 427–428.

Harrison, R. G. (1907). Observations on the living developing nerve fiber. *Proc. Soc. Exp. Biol. Med.* **4**, 140–143.

Holmes, B. E., and Watchorn, E. (1927). Studies in the metabolism of tissues growing *in vitro*. I. Ammonia and urea production by kidney. *Biochem. J.* **21**, 327–334.

Konigsberg, I. R. (1963). Clonal analysis of myogenesis. *Science* **140**, 1273–1284.

Krontowski, A. A., and Bronstein, J. A. (1926). Stoffwechselstudien an Gewebskulturen. I. Mikrochemische Untersuchungen des Zucherverbrauchs durch explantate aus normalen Geweben und durch Krebsexplantate. *Arch. Exp. Zellforsch. Besonders Gewebezuecht.* **3**, 32–57.

Krontowski, A. A., and Jazimirska-Krontowska, M. C. (1926). Stoffwechselstudien an Gewebskulturen. II. Über Zucherverbrauch durch Gewebskulturen eines mittels Passagen nach Carrel *in vitro* gezuchteten reinen Fibroblastenstammes. *Arch. Exp. Zellforsch. Besonders Gewebezwecht.* **5**, 114–124.

Laser, H. (1933). Der Stoffwechsel von Gewebskulturen und ihr Verhalt in der Anaerobiose. *Biochem. Z.,* **264**, 72–86.

Lewis, M. R. (1922). Importance of dextrose in the medium for tissue cultures. *J. Exp. Med.* **35**, 317–322.

Lewis, M. R., and Lewis, W. H. (1911a). The cultivation of tissues from chick embryos in solutions of NaCl, $CaCl_2$, KCl and NaHCO. *Anat. Rec.* **5**, 277–293.

Lewis, M. R., and Lewis, W. H. (1911b). The growth of embryonic chick tissues in artificial media, agar and bouillon. *Bull. Johns Hopkins Hosp.* **22**, 126–127.

Pomerat, C. M., and Willmer, E. N. (1939). Studies on the growth of tissues *in vitro*. VII. Carbohydrate metabolism and mitosis. *J. Exp. Biol.* **16**, 232–249.

Roux, W. (1885). Beiträge zur Entwicklungsmechanik des Embryo. *Z. Biol.* (*Munich*) **21**, 411–524.

Sato, G., and Buonassisi, V. (1964). Hormone secreting cultures of endocrine tumor origin. *Nat. Cancer Inst., Monogr.* **13**, 81–91.

Sato, G. H., and Yasumura, Y. (1966). Retention of differentiated function in dispersed cell culture. *Trans. N.Y. Acad. Sci.* [2] **28**, 1063–1079.

Temin, H. M., and Rubin, H. (1958). Characteristics of an assay for Rous sarcoma virus and Rous sarcoma cells in tissue culture. *Virology* **6**, 669–688.

Thomson, R. Y., Paul, J., and Davidson, J. N. (1956). Metabolic stability of DNA in fibroblast cultures. *Biochim. Biophys. Acta* **22**, 581–583.

Thomson, R. Y., Paul, J., and Davidson, J. N. (1958). The metabolic stability of the nucleic acids in cultures of a pure strain of mammalian cells. *Biochem. J.* **69**, 553–561.

Vogelaar, J. P. M., and Erlichman, E. (1933). A feeding solution for cultures of human fibroblasts. *Amer. J. Cancer* **18**, 28–38.

Warburg, O. (1930). "The Metabolism of Tumours." Constable, London.

Warburg, O., and Kubowitz, F. (1927). Stoffwechsel wachsender Zellen. *Biochem. Z.* **189**, 242–248.

White, P. R. (1949). Prolonged survival of excised animal tissue *in vitro* in nutrients of known constitution. *J. Cell. Comp. Physiol.* **34**, 221–241.

2

CONSTRUCTION OF TISSUE CULTURE MEDIA

Charity Waymouth

I. Brief History

The study of cell behavior apart from the normal tissue relationships in the parent organism was first practiced by Arnold (1887), who isolated leukocytes from the dorsal lymph sac of the frog in lymph-soaked elder pith and observed their movements microscopically for 4 or 5 days (see also Hughes, 1952). The detailed study of living nerve fibers made by Harrison (1906–1907, 1908), which has earned him the title of "father" of the tissue culture technique, employed frog lymph as supporting and nutrient medium (see also Wilens, 1969). Harrison's near contemporaries, Warren and Margaret Lewis, foresaw the importance of controlling and defining the environment of cells *in vitro* (M. R. Lewis and Lewis, 1911a,b; W. H. Lewis and Lewis, 1912). The interacting effects of cells and medium upon each other in a dynamic system was emphasized by Carrel (1913a). Burrows, a student of Harrison, introduced the use of clotted blood plasma as a support for explanted tissues from avian and mammalian sources (Burrows, 1910, 1911) and, in collaboration with Carrel (1912, 1913b; Burrows, 1916–1917), developed the method of growing cultures in plasma and embryonic extracts. Carrel regarded the coagulum of chicken plasma mainly as a physical substratum (Carrel, 1928) and the extracts of embryonic tissues as the source of nutrients and "growth stimulants." Much effort was expended in searching for "the growth-promoting substance" in embryonic extracts before it was realized that growth results from the favorable combination of a number of synthetic processes, fed by many substrates and intricately interlinked. Biological media remained the standard environment for tissue culture until the roles of the individual components (amino acids, vitamins, energy sources, etc.) began to be recognized. Even today, unknown components in plasma and tissue extracts are still being sought. Defined media, adequate for certain cell types, may require, when used for other cells, supplementation with small amounts of serum as a source of lipids, vitamins, hormones, trace metals, etc., not otherwise provided for, and perhaps to supply proteins and/or peptides as nutritional supplements.

Biochemistry was a very primitive science when M. R. Lewis and Lewis (1911a) wrote, "It is to be hoped that an artificial medium will be found as satisfactory as the plasma, for the advantages are obvious if one can work with a known medium in the investigation of the many new problems, which suggest themselves." Most of the vitamins and some of the amino acids were then unknown; few of the hormones had been recognized or characterized. Cell behavior and replication were at the descriptive stage and hardly understood at all in chemical terms. It is not therefore surprising that, at least into the late 1930s, biological nutrients such as lymph, blood plasma or serum, tissue and yeast extracts, meat digests, and peptones held the field as the principal components of media for cells and tissues in culture.

Early attempts by physiologists and pharmacologists (Ringer, 1880, 1883a,b, 1886, 1895; Locke, 1895a,b, 1900; Tyrode, 1910) to substitute simple salt solutions, or salt solutions with glucose, for serum as media for the study of isolated organs provided empirical knowledge of the major ionic requirements, and of the osmotic tolerances, of tissues (for review, see Waymouth, 1970). The "physiological salt solutions" became the basis for development of defined media designed on the imitative principle, that is, simulating those components and characteristics of serum which were regarded as the most important for sustaining life.

In the 1920s, physiological salt solutions designed primarily for other purposes (e.g., Tyrode's solution) began to be replaced by solutions formulated specifically for tissue culture work, e.g., those of Pannett and Compton (1924) and Drew (1927–1928) and still later those of Gey and Gey (1936), Earle (1943), and Hanks (1948; Hanks and Wallace, 1949).

Analysis of serum and other biological fluids was a popular activity among clinicians and others in the first half of the twentieth century, and, as specific substances were identified, artificial media were made to incorporate known serum or tissue components such as vitamins, hormones, and amino acids, as well as salts and glucose. Examples of these semidefined artificial media are those of Vogelaar and Erlichman (1933), Baker (1936), Baker and Ebeling (1938, 1939), and H. Wilson et al. (1942). Attempts to replace serum by simpler, though still undefined, components resulted in the inclusion of various kinds of protein hydrolysates (Baker, 1936; H. Wilson et al. 1942; Waymouth, 1956; G. A. Fischer, 1958; Hsu and Merchant, 1961). Studies are still being made of the nutritional roles of peptones and tissue and milk hydrolysates (Floss, 1964; Pumper et al., 1965; Amborski and Moskowitz, 1968) and of specific peptides (Ito and Moore, 1969). Semidefined media with fewer and fewer serum and tissue fractions led step by step to fully chemically defined media, among the earliest of which were those of White (1946, 1949) and Morgan et al. (1950). The nutrition of animal cells and the history of medium design have been reviewed frequently (Waymouth, 1952, 1954, 1960, 1965a, 1967; Stewart and Kirk, 1954; Geyer, 1958; Morgan, 1950, 1958; Swim, 1967). Recent progress will be documented here.

Media intended for long-term cultivation must necessarily be more complete and elaborate than those for short-term (1 week or less) studies. Some of the B vitamins, for example, may be omitted from short-term media, since their absence may not produce significant metabolic or growth deficiencies for about 2 or 3 weeks (Swim and Parker, 1958a; Dupree et al., 1962; Sanford et al., 1963; Swim, 1967). Trowell's medium T8 (Trowell, 1959) for organ culture is an example. This medium contains, of the B vitamins, only p-aminobenzoic acid and thiamine. Trowell's cultures were normally carried for 4–6 days. For purposes where optimal growth and metabolism of the cells are not the goals, e.g., in some viral studies, a very simple medium, such as that of Dubreuil and Pavilanis (1958), may be sufficient to keep the system functioning for several weeks. However, it should be understood that even short exposure of cells to nonnutrient saline solutions may be highly deleterious (Gey and Gey, 1936; Hanks, 1955). For example, Cleaver

(1965) has shown that exposure of strain L mouse cells to Tyrode's solution causes DNA and RNA synthesis to cease within 20 minutes. Prudence therefore suggests that the washing, diluting, and dispersing of cells should be done in nutritionally adequate media.

II. Variability, Purity, and Methods of Sterilization

One of the advantages claimed for chemically defined media over biological media is constancy of composition. Sera vary from sample to sample, and especially with the age, sex, health, and nutritional state of the donor. Serum can carry viruses, mycoplasmas, and drugs. Hormones, vitamins, lipids, and trace metals are among the components, variations in which may significantly affect cultures.

The reproducibility of a chemically defined medium from batch to batch is not guaranteed because a standard formula is presumably followed on each occasion. Variations in the purity of the chemicals, or in the quality of the water, used in different samples may affect the quality of the final product. Commercial suppliers of media are increasingly aware of these problems. Routine quality control to check biological and synthetic media for growth-promoting ability, for cytopathic properties, and for constancy of pH, tonicity, etc., should eliminate samples which do not meet strict standards. Pooled batches of serum minimize individual variations. Fetal calf sera are commonly used in tissue culture media, and many of the physical and chemical parameters of different samples of fetal calf serum were measured by Olmsted (1967). The variability is wide, e.g., Na^+, 268–460 mg/100 ml (61 samples); Cl^-, 98.9–140.0 mEq/liter (54 samples); Ca^{2+}, 3.1–8.8 mEq/liter (53 samples); acid-soluble phosphate, 4.13–11.84 mg/100 ml (53 samples); and glucose, 57.5–318.9 mg/100 ml (31 samples).

Basic to the preparation of chemically defined media is a good supply of distilled water of high resistivity. Passage through ion exchange resins is not, by itself, sufficient, both because nonionized materials are not removed and because substances absorbed on the resins may be released. Multiple (resin + charcoal + filtration) systems may possibly produce adequately pure water for tissue culture, but the proven system of distillation from glass or quartz, or Monel, is to be preferred until other methods have been thoroughly tested.

Chemicals of the highest purity should be used. After these precautions have been taken, the use of pure water and chemicals may be vitiated by improper methods of sterilization. Seitz asbestos filter pads release large amounts of salts into the filtrate, even after repeated washings. Sintered glass and porcelain filters release significant amounts of soluble material after each successive autoclaving, which can be largely removed by washing the filter with water or with a portion of the medium before collecting the filtrate. Cellulose membranes, now the sterilizing equipment used in most laboratories, should also be prewashed immediately before the collection of sterile samples to remove traces of plasticizers which may be cytotoxic. Adsorption of nutrients on filter materials may also occur; e.g., the glycoprotein hormone erythropoietin is strongly adsorbed on cellulose membrane filters and must be sterilized with pure silver membrane filters (Lowy and Keighley, 1968). Certain types of plastic tubing are potential sources of toxic substances. Borosilicate glass and polycarbonate, are materials of choice for storage of tissue culture media.

Media prepared in the form of powders, ball-milled to ensure thorough homogeneity (Swim and Parker, 1958b; Hayflick et al., 1964; Greene et al., 1965; Young et al., 1966), have the advantage of being capable of prolonged storage and of enabling samples from a single large batch to be used in long-term or replicate experiments. A degree of uniformity unattainable from successive batches of conventionally prepared media is possible.

Autoclaving has not been commonly used for sterilizing tissue culture media, either for protein-containing media, because of protein denaturation, or for chemically defined media, because of the instability of glutamine and of many of the B vitamins at physiological pH. Where stock solutions are prepared separately and combined to make up a complete medium, some of the component solutions (e.g., amino acids and salts) can be sterilized by heat with little risk of change. Interest in heat sterilization of tissue culture media has, however, grown substantially since the description of autoclavable media by Nagle (1968) and Yamane et al. (1968). Nagle's medium was used for almost 1 year for cultivation in suspension of mouse L cells, HeLa cells, and cat kidney cells. Only glutamine and bicarbonate are added after the rest of the medium has been sterilized in 121°C for 15 minutes. Glutamine is either sterilized by filtration or autoclaved dry. Sodium bicarbonate is sterilized separately and added to pH 7.2–7.4. The medium of Yamane et al. (1968), a modified Eagle's minimum essential medium (Eagle, 1959), is designed to be prepared as a dry powder, also without glutamine and bicarbonate. This powder contains succinic acid and sodium succinate, to maintain a pH of 4.0–4.5 during autoclaving of the solution. The effectiveness of the medium is sharply reduced if it is autoclaved below pH 4.0 and is almost completely lost if it is autoclaved above pH 5.0. The range 4.0–4.5 is presumably critical for the stability of the heat-labile B vitamins. Yamane *et al.* (1968) grew mouse strain L cells and baby hamster cells in their medium supplemented with serum, and L cells for short periods in the unsupplemented medium. Since this medium does not contain biotin, which is essential for the continuous growth of L cells (Haggerty and Sato, 1969), it is not to be expected that it could support long-term growth of these cells. Autoclavable media containing the succinate buffer recommended by Yamane, but otherwise similar to the more complete media used in the author's laboratory (Waymouth, 1965b), are now being used for continuous growth of strain L cells (Waymouth, 1972). It is noteworthy, since heat-sterilized media are coming into general use, that a material isolated from an autoclaved mixture of glucose and phosphate is active at low concentrations (optimum, 100 μg/ml) as a growth stimulant for HeLa cells (Sergeant and Smith, 1960).

III. Design of Chemically Defined Media

A. *Introduction*

Empiricism has been the chief method by which chemically defined media have been designed and tested. While the broad basic requirements for salts, amino acids, vitamins and/or their respective coenzymes, and energy sources are common to all cells, the proper proportions of these components, and the requirements for hormones or for substances such as trace metals or lipids, may be very specific. The task of determining optimal concentrations and appropriate relationships among the components in multivariant systems which permit almost infinite scope for quantitative modulations still continues. The principles and practice of the design of synthetic media have been reviewed (Waymouth, 1965a). About 60 chemically defined media described between 1932 and 1962 are listed therein. The complete compositions of 8 synthetic media (then believed to be those in most common use) have been tabulated and about 50 others cited (Waymouth, 1968). Recent formulations are referred to in Section IV.

Some of the most important recent advances in medium design have come from the relatively late realization of the hopes expressed long ago by Carrel and Burrows (1911) for media to regulate morphological development and by the Lewises (W. H. Lewis and Lewis, 1912) for selective media for particular cell types. These phases of medium design are not far advanced but clearly are where important developments may be expected in the near future. Many selective media, and media supporting organotypic development, incorporate hormones specifically active on the target cells (see Section III,B,6). Selective effects upon apparently similar tissues may be produced by substances other than hormones. Lucas (1969), for example, has found a very specific and selective effect of the protease-inhibitory leucine isomer, 6-aminohexanoic acid (ϵ-amino-n-caproic acid), in the cultivation of rat sublingual glands in Trowell's medium T8. Survival of these mucus-secreting glands is improved by addition of 0.2 M of this compound, whereas the protease-secreting salivary glands (submandibular and parotid) become necrotic in this concentration of 6-aminohexanoic acid.

It is important to reemphasize that media are usually developed for particular types of culture—not only for cells of a single species, usually from a single tissue or organ and perhaps from an inbred or mutant animal, but also for the study of a particular variable. The variables may be maximum or minimum growth; biochemical, physiological, or pathological functions; provision of a vehicle for viruses; or the study of oncogenesis or cytogenetics. It should be clear, from the diversity of uses to which tissue and organ cultures are put, that each problem demands careful selection of the proper medium (Evans *et al.*, 1964). The numerous media now available make possible, and require of the investigator, judicious selection of the most suitable medium for the problem under study. A medium designed for other cells in other circumstances is unlikely to be the best for his purpose.

B. Particular Nutrients

1. Major Ions

The major ions (Na^+, K^+, Ca^{2+}, Mg^{2+}, Cl^-, PO_4^{3-}, and HCO_3^-) were for a long time thought of mainly in the context of providing a salt mixture similar in composition to that of serum. The emphasis was on the osmotic role of the ions, in spite of Ringer's early conclusions about the importance of proper concentrations of specific ions in controlling biological functions (Ringer, 1880, 1883a,b, 1886, 1895). The integration of ionic functions into the flow of biological syntheses and in control mechanisms has only recently begun to be considered in relation to the design of media. Ion transport mechanisms are closely related to the uptake of nutrients (amino acids, sugars). The presence of highly specific recognition sites for particular ions and ways in which ions and ion fluxes affect surface charge and major synthetic pathways all indicate that more attention needs to be paid to the active functions of ions, e.g., in controlling nucleic acid synthesis and in triggering mitosis (Perris and Whitfield, 1967; Perris et al., 1968; Morton, 1968; Cone, 1969).

The effects of total $NaCl + KCl$ and of the ratio of KCl to NaCl on the growth of ERK cells was examined by Pirt and Thackeray (1964a,b). These cells were inhibited by the "normal" level of 150-mM $NaCl + KCl$; 50–100 mM and an osmolality of 303 mOsm with KCl/NaCl ratios of 1:32 to 1:4 were acceptable for growth. There are optimal levels of Na^+ and K^+ regulating amino acid uptake by Ehrlich ascites cells (Riggs et al., 1958) and LM mouse cells (Kuchler, 1967). The concentrations of K^+ and Ca^{2+} affect membrane potentials (Borle and Loveday, 1968). Optimal levels of K^+ and Ca^{2+} are also important for the development of preimplantation mouse embryos in vitro (Brinster, 1970; Wales, 1970). The ability of chick heart cells to initiate contractions is highly sensitive to the external K^+ concentration (De Haan, 1967, 1970). Potassium certainly plays a part in many biological functions. It is, for example, necessary for the synthesis of influenza virus in chick chorioallantoic membrane cultures (Levine et al., 1956a,b) and for controlling the synthesis of DNA in phytohemagglutinin-treated lymphocytes in vitro (Quastel and Kaplan, 1970). Chick ganglia produce a maximum number of neurons in a medium containing 40-mM K^+ (Scott and Fisher, 1970). Organ cultures of fetal mouse kidney develop abnormally in low K^+ media. At least 9 mEq/liter is required for kidneys from young (10–14-day) fetuses, and at least 6 mEq/liter for 14–18-day fetal kidneys (Crocker and Vernier, 1970). This study showed that the serum of the 16-day mouse fetus has about twice the maternal concentration of K^+, i.e., about 9–10 mEq/liter (670–750 mg/liter of KCl). Abnormalities in kidney development occurred in vitro with "normal" (maternal) levels of K^+, namely 4–6 mEq/liter lower than in fetal serum.

The potassium in fetal rat plasma, like that in the mouse, is nearly twice as high (about 9 mM) as in maternal plasma (Christianson and Jones, 1957; Adolph and Hoy, 1963). In fetal calf serum it is even higher (16 mM) (Scott and Fisher, 1970), which may be one of the causes of the effectiveness of such sera in tissue culture media. Testicular and follicular fluids are also higher in K^+ (and lower in Na^+) than serum (see Willmer, 1970).

Landschütz ascites tumor cells require an adequate level of Mg^{2+} in order to accumulate K^+ (Ryan et al., 1969), although the depletion of cell K^+ caused by low external concentrations retards growth only when the K^+ loss is accompanied by significant gain in intracellular Na^+.

In sarcoma 180 mouse cells growth, and protein and DNA synthesis are reversibly depressed when external K^+ is reduced (Lubin, 1967).

Mouse strain L cells can survive and grow for over 6 months in media without any added Ca^{2+} or Mg^{2+} (Yang and Morton, 1970), although omission of the Ca^{2+} causes major (but reversible) morphological changes. In that case, mitosis can presumably proceed with very low levels of Ca^{2+}. However, in cultures of rat bone marrow cells, mobilization of Ca^{2+} is suggested as the mechanism of action of the high concentrations of CO_2 which Morton (1967) found to stimulate cell division. Ionized calcium is here believed to be the immediate stimulus to mitosis (Perris and Whitfield, 1967; Morton, 1968; Perris et al., 1968). Poly-L-lysine and other polycations may stimulate mitosis by releasing Ca^{2+} from its binding sites (Whitfield et al., 1968).

2. Trace Metals

Some of the early media, e.g., those of A. Fischer et al. (1948) and Graff and McCarty (1957, 1958), contained deliberately added trace metals (Fe, Mn, Cu, Zn, and Co) in addition to probable traces included adventitiously as contaminants of other components. That these fortuitous traces might affect growth and metabolism was recognized by Healy et al. (1952), who noted the relatively high levels of trace metals in natural media. Shooter and Gey (1952) attempted to discover the requirements of rat cells for trace metals. They treated a biological medium with chelating agents to remove cations, and added back cation mixtures to restore the growth-supporting ability of the medium. By this means they demonstrate requirements of the rat cells for Fe, Cu, Co, Mn, Zn, and Mo.

Many media have been supplemented with Fe (Hetherington and Shipp, 1935; White, 1946; Morgan et al., 1950; Ham, 1960, 1962), with Fe and Zn (Ham, 1963a, 1965), or with mixtures of several trace metals, e.g., Graff and McCarty (1957, 1958). Our own media, since 1962 (Kitos et al., 1962), have contained all the ions recommended by Shooter and Gey, in similar proportions (Kitos et al., 1962) or in different combinations (Waymouth, 1965b). Thomas and Johnson (1967) examined the trace metals to be found in their chemically defined medium with and without supplementation with methylcellulose (Methocel MC or Methocel HG, Dow Chemical Co.). The methylcellulose contributed a large proportion of the iron found in the medium and some Zn and Cu. For strain L mouse cells, the optimal concentration of iron was found to be 0.6–1.4 μM, and lower if Cu was also provided at 0.4 μM. Mn was inhibitory, but this inhibition could be reversed by Fe. 0.6-μM Zn gave maximum growth. Pirt and Thackeray (1964b) found 5-μM Mn to improve the growth of ERK cells but not of L cells.

Higuchi (1970) has also examined the requirements of strain L cells for Fe and Zn, which, under his conditions—notably absence of Cu and Mn—were 4.0-μM Fe and 0.5-μM Zn. Fe, Cu, and Zn are included in the defined medium of Parsa et al. (1970) for organ culture. Parker (Healy and Parker, 1966a) no longer adds Fe or other trace metals to his medium CMRL1415, although Fe was present in the earlier media 199 (Morgan et al., 1950), 703 (Healy et al., 1954), and 858 (Healy et al., 1955).

Birch and Pirt (1969) include Fe, Mn, Zn, and Cu in their media for strain LS (a subline of strain L adapted to grow in suspension). The same group of cations is supplied by Ling et al. (1968), and, in addition, molybdate. They stabilize the trace metals by chelation with aspartic acid. Cobalt is provided as cobalamine.

3. ENERGY SOURCES

a. Carbohydrates. D-Glucose, as the major sugar in avian and mammalian serum, has been included in most chemically defined media. Some other sugars will support growth; e.g., Harris and Kutsky (1953) showed that chick fibroblasts can use D-mannose, D-fructose, and D-maltose but not D-sucrose, D-lactose, D-galactose, L-glucose, or pentoses. Several human cell lines can use mannose, fructose, and galactose (Eagle et al., 1958). A large number of sugars was tested on fresh chick embryonic heart cultures by Morgan and Morton (1960), their list of active sugars differing from that of Harris and Kutsky (1953) in that galactose was found to be effective, as well as mannose, fructose, β-glucose, glucose-1-phosphate, and glucose-6-phosphate. The disaccharides, maltose and turanose, but not pentoses, lactose, sucrose, or various citric acid cycle intermediates, could substitute for D-glucose. As noted in Section II, galactose has been used in complete or partial replacement of glucose in an attempt to reduce the amount of lactic acid produced (Leibovitz, 1963). Paul (1965) reviews this subject and tabulates those carbohydrates that have been tested as sole carbohydrate sources for cells in culture.

b. Other Carbon Sources: Keto Acids, Carboxylic Acids, Purines, and Pyrimidines. Gwatkin and Siminovitch (1960) found that it is possible, in serum-supplemented medium CMRL1066, for strain L cells or for HeLa cells, to replace bicarbonate as a source of CO_2 by 2.5-mM neutralized oxalacetic acid. Kelley et al. (1960), in their medium SRI-8, replaced bicarbonate by 1.14-mM oxalacetate, simultaneously increasing the NaH_2PO_4 by 10-fold, making it possible to grow HEp2 cells in an open system. Moore's bicarbonate-free medium RPMI906 also contains 1.14-mM oxalacetate and high NaH_2PO_4 (Moore et al., 1963, 1966). The 7Cs medium of Ling et al. (1968) includes 0.5-mM oxalacetate, is buffered principally with β-glycerophosphate, and contains only 2.7-mM bicarbonate.

A combination of ribonucleosides and 5-mM oxalacetate was found to substitute for CO_2 in HeLa and in human conjunctival cell cultures in a tris-buffered medium and a system employing a CO_2 trap (Chang et al., 1961). Runyan and Geyer (1967), using mouse strain L cells under similar conditions, found that partial replacement of CO_2 was more successful on addition to a hypoxanthine-containing medium of 10 μg/ml each of adenosine, cytidine, guanosine, and uridine than on addition of this mixture together with 1.5- or 5.0-mM oxalacetate. Other combinations of nucleosides and nucleotides were also effective. The growth response achieved with the mixture of nucleosides was also not enhanced by supplementation with malonate, pyruvate, asparagine, carbamyl phosphate + oxalacetate, or formate.

Keto acids have nevertheless been frequently included in tissue culture media. Sinclair (1966) used α-ketoglutarate of 0.25 mM. The most commonly used keto acid, now found in many chemically defined media, is pyruvate. An early use of pyruvate is that of Levine et al. (1956a,b), who found it to be essential, in a simple medium based on Hanks' solution, for synthesis of influenza virus in culture, and they included it at about 9 mM. Neuman and McCoy (1958) tested three keto acids, pyruvic, oxalacetic, and α-ketoglutaric, in media for the culture of the Walker 256 rat tumor cells and found 0.5 mM of any of the three to be beneficial. Ham (1960, 1962, 1963a, 1965) has included pyruvate in his media F10 and F12 at 1.0 mM. The media used by Higuchi (1963, 1970) also contain pyruvate at 1.0 mM. Herzenberg and Roosa (1959) demonstrated a requirement for pyruvate for mouse lymphoma 388 cells and added it at 1.0 mM. Nagle included pyruvate at 1.0 or 0.6 mM in media for mammalian cells, both in conventionally prepared (Nagle et al., 1963) and in autoclavable (Nagle, 1968) media. Leibovitz (1963), in his bicarbonate-free, galactose-containing medium L-15, uses 5.0-mM pyruvate, while Baugh et al. (1967a,b), in their galactose-containing media, include 4.55-mM pyruvate, the same level as in Rappaport's medium SM-3 (Rappaport et al., 1960). Healy and Parker (1966a), in medium CMRL1415, employ 2.0 mM. E. Steinberger et al. (1964) include 1.0-mM pyruvate in a medium for organ culture of testis, while Lostroh (1966) has very high levels (27 mM of each) of both pyruvate and succinate in a medium for mouse uteri. Media for mouse eggs, on the other hand, contain low levels of pyruvate, e.g., 0.25 or 0.5 mM (Brinster, 1970) or 0.32 mM (Whitten, 1971). In media containing high levels of glutamine and glutamic acid—e.g., Waymouth's MD705/1 and MAB87/3 (Kitos et al., 1962; Waymouth, 1965b); Leibovitz's L-15 (Leibovitz, 1963); Healy and Parker's CMRL1415 (Healy and Parker, 1966a); Moore's RPMI906, 1595, 1603, and 1640 (Moore et al., 1966, 1967; Moore and Kitamura, 1968); and Ling's 7Cs medium (Ling et al., 1968)—cells containing active glutamate-oxalacetate transaminase or glutamate-pyruvate transaminase are able to form keto acids endogenously (Kitos and Waymouth, 1966).

As mentioned in Section II, Yamane et al. (1968) demonstrated that it is possible to autoclave most of the components of a defined medium for mammalian cells by buffering the solution with a succinate buffer of pH 4.0–4.5 and adjusting the pH with bicarbonate after autoclaving. The total succinate level in their medium is 1.19 mM. Acetate is included in medium NCTC135 and many of its precursors (Evans et al., 1964) at 0.27 mM, but, although it was a component of Parker's media 199 (Morgan et al., 1950) through 858 (Healy et al., 1955) at 0.27 mM and of the modification by Biggers and Lucy (1960) of 858 at 0.4 mM, it is omitted from CMRL1066 (Parker, 1961) and CMRL1415 (Healy and Parker, 1966a), which contains nucleosides at about the levels recommended by Runyan and Geyer (1967).

Most cells in culture produce lactate from glucose by glycolysis. Mouse strain L cells convert glucose to lactate, glutamate, alanine, glycine, and serine (Kitos and Waymouth, 1964). Lactate is not, therefore, generally supplied exogenously. However, Biggers et al. (1961) include 1.8-mM calcium lactate in their medium for growth of avian and mammalian tibiae as a convenient source of calcium. Media for mammalian ova also contain lactate, 20 mM in Brinster's medium BMOC-2 (Brinster, 1970) (not 25 mM as indicated in his Fig. 6), and 21.72 mM in Whitten's medium (Whitten, 1971). Pyruvate and oxalacetate, but not lactate or phosphoenolpyruvate, permit cleavage from one cell to two cells in the mouse oocyte (Biggers et al., 1967). Pyruvate, lactate, oxalacetate, and phosphoenolpyruvate were all found by Brinster (1965, 1970) to be suitable energy sources for the development of mouse embryos from the two-cell stage; malate, fumarate, succinate, isocitrate, citrate, acetate, cis-aconitate, and α-ketoglutarate were not, although malate, citrate, and acetate were able to support eight-cell embryo development (fumarate, isocitrate, and cis-aconitate were not tested). (For further discussion, see Chapter 7, Volume 2.) The energy-producing pathways and regulating mechanisms, as they apply to cell cultures, were reviewed by Paul (1965).

4. AMINO ACIDS

A great amount of investigation has been made into the minimal and maximal amino acid requirements for cells in culture. Fashions have swung from the use of minimal numbers and amounts, in media supplemented with dialyzed plasma (A. Fischer, 1948; A. Fischer et al., 1948) or serum (Eagle, 1955), i.e., conditions which can compensate for some deficiencies (Chang, 1958), to maximal numbers (Parsa et al., 1970) and high concentrations (Ling et al., 1968). Much latitude is evidently compatible with satisfactory media. Useful combinations vary with the cell or organ and with the stage of growth and depend markedly on the composition of the rest of the medium. Some amino acids exert a sparing effect upon others, and interconversion and transamination may also occur. Information about amino acid utilization or liberation by L cells (Griffiths and Pirt, 1967) and by various cell strains (McCarty, 1962) shows highly characteristic patterns of utilization for each strain. This subject is dealt with by Patterson (Chapter 6, Volume 1).

5. VITAMINS

The requirements of certain cells in chemically defined media for particular nutrients are well exemplified by the report of Price et al. (1966) on the needs of five cell types for inositol and vitamin B_{12}. HeLa cells (S3 clone) and a Chinese hamster line require both vitamins, although a subline of the hamster cells could grow without inositol. Two lines of mouse liver cells from one clone and a green monkey kidney cell line require neither vitamin.

Vitamin requirements must, therefore, be examined for every culture system used. It may, however, be safely assumed that all, or almost all, cells need the major B vitamins (pantothenate or coenzyme A, of which it is a constituent; nicotinic acid or nicotinamide or the coenzymes of which they form a part; riboflavin or its coenzymes; thiamine or thiamine pyrophosphate; one of the pyridoxine group or the biologically active pyridoxal phosphate to which they can in most cell systems be converted; folic acid or folinic acid). As indicated above, inositol and vitamin B_{12}, which are also involved in coenzyme functions, and biotin, may not be required in every case. Strain 2071 mouse cells (a derivative of strain L, NCTC clone 929 cells) grow well if thymidine and/or vitamin B_{12} are omitted (Sanford et al., 1963). Para-aminobenzoic acid is sometimes supplied, either with folate or as a potential folate precursor. Lipoic acid is included in some media (Holmes, 1959; Holmes and Wolfe, 1961; Ham, 1963a, 1965; Ling et al., 1968). Choline plays an important part in most media (cf. Nagle, 1969; Birch and Pirt, 1969), probably mainly as a methyl donor.

The role of ascorbic acid in tissue culture media has been reviewed by Fell and Rinaldini (1965). The instability of this vitamin at physiological pH, due to its rapid oxidation, especially if antioxidants such as glutathione are not also present, makes evaluation of its functions difficult. However, although many media now do not include ascorbic acid, it may have an important function as a component of redox systems, especially in maintaining the redox potential at a level at which small numbers of cells can become established in culture.

Of the fat-soluble vitamins, vitamin A has been the most fully studied. It has profound effects on the differentiation and functions of epithelial tissues and of bone and cartilage. The subject is reviewed by Fell and Rinaldini (1965). Vitamins A, D, E, and K are included in medium NCTC135 (Evans et al., 1964) and in the medium of Parsa et al. (1970). Holmes and Wolfe (1961) included these four vitamins and also cholesterol. Cholesterol and vitamins A, D, E, and K, which were included in Parker's medium 858 (Healy et al., 1955), have been omitted from CMRL1066 and CMRL1415 (Healy and Parker, 1966a). Ling et al. (1968) have vitamins A, D, E, and cholesterol, but not vitamin K. The functions of all the vitamins in chemically defined media have been reviewed and evaluated by Sanford et al. (1963).

6. Hormones

Some of the first of the semidefined media contained hormones. The media of Baker (1936) for fibroblasts, epithelial cells, and monocytes contained insulin and thyroxine, and later versions (Baker and Ebeling, 1939) included adrenaline and crude adrenal cortical and pituitary preparations.

Insulin has been widely used in tissue culture media, where it probably has multiple functions, including roles in the stimulation of glucose utilization, uptake of other nutrients, and maintenance of differentiation. Its effects are at least in part membrane-mediated, since it stimulates pinocytosis (Barrnett and Ball, 1959, 1960; Paul and Pearson, 1960) and affects cell morphology (Waymouth and Reed, 1965). Maintenance of contractions in mouse heart cultures by insulin is related to dose (Wildenthal, 1970). Insulin-containing media have been used for cells in suspension (McCarty and Graff, 1959; Nagle et al., 1963; Tribble and Higuchi, 1963), for cells in monolayer (Paul, 1959; Crockett and Leslie, 1963; Vann et al., 1963; Hanss and Moore, 1964; Waymouth, 1965b), and for organ culture (Trowell, 1959, for many tissues; Elias,

1962; Elias and Rivera, 1959; Prop, 1960; Rivera and Bern, 1961; Moretti and De Ome, 1962; Lasfargues, 1962; Rivera, 1963; Ceriani, 1970, for mammary tissues; Gorham and Waymouth, 1965, for bone; Lostroh, 1963, 1966, for uteri). The effects of insulin on cell and organ cultures are reviewed by Lasnitzki (1965) and Paul (1965). Thyroxine has been tested in several culture systems but is not included in any recently published media except that of Garvey (1961) for liver reticulo-endothelial cells.

The steroid hormones, often in combination with insulin and other protein hormones, play important parts in stimulating and maintaining differentiation and function. Rat salivary glands, for example, require cortisol (or another adrenocorticosteroid) and insulin (Lucas *et al.*, 1970) in organ culture. Cell cultures of hepatocytes from 16-day fetal livers of C57BL/6J mice require an adrenocorticosteroid hormone for long-term culture in defined media (Waymouth, 1972). Prelactating mouse mammary tissues need cortisol (Elias and Rivera, 1959) or aldosterone (Rivera, 1963), prolactin, somatotropin and insulin (Rivera, 1964), and progesterone (Elias, 1957; Ceriani, 1970) for complete differentiation *in vitro*. Moreover, marked differences in responsiveness to prolactin and somatotropin were observed with mammary tissues from different inbred strains of mice (Rivera, 1964, 1966), demonstrating that not only specific tissues but also specific genetic constitution may need to be taken into consideration in designing a medium for a particular culture system. Not all organotypic culture systems require hormones. Morphogenesis of rat pancreatic anlagen, over a period of 9 days in organ culture, has been achieved by Parsa *et al.* (1970) in a highly complex defined medium not containing any hormones.

7. LIPIDS

There have been few unequivocal demonstrations of requirements of cells in tissue culture for lipids. Many cells, including mouse L cells, can synthesize lipids *de novo* from glucose and acetate (Bailey, 1966). Fatty acids can be taken up and metabolized but are not necessarily required (Bailey, 1967; Fillerup *et al.*, 1958; Howard and Kritchevsky, 1969). Cholesterol can be synthesized by cells *in vitro* (Berliner *et al.*, 1958), and mouse L5178Y cells take up cholesterol from a balanced salt solution + 5% fetal calf serum medium, the uptake being dependent on the proportion of phospholipids in the medium but independent of the viability of the cells (Rothblat *et al.*, 1968). Lipid deficiencies can be created in some cells, e.g., in MB III cells (Ling *et al.*, 1968) and in HeLa cells. The deficiency in HeLa cells can be prevented by the addition of protein-bound linoleic or arachadonic acids (Gershonson et al., 1967). Harary et al. (1966) demonstrated that rat heart cells in lipid-deficient media cease to beat. Serum lipids, or fatty acids, can restore their ability to do so. Where fatty acids are required, the limits of effective concentration appears to be narrow; e.g., 10–20 μg/ml of oleic acid can replace 1% calf serum as a growth supplement for the monkey kidney cell line LLC-MK$_2$, but more than 20 μg/ml is inhibitory (Jenkin and Anderson, 1970). Lipid metabolism in cells in culture is reviewed in a symposium edited by Rothblat and Kritchevsky (1967) and in Chapters 8 and 9, Volume 1.

8. MACROMOLECULAR ADJUVANTS

Various macromolecules have been added to media not containing serum for their "protective" effects upon cells. These effects, while possibly separable from nutritional effects, are not well understood. In suspension cultures, the functions may be partly mechanical, counteracting shearing damage. Many of these substances, however, probably act as carriers and chelators of nutrient substances (Gwatkin, 1960), or as bearers of charged groups, or may themselves, as in the case of certain purified proteins, in fact be nutritive (McCarty and Graff, 1959). Some affect cell attachment (see Section IV,B,1) and cell survival; others promote cell proliferation (Lieberman and Ove, 1957; Fisher et al., 1959; Holmes and Wolfe, 1961; Tozer and Pirt, 1964; Holmes, 1967; Morgan, 1970).

The adjuvants range from specific proteins or protein fractions from serum to several synthetic polymers [Pluronic F-68, polyvinylpyrrolidone (PVP), dextran, methylcellulose (Methocel), and Ficoll]. Healy and Parker (1966b) have used many of these supplements with their medium CMRL1415. Dextran alone was less effective than a low concentration of dextran together with a small amount of low-molecular-weight α-glycoprotein for primary mouse embryonic cell cultures. These authors review some of the previously reported uses of macromolecules as additions to synthetic media.

Methylcellulose (Methocel), of either high or low viscosity, was made popular by its use in suspension cultures by Kuchler et al. (1960), Bryant et al. (1961), Merchant and Hellman (1962), Nagle et al. (1963), and Bryant (1966, 1969). Methocel may contribute significant amounts of iron and other trace metals to the medium (Thomas and Johnson, 1967). Ling et al. (1968) used 0.1% levels of either Methocel (10 centipoises), PVP, or thiolated gelatin for the growth in suspension of MBIII cells (De Bruyn et al., 1949) in 7Cs medium.

C. Physical Factors and Buffers

The appropriate chemical composition of a medium is not by itself sufficient to provide a satisfactory total environment for cells. Various factors, some dependent on chemical components, must be taken into account. Among these are the oxidation-reduction potential, osmolality and ionic balance, pH and gas phase, temperature, and type of substrate (various types of glass and plastic and the presence of "protective" macromolecular adjuvants).

The importance of redox potentials in the control of cell functions, neglected since Havard and Kendal (1934) first drew attention to the effects of alteration in redox potential of a medium on the growth of cells in culture, has been reemphasized recently by Shipman (1969). Osmolality (Pirt and Thackeray, 1964a,b; Waymouth, 1970) has also been neglected as a factor controlling cell behavior. Membrane potential plays an important role in determining cellular events relating to mitosis (Cone, 1969). The pH and the gas phase, on the other hand, have been extensively investigated.

Because of their traditional basis as imitations of the blood serum, most complete media, as well as the balanced salt solutions (Tyrode, 1910; Gey and Gey, 1936; Earle, 1943; Hanks, 1948) which form their foundation (see Waymouth, 1965a, 1968; Morton, 1970), are buffered with phosphates and bicarbonate. The relative proportions of these components vary, but this system in some form is still the most popular. Because bicarbonate-buffered media require to be equilibrated with CO_2 and readily lose CO_2 when exposed to air, cultures in such media must either be kept in stoppered vessels or maintained in special incubators with a humidified CO_2-air atmosphere. Attempts to substitute nonvolatile buffers for bicarbonate, because of this inconvenience, have been numerous. McLimans et al. (1957) grew mouse L cells in large-scale submerged cultures in a medium containing 10% serum and no bicarbonate or CO_2 gassing, depending on phosphates alone for buffering. Pirt and Callow (1964) describe two nonbicarbonate media buffered only with phosphates. Glycylglycine (Swim and Parker, 1955; Goldstein et al., 1960) has not become popular. The tris(hydroxymethyl)aminomethane (tris) buffers of Gomori (1946) have been employed in balanced salt solutions, e.g., by Favour (1964) and Kern and Eisen (1959), and in tissue culture media (Swim and Parker, 1955; Runyan and Geyer, 1963; Martin, 1964; Waymouth, 1965c) but usually not for long-term cultures, although Hay and Paul (1967) have maintained strain HLM (human fetal liver) cells (Leslie et al., 1956) in a tris-citrate-buffered medium. When tris buffers are used, it is important to note the anomalous pH readings obtained with certain types of single-probe pH electrodes, because of reaction between tris and the linen fiber of the electrode (Ryan, 1969).

Bicarbonate is a metabolic requirement, at least for some cells (Harris, 1954; Swim and Parker, 1958c; Geyer and Chang, 1958; Runyan and Geyer, 1967), although in much smaller amounts than are needed for buffering. A low bicarbonate, phosphate-buffered medium based on medium 199 (Morgan et al., 1950) was devised by B. W. Wilson et al. (1966). The most radical efforts to eliminate bicarbonate from media were those of Kelley et al. (1960) and Leibovitz (1963). Kelley's medium SRI-8 and Leibovitz' medium L-15 contain a phosphate buffer and depend on increased amounts of amino acids (in L-15 especially high levels of the basic amino acids arginine and histidine) for buffering. Medium SRI-8 contains oxalacetate in place of bicarbonate, and L-15 contains galactose and pyruvate in place of glucose, as a means of reducing the amount of lactic acid formed from glucose by glycolysis. A modification, CMRL1415-ATM, of the CMRL1415 medium of Healy and Parker (1966a) lacks bicarbonate and contains galactose and pyruvate. Baugh et al. (1967b) found that addition of a very low concentration of glucose (100 mg/liter) to a galactose- and pyruvate-containing medium increased the growth of WI38 cells at pH 7.0–7.4, reduced growth at pH 6.8, and completely inhibited growth at pH 6.6.

Especially when bicarbonate-containing buffers are used, the overlying gas phase forms an important part of the total environment. Carbon dioxide not only maintains the equilibrium

$$HCO_3^- \rightleftharpoons CO_2 + OH^-,$$

but also enters the metabolic cycle as a nutrient, as noted above. To maintain an appropriate partial pressure of CO_2 at 37°C to keep the pH at 7.0–7.2, gas mixtures are often supplied, usually 5% CO_2 in air, although with their low bicarbonate medium CMRL1415, Healy and Parker (1966a) use 8% CO_2, and Bryant (1966, 1969) recommends 10% CO_2 in air for cells in suspension culture. An unusually high level of CO_2 (30%) has been found by Morton (1967) to stimulate mitosis in cultures of rat bone marrow.

Cells vary in their ability to grow anaerobically. Pace et al. (1962) found 95% O_2–5% CO_2 to be toxic to mouse L cells in monolayer. These cells grew in 0 to 50% O_2–5% CO_2–remainder N_2. The mouse liver cell line NCTC 1469 tolerated a narrower range of O_2 concentration (10–50%), HeLa cells died in 11 days in 50% O_2, and a human skin cell line NCTC 1769 grew well only between 10 and 20% O_2. High O_2 (95%) is toxic to rat testis in organ culture (A. Steinberger and Steinberger, 1966) and to mouse ascites tumor cells in monolayer culture (Morgan, 1970). Cell damage can be caused by sparging to control pO_2, and this can be prevented by addition of either 10% serum or 0.02% Pluronic F-68 (Kilburn and Webb, 1968). Pluronic F-68, a surfactant polymer, was introduced for the growth of cells in suspension on a rotary shaker by Swim and Parker (1960). The original organ culture technique introduced by Trowell (1952) called for the provision of an atmosphere of 100% oxygen, in accordance with his thesis that fragments of tissue are not easily penetrable by oxygen, which must therefore be supplied at high partial pressures. In later studies, Trowell employed 3% CO_2–97% O_2 (Trowell, 1955) or 5% CO_2–95% O_2 (Trowell, 1959). Trowell's technique is, however, far from uniformly successful for every tissue. Possibly some of this inequality may be attributable to the inability of some tissues to withstand very high levels of O_2. Armstrong and Elias (1968) found that rat eyes in a chemically defined medium degenerated at pH below 7.0 or above 8.0 and in 95% O_2–5% CO_2 or 95% N_2–5% CO_2, but remained healthy for 6 days in 95% air ($= 19\%$ v/v O_2)–5% CO_2. That some tissues do thrive in rather high O_2 is supported by the studies of Melcher and Hodges (1968), who found 40–80% O_2 to be optimal for culture of mouse mandibles. On the other hand, low oxygen (5% O_2–5% CO_2–90% N_2) is optimal for development of one-cell mouse embryos, for which 20% oxygen is toxic (Whitten, 1971). The report that growth of ERK cells is very sensitive to small changes in O_2 level (Cooper et al., 1958), so that a reduction from 19 to 17% v/v greatly increased the growth rate, was not confirmed by Pirt and Callow (1964). (For further discussion, see Chapter 5, Volume 1.)

Since the description by Good et al. (1966) of several nonvolatile and relatively nontoxic organic buffers, including N-2-hydroxypiperazine-N-2-ethanesulfonic acid (HEPES), N-tris(hydroxymethyl)methyl-2-aminoethanesulfonic acid (TES), and N-tris(hydroxymethyl)methylglycine (Tricine), several trials have been made of these buffers in tissue culture media. Williamson and Cox (1968) grew human, rabbit, and hamster cells for virus culture in TES or HEPES at 28 mM, with 15.6-mM bicarbonate in medium 199 or 10.4-mM bicarbonate in Eagle's minimal essential medium. Shipman (1969) has used HEPES at 10 mM without any bicarbonate, and Fisk and Pathak (1969) used HEPES at 14 mM with 13-mM $NaHCO_3$ in an open system. HEPES buffer, with a low concentration of bicarbonate, has been used successfully for the cultivation of mouse ova (Whitten, 1971). Gardner (1969) has maintained rat hepatoma cells for over 2 years in a modified Swim's S 103 medium (Swim and Parker, 1958a) buffered with 50-mM Tricine and containing only 5.95-mM bicarbonate. The growth of human diploid cells may be inhibited by Tricine-buffered media (Wood and Pinsky, 1970), but the inhibition may be reversed on transfer to bicarbonate-buffered media. Tris and many of the buffers of Good et al. (1966) have been found (Gregory and Sajdera, 1970) to interfere with the Lowry method of protein determination (Lowry et al., 1951), commonly used for cell cultures, a problem which may be circumvented by using a microbiuret method (Turner and Manchester, 1970).

A novel approach to buffering was made by Matsumura et al. (1968), who successfully used a cross-linked polymethacrylic acid (Amberlite IRC 50) to buffer media for mouse strain L cells. The use of buffer systems other than bicarbonate-phosphate, with the inclusion either of small amounts of bicarbonate or of substances which generate CO_2 endogenously, can be expected to commend itself to cell culturists and become more popular.

IV. Current Media

A. Recently Published Formulas (General)

Modifications of published medium formulations are apt to be adopted, by the original authors and by others, either as improvements on the media for their primary use or to adapt them for other types of culture. The survey compiled by Morton (1970), reviewing 23 commercially available tissue culture media and the methods of preparation recommended by their authors, is therefore a useful guide to current practice.

Since chemically defined media were last reviewed and tabulated by the author (Waymouth, 1965a, 1968), several new formulas have been devised. The medium CMRL1415 of Healy and Parker (1966a) is a development from the series of media from Parker's laboratory; the first medium of the series (medium 199, Morgan et al., 1950) and at least one of its later elaborations (CMRL1066, Parker, 1961) are still in common use. The new medium CMRL1415 is designed for primary mouse embryonic cell cultures, and it differs from CMRL1066 by having a lower bicarbonate concentration (depending for buffering on higher basic phosphate and on the free bases of the basic amino acids) and by omission of 13 of the ingredients of CMRL1066. Part of the glucose is replaced by galactose, and pyruvate is included. A modification, CMRL1415-ATM, lacking bicarbonate, is used in unsealed cultures in free gas exchange with water-saturated air.

Moore *et al.* (1963, 1966, 1967; Moore and Kitamura, 1968) and Iwakata and Grace (1964) have designed a series of media, RPMI906, 1311, 1595, 1629, 1630, and 1640, for a variety of different uses with normal and leukemic human and mouse cells in monolayer or suspension culture. Medium 906 contains oxalacetate, high phosphate, and low glucose; 1311 contains trace metals and insulin; 1595 contains trace metals and nucleosides; 1603 includes iron; 1629 is a modification of McCoy's medium 5A (McCoy *et al.*, 1959a) supplemented with Bactopeptone (for the history of medium 5A, see Morton, 1970); 1630 and 1634 contain high phosphate and no bicarbonate; 1634 also contains high glutathione; and 1640 has a high inositol content. These are the major distinguishing features; all contain many other variables.

A simplified maintenance medium containing galactose instead of glucose and only the two amino acids L-cystine and L-glutamine has been described by Jacobs (1966). This medium is suitable for maintaining human diploid cell strains or, supplemented with 10% serum, for growth (47 passages). The vitamin components are said to be based on Eagle's basal medium (Eagle, 1955), but in fact differ from that medium in the concentrations of all the vitamins listed; also, the medium contains inositol, which is not in Eagle's basal medium.

The 7Cs medium of Ling *et al.* (1968) is characterized by its high level of amino acids, designed to be closer to the amino acid composition of skeletal muscle than to that of serum, and by the inclusion of trace metals, fat-soluble vitamins and lipids, oxalacetate, thymidine, and deoxycytidine and of only a small amount of bicarbonate (sufficient for the metabolic requirements of the cells), buffering being accomplished by Na-β-glycerophosphate instead of NaHCO$_3$.

A technically useful advance has been made from the identification and description of conditions under which media containing heat-labile components may be autoclaved (see Section II). Nagle (1968, 1969) and Yamane *et al.* (1968) formulated media in which all components except glutamine and bicarbonate can be autoclaved. A medium similar to those of Nagle, but filter-sterilized, incorporating the high choline levels recommended by Nagle (1969) and containing Fe and Zn, has been described by Higuchi (1970).

The medium F10 of Ham (1963a) is distinctive in its inclusion of a high level of arginine and contains Fe, Cu, Zn, and pyruvate. The later modification, F12 (Ham, 1965), supplies much higher amounts of choline and inositol, lower amounts of some of the other B vitamins, and increased amounts of several amino acids and includes linoleic acid and putrescine. A new medium for organ culture, in which differentiation and function of rat pancreas explants continues for 9 days, has been described by Parsa *et al.* (1970). It contains a very large number of components, including trace metals, a very complete (25) amino acid complement, glucosamine, glycuronolactone, glucuronate, acetate, pyruvate, mucate, linoleic and oleic acids, water- and fat-soluble vitamins, lipoic acid, coenzymes, and nucleosides and nucleotides. It is unusual among media designed for study of differentiated function in that it contains no hormones.

B. Media for Special Purposes

1. For Cloning and Plating

The success of early cloning experiments depended on the use of "conditioned medium" from another culture (Sanford *et al.*, 1948) or of a "feeder layer" of nonproliferating cells (Puck and Marcus, 1955). The principle in both instances is the same, namely that the synthetic capacities of the conditioning or feeder cells produce nutritionally useful metabolites which compensate for inadequacies of the conditioned medium. With improvements in medium formulation, conditioned media and feeder layers have become less necessary and cloning (i.e., isolation of colonies from single cells) and plating (i.e., production of cultures by inoculating small numbers of cells) are now done in defined media with small supplements of serum (Ham and Sattler, 1968; Sanford and Westfall, 1969) or in completely defined media (Ham, 1964, 1965). Rubin (1966) has isolated from conditioned media, by high-speed centrifugation, a large-molecular-weight fraction which enables chick embryo cells to be grown at low population density.

Several substances have been proposed as replacements for serum in plating media. Among these are the acidic α_1-glycoprotein fetuin, first described by Pederson (1944), used in plating media by Fisher *et al.* (1958) and characterized by Spiro (1960). Fisher *et al.* (1958, 1959) found a mixture of fetuin and serum albumin superior to fetuin alone. In a medium also containing fetuin, Ham (1963b) substituted linoleic acid for serum albumin in a medium for plating Chinese hamster cells. Cholesterol has also been implicated as an attachment factor (Holmes *et al.*, 1969). Serine was proposed as an adjuvant to improve plating efficiency for various cell lines by Lockart and Eagle (1959). Later, Eagle and Piez (1962) provided a rationale for this by drawing attention to the nutritional requirements exhibited by some cells at low, but not at high, population densities for asparagine, cystine, glutamine, homocystine, inositol, pyruvate, and serine. Lieberman and Ove (1957,

1958) identified a glycoprotein "attachment factor" in serum but reported its nonidentity with fetuin (Lieberman *et al.*, 1959). They noted that basic peptides such as salmine are active in attachment, perhaps by changing the surface charge of the cells (Lieberman and Ove, 1958).

Two different fractions were obtained from chick embryo extract by Coon and Cahn (1966), one of which, of low molecular weight, supports differentiation of clones of chick retina and cartilage cells; the other, of high molecular weight, stimulates growth and plating efficiency but inhibits differentiation. Plating efficiency is a function not only of optimal nutrition, but also of properties of the cell surface which favor attachment. These may depend on membrane potential, which in turn is affected by the presence of ions and binding sites on the cells and in the environment. It is of interest that many of the "attachment factors" appear to have high anion-binding capacities (Lieberman and Ove, 1958; Spiro, 1960; Ham, 1964).

2. For Isolation of Variants and Mutants

The technique of plating out cells grown in the presence or absence of specific nutrients or specific antimetabolites has made possible the isolation of variant and mutant cell lines with different metabolic characteristics. Stable clonal sublines of HeLa cells, showing extreme differences in plating efficiency in the absence of a feeder layer, were isolated by Puck and Fisher (1956). Variants from conjunctival and HeLa cells capable of growing with only xylose, ribose, or lactate as the main energy source were isolated by Chang (1957). Maxwell *et al.* (1959) and McCoy *et al.* (1959b) discovered variants of Jensen rat sarcoma cells with and without a requirement of asparagine. Roosa and Herzenberg (1959) were among the first to use antimetabolites for the selection of variant cells.

Selective media have been devised for the isolation of drug-resistant or drug-sensitive mutants or of cells which either lack or carry certain key metabolic enzymes. For example, the HAT medium, containing 5 μg/ml of hypoxanthine, 0.02 μg/ml of aminopterin, and 5 μg/ml of thymidine in Eagle's basal medium + 10% horse serum (Szybalski *et al.*, 1962), is used to isolate revertant cells from populations lacking the enzymes guanine-hypoxanthine phosphoribosyltransferase (and are therefore resistant to the hypoxanthine analogue 8-azaguanine) or thymidine kinase (and are resistant to the thymidine analogue 5-bromodeoxyuridine) or of hybrids produced from two cells, each of which carries one of the enzymes (Szybalski *et al.*, 1962; Littlefield, 1964; Littlefield and Goldstein, 1970)

V. Conclusion

The design of media remains in part an empirical science and in part an art assisted by experience and intuition. Insofar as generalizations are possible, it may be said that just as every cell possesses unique genetic and epigenetic functional capacities, so every medium designed for particular cells must come as close as our knowledge makes possible to providing the nutritional and physical milieu within which these capacities can be expressed. This ideal is seldom reached, but we can now come sufficiently close to it to enable us to learn much about how cells work, both as individual entities and as parts of organs.

REFERENCES

Adolph, E. F., and Hoy, P. A. (1963). Regulation of electrolyte composition of fetal rat plasma. *Amer. J. Physiol.* 204, 392–400.
Amborski, R. L., and Moskowitz, M. (1968). The effects of low molecular weight materials derived from animal tissues on the growth of animal cells in vitro. *Exp. Cell Res.* 53, 117–128.
Armstrong, R. C., and Elias, J. J. (1968). Development of embryonic rat eyes in organ culture. I. Effect of glutamine on normal and abnormal development in a chemically defined medium. *J. Embryol. Exp. Morphol.* 19, 397–405.
Arnold, J. (1887). Ueber Theilungsvorgänge an der Wanderzellen; ihre progressiven und regressiven Metamorphosen. *Arch. Mikrosk. Anat.* 30, 205–310.
Bailey, J. M. (1966). Lipid metabolism in cultured cells. VI. Lipid biosynthesis in serum and synthetic growth media. *Biochim. Biophys. Acta* 125, 226–236.
Bailey, J. M. (1967). Cellular lipid nutrition and lipid transport. *Wistar Inst. Symp. Monogr.* 6, 85–113.
Baker, L. E. (1936). Artificial media for the cultivation of fibroblasts, epithelial cells and monocytes. *Science* 83, 605–606.
Baker, L. E., and Ebeling, A. H. (1938). Maintenance of fibroblasts in artificial and serumless media. *Proc. Soc. Exp. Biol. Med.* 39, 291–294.
Baker, L. E., and Ebeling, A. H. (1939). Artificial maintenance media for cell and organ cultivation. I. The cultivation of fibroblasts in artificial and serumless media. *J. Exp. Med.* 69, 365–378.
Barnett, R. J., and Ball, E. G. (1959). Morphological and metabolic changes produced in rat adipose tissue *in vitro* by insulin. *Science* 129, 1282.
Barnett, R. J., and Ball, E. G. (1960). Metabolic and ultrastructural changes induced in adipose tissue by insulin. *J. Biophys. Biochem. Cytol.* 8, 83–101.
Baugh, C. L., Fitzgerald, J. M., and Tytell, A. A. (1967a). Growth of the L-cell in galactose medium. *J. Cell. Physiol.* 69, 259–261.
Baugh, C. L., Lecher, R. W., and Tytell, A. A. (1967b). The effect of pH on the propagation of the diploid cell WI-38 in galactose medium. *J. Cell. Physiol.* 70, 225–228.
Berliner, D. L., Swim, H. E., and Dougherty, T. F. (1958). Synthesis of cholesterol

by a strain of human uterine fibroblasts propagated *in vitro. Proc. Soc. Exp. Biol. Med.* **99**, 51–53.

Biggers, J. D., and Lucy, J. A. (1960). Composition and preparation of small batches of a modification of medium 858 (denoted BL1). *J. Exp. Zool.* **144**, 253–256.

Biggers, J. D., Gwatkin, R. B. L., and Heyner, S. (1961). Growth of embryonic avian and mammalian tibiae on a relatively simple chemically defined medium. *Exp. Cell Res.* **25**, 41–58.

Biggers, J. D., Whittingham, D. G., and Donahue, R. P. (1967). The pattern of energy metabolism in the mouse oöcyte and zygote. *Proc. Nat. Acad. Sci. U.S.* **58**, 560–567.

Birch, J. R., and Pirt, S. J. (1969). The choline and serum protein requirements of mouse fibroblast cells (strain LS) in culture. *J. Cell Sci.* **5**, 135–142.

Borle, A. B., and Loveday, J. (1968). Effects of temperature, potassium, and calcium on the electrical potential difference in HeLa cells. *Cancer Res.* **28**, 2401–2405.

Brinster, R. L. (1965). Studies on the development of mouse embryos *in vitro.* IV. Interaction of energy sources. *J. Reprod. Fert.* **10**, 227–240.

Brinster, R. L. (1970). *In vitro* cultivation of mammalian ova. *Advan. Biosci.* **4**, 199–232.

Bryant, J. C. (1966). Mammalian cells in chemically defined media in suspension cultures. *Ann. N.Y. Acad. Sci.* **139**, 143–161.

Bryant, J. C. (1969). Methylcellulose effect on cell proliferation and glucose utilization in chemically defined medium in large stationary cultures. *Biotechnol. Bioeng.* **11**, 155–179.

Bryant, J. C., Evans, V. J., Schilling, E. L., and Earle, W. R. (1961). Effect of chemically defined medium NCTC 109 supplemented with Methocel and of silicone coating the flasks on strain 2071 cells in suspension culture. *J. Nat. Cancer Inst.* **26**, 239–252.

Burrows, M. T. (1910). The cultivation of tissues of the chick embryo outside the body. *J. Amer. Med. Ass.* **55**, 2057–2058.

Burrows, M. T. (1911). The growth of tissues of the chick embryo outside the animal body, with special reference to the nervous system. *J. Exp. Zool.* **10**, 63–83.

Burrows, M. T. (1916–1917). Some factors regulating growth. *Anat. Rec.* **11**, 335–339.

Carrel, A. (1912). On the permanent life of tissues outside of the organism. *J. Exp. Med.* **15**, 516–528.

Carrel, A. (1913a). Contributions to the study of the mechanisms of the growth of tissues. The effects of the dilution of the medium on the growth of connective tissue. *J. Exp. Med.* **18**, 287–299.

Carrel, A. (1913b). Artificial activation of the growth *in vitro* of connective tissue. *J. Exp. Med.* **17**, 14–19.

Carrel, A. (1928). Modern techniques of tissue culture and results. *Arch. Exp. Zellforsch. besonders Gewebezücht.* **6**, 70–81.

Carrel, A., and Burrows, M. T. (1911). On the physiocochemical regulation of the growth of tissues. The effects of the dilution of the medium on the growth of the spleen. *J. Exp. Med.* **13**, 562–570.

Ceriani, R. L. (1970). Fetal mammary gland differentiation *in vitro* in response to hormones. I. Morphological findings. *Develop. Biol.* **21**, 506–529.

Chang, R. S. (1957). Isolation of nutritional variants from conjunctival and HeLa cells. *Proc. Soc. Exp. Biol. Med.* **96**, 818–820.

Chang, R. S. (1958). Differences in inositol requirements of several strains of HeLa, conjunctival and amnion cells. *Proc. Soc. Exp. Biol. Med.* **99**, 99–102.

Chang, R. S., Liepens, H., and Margoliash, M. (1961). Carbon dioxide requirements and nucleic acid metabolism of HeLa and conjunctival cells. *Proc. Soc. Exp. Biol. Med.* **106**, 149–152.

Christianson, M., and Jones, I. C. (1957). The interrelationships of the adrenal glands of mother and foetus in the rat. *J. Endocrinol.* **15**, 17–42.

Cleaver, J. E. (1965). Effect of physiological saline on synthesis of deoxyribonucleic acid and ribonucleic acid in mammalian cells. *Nature (London)* **206**, 401–403.

Cone, C. D. (1969). Electroosmotic interactions accompanying mitosis initiation in sarcoma cells *in vitro*. *Trans. N.Y. Acad. Sci.* [2] **31**, 404–427.

Coon, H. G., and Cahn, R. D. (1966). Differentiation *in vitro:* Effects of Sephadex fractions of chick embryo extract. *Science* **153**, 1116–1119.

Cooper, P. D., Burt, A. M., and Wilson, J. N. (1958). Critical effect of oxygen tension on rate of growth of animal cells in continuous suspended culture. *Nature (London)* **182**, 1508–1509.

Crocker, F. S., and Vernier, R. L. (1970). Fetal kidney in organ culture: Abnormalities of development induced by decreased amounts of potassium. *Science* **169**, 485–487.

Crockett, R. L., and Leslie, I. (1963). Utilization of ^{14}C-labeled glucose by human cells (strain HLM) in tissue culture. *Biochem. J.* **89**, 516–525.

De Bruyn, W. M., Korteweg, R., and Kits van Waveren, E. (1949). Transplantable mouse lymphosarcoma T86157(MB) studied *in vivo, in vitro*, and at autopsy. *Cancer Res.* **9**, 282–293.

De Haan, R. L. (1967). Regulation of spontaneous activity and growth of embryonic chick heart cells in tissue culture. *Develop. Biol.* **16**, 216–249.

De Haan, R. L. (1970). The potassium-sensitivity of isolated embryonic heart cells increases with development. *Develop. Biol.* **23**, 226–240.

Drew, A. H. (1927–1928). Notes on the cultivation of tumours *in vitro*. *Arch. Exp. Zellforsch. besonders Gewebezücht.* **5**, 128–130.

Dubreuil, R., and Pavilanis, V. (1958). Emploi d'un milieu synthétique simplifié pour la culture *in vitro* de cellules animales et de la poliomyélite. *Can. J. Microbiol.* **4**, 543–550.

Dupree, L. T., Sanford, K. K., Westfall, B. B., and Covalsky, A. B. (1962). Influence of serum protein on determination of nutritional requirements of cells in culture. *Exp. Cell Res.* **28**, 381–405.

Eagle, H. (1955). Nutrition needs of mammalian cells in tissue culture. *Science* **122**, 501–504.

Eagle, H. (1959). Amino acid metabolism in mammalian cell cultures. *Science* **130**, 432–437.

Eagle, H., and Piez, K. A. (1962). The population-dependent requirement by cultured mammalian cells for metabolites which they can synthesize. *J. Exp. Med.* **116**, 29–43.

Eagle, H., Barban, S., Levy, M., and Schulze, H. O. (1958). The utilization of carbohydrates by human cell cultures. *J. Biol. Chem.* **233**, 551–558.

Earle, W. R. (1943). Production of malignancy *in vitro*. IV. The mouse fibroblast cultures and changes seen in the living cells. *J. Nat. Cancer Inst.* **4**, 165–212.

Elias, J. J. (1957). Cultivation of adult mouse mammary gland in hormone-enriched synthetic medium. *Science* **126**, 842–844.

Elias, J. J. (1962). Response of mouse mammary duct end-buds to insulin in organ culture. *Exp. Cell Res.* **27**, 601–604.

Elias, J. J., and Rivera, E. (1959). Comparison of the responses of normal, precancerous, and neoplastic mouse mammary tissues to hormones *in vitro. Cancer Res.* **19**, 505–511.

Evans, V. J., Bryant, J. C., Kerr, H. A., and Schilling, E. L. (1964). Chemically defined media for cultivation of long-term cell strains from four mammalian species. *Exp. Cell Res.* **36**, 439–474.

Favour, C. B. (1964). Antigen-antibody reactions in tissue culture. In "Immunological Methods" (J. F. Ackroyd, ed.), pp. 195–223. Davis, Philadelphia, Pennsylvania.

Fell, H. B., and Rinaldini, L. (1965). The effects of vitamins A and C on cells and tissues in culture. Chapter 17, pp. 659–699. In "Cells and Tissues in Culture" (E. N. Willmer, ed.), Vol. 1, Academic Press, New York.

Fillerup, D. L., Migliore, J. V., and Mead, J. F. (1958). The uptake of lipoproteins by ascites tumor cells. The fatty acid-albumin complex. *J. Biol. Chem.* **233**, 98–101.

Fischer, A. (1948). Amino-acid metabolism of tissue cells *in vitro. Biochem. J.* **43**, 491–497.

Fischer, A., Astrup, T., Ehrensvärd, G., and Øhlenschläger, V. (1948). Growth of animal tissue cells in artificial media. *Proc. Soc. Exp. Biol. Med.* **67**, 40–46.

Fischer, G. A. (1958). Studies of the culture of leukemic cells *in vitro. Ann. N.Y. Acad. Sci.* **76**, 673–680.

Fisher, H. W., Puck, T. T., and Sato, G. (1958). Molecular growth requirements of single mammalian cells: The action of fetuin in promoting cell attachment to glass. *Proc. Nat. Acad. Sci. U.S.* **44**, 4–10.

Fisher, H. W., Puck, T. T., and Sato, G. (1959). Molecular growth requirements of single mammalian cells: III. Quantitative colonial growth of single S3 cells in a medium containing synthetic small molecular constituents and two purified protein fractions. *J. Exp. Med.* **109**, 649–660.

Fisk, A., and Pathak, S. (1969). HEPES-buffered medium for organ culture. *Nature (London)* **224**, 1030–1031.

Floss, D. R. (1964). The determination of the optimal concentration of lactalbumin hydrolyzate for the cultivation of a human cell line *in vitro. Exp. Cell Res.* **34**, 603–605.

Gardner, R. S. (1969). The use of tricine buffer in animal tissue cultures. *J. Cell Biol.* **42**, 320–321.

Garvey, J. S. (1961). Separation and *in vitro* culture of cells from liver tissue. *Nature (London)* **191**, 972–974.

Gershonson, L. E., Mead, J. F., Harary, I., and Haggerty, D. F. (1967). Studies on the effects of essential fatty acids on growth rate, fatty acid composition, oxidative phosphorylation and respiratory control of HeLa cells in culture. *Biochim. Biophys. Acta* **131**, 42–49.

Gey, G. O., and Gey, M. K. (1936). The maintenance of human normal cells and tumor cells in continuous culture. I. Preliminary report: Cultivation of mesoblastic tumors and normal tissue and notes on methods of cultivation. *Amer. J. Cancer* **27**, 45–76.

Geyer, R. P. (1958). Nutrition of mammalian cells in tissue culture. *Nutr. Rev.* **16**, 321–323.

Geyer, R. P., and Chang, R. S. (1958). Bicarbonate as an essential for human cells *in vitro. Arch. Biochem. Biophys.* **73**, 500–506.

Goldstein, L., Cailleau, R., and Crocker, T. T. (1960). Nuclear-cytoplasmic relationships in human cells in tissue culture. II. The microscopic behavior of enucleate human cell fragments. *Exp. Cell Res.* 19, 332–342.

Gomori, G. (1946). Buffers in the range of pH 6.5 to 9.6. *Proc. Soc. Exp. Biol. Med.* 63, 33–34.

Good, N. E., Winget, G. D., Winter, W., Connolly, T. N., Izawa, S., and Singh, R. M. M. (1966). Hydrogen ion buffers for biological research. *Biochemistry* 5, 467–477.

Gorham, L. W., and Waymouth, C. (1965). Differentiation *in vitro* of embryonic cartilage and bone in a chemically defined medium. *Proc. Soc. Exp. Biol. Med.* 119, 287–290.

Graff, S., and McCarty, K. S. (1957). Sustained cell culture. *Exp. Cell Res.* 13, 348–357.

Graff, S., and McCarty, K. S. (1958). Energy costs of growth in continuous metazoan cell cultures. *Cancer Res.* 18, 741–746.

Greene, A. E., Silver, R. K., Krug, M. D., and Coriell, L. L. (1965). A premixed powder for preparation of tissue culture media. *Proc. Soc. Exp. Biol. Med.* 118, 122–128.

Gregory, J. D., and Sajdera, S. W. (1970). Interference in the Lowry method for protein determination. *Science* 169, 97–98.

Griffiths, J. B., and Pirt, S. J. (1967). The uptake of amino acids by mouse cells (strain LS) during growth in batch culture and chemostat culture: The influence of cell growth rate. *Proc. Roy. Soc., Ser. B* 168, 421–438.

Gwatkin, R. B. L. (1960). Are macromolecules required for growth of single isolated mammalian cells? *Nature (London)* 196, 984–985.

Gwatkin, R. B. L., and Siminovitch, L. (1960). Multiplication of single cells in a non-bicarbonate medium. *Proc. Soc. Exp. Biol. Med.* 103, 718–721.

Haggerty, D. F., and Sato, G. (1969). The requirement for biotin in mouse fibroblast L-cells cultured on serumless medium. *Biochem. Biophys. Res. Commun.* 34, 812–815.

Ham, R. G. (1960). Small molecule requirements for clonal growth of single diploid Chinese hamster cells. *Fed. Proc., Fed. Amer. Soc. Exp. Biol.* 19, 387.

Ham, R. G. (1962). Clonal growth of diploid Chinese hamster cells in a synthetic medium supplemented with purified protein fractions. *Exp. Cell Res.* 28, 489–500.

Ham, R. G. (1963a). An improved nutrient solution for diploid Chinese hamster and human cell lines. *Exp. Cell Res.* 29, 515–526.

Ham, R. G. (1963b). Albumin replacement by fatty acids in clonal growth of mammalian cells. *Science* 140, 802–803.

Ham, R. G. (1964). Putrescine and related amines as growth factors for a mammalian cell line. *Biochem. Biophys. Res. Commun.* 14, 34–38.

Ham, R. G. (1965). Clonal growth of mammalian cells in a chemically defined, synthetic medium. *Proc. Nat. Acad. Sci. U.S.* 53, 288–293.

Ham, R. G., and Sattler, G. L. (1968). Clonal growth of differentiated rabbit cartilage cells. *J. Cell. Physiol.* 72, 109–114.

Hanks, J. H. (1948). The longevity of chick tissue cultures without renewal of medium. *J. Cell. Comp. Physiol.* 31, 235–260.

Hanks, J. H. (1955). Balanced salt solutions, inorganic requirements and pH control. *In* "An Introduction to Cell and Tissue Culture" (J. H. Hanks *et al.*, eds.), Chapter 3, pp. 5–8. Burgess, Minneapolis, Minnesota.

Hanks, J. H., and Wallace, R. E. (1959). Relation of oxygen and temperature in the preservation of tissues by refrigeration. *Proc. Soc. Exp. Biol. Med.* **71**, 196–200.

Hanss, J., and Moore, G. E. (1964). Studies of culture media for the growth of human tumor cells. *Exp. Cell Res.* **34**, 243–256.

Harary, I., McCarl, R., and Farley, B. (1966). Studies *in vitro* on single beating rat heart cells. IX. The restoration of beating by serum lipids and fatty acids. *Biochem. Biophys. Acta* **115**, 15–22.

Harris, M. (1954). The role of bicarbonate for outgrowth of chick heart fibroblasts *in vitro. J. Exp. Zool.* **125**, 85–98.

Harris, M., and Kutsky, R. (1953). Utilization of added sugar by chick heart fibroblasts in dialyzed media. *J. Cell. Comp. Physiol.* **42**, 449–470.

Harrison, R. G. (1906–1907). Observations on the living developing nerve fiber. *Proc. Soc. Exp. Biol. Med.* **4**, 140–143.

Harrison, R. G. (1908). Embryonic transplantation and development of the nervous system. *Anat. Rec.* **2**, 385–410.

Havard, R. E., and Kendal, L. P. (1934). The effect of the oxidation-reduction potential of the medium on the growth of tissue cultures. *Biochem. J.* **28**, 1121–1130.

Hay, R. J., and Paul, J. (1967). Factors influencing glucose flux and the effect of insulin in cultured human cells. *J. Gen. Physiol.* **50**, 1663–1680.

Hayflick, L., Jacobs, P., and Perkins, F. (1964). A procedure for the standardization of tissue culture media. *Nature (London)* **204**, 146–147.

Healy, G. M., and Parker, R. C. (1966a). An improved chemically defined basal medium (CMRL-1415) for newly explanted mouse embryo cells. *J. Cell Biol.* **30**, 531–538.

Healy, G. M., and Parker, R. C. (1966b). Cultivation of mammalian cells in defined media with protein and nonprotein supplements. *J. Cell Biol.* **30**, 539–553.

Healy, G. M., Morgan, J. F., and Parker, R. C. (1952). Trace metal content of some natural and synthetic media. *J. Biol. Chem.* **198**, 305–312.

Healy, G. M., Fisher, D. C., and Parker, R. C. (1954). Nutrition of animal cells in tissue culture. IX. Synthetic medium No. 703. *Can. J. Biochem. Physiol.* **32**, 327–337,

Healy, G. M., Fischer, D. C., and Parker, R. C. (1955). Nutrition of animal cells in tissue culture. X. Synthetic medium No. 858. *Proc. Soc. Exp. Biol. Med.* **89**, 71–77.

Hetherington, D. C., and Shipp, M. E. (1935). The effect of cupric, manganous and ferric chlorides upon cardiac explants in tissue culture. *Biol. Bull.* **68**, 215–230.

Herzenberg, L. A., and Roosa, R. A. (1959). Serine, glycine, alanine, and certain α-keto acids as alternative growth factors for a mouse lymphoid neoplasm in cell culture. *Fed. Proc., Fed. Amer. Soc. Exp. Biol.* **18**, 401.

Higuchi, K. (1963). Studies on the nutrition and metabolism of animal cells in serum-free media. I. Serum-free monolayer cultures. *J. Infec. Dis.* **112**, 213–220.

Higuchi, K. (1970). An improved chemically defined culture medium for strain L mouse cells based on growth responses to graded levels of nutrients including iron and zinc. *J. Cell. Physiol.* **75**, 65–72.

Holmes R,. (1959). Long-term cultivation of human cells (Chang) in chemically defined medium and effect of added peptone fractions. *J. Biophys. Biochem. Cytol.* **6**, 535–536.

Holmes, R. (1967). Preparation from human serum of an alpha-one protein which induces the immediate growth of unadapted cells *in vitro. J. Cell Biol.* **32**, 297–308.

Holmes, R., and Wolfe, S. W. (1961). Serum fractionation and the effects of bovine serum fractions on human cells grown in a chemically defined medium. *J. Biophys. Biochem. Cytol.* **10**, 389–401.

Holmes, R., Helms, J., and Mercer, G. (1969). Cholesterol requirement of primary diploid human fibroblasts. *J. Cell Biol.* **42**, 262–271.

Howard, B. V., and Kritchevsky, D. (1969). The source of cellular lipid in the human diploid cell strain WI-38. *Biochim. Biophys. Acta* **187**, 293–301.

Hsu, T. C., and Merchant, D. J. (1961). Mammalian chromosomes *in vitro.* XIV. Genotypic replacement in cell populations. *J. Nat. Cancer Inst.* **26**, 1075–1083.

Hughes, A. (1952). "The Mitotic Cycle." Butterworth, London.

Ito, T., and Moore, G. E. (1969). The growth-stimulating activity of peptides on human hematopoietic cell cultures. *Exp. Cell Res.* **56**, 10–14.

Iwakata, J. T., and Grace, J. T. (1964). Cultivation *in vitro* of myeloblasts from human leukemia. *N.Y. State J. Med.* **64**, 2279–2282.

Jacobs, J. P. (1966). A simple medium for the propagation and maintenance of human diploid cell strains. *Nature (London)* **210**, 100–101.

Jenkin, H. M., and Anderson, L. E. (1970). The effect of oleic acid on the growth of monkey kidney cells (LLC-MK₂). *Exp. Cell Res.* **59**, 6–10.

Kelley, G. G., Adamson, D. J., and Vail, M. H. (1960). Further studies on the growth of human tissue cells in the absence of an added gas phase. *Amer. J. Hyg.* **72**, 275–278.

Kern, M., and Eisen, H. N. (1959). The effect of antigenic stimulation on incorporation of phosphate and methionine into proteins of isolated lymph node cells. *J. Exp. Med.* **110**, 207–219.

Kilburn, D. G., and Webb, F. C. (1968). The cultivation of animal cells at controlled dissolved oxygen partial pressure. *Biotechnol. Bioeng.* **10**, 801–814.

Kitos, P. A., and Waymouth, C. (1964). Glucose metabolism by mouse cells (NCTC clone 929) under conditions of defined nutrition. *Exp. Cell Res.* **35**, 108–118.

Kitos, P. A., and Waymouth, C. (1966). The metabolism of L-glutamate and L-5-carboxy-pyrrolidone by mouse cells (NCTC clone 929) under conditions of defined nutrition. *J. Cell. Physiol.* **67**, 383–398.

Kitos, P. A., Sinclair, R., and Waymouth, C. (1962). Glutamine metabolism by animal cells growing in a synthetic medium. *Exp. Cell Res.* **27**, 307–316.

Kuchler, R. J. (1967). The role of sodium and potassium in regulating amino acid accumulation and protein synthesis in LM-strain mouse fibroblasts. *Biochim. Biophys. Acta* **136**, 473–483.

Kuchler, R. J., Marlowe, M. L., and Merchant, D. J. (1960). The mechanism of cell binding and cell-sheet formation in L strain fibroblasts. *Exp. Cell Res.* **20**, 428–437.

Lasfargues, E. Y. (1962). Concerning the role of insulin in the differentiation and functional activity of mouse mammary tissues. *Exp. Cell Res.* **28**, 531–542.

Lasnitzki, I. (1965). The action of hormones on cell and organ cultures. *In* "Cells and Tissues in Culture" (E. N. Willmer, ed.), Vol. 1, Chapter 16, pp. 591–658. Academic Press, New York.

Leibovitz, A. (1963). The growth and maintenance of tissue-cell cultures in free gas exchange with the atmosphere. *Amer. J. Hyg.* **78**, 173–180.

Leslie, I., Fulton, W. C., and Sinclair, R. (1956). Biochemical tests for malignancy applied to a new strain of human cells. *Nature (London)* **178**, 1179–1180.

Levine, A. S., Bond, P. H., and Rouse, H. C. (1956a). Modification of viral synthesis in tissue culture by substituting pyruvate for glucose in the medium. *Proc. Soc. Exp. Biol. Med.* **93**, 233–235.

Levine, A. S., Bond, P. H., Scala, A. R., and Eaton, M. D. (1956b). Studies on the relationship of potassium to influenza virus multiplication. *J. Immunol.* **76**, 386–392.

Lewis, M. R., and Lewis, W. H. (1911a). The cultivation of tissues from chick embryos in solutions of NaCl, CaCl₂, KCl and NaHCO₃. *Anat. Rec.* **5**, 277–293.

Lewis, M. R., and Lewis, W. H. (1911b). On the growth of embryonic chick tissues in artificial media, nutrient agar and bouillon. *Bull. Johns Hopkins Hosp.* **22**, 126–127.

Lewis, W. H., and Lewis, M. R. (1912). The cultivation of chick tissues in media of known chemical composition. *Anat. Rec.* **6**, 207–211.

Lieberman, I., and Ove, P. (1957). Purification of a serum protein required by a mammalian cell in tissue culture. *Biochim. Biophys. Acta* **25**, 449–450.

Lieberman, I., and Ove, P. (1958). A protein growth factor for mammalian cells in culture. *J. Biol. Chem.* **233**, 637–642.

Lieberman, I., Lamy, F., and Ove, P. (1959). Non-identity of fetuin and protein growth (flattening) factor. *Science* **129**, 43–44.

Ling, C. T., Gey, G. O., and Richters, V. (1968). Chemically characterized concentrated corodies for continuous cell culture (the 7C's media). *Exp. Cell Res.* **52**, 469–489.

Littlefield, J. W. (1964). Selection of hybrids from matings of fibroblasts *in vitro* and their presumed recombinants. *Science* **145**, 709–710.

Littlefield, J. W., and Goldstein, S. (1970). Some aspects of somatic cell hybridization. *In Vitro* **6**, 21–31.

Lockart, R. Z., and Eagle, H. (1959). Requirements for growth of single human cells. *Science* **129**, 252–254.

Locke, F. S. (1895a). Artificial fluids as uninjurious as possible to animal tissues. *Boston Med. Surg. J.* **134**, 173.

Locke, F. S. (1895b). Towards the ideal artificial circulating fluid for the isolated frog's heart. *J. Physiol. (London)* **18**, 332–333.

Locke, F. S. (1900). Die Wirkung der Metalle das Blutplasmas und verschiedener Zucker auf des isoliertes Saügetierherz. *Zentralbl. Physiol.* **14**, 670.

Lostroh, A. J. (1963). Effects of insulin and of DL-aldosterone on protein synthesis by mouse uteri in organ culture. *Exp. Cell Res.* **32**, 327–332.

Lostroh, A. J. (1966). Effect of glucose, pyruvate, and insulin on protein synthesis in explanted mouse uteri. *Amer. J. Physiol.* **211**, 809–814.

Lowry, O. H., Rosebrough, N. J., Farr, A. L., and Randall, R. J. (1951). Protein measurement with the Folin phenol reagent. *J. Biol. Chem.* **193**, 265–275.

Lowy, P. H., and Keighley, G. (1968). Inactivation of erythropoietin by Koshland's tryptophan reagent and by membrane filtration. *Biochim. Biophys. Acta* **160**, 413–419.

Lubin, M. (1967). Intracellular potassium and macromolecular synthesis in mammalian cells. *Nature (London)* **213**, 451–453.

Lucas, D. R. (1969). The effect of hydrocortisone, oxygen tension and other factors on the survival of the submandibular, sublingual, parotid and exorbital lacrimal glands in organ culture. *Exp. Cell Res.* **55**, 229–242.

Lucas, D. R., Peakman, E. M., and Smith, C. (1970). The effect of insulin, steroid and other hormones on the survival of the rat salivary glands in organ culture. *Exp. Cell Res.* **60**, 262–268.

McCarty, K. S. (1962). Selective utilization of amino acids by mammalian cell cultures. *Exp. Cell Res.* **27**, 230–240.

McCarty, K. S., and Graff, S. (1959). Some aspects of nitrogen metabolism in strain L cell cultures. *Exp. Cell Res.* **16**, 518–526.

McCoy, T. A., Maxwell, M., and Kruse, P. F. (1959a). Amino acid requirements of the Novikoff hepatoma *in vitro*. *Proc. Soc. Exp. Biol. Med.* **100**, 115–118.

McCoy, T. A., Maxwell, M., Irvine, E., and Sartorelli, A. C. (1959b). Two nutritional variants of cultured Jensen sarcoma cells. *Proc. Soc. Exp. Biol. Med.* **100**, 862–865.

McLimans, W. F., Giardinello, F. E., Davis, E. V., Kucera, J., and Rake, G. W. (1957). Submerged culture of mammalian cells: The five liter fermentor. *J. Bacteriol.* **74**, 768–774.

Martin, G. (1964). Use of tris(hydroxymethyl)aminomethane buffers in cultures of human diploid fibroblasts. *Proc. Soc. Exp. Biol. Med.* **116**, 167–171.

Matsumura, T., Takaoka, T., and Katsuta, H. (1968). A polyelectrolyte buffer system for bacterial and mammalian cell culture. *Exp. Cell Res.* **53**, 337–347.

Maxwell, M., Orr, G. R., and McCoy, T. A. (1959). Biochemical studies of the JA-2 strain of the Jensen sarcoma *in vitro*. *Proc. Amer. Ass. Cancer Res.* **3**, 40.

Melcher, A. H., and Hodges, G. M. (1968). In vitro culture of an organ containing mixed epithelial and connective tissues on a chemically defined medium. *Nature (London)* **219**, 301–302

Merchant, D. J., and Hellman, K. B. (1962). Growth of LM strain of mouse cells in a chemically defined medium. *Proc. Soc. Exp. Biol. Med.* **110**, 194–198.

Moore, G. E., and Kitamura, H. (1968). Cell line derived from a patient with multiple myeloma. *N.Y. State J. Med.* **68**, 2054–2060.

Moore, G. E., Mount, D. D., Tara, G., and Schwartz, N. (1963). Growth of human tumor cells in suspension culture. *Cancer Res.* **23**, 1735–1741.

Moore, G. E., Sandberg, A. A., and Ulrich, K. (1966). Suspension cell culture and *in vivo* and *in vitro* chromosome constitution of mouse leukemia. *J. Nat. Cancer Inst.* **36**, 405–421.

Moore, G. E., Gerner, R. E., and Franklin, H. A. (1967). Culture of normal human leukocytes. *J. Amer. Med. Ass.* **199**, 519–524.

Moretti, R. L., and De Ome, K. M. (1962). Effect of insulin on glucose uptake by normal and neoplastic mouse mammary tissues in organ culture. *J. Nat. Cancer Inst.* **29**, 321–329.

Morgan, J. F. (1958). Tissue culture nutrition. *Bacteriol. Rev.* **22**, 20–45.

Morgan, J. F. (1950). In "Methods of Tissue Culture" (R. C. Parker, ed.), 2nd ed., Chapter VIII, pp. 115–131. Harper (Hoeber), New York.

Morgan, J. F. (1970). Nutrition of mouse ascites tumor cells in primary culture. I. Large molecular substances and conditioned media. *J. Nat. Cancer Inst.* **44**, 623–631.

Morgan, J. F., and Morton, H. J. (1960). Carbohydrate utilization by chick embryonic heart cultures. *Can. J. Biochem. Physiol.* **38**, 69–78.

Morgan, J. F., Morton, H. J., and Parker, R. C. (1950). Nutrition of animal cells in tissue culture. I. Initial studies on a synthetic medium. *Proc. Soc. Exp. Biol. Med.* **73**, 1–8.

Morton, H. J. (1967). Role of carbon dioxide in erythropoiesis. *Nature* (*London*) **215**, 1166–1167.

Morton, H. J. (1968). Effect of calcium on bone marrow mitosis *in vitro*. *Proc. Soc. Exp. Biol. Med.* **128**, 112–116.

Morton, H. J. (1970). A survey of commercially available tissue culture media. *In Vitro* **6**, 80–108.

Nagle, S. C. (1968). Heat-stable chemically defined medium for growth of animal cells in suspension. *Appl. Microbiol.* **16**, 53–55.

Nagle, S. C. (1969). Improved growth of mammalian and insect cells in media containing increased levels of choline. *Appl. Microbiol.* **17**, 318–319.

Nagle, S. C., Tribble, H. R., Anderson, R. E., and Gary, N. D. (1963). A chemically defined medium for growth of animal cells in suspension. *Proc. Soc. Exp. Biol. Med.* **112**, 240–346.

Neuman, R. E., and McCoy, T. A. (1958). Growth-promoting properties of pyruvate, oxalacetate, and α-ketoglutarate for isolated Walker carcinosarcoma 256 cells. *Proc. Soc. Exp. Biol. Med.* **98**, 303–306.

Olmsted, C. A. (1967). A physico-chemical study of fetal calf sera used as tissue culture nutrient correlated with biological tests for toxicity. *Exp. Cell Res.* **48**, 283–299.

Pace, D. M., Thompson, J. R., and Van Camp, W. A. (1962). Effects of oxygen on growth in several established cell lines. *J. Nat. Cancer Inst.* **28**, 897–909.

Pannett, C. A., and Compton, A. (1924). The cultivation of tissues in saline embryonic juice. *Lancet* **1**, 381–384.

Parker, R. C., ed. (1961). "Methods of Tissue Culture," 3rd ed. Harper (Hoeber), New York.

Parsa, I., Marsh, W. H., and Fitzgerald, P. J. (1970). Chemically defined medium for organ culture differentiation of rat pancreas anlage. *Exp. Cell Res.* **59**, 171–175.

Paul, J. (1959). Environmental influences on the metabolism and composition of cultured cells. *J. Exp. Zool.* **142**, 475–505.

Paul, J. (1965). Carbohydrate and energy metabolism. *In* "Cells and Tissues in Culture," (E. N. Willmer, ed.), Vol. 1, Chapter 7, pp. 239–276. Academic Press, New York.

Paul, J., and Pearson, E. S. (1960). The action of insulin on the metabolism of cell cultures. *J. Endocrinol.* **21**, 287–294.

Pederson, K. O. (1944). Fetuin, a new globulin isolated from serum. *Nature* (*London*) **154**, 575.

Perris, A. D., and Whitfield, J. F. (1967). Calcium and the control of mitosis in the mammal. *Nature* (*London*) **216**, 1350–1351.

Perris, A. D., Whitfield, J. F., and Tolg, P. K. (1968). Role of calcium in the control of growth and cell division. *Nature* (*London*) **219**, 527–529.

Pirt, S. J., and Callow, D. S. (1964). Continuous-flow culture of the ERK and L types of mammalian cells. *Exp. Cell Res.* **33**, 413–421.

Pirt, S. J., and Thackeray, E. J. (1964a). Environmental influences on the growth of ERK mammalian cells in monolayer culture. *Exp. Cell Res.* **33**, 396–405.

Pirt, S. J., and Thackeray, E. J. (1964b). Environmental influences on growth of L and ERK mammalian cells in shake-flask cultures. *Exp. Cell Res.* **33**, 406–412.

Price, F. M., Kerr, H. A., Andresen, W. F., Bryant, J. C., and Evans, V. J. (1966).

Some *in vitro* requirements of cells of C3H mouse, Chinese hamster, green monkey, and human origin. *J. Nat. Cancer Inst.* 37, 601–617.

Prop, F. J. A. (1960). Development of alveoli in organ cultures of total mammary glands of six weeks old virgin mice. *Exp. Cell Res.* 20, 256–258.

Puck, T. T., and Fisher, H. W. (1956). Genetics of somatic mammalian cells. I. Demonstration of the existence of mutants with different growth requirements in a human cancer cell strain (HeLa). *J. Exp. Med.* 104, 427–434.

Puck, T. T., and Marcus, P. I. (1955). A rapid method for viable cell titration and clone production with HeLa cells in tissue culture: The use of X-irradiated cells to supply conditioning factors. *Proc. Nat. Acad. Sci. U.S.* 41, 432–437.

Pumper, R. W., Yamashiroya, H. M., and Molander, L. T. (1965). Growth of mammalian cells in a heat-stable dialysable medium. *Nature (London)* 207, 662–663.

Quastel, M. R., and Kaplan, J. G. (1970). Significance of the early stimulation by phytohemagglutinin of potassium transport in lymphocytes *in vitro. J. Cell Biol.* 47, 164a.

Rappaport, C., Poole, J. P., and Rappaport, H. P. (1960). Studies on properties of surfaces required for growth of mammalian cells in synthetic media. *Exp. Cell Res.* 20, 465–510.

Riggs, T. R., Walker, L. M., and Christensen, H. N. (1958). Potassium migration and amino acid transport. *J. Biol. Chem.* 233, 1479–1484.

Ringer, S. (1880). Concerning the influence exerted by each of the constituents of the blood on the contraction of the ventricle. *J. Physiol. (London)* 3, 380–393.

Ringer, S. (1883a). A further contribution regarding the influence of the different constituents of the blood on the contraction of the heart. *J. Physiol. (London)* 4, 29–32.

Ringer, S. (1883b). A third contribution regarding the influence of the inorganic constituents of the blood on the ventricular contraction. *J. Physiol. (London)* 4, 222–225.

Ringer, S. (1886). Further experiments regarding the influence of small quantities of lime, potassium and other salts on muscular tissue. *J. Physiol. (London)* 7, 291–308.

Ringer, S. (1895). Further observations regarding the antagonism between calcium salts and sodium potassium and ammonium salts. *J. Physiol. (London)* 18, 425–429.

Rivera, E. M. (1963). Hormonal requirements for survival and growth of mouse primary mammary ducts in organ culture. *Proc. Soc. Exp. Biol. Med.* 114, 735–738.

Rivera, E. M. (1964). Differential responsiveness to hormones of C3H and A mouse mammary tissues in organ culture. *Endocrinology* 74, 853–864.

Rivera, E. M. (1966). Strain differences in mouse mammary tissue sensitivity to prolactin and somatotrophin in organ culture. *Nature (London)* 209, 1151–1152.

Rivera, E. M., and Bern, H. A. (1961). Influence of insulin on maintenance and secretory stimulation of mouse mammary tissues by hormones in organ-culture. *Endocrinology* 69, 340–353.

Roosa, R. A., and Herzenberg, L. A. (1959). The selection of variants resistant to folic and antagonists and 5-fluorouridine in cell cultures of a lymphocytic neoplasm. *Proc. Amer. Ass. Cancer Res.* 3, 58.

Rothblat, G. H., and Kritchevsky, D., eds. (1967). "Lipid Metabolism in Tissue

Culture Cells," Wistar Inst. Symp., Monograph No. 6. Wistar Inst. Anat. Biol., Philadelphia, Pennsylvania.

Rothblat, G. H., Buchko, M. K., and Kritchevsky, D. (1968). Cholesterol uptake by L5178Y tissue culture cells: Studies with delipidized serum. *Biochim. Biophys. Acta* 164, 327–338.

Rubin, H. (1966). A substance in conditioned medium which enhances the growth of small numbers of chick embryo cells. *Exp. Cell Res.* 41, 138–148.

Runyan, W. S., and Geyer, R. P. (1963). Growth of L cell suspensions on a Warburg apparatus. *Proc. Soc. Exp. Biol. Med.* 112, 1027–1030.

Runyan, W. S., and Geyer, R. P. (1967). Partial replacement of CO_2 in strain L and HeLa cultures. *Proc. Soc. Exp. Biol. Med.* 125, 1301–1304.

Ryan, M. F. (1969). Unreliable results. *Science* 165, 851.

Ryan, M. P., Smyth, H., and Hingerty, D. (1969). Effects of Mg^{++} and K^+ deficiencies on composition and growth of ascites tumour cells *in vivo*. *Life Sci.* 8, 485–494.

Sanford, K. K., and Westfall, B. B. (1969). Growth and glucose metabolism of high and low tumor-producing clones under aerobic and anaerobic conditions *in vitro*. *J. Nat. Cancer Inst.* 42, 953–959.

Sanford, K. K., Earle, W. R., and Likely, G. D. (1948). The growth *in vitro* of single isolated tissue cells. *J. Nat. Cancer Inst.* 9, 229–246.

Sanford, K. K. Dupree, L. T., and Covalesky, A. B. (1963). Biotin, B_{12}, and other vitamin requirements of a strain of mammalian cells grown in chemically defined medium. *Exp. Cell Res.* 31, 345–375.

Scott, B. S., and Fisher, K. S. (1970). Potassium concentration and number of neurons in cultures of dissociated ganglia. *Exp. Neurol.* 27, 16–22.

Sergeant, T. P., and Smith, S. (1960). A growth-stimulating factor for an epithelial cell line in a reduced serum medium. *Science* 131, 606–607.

Shipman, C. (1969). Evaluation of 4-(2-hydroxyethyl)-1-piperazineëthane-sulfonic acid (HEPES) as a tissue culture buffer. *Proc. Soc. Exp. Biol. Med.* 130, 305–310.

Shooter, R. A., and Gey, G. O. (1952). Studies of the mineral requirements of mammalian cells. *Brit. J. Exp. Pathol.* 33, 98–103.

Sinclair, R. (1966). Steady-state suspension culture and metabolism of strain L mouse cells in simple defined medium. *Exp. Cell Res.* 41, 20–33.

Spiro, R. G. (1960). Studies on fetuin, a glycoprotein of fetal serum. *J. Biol. Chem.* 235, 2860–2869.

Steinberger, A., and Steinberger, E. (1966). Stimulatory effects of vitamins and glutamine on the differentiation of germ cells in rat testes organ culture grown in chemically defined media. *Exp. Cell Res.* 44, 429–435.

Steinberger, E., Steinberger, A., and Perloff, W. H. (1964). Studies on growth in organ culture of testicular tissue from rats of various ages. *Anat. Rec.* 148, 581–589.

Stewart, D. C., and Kirk, P. L. (1954). The liquid medium in tissue culture. *Biol. Rev.* 29, 119–153.

Swim, H. E. (1967). Nutrition of cells in culture—A review. *Wistar Inst. Symp., Monogr.* 6, 1–16.

Swim, H. E., and Parker, R. F. (1955). Non-bicarbonate buffers in cell culture media. *Science* 122, 466.

Swim, H. E., and Parker, R. F. (1958a). Vitamin requirements of uterine fibroblasts, strain U12-79; their replacement by related compounds. *Arch. Biochem. Biophys.* 78, 46–53.

Swim, H. E., and Parker, R. F. (1958b). Stable tissue culture media prepared in dry form. *J. Lab. Clin. Med.* **52**, 309–311.

Swim, H. E., and Parker, R. F. (1958c). The role of carbon dioxide as an essential nutrient for six permanent strains of fibroblasts. *J. Biophys. Biochem. Cytol.* **4**, 525–528.

Swim, H. E., and Parker, R. F. (1960). Effect of Pluronic F68 on growth of fibroblasts in suspension on rotary shaker. *Proc. Soc. Exp. Biol. Med.* **103**, 252–254.

Szybalski, W., Szybalski, E. H., and Ragni, G. (1962). Genetic studies with human cell lines. *Nat. Cancer Inst., Monogr.* **7**, 75–89.

Thomas, J. A., and Johnson, M. J. (1967). Trace-metal requirements of NCTC clone 929 strain L cells. *J. Nat. Cancer Inst.* **39**, 337–345.

Tozer, B. T., and Pirt, S. J. (1964). Suspension cultures of mammalian cells and macromolecular growth-promoting fractions of calf serum. *Nature (London)* **201**, 375–378.

Tribble, H. R., and Higuchi, K. (1963). Studies on nutrition and metabolism of animal cells in serum-free media. II. Cultivation of cells in suspension. *J. Infec. Dis.* **112**, 221–225.

Trowell, O. A. (1952). The culture of lymph nodes *in vitro*. *Exp. Cell Res.* **3**, 79–107.

Trowell, O. A. (1955). The culture of lymph nodes in synthetic media. *Exp. Cell Res.* **9**, 258–276.

Trowell, O. A. (1959). The culture of mature organs in a synthetic medium. *Exp. Cell Res.* **16**, 118–147.

Turner, L. V., and Manchester, K. L. (1970). Interference of HEPES with the Lowry method. *Science* **170**, 649.

Tyrode, M. V. (1910). The mode of action of some purgative salts. *Arch. Int. Pharmacodyn. Ther.* **20**, 205–223.

Vann, L. S., Nerenberg, S. T., and Lewin, C. J. (1963). Glucose uptake by HeLa cells as influenced by insulin. *Exp. Cell Res.* **32**, 358–367.

Vogelaar, J. P. M., and Erlichman, E. (1933). A feeding solution for cultures of human fibroblasts. *Amer. J. Cancer* **18**, 28–48.

Wales, R. G. (1970). Effects of ions on the development of preimplantation mouse embryos *in vitro*. *Aust. J. Biol. Sci.* **23**, 421–429.

Waymouth, C. (1952). Nature of the stimulus to mitosis. In "The Mitotic Cycle" (A. Hughes), Chapter 5, pp. 163–183. Butterworth, London.

Waymouth, C. (1954). The nutrition of animal cells. *Int. Rev. Cytol.* **3**, 1–68.

Waymouth, C. (1956). A serum-free nutrient solution sustaining rapid and continuous proliferation of strain L (Earle) mouse cells. *J. Nat. Cancer Inst.* **17**, 315–327.

Waymouth, C. (1960). Growth in tissue culture. In "Fundamental Aspects of Normal and Malignant Growth" (W. W. Nowinski, ed.), Chapter 6, pp. 546–587. Elsevier, Amsterdam.

Waymouth, C. (1965a). Construction and use of synthetic media. In "Cells and Tissues in Culture" (E. N. Willmer, ed.), Vol. 1, Chapter 3, pp. 99–142. Academic Press, New York.

Waymouth, C. (1965b). The cultivation of cells in chemically defined media and the malignant transformation of cell *in vitro*. In "Tissue Culture" (C. V. Ramakrishnan, ed.), pp. 168–179. Junk Publ., The Hague.

Waymouth, C. (1965c). A chemically defined medium for study of retinal pigmentation. *Develop. Biol.* **12**, 115–116.

Waymouth, C. (1967). Somatic cells *in vitro:* Their relationship is progenitive cells and to artificial milieux. *Nat. Cancer Inst., Monogr.* **26**, 1–21.

Waymouth, C. (1968). Tissue culture media. Animal cells. *In* "Metabolism" (P. L. Altman and D. S. Dittmer, eds.), pp. 180–187. FASEB, Washington, D.C.

Waymouth, C. (1970). Osmolality of mammalian blood and of media for culture of mammalian cells. *In Vitro* **6**, 109–127.

Waymouth, C. (1972). Unpublished data.

Waymouth, C., and Reed, D. E. (1965). A reversible morphological change in mouse cells (Strain L, clone NCTC929) under the influence of insulin. *Tex. Rep. Biol. Med.* **23**, 413–419.

White, P. R. (1946). Cultivation of animal tissues *in vitro* in nutrients of precisely known constitution. *Growth* **10**, 231–289.

White, P. R. (1949). Prolonged survival of excised animal tissues *in vitro* in nutrients of precisely known constitution. *J. Cell. Comp. Physiol.* **34**, 221–241.

Whitfield, J. F., Perris, A. D., and Youdale, T. (1968). The role of calcium in the mitotic stimulation of rat thymocytes by detergents, agmatine and poly-L-lysine. *Exp. Cell Res.* **53**, 155–165.

Whitten, W. K. (1971). Nutrient requirements for the culture of preimplementation embryos in vitro. *Advan. Biosci.* **6**, 129–141.

Wildenthal, K. (1970). The influence of insulin in promoting the survival and beating of intact foetal mouse hearts in organ culture. *J. Physiol. (London)* **207**, 33P–34P.

Wilens, S. (1969). "Ross G. Harrison. Organization and Development of the Embryo" Yale Univ. Press, New Haven, Connecticut.

Williamson, J. D., and Cox, P. (1968). Use of a new buffer in the culture of animal cells. *J. Gen. Virol.* **2**, 309–312.

Willmer, E. N. (1970). "Cytology and Evolution," 2nd ed. Academic Press, New York.

Wilson, B. W., Peterson, D. W., Stinnett, H. O., Nelson, T. K., and Hamilton, W. H. (1966). Growth and metabolism of normal and genetically dystrophic chick embryo cells on a phosphate buffered medium. *Proc. Soc. Exp. Biol. Med.* **121**, 954–958.

Wilson, H., Jackson, E. B., and Brues, A. M. (1942). The metabolism of tissue cultures. I. Preliminary studies on chick embryo. *J. Gen. Physiol.* **25**, 689–703.

Wood, S., and Pinsky, L. (1970). Failure of tricine-buffered medium to support growth of human diploid fibroblasts. *In Vitro* **6**, 246.

Yamane, I., Matsuya, Y., and Jimbo, K. (1968). An autoclavable powdered culture medium for mammalian cells. *Proc. Soc. Exp. Biol. Med.* **127**, 335–336.

Yang, D. P., and Morton, H. J. (1970). Effect of Ca^{++} and Mg^{++} on morphology and growth pattern of L-M cells. *In Vitro* **6**, 224.

Young, F. B., Sharon, W. S., and Long, R. B. (1966). Preparation and use of dry powder tissue culture media. *Ann. N.Y. Acad. Sci.* **139**, 108–110.

THE ROLE OF SERUM IN THE CONTROL OF

MULTIPLICATION OF AVIAN AND

MAMMALIAN CELLS IN CULTURE

Howard M. Temin
Robert W. Pierson, Jr.
Norman C. Dulak

50   H. M. Temin, R. W. Pierson, Jr., and N. C. Dulak

I. Introduction

Macromolecular components always appear to be part of the medium of cells in culture. This statement is obviously true when serum is a component of the medium that is added to the cells. It is also true if casein, embryo extract, or other proteins are added. However, some cells multiply in cell culture in the absence of added serum or other protein (macromolecular) supplements (Evans et al., 1964; Ham, 1965). In these circumstances, although the medium as it was placed on the cells did not contain macromolecular components, the medium soon contained macromolecular components released from live or from dead cells (see below).

The role of macromolecular factors, either those added to the medium or those produced by cells, has not yet been clearly defined. In general terms, these factors appear to be required for cell multiplication. On a molecular level, little more seems known now than was known several years ago (Lucy, 1960; Harris, 1964).

Direct experiments on the role of macromolecular factors in cell culture medium could be performed only after fairly complete understanding was gained of the nature of the low-molecular-weight components of the medium (see Chapter 2, Volume 1). In fact, it appears that one role of serum in cell culture medium may be to act as a carrier of some low-molecular-weight components. Experiments suggesting that serum carried substances (unspecified) affecting cell multiplication which could pass through a dialysis bag were described by Metzgar and Moskowitz (1960) and Eagle (1960). These substances carried by serum presumably include vitamins, amino acids, lipids, and choline since with certain cells these compounds have been shown to replace much of the requirement for serum (Dupree et al., 1962; Eagle et al., 1960; Bailey, 1967; Birch and Pirt, 1969).

In evaluating the significance of the replacement of much of the requirement for serum by low-molecular-weight compounds, we must realize that these replacement experiments were all carried out with permanent cell lines, usually lines such as L cells or HeLa cells which have been cultured for many years. Since serum proteins must be produced by some cells in an animal, and since all cells in an organism appear to have the same genome, cells in culture must have the genetic potential to make serum proteins. Furthermore, cells that have been grown in culture for many years have usually grown under conditions in which serum is the limiting factor in their multiplication. Therefore, the cells will have been selected for a decreased requirement for serum and/or an increased production of substances which can replace serum factors necessary for cell multiplication. So, it is not surprising that the main requirement for serum of cells such as L or HeLa cells can be replaced by low-molecular-weight factors. Presumably, these factors are adsorbed to serum proteins and are released in culture. Even these types of cells, L and HeLa, seem to multiply more rapidly when serum is present along with the low-molecular-weight factors.

The ability of established cell lines to synthesize substances that can replace serum factors necessary for cell multiplication is illustrated by the behavior of a line of cells obtained originally by Hayden Coon as a clone from rat liver (BRL cells). These cells are epithelial in appearance and were found to multiply in the absence of serum; that is, they multiplied in a modified Eagle's medium with 20% tryptose phosphate broth and no serum. In this case, however, the presence of serum caused an actual decrease in the rate of multiplication of these BRL cells. When medium containing no added serum that had been on these BRL cells for 1 or 2 days was added to cultures of stationary rat embryo fibroblasts, which require serum for multiplication, it behaved like serum; that is, the medium from the cultures of BRL cells stimulated DNA synthesis in the fibroblasts (see Section II,D). Furthermore, a nondialyzable factor with the same multiplication-stimulating activity could be isolated from the BRL medium which did not contain any added serum (Dulak and Temin, 1972). Since these BRL cells had been kept in medium without serum for many passages, the nondialyzable factor with multiplication-stimulating activity must have been produced by the cells in culture.

We conclude, therefore, that although the supply by serum of low-molecular-weight factors may be important for some cells growing in some media, macromolecular factors must have an additional role in the multiplication of cells in culture. This topic is the primary focus of this chapter. We must keep in mind, however, the possibility that some of the active fractions isolated from serum (Section III) may be active because they carry low-molecular-weight factors. [See Birch and Pirt (1969) for further consideration of this point.]

In most early studies of the role of serum, comparison was made of the effects of medium with and medium without serum. Since comparison was not made of the effects of medium with different levels of serum, the role of serum in the control of the extent of multiplication of cells in culture was ignored. Cell multiplication was believed to be controlled by the ill-defined phenomenon of "contact inhibition of cell multiplication."

In 1954 Abercrombie and Heaysman observed that the movement of fibroblasts in cell culture appeared to be controlled by contact relations between cells—in particular, fibroblasts did not seem to migrate over other fibroblasts. To describe this phenomenon, they introduced the phrase "contact inhibition of cell movement."

Cultures of nonmultiplying or stationary fibroblasts are often monolayer. The monolayering appears to be the result of contact inhibition of movement. The cessation of cell multiplication in these cultures was, therefore, also assumed to be the result of contact inhibition. However, two types of observations led to questioning of this hypothesis. One was direct measurement of contact inhibition of cell movement and extent of cell multiplication control in two related cell populations. Macieira-Coelho (1967b) showed that human fibroblasts transformed by Rous sarcoma virus had increased contact inhibition of movement and decreased cell multiplication control as compared to untransformed human fibroblasts. The second was the study of the effects of constant renewal of serum-containing medium by either perfusion (Kruse and Miedema, 1965) or by frequent medium changes (Temin, 1965). These studies showed that cultures of contact-inhibited cells could grow to much higher population densities than usual if sufficient fresh medium with serum was added. The latter experiments shifted attention to the role of the medium in controlling cell multiplication. Experiments were performed showing that adding serum to cultures of stationary cells with fresh or with depleted medium resulted in stimulation of cell multiplication (Todaro et al., 1965; Temin, 1966; Yoshikura and Hirokawa, 1968). Other experiments showed that the final cell population density was directly proportional to the amount of serum in the medium (Temin, 1966, 1968a; Todaro et al., 1967). These experiments led to the hypothesis that serum contains some specific factors which are used in cell multiplication, and that it is the availability of these or related factors which usually controls cell multiplication.

There are factors in addition to the availability of serum factors which are involved in the control of cell multiplication. Perhaps most important are the toxic or inhibitory factors produced by cells in culture (Bürk, 1966; Rubin, 1966; Temin, 1966, 1967b, 1968a,b; Yeh and Fisher, 1969; Kotler, 1970; Gori, 1970). Furthermore, the response of cells to serum is influenced by the history of the cells and by their surroundings (Clarke et al., 1970; Dulbecco, 1970a; Temin, 1971).

In the remainder of this chapter, we shall first consider the various assays which have been used in demonstrating a biological effect of serum, and then we shall consider the attempts which have been made with these assays to isolate active macromolecular factors from serum.

II. Biological Effects of Serum

A. Cell Attachment and Detachment

Most cells in culture grow attached to a solid substratum, either glass or plastic. Most cells in culture are unable to multiply in suspension, although some cells are routinely propagated in suspension. [This difference may relate to serum requirements (see Section II,D).] Cells which do not multiply in suspension must first attach to a substratum before starting cell multiplication. Attachment of cells can be defined experimentally by a determination of cells remaining on the substratum after inversion of a culture chamber filled with liquid.

Harris (1964) discussed at length the studies on cell attachment. Several major generalizations were made. (1) Most cells can attach to a clean plastic glass surface in serum-free chemically defined medium. (2) Such attachment is rapid, does not require divalent cations, and is often temporary. (3) Soft glass with a high number of accessible sites for sodium ion exchange provides a better surface for attachment than Pyrex surfaces. (4) Serum, or a serum factor, perhaps fetuin or similar protein, in otherwise defined culture medium, promotes cell attachment. Attachment in the presence of serum is usually more permanent in nature than attachment without serum, and it does require divalent cations. (5) Attachment in the presence of serum occurs more slowly than without serum. More recently it has been shown that serum lipoproteins form a layer on the surface of the culture dish which the cells must circumvent in order to attach (Myllylä et al., 1966; Häyry et al., 1966).

The role of serum in this attachment is unclear. Curtis (1964) inferred from studies using interference microscopy that there was no protein material in the distance of 100–200Å that separated the cells from the substrate. Therefore, the serum may not be providing a direct cell-glass glue.

Recently Moore (1972) has studied attachment to plastic petri dishes of uninfected and avian sarcoma-virus-transformed rat embryo cells. Attached cells were scored by inverting the dishes 20–24 hours after the cells were plated. He found that uninfected cells in Eagle's medium attached to the plastic substratum in the absence of serum, while the avian sarcoma-virus-transformed rat cells did not. The avian sarcoma-virus-transformed rat cells would attach to the plastic in the presence of calf serum or of conditioned medium from untransformed rat cells. Furthermore, the avian sarcoma-virus-transformed rat cells would attach to the plastic in Eagle's medium if the plastic had been treated with calf serum or conditioned medium from untransformed rat cells. If a glass substrate was used, the conditioned medium was not able to promote attachment of the avian sarcoma-virus-transformed rat cells. When normal cells were plated on culture dishes that had been used to plate transformed rat cells that did not attach, the normal rat cells did not attach either. These experiments indicate that normal rat cells produce an attachment factor distinct from that in serum and that transformed rat cells produce an attachment-inhibiting factor. These appear to be tightly bound to the culture dish.

Attachment provides a good assay for isolation of active components from serum (see Section III,A). However, the molecular nature of the process and its relation to other aspects of cell growth is unclear at present.

It also appears that serum is needed to prevent cell detachment in some cases. Temin (1968a) noted that some chicken fibroblasts detached within a day after the removal of serum. With 3T3 cells, Dulbecco (1970b) found that the rate of cell detachment was inversely proportional to the concentration of serum, and directly proportional to temperature. Temin (1970a) found that rat embryo fibroblasts did not detach more rapidly in the absence of serum. Interpretation of these studies of cell detachment is complicated by the death of cells in medium without serum (see Section II,G), which also leads to detachment.

B. Cell Stretching on Solid Substratum

Cells in suspension are round. When they first attach to a solid substratum they are round. Then they lose their round shape, stretch, flatten, and attach more broadly to the substratum. Even cells whose usual shape is round, such as chicken cells converted by *morph*[r] Rous sarcoma virus, are more stretched and flattened soon after initial attachment. Since this stretching process is continuous, subjective criteria must be used in scoring stretching experimentally.

The most detailed studies of the kinetics of cell attachment and stretching were reported by A. C. Taylor (1961). He studied a line of human conjunctiva cells plated on glass. Serum was found to retard stretching strongly. Possibly this serum effect, like the effect of serum in retarding cell attachment, is the result of formation of a layer of lipoprotein on the glass, mechanically blocking the spreading. However, even though the presence of serum in the medium retarded the stretching of the human conjunctiva cells, serum seemed necessary to maintain these cells in the stretched or flattened shape. In the absence of serum, the conjunctiva cells began to round up and detach from the glass after 8 hours (see Section II,A). Furthermore, cells pretreated with serum before plating in serum-free medium remained stretched longer than cells plated without the pretreatment with serum. These results suggest that serum acts directly on these cells and not on the glass.

Fetuin, a glycoprotein from fetal bovine serum (see Section III,A), does not affect stretching on glass of conjunctiva cells at a concentration of 8 μg/ml, and retards stretching of these cells at a concentration of 1 mg/ml (A. C. Taylor, 1961). However, fetuin seems to accelerate attachment and stretching of a number of types of epithelial cells (Fisher et al., 1958; Lieberman and Ove, 1958).

Other cells, such as chicken and rat embryo fibroblasts, do not require serum for attachment or stretching on a plastic surface. In this laboratory, after transfer we routinely plate these cells in a serum-free Eagle's medium.

C. DNA Synthesis

As discussed in Section I, serum appears to contain some specific factors which are used in cell multiplication and whose availability usually controls cell multiplication. One direct measure of cell multiplication is the rate or amount of DNA synthesis. DNA synthesis is easily determined by the incorporation of radioactive thymidine measured either as a rate of incorporation or as a percent of nuclei labeled. In our experience with normal and RNA tumor-virus-converted chicken and rat embryo fibroblasts, the results of the two types of measurement usually vary together (Temin, 1969). This result demonstrates that the effects of serum on rates of transport (see Section II,F) are not important in these measurements of DNA synthesis. The effects on transport seem to be very rapid, whereas those on DNA synthesis occur after a lag of several hours. However, with any cell system it is desirable to validate experiments measuring the rate of thymidine incorporation with experiments using autoradiography.

The rate of incorporation of thymidine was first used as a quantitative measure of the activity of different fractions of serum by Todaro *et al.* (1967). Plots of the amount of serum added against the rate of DNA synthesis or number of cells synthesizing DNA—dose-response curves—gave linear curves for a variety of normal and transformed fibroblasts from chicken, rat, hamster, and mouse (Temin, 1968b, 1969, 1970a; Clarke *et al.*, 1970; Dulbecco, 1970a). However, these linear curves do not appear to indicate that a single unit of some factor in serum is sufficient to stimulate stationary cells to enter into DNA synthesis. Recent work with chicken embryo fibroblasts has indicated that the stimulation of DNA synthesis by serum is a complex reaction which requires the interaction of serum with cells over a period of a few hours (Temin, 1971). Therefore, the approximately linear dose-response curves do not indicate a "one-hit" process. However, the dose-response curves can still be used as a quantitative measure of the activity of serum or fractions from serum (see Section III,B).

Two major features of the stimulation of DNA synthesis by serum have been studied: the events between the time of addition of serum and the start of DNA synthesis, and the factors affecting the number of cells responding to the addition of serum by starting DNA synthesis.

In most of the stationary cell culture systems studied, serum appeared to act on G1 phase cells to stimulate them into DNA synthesis (Todaro *et al.*, 1965; Yoshikura and Hirokawa, 1968; Temin, 1968b). But with human embryo fibroblasts, cells stopped in G2 are also found (Maciera-Coelho, 1967c). These assignments were supported by study of the time of DNA synthesis and mitosis and by direct study of whether most mitotic cells had engaged in DNA synthesis before mitosis and after the addition of serum.

Furthermore, there was a delay of some hours—4 or 5 hours for chicken fibroblasts, about 12 hours for rat fibroblasts, and about 15 hours for mouse fibroblasts—before the start of DNA synthesis. Serum need not be present throughout this period: In the case of chicken fibroblasts about 2 hours before a cell started DNA synthesis it was irreversibly committed to start DNA synthesis even if serum was removed (Temin, 1971). A few hours of exposure of cells to serum before they were committed to enter DNA synthesis was sufficient to stimulate cells to start and to complete an entire round of DNA synthesis. In the case of chicken fibroblasts, this time was not decreased by a 10-fold increase in the concentration of serum or by different means of preparing the stationary cells (Temin, 1971). The effects of exposure to serum for periods too short to cause cells to become committed to start DNA synthesis disappeared after overnight incubation in serum-free medium.

In a population of stationary chicken cells, there is a good deal of

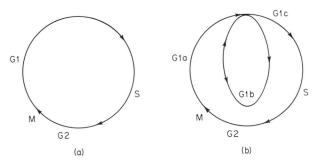

Fig. 1. Models of cell cycle.

heterogeneity in the time of exposure to serum required to commit the cells to enter DNA synthesis. The longer the exposure, the more cells are so committed. (This relationship follows complex kinetics.) From consideration of these kinds of results, Temin (1971) proposed that the cell cycle be described as indicated in Fig. 1(b) rather than as classically drawn in Fig. 1(a). In this formulation, serum would act in some complex fashion on cells in Glb until they entered Glc. An important consequence of this new formulation is that stationary cells are in a phase of the cell cycle entirely different from the pre-DNA synthesis phase of exponentially multiplying cells.

In the time between the addition of serum and the start of DNA synthesis, many cellular processes are stimulated (see Section II,F). Some of these may be irrelevant to the stimulation of DNA synthesis by serum. For example, the addition of serum to stationary 3T3 cells increases the rate of transport of uridine (Cunningham and Pardee, 1969). However, dialyzed serum stimulates DNA synthesis in stationary chicken fibroblasts in a balanced salt solution (Temin, 1971). Therefore it is hard to see how increased transport leads to stimulation of DNA synthesis.

Other processes that are stimulated by serum, such as RNA and protein synthesis and glycolysis, are probably important for stimulation of DNA synthesis. Addition of inhibitors of these syntheses or of glycolysis prevents the stimulation of DNA synthesis by serum (Wiebel and Baserga, 1969; Temin, 1971). This result is not surprising in light of studies showing that RNA and protein synthesis are necessary for normal S phase DNA synthesis (Mueller, 1969).

None of these studies define a primary role of serum in initiation of DNA synthesis. Furthermore, they do not tell if there is in fact a primary role. The availability of purified multiplication-stimulating factors from serum should aid in answering these questions (see Section III).

The other major feature of stimulation of DNA synthesis in stationary cells by serum which has been studied is the factors affecting the number of cells responding to the addition of serum by starting DNA synthesis. A striking illustration of the difference between related cells is seen with untransformed and RNA tumor-virus-transformed chicken or mouse cells (Temin, 1968b; Clarke *et al.*, 1970). Withdrawal of serum leads to a rapid cessation of DNA synthesis in the untransformed cells, as cells cannot be committed to DNA synthesis in the absence of serum. However, with the transformed cells DNA synthesis continues after the withdrawal of serum.

For any one type of cell prepared in any one way and exposed for the same period of time to different amounts of serum, there is a direct proportionality between the amount of serum and the number of cells stimulated to enter DNA synthesis; that is, there is a linear dose-response curve (discussed above). However, the sensitivity to serum of the same cells prepared in different ways or of different cells prepared in the same way varies widely (Clarke *et al.*, 1970; Dulbecco, 1970a; Temin, 1968b, 1970a, 1971). The most striking difference is the 60-fold decreased sensitivity of BHK cells to serum when they are placed in suspension as compared to the same cells attached to a solid substratum (Clarke *et al.*, 1970). Cell density appears to affect the sensitivity of mouse and hamster cells, but not rat cells. Wounding of the cell sheet, the type of medium present before the addition of serum, and trypsinization have all been reported to influence the amount of stimulation of DNA synthesis by serum. In addition, cells produce factors which enhance or depress stimulation (Temin, 1969, 1970a,b). Because of the multitude of factors which influence the efficiency of stimulation, it is somewhat difficult to draw conclusions from differences seen between cells treated in different ways.

This problem is even greater in interpreting experiments comparing two different types of cells. Until a fairly complete knowledge is gained of the significance of these factors for each of the cells compared, caution must be exercised in drawing conclusions.

D. Mitosis

In the cell cycle (Fig. 1), mitosis follows the completion of DNA synthesis. As discussed in Section II,C, serum actually acts on G1 phase cells to stimulate them into DNA synthesis. After completion of the

DNA synthesis stimulated by serum, mitosis usually results. This mitosis is another easily measured parameter.

Mitoses are usually enumerated either by fixing and staining cells and counting the mitotic index or by adding colchicine or one of its derivatives and accumulating cells in mitosis. Since mitotic cells are known to detach more easily than interphase cells (Axelrad and McCulloch, 1958; Terasima and Tolmach, 1963) and serum appears to affect the rate of cell detachment (see Section II,A), care must be exercised to determine if the effects of the presence or absence of serum are on mitosis or on cell detachment.

As expected from the results measuring stimulation of DNA synthesis by serum, addition of serum to stationary cells leads to an increase in mitosis. The number of cells entering mitosis seems to vary directly with the amount of serum added (Temin, 1969). However, some types of cells, especially transformed cells, may show a high percentage of cells synthesizing DNA and few cells entering mitosis (Dulbecco, 1970a; Macieira-Coelho, 1967a). This seeming paradox probably relates both to a serum requirement for mitosis separate from that for DNA synthesis and the high rate of cell detachment in cultures of transformed cells.

In agreement with the hypothesis of a separate serum requirement for mitosis is the finding that some stationary cells may be in the G2 phase of the cell cycle (Macieira-Coelho, 1967c). This hypothesis is also supported by study of the effects of pulses of serum on DNA synthesis and mitosis in chicken fibroblasts (Temin, 1971). It was found that in the absence of serum 25% of the cells which had been stimulated by a pulse of serum to enter DNA synthesis did not continue into mitosis. Therefore, at least for some cells, serum must have a function in addition to stimulation of DNA synthesis and prevention of cell detachment for these cells to enter mitosis. It is not known if this function is quantitatively or qualitatively related to the stimulation of DNA synthesis. Use of purified serum factors should also help resolve this problem.

E. Cell Multiplication

Since serum stimulates DNA synthesis and mitosis, it will usually cause an increase in cell number. Such an increase is not found if there is a high rate of cell detachment (Macieira-Coelho, 1967a; Temin, 1968b, 1970a). Total cell number is estimated directly by hemocytometer or electronic cell counts or by measurements of amounts of DNA, RNA, or protein. We have found with chicken embryo fibroblasts that all these values usually vary coordinately. However, there may be a twofold

variation in cell size from stationary to exponentially multiplying populations. Cell detachment is measured by comparing the number of cells
synthesizing DNA and increase in cell number or by measuring loss
of prelabeled DNA from a culture or by observing actual decreases
in the number of cells in a culture. Cell detachment has appeared to
be of great magnitude only in cultures of transformed cells and in cultures of untransformed cells with little or no serum (see Section II,A).

A common way to express the effects of serum on increase in number
of cells is to plot serum concentration against the final density of the
population of cells. This final density has been called saturation density
or stationary population density. It has been expressed as cells per culture dish, cells per unit area, protein per culture dish, or monolayer equivalents per culture dish, all of which are interconvertible if one knows
the conversion factors. Because this cell density is influenced by the
medium and the rate of cell detachment, it is not accurate to call it
either a stationary population or a saturation population density unless
it has been shown that the cells are in stationary phase, for the former,
and cannot be made to multiply more, for the latter.

When dose-response experiments were carried out (Temin, 1966,
1967b, 1968a; Todaro et al., 1967) it was found that the stationary
population density of any one type of cell increased with the serum
concentration. Except at very high serum concentrations, when inhibitory
effects may begin to appear, this observation appears to be generally
true for all fibroblasts in culture.

The position of the dose-response curve for serum concentration versus
stationary population density can be interpreted as showing the magnitude of the serum requirement for cell multiplication. From comparisons
of dose-response curves, it appears that there are differences between
fibroblasts from different species of animals in their requirement for
serum for cell multiplication. For example, chicken fibroblasts have a
much lower serum requirement than duck fibroblasts (Temin, 1967a,b).

Quite interesting results have come from comparison of the serum requirement of untransformed and transformed fibroblasts (Temin, 1966,
1967b, 1968a; Holley and Kiernan, 1968; Clarke et al., 1970; Dulbecco,
1970a). It has been found with fibroblasts from chicken, duck, rat,
mouse, and hamster—transformed with RNA or DNA tumor viruses,
chemicals, or irradiation—that the transformation leads to a decreased
requirement for serum for cell multiplication. [Because of toxic factors
produced by some cells, this phenomenon could not be seen with these
cells until the toxic factor was removed (Kotler, 1970).]

Since an increase in cell number is brought about by stimulation
of DNA synthesis and mitosis and inhibition of cell detachment, there

could be several mechanisms for the decreased serum requirement for cell multiplication of transformed cells. In the case of chicken and rat cells converted by infection with RNA tumor viruses, the mechanism seemed to be an increased efficiency of utilization of the multiplication-stimulating activity of serum (Temin, 1969, 1970a,b). There were no differences in sensitivity to stimulation. It is not clear how general this mechanism is. Jainchill and Todaro (1970) reported that transformed mouse cells multiplied in a medium from which some limiting multiplica-tion-stimulating factors for untransformed mouse cells had been chemi-cally removed. Again, the availability of purified multiplication-stimulat-ing factors will be necessary to understand these problems.

F. Other Syntheses and Transport

1. RNA Synthesis and Transport of RNA Precursors

There is an increase in rate of incorporation of labeled uridine within minutes after the addition of serum to cultures of stationary cells (To-daro et al., 1965, 1967; Rhode and Ellem, 1968; Yoshikura and Hirokawa, 1968; Yeh and Fisher, 1969; Cunningham and Pardee, 1969). This effect was at first ascribed to an increase in the rate of RNA synthesis. How-ever, Cunningham and Pardee (1969) have shown that essentially all the increased labeling of RNA in 3T3 cells occurring within 15 minutes after the addition of serum can be accounted for by an increased rate of uptake of labeled uridine from the medium. Dialyzed serum stimu-lated the rate of uptake of uridine, guanosine, and cytosine. It did not increase the rate of uptake of adenosine. This increased rate of uptake increased the concentration of labeled precursor in the internal pool more rapidly than in control cultures, thereby increasing the relative rate of labeling of RNA. Consistent with this interpretation is the fact that addition of serum to stationary cultures increases the basal, unstimu-lated incorporation of a short pulse of radioactive uridine into each class of RNA to the same degree as any other class (Bloom et al., 1966).

In addition to this problem, there is another one. Becker and Levitt (1968) have observed that simply replacing the medium on a culture of chicken fibroblasts with fresh medium without any serum caused a stimulation of uptake of labeled uridine. Therefore, it is clear that experiments on immediate effects of serum on synthesis must be well controlled with respect to effects on transport and for effects of medium change.

2. PROTEIN SYNTHESIS AND AMINO ACID TRANSPORT

The addition of fresh, serum-containing medium to stationary fibroblasts caused an increase in the rate of incorporation of labeled amino
acids (Todaro et al., 1965). Less stimulation was found when fresh
serum alone was added (Wiebel and Baserga, 1969).

Wiebel and Baserga (1969) showed that a 1-hour pulse of labeled
leucine was an adequate measure of the rate of protein synthesis in
human fibroblasts since the labeled leucine came to equilibrium with
the intracellular leucine pool in 8 minutes. With this technique they
found a 1.4-fold increase in the rate of protein synthesis after the addition of serum to cultures of stationary human fibroblasts.

Presumably, some protein synthesis is essential for the serum-stimulated DNA synthesis (Section II,C). Addition of cycloheximide for 4
hours either concurrently with serum or after a 4-hour exposure to serum
prevented the serum-stimulated DNA synthesis. It is not known if the
required protein synthesis was that protein synthesis which is stimulated
by serum.

In the same study, Wiebel and Baserga (1969) observed alterations
in the size of various amino acid pools after addition of serum. Most
pools were depleted two- to four-fold within 165 minutes after addition
of serum. However, the pools of aspartic and glutamic acid were unal-
tered and that of serine was enlarged. The depletion of the leucine
pool was not blocked by cycloheximide. Presumably, therefore, the serum
effects on pool sizes and on protein synthesis are separable effects (see
Cunningham and Pardee, 1969).

Soeiro and Amos (1966) found that the microsomal fractions from
chicken fibroblasts deprived of serum for 24 hours had a significantly
reduced capacity to incorporate amino acids into protein as compared
to microsomes from cells grown with serum. This reduction was found
with either endogenous or exogenous template present. After the serum-
deprived cells were exposed to serum for 3 hours, the protein-synthesizing capacity of their microsomes returned to full activity. Addition of
insulin to the cells had the same effect (Schwartz and Amos, 1968).
However, incubation of the microsomes from serum-deprived cells with
serum or insulin did not affect their activity. These observations of Amos
and his co-workers are remarkably similar to the *in vivo* effects of insulin
on the protein-synthesizing capacity of microsomes from diabetic rats
(Wool and Cavicchi, 1967).

3. PHOSPHOLIPID METABOLISM

Peterson and Rubin (1969) showed with the use of labeled choline
that chicken fibroblasts continually release phospholipids to the medium.

Serum stimulated this release of phospholipid; the more serum, the higher the rate of release.

Cunningham and Pardee (1969) found that addition of serum to stationary 3T3 cells stimulated both the uptake and release of inorganic phosphate and incorporation of inorganic phosphate into phospholipid. The component of serum responsible for stimulation of phosphate metabolism could be separated by chromatography on Sephadex G-200 from the component of serum responsible for stimulation of DNA synthesis. Therefore, the effects on phosphate metabolism may not have been necessary for stimulation of DNA synthesis.

G. Other Biological Effects of Serum

1. Short-Term Survival

With several types of fibroblasts, it has been shown that incubation for a day or so in the absence of serum leads to cell death (Temin, 1968b; Dulbecco, 1970a). This death is independent of the cell detachment discussed in Section II,A. The dead cells are still attached, but are unable to respond to the readdition of serum with DNA synthesis and cell multiplication. This effect is especially noticeable with chicken fibroblasts infected and converted by the Schmidt-Ruppin strain of avian sarcoma virus. Usually by 24 hours after removal of serum these cells have died. The reasons for this cell death, and which component of serum prevents it, are not known. The untransformed chicken fibroblasts can survive for over a week in the absence of serum. Morgan (1970) showed that serum prolonged the survival of primary cultures of mouse ascites tumor cells.

2. Cloning Efficiency

Cells, even those taken directly from an animal, can grow at low cell densities so that clones develop. Careful nutritional studies by Ham and co-workers (Ham and Murray, 1967; Ham and Sattler, 1968) have shown that cloning efficiencies of 45% can be secured with cells from adult rabbits plated in F12 medium with 5% fetal bovine serum and 5 or 10% rabbit serum. Further study (Ham et al., 1970) has shown that the serum requirement can be reduced if the macromolecular fraction of chicken embryo extract is also added to the medium.

It is not yet known if the requirement for macromolecular factors for cloning is more stringent than that for multiplication in mass culture.

3. Transformation to Cell Lines

Sanford and her associates (Sanford, 1967; Jackson *et al.*, 1970) have studied the spontaneous transformation of mouse and rat embryo cells into neoplastic cells. They have found that transformation is much more rapid for cells growing in horse serum than for cells growing in fetal bovine serum, and more rapid for cells growing in serum than for cells growing in the absence of serum. Because of the possible implication of viruses in this change, it is unlikely that this effect of serum is related to the others discussed here.

4. Cell Differentiation

A number of systems have been developed in which cells carry out differentiated functions while multiplying in culture. In all cases the serum or other macromolecular factors in the medium must be selected with care. Other chapters in these volumes will consider these systems in detail.

5. Cell Movement

Raff and Houck (1969) reported that migration of a diploid human fibroblast into a wound in the monolayer did not require serum. But Clarke *et al.* (1970) reported that in the absence of serum very few cells moved from an adjacent cell layer into a wound area in monolayer cultures of BHK cells. They found that as the concentration of serum was increased, the number of migrating cells increased.

III. Isolation from Serum of Fractions Having Biological Effects on Cells in Culture

It is clear from Section II that there are many biological effects of serum on cells in culture. It is not clear from these studies how many of these are independent effects, not secondary to another effect of serum. It is also not clear from these studies how many specific factors there are in serum which have biological effects on cells in culture. To answer both these questions, it is necessary to have purified factors from serum. Such purified factors are also necessary if a completely defined medium is to be prepared. Therefore, several groups have tried to isolate fractions from serum with biological effects on cells in culture.

A. Globulin Fractions

1. FETUIN

Fetuin, a protein component of fetal bovine serum, first described by Pedersen (1944), has been the object of much study and considerable controversy. Fisher et al. (1958) reported that fetuin was active in promoting the attachment and stretching of HeLa cells in culture. They partially purified fetuin from fetal bovine serum by fractionation with ammonium sulfate. Fetuin was obtained as the fraction soluble in ammonium sulfate at 40% of saturation and insoluble at 45% of saturation. The fetuin was judged 95% homogeneous by analytical ultracentrifugation.

However, Lieberman et al. (1959) reported that fetuin purified on DEAE-cellulose columns had poor activity in promoting attaching and stretching of A1 cells from human appendix. Fractions from the columns, containing 74% of the recovered protein and having a sedimentation constant equal to that of fetuin, had only 6% of the recovered biological activity. The remainder of the activity recovered appeared in fractions which contained only 12% of the recovered protein. This protein appeared as two peaks in the ultracentrifuge. One had $s_{20,w} = 3$, equal to that of fetuin, and the other sedimented faster. It should be noted, however, that while the recovery of protein from the column was quantitative, only 40% of the applied activity was recovered.

Further evidence that fetuin may not possess attachment and stretching activity comes from the work of Spiro (1960), who purified fetuin by alcohol precipitation in the presence of zinc and barium ions. The purified material had no ability to cause the attachment to glass or stretching of HeLa cells, even at high concentrations of protein.

Marr et al (1962), in attempting to determine whether the growth-promoting activity of fetuin was a property of fetuin itself or of a smaller component carried by it, achieved further fractionation of active fetuin by subjecting fetuin prepared by ammonium sulfate precipitation to ultracentrifugation. Three fractions were obtained, a colorless liquid layer designated the "light" fraction, a yellow liquid layer, and a brown pellet. The last two fractions, designated the "heavy" fraction, were recombined and recentrifuged. The separation of the light fraction and the combining of the heavier fractions was repeated a number of times. Each time the three fractions were obtained, and the light fraction was discarded. The final heavy fraction and the initial light fraction were then analyzed by ultracentrifugation. It was found that the heavy fraction consisted mainly of an α-macroglobulin with $s_{20,w} = 20$. The light fraction was enriched in fetuin $s_{20,w} = 3$. Both fractions promoted the multiplication of HeLa cells in monolayer culture (measured by the increase in cell number). However, it was not possible to correlate the level of activity with the proportion of fetuin in the two preparations.

More recently, Puck and co-workers (1968) showed that fetuin lost its ability to stimulate single-cell cloning after purification on DEAE-cellulose or by alcohol precipitation by the method of Spiro (1960). They achieved a somewhat higher degree of purification of active fetuin by more exhaustive ammonium sulfate fractionation. Disc gel analysis of a standard preparation of fetuin revealed one major and seven minor protein bands, one of which had the mobility of albumin. The more highly purified active fetuin preparation showed no albumin band, while the remaining minor bands were greatly reduced. In terms of its ability to increase plating efficiency of single HeLa cells, the specific activity of the more highly purified preparation was indistinguishable from that of the standard preparation, suggesting it was unlikely that the activity was a property of any of the minor components. As little as 0.2 μg/ml of fetuin showed an effect on plating efficiency. Antibody prepared against the purified fetuin produced a strong precipitin reaction with the fetuin and reduced its ability to promote plating efficiency of cells to zero.

These results may all be reconciled if one assumes that fetuin can act as a carrier for a small polypeptide which is actually the molecule with biological activity (see Section III,B).

2. OTHER GLOBULIN FRACTIONS

Michl (1961) has isolated from calf serum a globulin which promotes the rapid attachment to glass of HeLa and monkey heart cells and stimulates the multiplication of HeLa cells. The globulin was isolated by a combination of ammonium sulfate and isoelectric precipitation. The resulting preparation was not homogeneous when examined in the ultracentrifuge, giving two peaks with sedimentation constants of $s_{20,w} = 3.1$ and 8.1.

Lieberman and Ove (1958) isolated a protein which induces the attachment to glass and stretching and multiplication of human appendix A1 cells and HeLa cells. The protein was prepared by the successive application to bovine serum of ammonium sulfate precipitation, alcohol fractionation, and DEAE-cellulose column chromatography. The purest preparation displayed a specific activity of 1152 units/mg of protein compared to 27.7 units/mg for whole bovine serum. An activity unit was defined as that amount of activity which induce the attachment to glass of 50% of the cells (determined without inversion of the culture chamber—the standard method discussed in Section IIA,B). In the ultra-centrifuge, the active material gave three peaks.

Katsuta *et al.* (1959) fractionated dialyzed horse serum by zone electrophoresis, cold ethanol precipitation, and ammonium sulfate precipitation and found that all fractions displayed some activity in stimulating the growth of Yoshida ascites tumor cells of the rat, measured as the increase in the number of cell nuclei. Most of the activity, however, was associated with the α-globulin and albumin fractions. Cold ethanol fractionations of human and horse sera by Bazely *et al.* (1954), of horse serum by Sanford *et al.* (1955), and of human and horse sera by Chang *et al.* (1959) also indicated that most growth-promoting activity was associated with the albumin and α-globulin fractions.

Cellulose powder column, curtain electrophoresis, and gel filtration in Sephadex G-200 were employed by Tozer and Pirt (1964) to isolate an α-globulin fraction from calf serum which enhanced cell multiplication. The fraction with highest activity was 70 times as effective as whole serum in causing the multiplication of cells in suspension. Because the cells were growing in suspension, cell attachment and stretching were not involved.

Curtain electrophoresis was also employed by Holmes and Wolfe (1960, 1961) for the preparation of cell-growth-promoting activities from dialyzed bovine serum. Two fractions were obtained, both of which promoted the survival and growth of Chang's endoepithelial cells and HeLa cells, measured as the increase in cell number. Only one of the fractions, however, caused appreciable cell spreading. Electrophoretic analysis showed that both fractions contained α-globulin proteins.

Healy and Parker (1966) demonstrated that a growth-promoting activity for newly explanted mouse embryo cells resided solely in the protein components of serum when they separated the protein constituents of fetal calf serum from the nonprotein, low-molecular-weight components by three different means—dialysis, total protein precipitation using sodium sulfate, and chromatography in Sephadex G-25. In each case when the separated protein and nonprotein fractions were combined with base medium and examined for their effects on mouse cells, all the growth-promoting activity was found to reside in the protein. The assessment of growth-promoting activity was made primarily by visual inspection; a marked increase in cell density was produced by the protein fractions compared to base medium alone or base medium in combination with the low-molecular-weight components of the serum.

Healy and Parker (1966) also effected a further fractionation of serum by subjecting the supernatant of Cohn's fraction V to gel filtration chromatography in Sephadex G-50 or G-75. In both cases, fractions active in the promotion of cell growth were obtained. Analysis of these fractions by starch gel electrophoresis showed that the active protein consisted of α_1-glycoprotein.

When the active protein from either of the Sephadex columns was tested in combination with an α_2-macroglobulin, prepared by the method of Schultze *et al.* (1955), there was a strong synergistic enhancement of cell growth. The role of the α_2-macroglobulin appeared to be nonspecific and probably protective in nature since dextran and Ficoll were also synergistic in combination with the α_1-glycoprotein. Such a protective function has also been ascribed to albumin (see below). The significance of the α_2-macroglobulin was further shown by starch gel electrophoretic comparison of horse serum and the same serum which had been used in a medium over newly explanted mouse cells. The serum that had been in contact with the mouse cells revealed a decreased intensity of the electrophoretic bands corresponding to α_1-acid glycoprotein and to the α_2-macroglobulin.

In later studies, Healy and Parker (1970) reported a relatively elaborate fractionation scheme in which bentonite absorption, potassium phosphate precipitation, glutamic acid precipitation, rivanol precipitation, and hydroxylapatite column chromatography were applied in succession to fetal bovine serum. A step-by-step balance sheet indicating protein and activity was not given. The final fraction at a protein concentration of 250 μg/ml reduced the doubling time for mouse embryo cells in a base medium from 4.75 to 1.33 days. The preparation was shown by electrophoretic analysis to contain three components.

An α-globulin was separated by Holmes (1967) from human serum. It could not be separated from either bovine or equine sera. The separation was effected by a glass bead column which was developed by the application of a stepwise pH gradient. The isolated globulin was active in stimulating the multiplication (measured as increase in cell protein) and spreading of HeLa, conjunctiva, and human heart cells in monolayer cultures. The growth and spreading of cells which had been adapted over a period of time to Holmes' base medium in the absence of serum were stimulated by as little as 0.4 μg/ml of the active α-globulin. Electrophoretic analysis of the globulin in discontinuous polyacrylamide gels revealed that it was different both from inactive fetuin prepared by the method of Spiro (1960) and Deutsch (1954) and from the active fetuin prepared according to Fisher *et al.* (1958). Similarly, the Holmes globulin was shown to be distinct from the heavy or light sedimenting fractions described by Marr *et al.* (1962). The biological effect of the Holmes α-globulin was not duplicated by insulin, carbamyl phosphate, putrescine, or linoleic acid.

Todaro *et al.* (1967) subjected calf serum to gel filtration in Sephadex G-200 and obtained approximately a three-fold purification of a factor, with apparent molecular weight of 100,000 to 150,000, which stimulated both DNA synthesis and mitosis in 3T3 cells.

Jainchill and Todaro (1970) prepared agamma calf serum by subjecting calf serum to alcohol precipitation. Medium containing this agamma serum supported the survival but not the division of 3T3 cells. Addition of the precipitated protein or of whole serum to the medium resulted in a rapid resumption of cell division.

3. How Many Globulins with Biological Activity Are There in Serum?

In all the cases discussed so far, the fractions with biological activity have been α-globulins. The fractionations of fetal bovine serum have shown that biological activity is associated with fetuin, the principal protein component of fetal bovine serum. In spite of efforts to resolve further these fetuin preparations, some protein with a sedimentation constant of fetuin has remained in all preparations with biological activity.

The active preparations of Healy and Parker (1966, 1970) were also obtained from fetal calf serum, but their relationship to fetuin is not clear. One of their preparations (Healy and Parker, 1966) showed three bands by electrophoresis under conditions in which standard fetuin preparations showed eight bands.

The active preparations of Michl (1961) from calf serum and Lieberman and Ove (1958) from beef serum could have contained fetuin. Although the amount of fetuin in calf serum declines rapidly from the time of birth, fetuin is still present in substantial amounts in 2-week-old calves, and it is still detectable in beef serum (Pedersen, 1944).

The preparations of Tozer and Pirt (1964) from calf serum, Bazely *et al* (1954) and Chang *et al.* (1959) from human and horse sera, and Sanford *et al.* (1955) from horse serum could not have contained fetuin. All these proteins were associated with the α-globulin or with the albumin fractions of serum, but no more can be said about their identity. Finally, Holmes (1967) demonstrated by electrophoretic comparison that his α-globulin, isolated from human serum, was distinct from fetuin prepared by a variety of procedures.

Therefore, at least two and possibly more α-globulins with biological activity must exist in serum. If the hypothesis of a small polypeptide with the actual biological activity on a larger inactive carrier is correct, these different globulins may represent the same active polypeptide with different carriers.

B. Insulin-like Activity

Crystalline and amorphous insulin are able to stimulate cell multiplication in several cell culture systems (see Waymouth and Reed, 1965, for references to earlier work; Temin, 1967a,b; Tiötta, 1968; Clarke et al., 1970). However, a quantitative comparison of the amount of immunoassayable insulin in serum and the multiplication-stimulating activity of insulin revealed that the amount of immunoassayable insulin in serum was at least two or three orders of magnitude less than that required to explain the multiplication-stimulating activity of serum (Temin, 1967a, 1968b).

Our attention then turned to nonsupressible insulin-like activity. This is a substance in serum that has activity in several insulin assays but is not neutralized by antibody to insulin (Bürgi et al., 1966). Samples of partially purified nonsuppressible insulin-like activity were able to stimulate DNA synthesis of stationary chicken fibroblasts; that is, they had multiplication-stimulating activity (Temin, 1967b, 1970a). We (Pierson and Temin, 1972) then developed a purification scheme for multiplication-stimulating activity in serum based on published methods for purification of serum insulin-like activity. Multiplication-stimulating activity was assayed by measuring the rate of incorporation of labeled thymidine (see Section II,C).

Whole calf serum was chromatographed on Dowex 50W by a procedure based on that of Antoniades and Gundersen (1962). Twenty to 30 percent of the multiplication-stimulating activity and less than 1% of the protein was bound to the resin and eluted at high pH. This separation was consistent with the observation that multiplication-stimulating activity was bound by agar and other polyanions (Temin, 1966). This active protein fraction was then dissolved in 1M acetic acid. Insoluble material was removed by centrifugation, and the supernatant was applied to a column of Biogel P-100 and eluted with 1M acetic acid. This procedure was adapted from Bürgi et al (1966). The slowly migrating proteins from the separation were then fractionated by preparative polyacrylamide gel electrophoresis in 8 M urea and 4 M acetic acid. The multiplication-stimulating activity was associated with the fastest migrating proteins. It was approximately 6000- or 8000-fold purified as compared to whole serum protein.

On gel electrophoresis fractionation this fraction with multiplication-stimulating activity ran as one band in the urea-acetic acid system, but as one major and two minor bands in an SDS-mercaptoethanol-urea system (Swank and Munkres, 1971). The activity of the purified material was stable to heating and to acid and was destroyed by treatment with crystalline trypsin. On the basis of its migration in Biogel and polyacrylamide gels, the purified multiplication-stimulating activity appeared to be slightly smaller than insulin. The purified multiplication-stimulating activity was about 5000 times less effective on a weight basis than insulin in the epididymal fat pad assay for insulin (Gliemann, 1967).

The properties of this multiplication-stimulating activity purified from calf serum, that is, its small size and extremely sticky character, may explain the disagreements discussed in Section III,A. When we chromatographed whole calf serum on Sephadex G-200, all protein fractions had about the same multiplication-stimulating activity. Therefore, it is reasonable to assume that in its native state this multiplication-stimulating activity is attached to one or more inactive carriers.

C. Relationship of Active Fractions from Serum to Other Pure Proteins of Known Biological Activity

A fair number of pure proteins with specific growth-stimulating activities are known. In addition to the polypeptide hormones, these are nerve growth factor, epidermal growth factor, erythropoietin, and colony-stimulating factor. Further work will be necessary to understand the relationship of these proteins to those discussed in Sections II,A and B.

1. NERVE GROWTH FACTOR

Nerve growth factor increases the size of the superior cervical ganglia in newborn mice, and, in organ culture, stimulates fiber outgrowth from explanted ganglia. (For a review of this topic, see Levi-Montalcini and Angeletti, 1968.) It has been isolated from the submaxillary gland of mice (Cohen, 1960) and from snake venom (Cohen, 1959). Nerve growth factor from mice is a protein containing less than 0.3% hexose (Cohen, 1960). Currently there is no agreement as to whether nerve growth factor (molecular weight, 140,000) is made up of three distinct but nearly inactive subunits (Smith et al., 1968) or whether nerve growth factor is a smaller, fully active substance of molecular weight

30,000 (Bocchini and Angeletti, 1969). A yet unconfirmed report suggests that nerve growth factor is a very minor and extraordinarily potent contaminant of preparations throught to be homogeneous (Schenkein *et al.*, 1968).

2. ERYTHROPOIETIN

Erythropoietin stimulates cell proliferation in blood cell-forming tissue. It is isolated from urine or plasma of anemic subjects and is probably a glycoprotein of molecular weight 50,000–70,000. Although it has been purified 10^6-fold from plasma it has not been shown to be homogeneous, primarily because only minute quantities have been available to investigators (see reviews, Krantz and Jacobson, 1970; Fisher, 1968).

3. EPIDERMAL GROWTH FACTOR

Epidermal growth factor stimulates precocious eye opening and eruption of teeth in newborn mice (Cohen, 1962) and stimulates DNA synthesis and mitosis in explants of mouse mammary carcinoma (Turkington, 1969). Epidermal growth factor has been isolated from the submaxillary gland of male mice in homogeneous form. It is a protein with a minimal molecular weight of about 14,500 (Taylor *et al.*, 1970) and does not appear to contain carbohydrate (Cohen, 1962). It is also called epithelial growth factor.

4. COLONY-STIMULATING FACTOR

The colony-stimulating factor stimulates colony formation in cell culture of spleen or bone marrow cells presumably by stimulating the multiplication of single cells (Stanley *et al.*, 1968). It has been partially purified from mouse serum (Stanley *et al.*, 1968) and from human urine (Metcalf and Stanley, 1969). Colony-stimulating factor from urine has been purified several hundredfold and appears to be associated with a glycoprotein, having an apparent molecular weight of 190,000 as determined by gel filtration. By the same index, however, colony-stimulating factor in urine concentrates behaves as a larger molecule (Stanley and Metcalf, 1969). This discrepancy and the lack of information on the number of components in the most purified samples leave open the possibilities that colony-stimulating factor may be made up of loosely associated active subunits or that it is a small molecule bound to various larger inert carrier molecules.

IV. Nonserum Proteins and Other Substances Able to Mimic Some Biological Effects of Serum

A. Low-Molecular-Weight Compounds

Low-molecular-weight compounds were discussed in Section I.

B. Albumin

While albumin has been shown to be of value for growth of certain cells, it is generally not essential. Fisher et al. (1959) found that albumin was needed in combination with fetuin in a base medium to facilitate colony formation from single HeLa cells. However, Ham (1965) later developed a medium for this purpose in which the albumin requirement was eliminated. Michl (1961) found that the inclusion of albumin with his active α-globulin fraction enhanced the multiplication of HeLa cells in culture, but the role of the albumin appeared to be protective and nonspecific, since methylcellulose could be used in its place. In view of its general binding capability, the albumin probably served to trap toxic substances in the medium. Moskowitz and Schenck (1965), using a dialysate of a porcine pancreas extract to stimulate the growth of mouse tumor cells and other established cell lines, noted that the dialysate also contained toxic substances. The addition of human serum albumin to the medium afforded protection from the toxic factors, but again the nonspecific nature of this effect was demonstrated by the fact that polyvinylpyrrolidone could replace the albumin with almost the same degree of effectiveness.

C. Insulin

Insulin was discussed in Section III, B.

D. Antiproteases

Wallis et al. (1969) proposed that serum acted to stimulate cell multiplication only because of its antiprotease activity. They found that a number of cells would grow in the absence of serum if the protease activity was removed by frequent washings. This hypothesis is supported by the antitrypsin activity of fetuin (Fisher et al., 1958) and multiplication-stimulating activity purified from calf serum (Pierson and Temin, 1972). However, we have not generally found antiproteases (egg white or soybean trypsin inhibitors) to have multiplication-stimulating activity for stationary chicken fibroblasts.

E. Hydrolytic Enzymes

Burger (1970) reported that low doses of trypsin stimulated cell division in stationary 3T3 cells. Vasiliev et al. (1970) reported that ribonuclease and hyaluronidase had the same effect. Rubin (1970a,b) reported that chicken cells, especially after infection with Rous sarcoma virus, contained and released a multiplication-stimulating factor for dense populations of chicken cells. He believed this factor to be a protease. All these experiments used cells in the presence of serum. An indirect effect of the protease may be involved (see Section II,C).

F. Conditioned Medium

Medium that has been in contact with cells for a day or so may have increased or decreased ability to support the multiplication of cells. The decrease is a result of depletion of needed factors or of production of inhibitory or toxic factors. The increased ability is a result of production of some needed or stimulatory factors. Medium with this effect is called conditioned. Another effect of conditioning could be removal of toxic factors from the medium. Although this conditioning effect has been known for many years (Sanford et al., 1948), even less is known about its basis than is known about serum. All of the discussion of serum is relevant here. In one case, using techniques described in Section III,B, we have purified a multiplication-stimulating activity from a line of BRL cells (Dulak and Temin, 1972). Therefore, this medium was conditioned because the cells produced a serum substitute.

G. Embryo Extract

Embryo extract was a component of most cell culture media a decade or so ago. It is now not used except for special cloning and differentiation studies (see Section II,C). No real fractionation has been carried out beyond separation of high- and low-molecular-weight components on Sephadex G-25 (Coon and Cahn, 1966; Ham et al., 1970).

H. Urine

Urine has been a source of multiplication-stimulating activity for 3T3 cells (Holley and Kiernan, 1968) and for colony-stimulating factor for blood cell precursors (Stanley and Metcalf, 1969).

I. Polyions

Agar and other large polyions have often been added to cell culture media. They undoubtedly play multiple roles. A recent paper (Mizrahi and Moore, 1970) gives references to earlier work with polyions and emphasizes the present lack of knowledge of their functions in cell culture.

V. Conclusion

The nature of the biological effects of serum on cells in culture and of the factors in serum which exert these effects is still little known. At present, factors with specific effects on cells in culture are being isolated from serum and characterized both in our lab and in that of Dr. R. Holley (pers. communication). With these purified factors it should be possible to understand better the phenomena discussed in this review.

ACKNOWLEDGMENTS

The work in our laboratory was supported by Program Project Research Grant CA-07175 and Training Grant CRTY-5002 from the National Cancer Institute. Howard M. Temin holds Research Career Development award 10K 3-CA8182 from the National Cancer Institute.

No claim is made to an exhaustive survey of the literature or a complete listing of all references bearing on these topics. This review was completed in Dec., 1970.

REFERENCES

Abercrombie, M., and Heaysman, J. E. M. (1954). Observations on the social behavior of cells in tissue culture. II. "Monolayering" of fibroblasts. *Exp. Cell Res.* 6, 293–306.
Antoniades, H. N., and Gundersen, K. (1962). Studies on the state of insulin in blood: materials and methods for the estimation of "free" and "bound" insulin-like activity in serum. *Endocrinology* 70, 95–98.

Axelrad, A. A., and McCulloch, E. A. (1958). Obtaining suspensions of animal cells in metaphase from cultures propagated on glass. *Stain Technol.* 33, 67–71.

Bailey, J. M. (1967). Cellular lipid nutrition and lipid transport. In "Lipid Metabolism in Tissue Culture Cells" (G. H. Rothblat and D. Kritchevsky, eds.), Wistar Inst. Symp., Monogr. No. 6, pp. 85–113. Wistar Inst. Press, Philadelphia, Pennsylvania.

Bazeley, P. L., Rotundo, R., and Buscheck, F. T. (1954). Plasma derivatives in tissue cultures intended for growth of poliomyelitis viruses. *Proc. Soc. Exp. Biol. Med.* 87, 420–424.

Becker, Y., and Levitt, J. (1968). Stimulation of macromolecular processes in BSC_1 cells due to medium replenishment. *Exp. Cell Res.* 51, 27–33.

Birch, J. R., and Pirt, S. J. (1969). The choline and serum protein requirements of mouse fibroblast cells (strain LS) in culture. *J. Cell Sci.* 5, 135–142.

Bloom, S., Todaro, G. J., and Green, H. (1966). RNA synthesis during preparation for growth in a resting population of mammalian cells. *Biochem. Biophys. Res. Commun.* 24, 412–417.

Bocchini, V., and Angeletti, P. U. (1969). The nerve growth factor: Purification as a 30,000-molecular-weight protein. *Proc. Nat. Acad. Sci. U.S.* 64, 787–794.

Burger, M. M. (1970). Proteolytic enzymes initiating cell division and escape from contact inhibition of growth. *Nature (London)* 227, 170–171.

Bürgi, H., Müller, W. A., Humbel, R. E., Labhart, A., and Froesch, E. R. (1966). Non-suppressible insulin-like activity of human serum. I. Physicochemical properties, extraction and partial purification. *Biochim. Biophys. Acta* 121, 349–359.

Bürk, R. R. (1966). Growth inhibition of hamster fibroblast cells. *Nature (London)* 212, 1261–1262.

Chang, R. S., Pennell, R. B., Keller, W., Wheaton, L., and Liepens, H. (1959). Macromolecular growth requirements of human cells in continuous culture. *Proc. Soc. Exp. Biol. Med.* 102, 213–217.

Clarke, G. D., Stoker, M. G. P., Ludlow, A., and Thornton, M. (1970). Requirement of serum for DNA synthesis in BHK-21 cells: Effects of density, suspension and virus transformation. *Nature (London)* 227, 798–801.

Cohen, S. (1959). Purification and metabolic effects of a nerve growth-promoting protein from snake venom. *J. Biol. Chem.* 234, 1129–1137.

Cohen, S. (1960). Purification of a nerve growth-promoting protein from the mouse salivary gland and its neuro-cytotoxic antiserum. *Proc. Nat. Acad. Sci. U.S.* 46, 302–311.

Cohen, S. (1962). Isolation of a mouse submaxillary gland protein accelerating incisor eruption and eyelid opening in the newborn animal. *J. Biol. Chem.* 237, 1555–1562.

Coon, H. G., and Cahn, R. D. (1966). Differentiation in vitro: Effects of Sephadex-fractions of chick embryo extract. *Science* 153, 1116–1119.

Cunningham, D. D., and Pardee, A. B. (1969). Transport changes rapidly initiated by serum addition to "contact inhibited" 3T3 cells. *Proc. Nat. Acad. Sci. U.S.* 64, 1049–1056.

Curtis, A. S. G. (1964). The mechanism of adhesion of cells to glass. A study by interference reflection microscopy. *J. Cell Biol.* 20, 199–215.

Deutsch, H. F. (1954). Fetuin. The mucoprotein of fetal calf serum. *J. Biol. Chem.* 208, 669–678.

Dulak, N. C., and Temin, H. M. (1972). In preparation.

Dulbecco, R. (1970a). Topoinhibition and serum requirement of transformed and untransformed cells. *Nature* (*London*) **227**, 802–806.

Dulbecco, R. (1970b). Behavior of tissue culture cells infected with polyoma virus. *Proc. Nat. Acad. Sci. U.S.* **67**, 1214–1220.

Dupree, L. T., Sanford, K. K., Westfall, B. B., and Covalesky, A. B. (1962). Influence of serum protein on determination of nutritional requirements of cells in culture. *Exp. Cell Res.* **28**, 381–405.

Eagle, H. (1960). The sustained growth of human and animal cells in a protein-free environment. *Proc. Nat. Acad. Sci. U.S.* **46**, 427–432.

Eagle, H., Oyama, V. I., and Piez, K. A. (1960). The reversible binding of half-cystine residues to serum protein, and its bearing on the cystine requirement of cultured mammalian cells. *J. Biol. Chem.* **235**, 1719–1726.

Evans, V. J., Bryant, J. C., Kerr, H. A., and Schilling, E. L. (1964). Chemically defined media for cultivation of long-term cell strains from four mammalian species. *Exp. Cell Res.* **36**, 439–474.

Fisher, J. W. (1968). "Erythropoietin." *Ann. N.Y. Acad. Sci.* **149**, 1–583.

Fisher, H. W., Puck, T. T., and Sato, G. (1958). Molecular growth requirements of single mammalian cells: The action of fetuin in promoting cell attachment to glass. *Proc. Nat. Acad. Sci. U.S.* **44**, 4–10.

Fisher, H. W., Puck, T. T., and Sato, G. (1959). Molecular growth requirements of single mammalian cells. III. Quantitative colonial growth of single S3 cells in a medium containing synthetic small molecular constituents and two purified protein fractions. *J. Exp. Med.* **109**, 649–660.

Gliemann, J. (1967). Assay of insulin-like activity by the isolated fat cell method. I. Factors influencing the response to crystalline insulin. II. The suppressible and non-suppressible insulin-like activity of serum. *Diabetologia* **3**, 382–394.

Gori, G. B. (1970). Medium pollution rates in cell cultures. *J. Nat. Cancer Inst.* **44**, 275–281.

Ham, R. G. (1965) Clonal growth of mammalian cells in a chemically defined, synthetic medium. *Proc. Nat. Acad. Sci. U.S.* **53**, 288–293.

Ham, R. G., and Murray, L. W. (1967). Clonal growth of cells taken directly from adult rabbits. *J. Cell. Physiol.* **70**, 275–280.

Ham, R. G., and Sattler, G. L. (1968). Clonal growth of differentiated rabbit cartilage cells. *J. Cell. Physiol.* **72**, 109–114.

Ham, R. G., Murray, L. W., and Sattler, G. L. (1970). Beneficial effects of embryo extract on cultured rabbit cartilage cells. *J. Cell. Physiol.* **75**, 353–360.

Harris, M. (1964). "Cell Culture and Somatic Variation." Holt, New York.

Häyry, P., Myllylä, G., Saxén, E., and Penttinen, K. (1966). The inhibitory mechanism of serum on the attachment of HeLa cells on glass. *Ann. Med. Exp. Biol. Fenn.* **44**, 166–170.

Healy, G. M., and Parker, R. C. (1966). Cultivation of mammalian cells in defined media with protein and nonprotein supplements. *J. Cell Biol.* **30**, 539–553.

Healy, G. M., and Parker, R. C. (1970). Growth-active globulins from calf serum tested on cultures of newly isolated mouse embryo cells. *Proc. Soc. Exp. Biol. Med.* **133**, 1257–1258.

Holley, R. W., and Kiernan, J. A. (1968). "Contact inhibition" of cell division in 3T3 cells. *Proc. Nat. Acad. Sci. U.S.* **60**, 300–304.

Holmes, R. (1967). Preparation from human serum of an alpha-one protein which induces the immediate growth of unadapted cells *in vitro*. *J. Cell Biol.* **32**, 297–308.

Holmes, R., and Wolfe, S. W. (1960). Effect of carboxymethylcellulose on the electrophoresis of serum proteins on paper. *Arch. Biochem. Biophys.* 87, 13–18.

Holmes, R., and Wolfe, S. W. (1961). Serum fractionation and the effects of bovine serum fractions on human cells grown in a chemically defined medium. *J. Biophys. Biochem. Cytol.* 10, 389–401.

Jackson, J. L., Sanford, K. K., and Dunn, T. B. (1970). Neoplastic conversion and chromosomal characteristics of rat embryo cells *in vitro. J. Nat. Cancer Inst.* 45, 11–21.

Jainchill, J. L., and Todaro, G. J. (1970). Stimulation of cell growth *in vitro* by serum with and without growth factor. *Exp. Cell Res.* 59, 137–146.

Katsuta, H., Takaoka, T., Hattori, K., Kawada, I., Kuwabara, H., and Kuwabara, S. (1959). The growth-promoting substances for Yoshida sarcoma cells in tissue culture. *Jap. J. Exp. Med.* 29, 297–309.

Kotler, M. (1970). Control of multiplication of uninfected mouse embryo fibroblasts and mouse embryo fibroblasts converted by infection with murine sarcoma virus (Harvey). *Cancer Res.* 30, 2493–2496.

Krantz, S. B., and Jacobson, L. O. (1970). "Erythropoietin and the regulation of erythropoiesis." Univ. of Chicago Press, Chicago.

Kruse, P. F., Jr., and Miedema, E. (1965). Production and characterization of multiple-layered populations of animal cells. *J. Cell Biol.* 27, 273–279.

Levi-Montalcini, R., and Angeletti, P. U. (1968). Nerve growth factor. *Physiol. Rev.* 48, 534–569.

Lieberman, I., and Ove, P. (1958). A protein growth factor for mammalian cells in culture. *J. Biol. Chem.* 233, 637–642.

Lieberman, I., Lamy, F., and Ove, P. (1959). Nonidentity of fetuin and protein growth (flattening) factor. *Science* 129, 43–44.

Lucy, J. A. (1960). The amino acid and protein metabolism of tissues cultivated *in vitro. Biol. Rev.* 35, 533–571.

Macieira-Coelho, A. (1967a). Relationships between DNA synthesis and cell density in normal and virus-transformed cells. *Int. J. Cancer* 2, 297–303.

Macieira-Coelho, A. (1967b). Dissociation between inhibition of movement and inhibition of division in RSV transformed human fibroblasts. *Exp. Cell Res.* 47, 193–200.

Macieira-Coelho, A. (1967c). Influence of cell density on growth inhibition of human fibroblasts *in vitro. Proc. Soc. Exp. Biol. Med.* 125, 548–552.

Marr, A. G. M., Owen, J. A., and Wilson, G. S. (1962). Studies on the growth-promoting glycoprotein fraction of foetal calf serum. *Biochim. Biophys. Acta* 63, 276–285.

Metcalf, D., and Stanley, E. R. (1969). Quantitative studies on the stimulation of mouse bone marrow colony growth *in vitro* by normal human urine. *Aust. J. Exp. Biol. Med. Sci.* 47, 453–466.

Metzgar, D. P., Jr., and Moskowitz, M. (1960). Separation of growth promoting activity from horse serum by dialysis. *Proc. Nat. Acad. Sci. U.S.* 104, 363–365.

Michl, J. (1961). Metabolism of cells in tissue culture *in vitro.* The influence of serum protein fractions on growth of normal and neoplastic cells. *Exp. Cell Res.* 23, 324–334.

Mizrahi, A., and Moore, G. E. (1970). Partial substitution of serum in hematopoietic cell line media by synthetic polymers. *Appl. Microbiol.* 19, 906–910.

Moore, E. G. (1972). Ph.D. Thesis University of Wisconsin, Madison.

Morgan, J. F. (1970). Nutrition of mouse ascites tumor cells in primary culture.

I. Large molecular substances and conditioned media. *J. Nat. Cancer Inst.* **44**, 623–631.

Moskowitz, M., and Schenck, D. M. (1965). Growth promoting activity for mammalian cells in fractions of tissue extracts. *Exp. Cell Res.* **38**, 523–535.

Mueller, G. C. (1969). Biochemical events in the animal cell cycle. *Fed. Proc., Fed. Amer. Soc. Exp. Biol.* **28**, 1780–1789.

Mÿllylä, G., Häyry, P., Penttinen, K., and Saxén, E. (1966). Serum lipoproteins in primary cell attachment and growth behavior of cells on glass. *Ann. Med. Exp. Biol. Fenn.* **44**, 171–176.

Pedersen, K. O. (1944). Fetuin, a new globulin isolated from serum. *Nature (London)* **154**, 575.

Peterson, J. A., and Rubin, H. (1969). The exchange of phospholipids between cultured chick embryo fibroblasts and their growth medium. *Exp. Cell Res.* **58**, 365–378.

Pierson, R. W., Jr., and Temin, H. M. (1972). The partial purification from calf serum of a fraction with multiplication-stimulating activity for chicken fibroblasts in cell culture and with non-suppressible insulin-like activity. *J. Cell. Physiol.*, in press.

Puck, T. T., Waldren, C. A., and Jones, C. (1968). Mammalian cell growth proteins. I. Growth stimulation by fetuin. *Proc. Nat. Acad. Sci. U.S.* **59**, 192–199.

Raff, E. C., and Houck, J. C. (1969). Migration and proliferation of diploid human fibroblasts following "wounding" of confluent monolayers. *J. Cell. Physiol.* **74**, 235–244.

Rhode, S. L., III, and Ellem, K. A. O. (1968). Control of nucleic acid synthesis in human diploid cells undergoing contact inhibition. *Exp. Cell Res.* **53**, 184–204.

Rubin, H. (1966). The inhibition of chick embryo cell growth by medium obtained from cultures of Rous sarcoma cells. *Exp. Cell Res.* **41**, 149–161.

Rubin, H. (1970a). Overgrowth-stimulating factor released from Rous sarcoma cells. *Science* **167**, 1271–1272.

Rubin, H. (1970b). Overgrowth-stimulating activity of disrupted chick embryo cells and cells infected with Rous sarcoma virus. *Proc. Nat. Acad. Sci. U.S.* **67**, 1256–1263.

Sanford, K. K. (1967). "Spontaneous" neoplastic transformation of cells *in vitro:* Some facts and theories. *Nat. Cancer Inst., Mono* **26**, 387–418.

Sanford, K. K., Earle, W. R., and Likely, G. D. (1948). The growth *in vitro* of single isolated tissue cells. *J. Nat. Cancer Inst.* **9**, 229–246.

Sanford, K. K., Westfall, B. B., Fioramonti, M. C., McQuilkin, W. T., Bryant, J. C., Peppers, E. V., Evans, V. J., and Earle, W. R. (1955). The effect of serum fractions on the proliferation of strain L mouse cells *in vitro*. *J. Nat. Cancer Inst.* **16**, 789–802.

Schenkein, I., Levy, M., Bueker, E. D., and Tokarsky, E. (1968). Nerve growth factor of very high yield and specific activity. *Science* **159**, 640–643.

Schultze, H. E., Göllner, I., Heide, K., Schönenberger, M., and Schwick, G. (1955). Zur kenntnis der α-globuline des menschlichen normalserums. *Z. Naturforsch. B* **10**, 463–473.

Schwartz, A. G., and Amos, H. (1968). Insulin dependence of cells in primary culture: Influence on ribosome integrity. *Nature (London)* **219**, 1366–1367.

Smith, A. P., Varon, S., and Shooter, E. M. (1968). Multiple forms of the nerve growth factor protein and its subunits. *Biochemistry* **7**, 3259–3268.

Soeiro, R., and Amos, H. (1966). Arrested protein synthesis in polysomes of cultured chick embryo cells. *Science* **154**, 662–665.

Spiro, R. G. (1960). Studies on fetuin, a glycoprotein of fetal serum. I. Isolation, chemical composition, and physicochemical properties. *J. Biol. Chem.* **235**, 2860–2869.

Stanley, E. R., and Metcalf, D. (1969). Partial purification and some properties of the factor in normal and leukaemic human urine stimulating mouse bone marrow colony growth *in vitro*. *Aust. J. Exp. Biol. Med. Sci.* **47**, 467–483.

Stanley, E. R., Robinson, W. A., and Ada, G. L. (1968). Properties of the colony stimulating factor in leukaemic and normal mouse serum. *Aust. J. Exp. Biol. Med. Sci.* **46**, 715–726.

Swank, R. T., and Munkres, K. D. (1971). Molecular weight analysis of oligopeptides by electrophoresis in polyacrylamide gel with sodium dodecylsulfate. *Anal. Biochem.* **39**, 462–477.

Taylor, A. C. (1961). Attachment and spreading of cells in culture. *Exp. Cell Res., Suppl.* **8**, 154–173.

Taylor, J. M., Cohen, S., Mitchell, W. M. (1970). Epidermal growth factor: High and low molecular weight forms. *Proc. Nat. Acad. Sci. U.S.* **67**, 164–171.

Temin, H. M. (1965). The mechanism of carcinogenesis by avian sarcoma viruses. I. Cell multiplication and differentiation. *J. Nat. Cancer Inst.* **35**, 679–693.

Temin, H. M. (1966). Studies on carcinogenesis by avian sarcoma viruses. III. The differential effect of serum and polyanions on multiplication of uninfected and converted cells. *J. Nat. Cancer Inst.* **37**, 167–175.

Temin, H. M. (1967a). Studies on carcinogenesis by avian sarcoma viruses. VI. Differential multiplication of uninfected and of converted cells in response to insulin. *J. Cell. Physiol.* **69**, 377–384.

Temin, H. M. (1967b). Control by factors in serum of multiplication of uninfected cells and cells infected and converted by avian sarcoma viruses. *In* "Growth Regulating Substances for Animal Cells in Culture" (V. Defendi and M. Stoker, eds.), Wistar Inst. Symp., Monogr. No. 7, pp. 103–114. Wistar Inst. Press, Philadelphia, Pennsylvania.

Temin, H. M. (1968a). Studies on carcinogenesis by avian sarcoma viruses. VIII. Glycolysis and cell multiplication. *Int. J. Cancer* **3**, 273–282.

Temin, H. M. (1968b). Carcinogenesis by avian sarcoma viruses. X. The decreased requirement for insulin-replaceable activity in serum for cell multiplication. *Int. J. Cancer* **3**, 771–787.

Temin, H. M. (1969). Control of cell multiplication in uninfected chicken cells and chicken cells converted by avian sarcoma viruses. *J. Cell. Physiol.* **74**, 9–16.

Temin, H. M. (1970a). Control of multiplication of uninfected rat cells and rat cells converted by murine sarcoma virus. *J. Cell. Physiol.* **75**, 107–120.

Temin, H. M. (1970b). Control of cell multiplication in normal cells and cells converted by infection with RNA sarcoma viruses. *In* "The Biology of Large RNA Viruses" (R. D. Barry and B. W. J. Mahy, eds.), pp. 687–711. Academic Press, New York.

Temin, H. M. (1971). Stimulation by serum of multiplication of stationary chicken cells. *J. Cell Physiol.* **78**, 101–170.

Terasima, T., and Tolmach, L. J. (1963). Growth and nucleic acid synthesis in synchronously dividing populations of HeLa cells. *Exp. Cell Res.* **30**, 344–362.

Tiötta, E. (1968). Decreased insulin growth dependence in transformed hamster cells. *Arch. Gesamte Virusforsch.* **25**, 363–364.

Todaro, G. J., Lazar, G. K., and Green, H. (1965). The initiation of cell division in a contact-inhibited mammalian cell line. *J. Cell. Comp. Physiol.* **66**, 325–333.

Todaro, G. J., Matsuya, Y., Bloom, S., Robbins, A., and Green, H. (1967). Stimulation of RNA synthesis and cell division in resting cells by a factor present in serum. *In* "Growth Regulating Substances for Animal Cells in Culture" (V. Defendi and M. Stoker, eds.), Wistar Inst. Symp., Monogr. No. 7, pp. 87–98. Wistar Inst. Press, Philadelphia, Pennsylvania.

Tozer, B. T., and Pirt, S. J. (1964). Suspension culture of mammalian cells and macromolecular growth-promoting fractions of calf serum. *Nature (London)* **201**, 375–378.

Turkington, R. W. (1969). Stimulation of mammary carcinoma cell proliferation by epithelial growth factor *in vitro*. *Cancer Res.* **29**, 1457–1458.

Vasiliev J. M., Gelfand, I. M., Guelstein, V. I., and Fetisova, E. K. (1970). Stimulation of DNA synthesis in cultures of mouse embryo fibroblast-like cells. *J. Cell. Physiol.* **75**, 305–313.

Wallis, C., Ver, B., and Melnick, J. L. (1969). The role of serum and fetuin in the growth of monkey kidney cells in culture. *Exp. Cell Res.* **58**, 271–282.

Waymouth, C., and Reed, D. E. (1965). A reversible morphological change in mouse cells (strain L, clone NCTC 929) under the influence of insulin. *Tex. Rep. Biol. Med.* **23**, Suppl. 1, 413–419.

Wiebel, F., and Baserga, R. (1969). Early alteration in amino acid pools and protein synthesis of diploid fibroblasts stimulated to synthesize DNA by addition of serum. *J. Cell. Physiol.* **74**, 191–202.

Wool, I. G., and Cavicchi, P. (1967). Protein synthesis by skeletal muscle ribosomes. Effect of diabetes and insulin. *Biochemistry* **6**, 1231–1242.

Yeh, J., and Fisher, H. W. (1969). A diffusible factor which sustains contact inhibition of replication. *J. Cell Biol.* **40**, 382–388.

Yoshikura, H., and Hirokawa, Y. (1068). Induction of cell replication. *Exp. Cell Res.* **52**, 439–444.

Yoshikura, H., Hirokawa, Y., and Yamada, M. (1967). Synchronized cell division induced by medium change. *Exp. Cell Res.* **48**, 226–228.

4

SOME ASPECTS OF THE ENERGY METABOLISM OF MAMMALIAN CELLS

Charles T. Gregg

> Energy is the only life, and is from the
> Body; and Reason is the bound or out-
> ward circumference of Energy. *Energy
> is Eternal Delight.*
>
> WILLIAM BLAKE, *The Marriage of
> Heaven and Hell* (ca. 1793)

I. Introduction

A. Introductory Remarks

1. SCOPE OF THE CHAPTER

Any cell maintains itself by extracting energy from its environment and transforming it into chemical energy to drive biosynthetic reactions, maintain concentration gradients, do mechanical work, or carry out the myriad other energy-requiring reactions of which various specialized cells are capable (e.g., the transmission of electric energy by nerve cells). Thus, energy metabolism, broadly conceived, is an essential part of every aspect of cellular activity. A complete account would treat in detail all the reactions by which chemical energy is extracted from the cell's environment and those in which the cell expends this energy. Owing to space limitations,* nothing of the sort is attempted here.

This chapter has two aims: first, to help cellular biologists to better understand the literature in this area and perhaps to see its relevance to their own work. Hence, current theories on the mechanism by which ATP formation is coupled to electron transport are discussed. This is an area to which much current effort is devoted; additionally, inclusion of some details of the chemical coupling hypothesis provides a convenient framework in which to discuss the mode of action of the most common inhibitors. It is the author's feeling that inability to understand the significance of various inhibitors used in this area is a principal stumbling block for nonspecialists trying to follow the literature.

There is another goal. An area in which the interests of cell culturists and biochemists interested in energy metabolism impinge is that of malignant transformation. Much cell biology literature is concerned with malignant cells studied in vitro. Correspondingly, the altered energy metabolism of malignant cells has played a central role in theories of the biochemistry of malignant transformation for nearly 50 years. Thus, throughout this chapter aspects of the energy metabolism of malignant cells are discussed with the hope of enhancing the reader's grasp of this important area of the biology of cells in culture and its relation to a major clinical problem.

Wherever possible, references are to reviews rather than to original work; some slighting of major contributors to this field is thereby inevitable. The energy metabolism of cells in culture was carefully reviewed earlier by Paul (1965) and the energy metabolism of tumors by Wenner (1967). In this discussion the emphasis is on reactions by which glucose is utilized as fuel by mammalian cells. The use of other carbon sources such as endogenous cellular lipids or amino acids from the medium was discussed by Paul (1965, p. 249).

I assume that the oxygen consumption of mammalian cells in culture arises from the oxidation of NADH or reduced flavoprotein via the mitochondrial respiratory chain. There is considerable experimental evidence to support this view, but, under some conditions, the microsomal electron

* Dr. Rothblat and Dr. Cristofalo (1971) quote Proverbs 17:9, "He that hath knowledge spareth his words."

transport system may be quantitatively important. The contribution of this system to the oxygen consumption of liver cells was recently reviewed (Estabrook et al., 1970); no comparable data are available for cells in culture. Sato and Hagihara (1970) find no cytochrome b_5 or cytochrome P-450 in a variety of ascites hepatomas or Ehrlich or Sarcoma 180 ascites cells. Since these are essential components of the microsomal electron transport chain, this system may be less significant in ascites cells (and, by extension, in cells in culture generally) than in liver. The role of nuclei in cellular respiration is discussed in Section II,C.

2. The Importance of Controlled Cultured Conditions for the Study of Energy Metabolism

The cell in culture is an adaptable organism; otherwise it would not survive the environmental insults heaped on it by callous investigators. This adaptability arises, in part, from the cells' ability to alter its energy metabolism in consonance with environmental changes. A few examples will indicate the nature of the response. With HLM cells, Paul (1965, p. 258) showed that a change in medium pH from 7.2 to 6.8 caused a 10-fold decrease in lactate formed per mole of glucose utilized. Such a pH change could readily occur over several hours with a high concentration of a rapidly metabolizing cell such as the mouse lymphoma, L5178Y (Gregg, 1970). Racker (1965) showed that a 4-fold change in residual medium glucose (from 0.8 to 0.2 mM) caused a 25-fold drop in the glycogen content of HeLa cells, while the level of phosphorylase a decreased 70-fold. The specific activities of aldolase and lactic dehydrogenase in mouse LS cells increased 65 and 97%, respectively, when the medium oxygen tension was raised from 32 to 48 mm Hg (Self et al., 1968), while the maximum cell yield doubled when the oxygen level was reduced from 140 to 48 mm Hg (Kilburn et al., 1969).

The latter studies raise an important point on evaluation of enzyme levels in cells, namely the difficulty of deciding whether an increased level of a particular enzyme is due to enzyme synthesis de novo or to removal or supply of inhibitors or activators, respectively. Self et al. (1968) conclude that the increase in aldolase and lactic dehydrogenase they observed in these cells was due to de novo synthesis. Their data also show substantial fluctuations of enzyme level in a random cell population as growth proceeds (and the oxygen tension in the medium drops). This raises a question concerning the many assays reported on variation of enzyme levels in synchronized cell populations. Self et al. (1968) have shown considerable variation of enzyme level in *random* populations due to the various perturbations of dilution, decreasing oxygen tension, etc. Thus studies on synchronized cell populations in which these factors are not adequately controlled are difficult to interpret.

Typically, cells in spinner culture grow in a medium of high glucose and low oxygen concentration with no environmental controls except temperature and stirring rate. Small laboratory fermentors provide the additional important capacities to measure and control pH and oxygen tension; studies done in the author's laboratory were carried out entirely on cells raised under these conditions. This is a step forward, but measurement of any cell's capacity for glycolysis and respiration reflects the prehistory of the cell at least as much as its intrinsic capacity to carry out the two reaction sequences. Thus the message of this section is that real understanding of the relation between cellular ability and capacity must await studies carried out on cells grown under conditions of constant pH, oxygen tension, glucose concentration, etc. (i.e., cells grown in a chemostat). (For further discussion, see Chapter 2, Volume 2.)

3. The Advantages of Suspension-Cultured Cells for the Study of Energy Metabolism and Malignancy

The earliest investigations on energy metabolism of malignant cells were made using tissue slices or minces of solid tumors. This material has many disadvantages. There are frequently necrotic areas in solid tumors and, depending on the tumor's vascularization, there may be steep gradients of oxygen, nutrient, and metabolite concentration. Moreover, such tumors are in general not pure cultures but contain both blood vessels and connective tissue.

A great advance was the discovery of transplantable ascites tumors (Loewenthal and Jahn, 1932) and their introduction into the field of tumor biochemistry by Klein (1951), who showed that they were relatively homogeneous, easily obtained, reproducible, and readily quantifiable (by pipetting, for example). The importance of this development is attested to by the fact that most of our information on the metabolism of malignant cells is derived from studies on ascites tumor preparations.

The experimental convenience of ascites tumor cells should not cause their disadvantages to be overlooked. I discussed earlier the importance of knowledge of and control over the conditions of cell growth. With ascites cells the experimenter has neither. To maximize yield, cells are commonly harvested from a desperately ill animal hours, or at the most days, away from death. The ascites fluid is severely depleted of both oxygen and glucose (Del Monte, 1967; Nakamura and Hosada, 1968). [This is also true of rapidly growing cells in spinner culture (Gregg, 1970).] And, although ascites cells are frequently described as "pure" or "nearly pure" cultures, quantitative studies indicate that they are contaminated with 20% or more of phagocytes and other host cells as well as red blood cells (Klein, 1951; Koprowska, 1956). Methods exist for the complete removal of contaminating cells (Lindahl and Klein, 1955; Hawtrey, 1958), but these methods are either seldom used or poorly understood. An example of the latter is the removal of red blood cells by differential lysis with distilled water. The effect of this treatment on the subsequent metabolism of ascites cells has never been carefully studied, and the remaining contamination is seldom evaluated microscopically. An ascites suspension judged "free" of red blood cells by visual examination may contain up to 10% erythrocytes (LaBauve and Gregg, 1970).

Monolayer cultures, while homogeneous and free of contaminants, are difficult to grow in large quantity, are easily damaged in removal from the glass surface (Luthardt and Fischer, 1964; Günther and Goecke, 1966), and are grown under conditions which are difficult to control. Thus the suspension-cultured cell grown under carefully controlled conditions is the material of choice for such studies. There is a further caveat if the investigator wishes to study the biochemistry of malignancy. There is an important distinction between a malignant population of cells and a population of malignant cells. Most lines of cells in culture fall into the first class. If a million such cells are injected into a susceptible animal, a tumor will arise because in that number of cells a few are malignant in the sense used here, that of giving rise to progressive, transplantable, and ultimately fatal tumors (Jarrett and Macpherson, 1968; Sanford et al., 1969).

Biochemical studies on malignancy cannot be done on such a malignant population because the investigator has no way of distinguishing the metabolic properties of the malignant minority of cells from the vastly greater nonmalignant majority. For such studies a population of malignant cells is required (i.e., a population in which the majority of cells is capable of inducing a fatal tumor in a suitable host). At the present time, the number of cell lines which qualify under this restrictive definition is very small, since few studies have been made with lines of cells in culture in which only a small number of cells have been administered to a suitable host. The methodology of titrations to establish degree of malignancy is discussed in detail by Hewitt

(1958).* Clearly, the word "malignant" has an operational definition with meaning only when the experimental conditions and criteria for tumorigenicity are clearly defined.

B. The Gross Anatomy of Energy Metabolism

1. GLYCOLYSIS

In discussions of energy metabolism glycolysis is usually defined, following Warburg (1930b), as the production of *lactic acid* from glucose. This restrictive definition is unfortunate, but it is thoroughly entrenched in the literature. The complete glycolytic pathway is shown in Fig. 1; details of the isomerase, mutase, and aldolase reactions are omitted. The remaining steps will be briefly discussed as to their role in energy metabolism. For a more nearly complete analysis, the excellent review on the mechanism and control of glycolysis by Rose and Rose (1969) is recommended.

The four kinase steps involving ATP and ADP all have equilibria far to the right. All four kinase enzymes require Mg^{2+}, as is usual for enzymes catalyzing reactions involving nucleotides. Thus the intracellular concentration and distribution of Mg^{2+} may play an important role in controlling glycolysis, respiration, and other nucleotide-dependent reactions. Hexokinase is inhibited allosterically by G-6-P; this inhibition is reversed by P_i. Free ATP inhibits the reaction competitively with MgATP. In mammalian cells the reaction is essentially irreversible, as is the phosphofructokinase reaction. Phosphofructokinase is one of the most complex enzymes in intermediary metabolism; a large number of normal metabolites affect its activity at physiological concentrations. The enzyme is stimulated by P_i, AMP, and ADP and is inhibited by high concentrations of ATP and by citrate. In Ehrlich ascites cells a linear, reciprocal correlation exists between intracellular citrate level and the utilization of glucose (Suzuki *et al.*, 1968). The enzyme molecule contains essential —SH groups, which are not at the catalytic site but must be reduced for full enzymatic activity. Thus the enzyme is also sensitive to the redox potential of the cytosol as reflected in the —SH/—S—S— ratio. As Rose and Rose point out (1969, p. 111), this enzyme is present in tissues at very high levels. Therefore, the conditions under which

---

* The LD_{50} for intraperitoneal injection of L5178Y cells in culture into DBA/2J (F₁ hybrid) mice is approximately one cell (Gregg, 1970). Although such data are necessarily imprecise, these lymphoma cells clearly constitute a population of malignant cells.

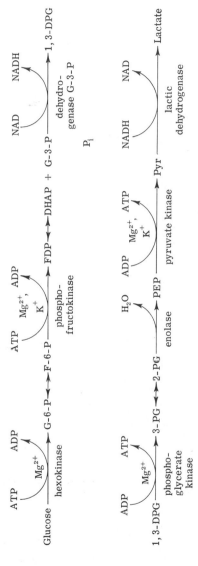

Fig. 1. Glycolytic pathway showing the steps involved in energy metabolism. Details of the isomerase, mutase, and aldolase reactions are omitted. Nonstandard abbreviations used are G-6-P and F-6-P, glucose and fructose-6-phosphate, respectively; FDP, fructose-1,6-diphosphate; DHAP, dihydroxyacetone phosphate; G-3-P, glyceraldehyde-3-phosphate; 1,3-DPG, 1,3-diphosphoglyceric acid; 3-PG and 2-PG, 3-phospho- and 2-phosphoglyceric acid, respectively; PEP, phosphoenol-pyruvate; and Pyr, pyruvate.

kinetic parameters are usually measured (very low enzyme levels relative to substrates and activators or inhibitors) may give results not readily applicable to the physiological situation.

This also applies to 3-phosphoglycerate dehydrogenase. It, too, is present in cells in amounts greater than either its substrate or product, and results obtained under the usual assay conditions must be cautiously interpreted. Since P_i takes part stoichiometrically in the reaction, 3-PG dehydrogenase was originally thought to be the step responsible for stimulation of glycolysis by P_i. The current view (Rose and Rose, 1969, pp. 116–117) is that this control is exerted through hexokinase and phosphofructokinase. The K_m for P_i for 3-phosphoglycerate dehydrogenase is about 0.3 mM.[*] Both products (NADH and 1,3-DPG) inhibit competitively with substrate. The K_m for NAD is high relative to the K_i for NADH: 90 μM versus 0.3 μM; thus, product inhibition may be a major factor in influencing the reaction rate. Like phosphofructokinase, 3-PG dehydrogenase is an —SH enzyme and thus may be sensitive to the redox potential of the cytosol.[†]

By carrying the reactions of glycolysis to this point the cell has consumed 2 moles of ATP, 2 moles of P_i, and 2 moles of NAD per mole of glucose utilized. For glycolysis to continue these molecules must be continually regenerated or supplied. This is partially accomplished in subsequent glycolytic reactions. In the next step the 2 moles of ATP originally supplied are regenerated in the phosphoglycerate kinase step (K_m for MgADP, about 0.1 mM).

Enolase, which removes 1 mole of water from 2-phosphoglycerate to form phosphoenolpyruvate, is inhibited by F^- in the presence of phosphate. The true inhibitor is presumably an F-P complex capable of reaction with the Mg^{2+} on which the activity of the enzyme depends. Pyruvate kinase forms 2 moles of ATP for every mole of glucose utilized and thus leads to a net gain for the glycolytic sequence of 2ATP. This reaction, like the phosphofructokinase reaction, also requires K^+. Although, like the phosphoglycerate kinase reaction, pyruvate kinase is readily reversible, equilibrium for both reactions strongly favors ATP formation. Pyruvate kinase is also inhibited by ATP and P_i, with important consequences for metabolic control (J. R. Williamson, 1965). The

---

[*] The concentration of substrate or cofactor required to give half-maximum velocity of an enzyme reaction is the K_m. A concentration of $10 \times K_m$ gives a velocity of about $0.9V_{max}$, where V_{max} is the maximum velocity. Thus the lower the K_m, the more effective the substrate or cofactor in stimulating the enzyme. The K_i has an analogous meaning for inhibitors; i.e., a concentration of inhibitor equal to the K_i will reduce the enzyme reaction velocity one half.

[†] For a brief but fascinating account of the 10 years of work required to determine the mechanism of the oxidative reaction of glycolysis, see Oesper (1968).

liver enzyme is stimulated by FDP and can change its concentration some 10-fold in response to changes in dietary carbohydrate level (Rose and Rose, 1969, p. 141), the only glycolytic enzyme so affected. Pyruvate kinase is inhibited by Ca^{2+} (by its antagonism with K^+). Ehrlich ascites cells are relatively impermeable to Ca^{2+}, and it was suggested that this plays a role in the high glycolytic rates of which these cells are capable (Bygrave, 1966). The K_m for ADP is approximately equal to the K_i for ATP for the rabbit muscle enzyme; both are about 0.3 mM.

The glycolytic enzymes are usually found in the "soluble" portion of the cell [i.e., the supernatant fluid remaining after centrifugation of a homogenate at 144,000 $\times$ g for 30–120 minutes (De Duve et al., 1962; Rose and Rose, 1969, p. 151)]. The best documented exception is hexokinase, which is found partially associated with mitochondria in tumor, brain, skeletal muscle, and plants and also with "microsomes" of skeletal muscle (Rose and Rose, 1969, p. 151). Other reports suggest that glycolytic enzymes are membrane-bound in various tissues (D. E. Green et al., 1965). The significance of these results remains to be evaluated.

The mechanism of ATP formation in glycolysis is fundamentally different from that accompanying terminal oxidation. One difference is that in glycolysis phosphate is transferred from a phosphorylated substrate directly to ADP. This is not true in respiratory-chain phosphorylations (see Section I,B,2). This distinction is preserved in the nomenclature in which glycolytic formation of ATP is called substrate-level phosphorylation to distinguish it from that occurring in terminal oxidation.[*] Substrate-level phosphorylations are also not inhibited by uncoupling agents or by other compounds which inhibit respiratory-chain phosphorylation (see Section I,B,2).

The phosphorylations accompanying oxidation and reduction of respiratory-chain components are often referred to as oxidative phosphorylations with the implication that substrate-level phosphorylation is not oxidative. The terminology arose from the fact that glycolytic substrate-level phosphorylations continue under anaerobic conditions, while respiratory-chain phosphorylations are oxygen-dependent. However, mechanistically, the distinction is dubious. The high-energy phosphate compound 1,3-diphosphoglyceric acid, which transfers phosphate to ADP in the phosphoglycerate kinase step, is formed in the only clearly oxidative step in glycolysis, that catalyzed by glyceraldehyde-3-phosphate dehydrogenase. Similarly, the substrate-level step of the Krebs cycle is oxygen-dependent.

[*] There is also a substrate-level phosphorylation in the Krebs cycle, which will be discussed later (Section I,B,2).

2. TERMINAL OXIDATION

The reactions of terminal oxidation are localized in the mitochondria. Here pyruvate is oxidized via the Krebs tricarboxylic acid cycle and the electron transport chain to CO_2 and H_2O. The Krebs cycle, narrowly considered as part of the energy-producing system, obeys the following reaction[*]:

$$Ac \sim SCoA + 3 H_2O + 3 NAD^+ + fp + GDP + P_i$$

$$\downarrow \qquad\qquad (1)$$

$$2 CO_2 + 3 NADH + 3 H^+ + fp\text{-}H_2 + GTP + HSCoA$$

One mole of GTP per mole of Ac $\sim$ CoA is produced by substrate-level phosphorylation in the α-ketoglutarate dehydrogenase complex. This may be used directly as an energy source (for protein synthesis or fatty acid activation, for example) or converted to ATP via the nucleoside diphosphokinase reaction:

$$GTP + ADP \rightleftharpoons GDP + ATP$$

However, the bulk of chemical energy conserved by the cell in oxidizing glucose to CO_2 and H_2O results from the phosphorylation reactions coupled to transfer of electrons down the respiratory chain from either NADH or reduced flavoprotein. In Eq. (2) are included the three known sites at which electron transport is coupled to ATP formation.[†] In the literature they are referred to as sites I, II, and III (from left to right). While it is known that site I is between NADH and cyt b (cytochrome b), site II between cyt b and cyt c_1, and site III between cyt c and O_2, the locations cannot now be more precisely specified. Sites I and II, for example, may be on the oxygen side of nhFe rather than on the substrate side as shown here. Equation (2) contains only the minimum number of generally accepted components (although inclusion of UQ is not universally accepted). Other electron carriers may exist within the chain (cf. Chance *et al.*, 1967; Galeotti *et al.*, 1969).

The properties of the individual coupling sites are discussed in some detail by Racker (1970), the site and mechanism of uncouplers and inhibitors by Boyer (1968), the properties of the various cytochromes

---

[*] The Krebs cycle has a much broader function in producing intermediates for a variety of biosynthetic reactions in microorganisms growing on 1- or 2-carbon substrates. In mammalian tissue, it functions largely as part of an energy-yielding pathway (Garland *et al.*, 1967).

[†] Nonstandard abbreviations used are fp₁, flavoprotein 1, NADH dehydrogenase, fp₂, succinic dehydrogenase; nhFe, nonheme iron; UQ, ubiquinone (coenzyme Q); cyt, cytochrome; TTB, 4,4,4-trifluoro-(2-thienyl)-1,3-butanedione.

by Gibson (1968), and the stoichiometry of the respiratory carriers by Klingenberg (1968). Lardy and Ferguson (1969) review the more controversial aspects of electron transport and respiratory-chain phosphorylation. As these authors point out, their 383 references are only a fraction of the papers published in this area over a 2-year period.

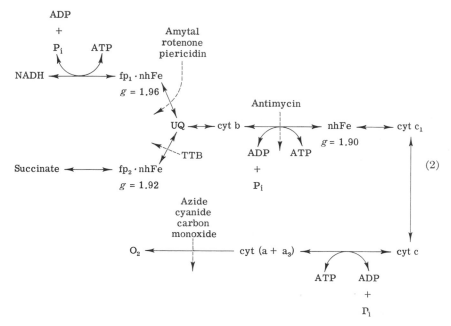

$$(2)$$

Here only the broadest features of this subject will be discussed. According to Eq. (2), oxidation of 1 mole of NADH via the respiratory chain should consume 1 gm atom of oxygen and yield 3 moles of ATP.* Thus the familiar P/O ratio should be 3.0. The P/O ratio for oxidation of 1 mole of succinate, according to Eq. (2), should be 2.0. Therefore, the P/O ratio for oxidation of 1 mole of pyruvate via the Krebs cycle should be 3.0 since, although the succinate oxidation step yields a P/O ratio of 2.0, the oxidation of α-ketoglutarate has a P/O ratio of 4.0 because of the contribution of 1 mole of GTP from substrate-level phosphorylation. There was a recent flurry of reports of P/O ratios much greater than 3.0 or 4.0. Most of these arose from ignorance of elementary manometric techniques; the few remaining reports are unconfirmed (for a more detailed discussion, see Slater, 1966). From Eq. (2) we can conclude that complete oxidation of 2 moles of pyruvate to CO_2 and

* The expression "μ atoms O_2" frequently appears in the literature. No such unit exists. The correct expression is "μg atoms O_2" or "μg A O_2."

H_2O yields 30 moles of ATP, remembering that 1 mole of NADH is formed in the conversion of 1 mole of pyruvate to acetyl CoA. Formation of 2 moles of pyruvate from glucose yields either 2 or 6 moles of ATP, depending on the fate of the cytoplasmic NADH formed during glycolysis (see Section II,A,2).

Classically, the enzymologist faced with a metabolic sequence to unravel has isolated, purified, and characterized the individual enzymes involved in the sequence, finally reconstituting the entire system from purified materials. That we know so little about the mechanism of electron transport and respiratory-chain phosphorylation after 30 years of work shows how abjectly this approach has failed with mitochondrial multienzyme systems firmly bound to a lipoprotein membrane (but see Racker, 1970). Of the cytochromes, only cytochrome c has been obtained in homogeneous form (Gibson, 1968), and the enzymes involved in ATP formation are much less well characterized than the respiratory-chain components. Consequently, use of specific inhibitors, combined with highly evolved spectrophotometric and polarographic techniques, was essential in furnishing what knowledge we have in this area. Much of the methodology dates from the pioneering work of Chance and Williams (1956). A few examples will illustrate the procedure.

If antimycin is added to a mitochondrial or cellular preparation (Table I and Table IV) oxidizing an NAD-linked substrate (e.g., pyruvate), oxygen consumption can be completely inhibited as shown by polarographic measurement with a variety of oxygen electrodes. Oxygen consumption due to succinate oxidation is similarly eliminated. Applying spectrophotometric techniques, one can show that the respiratory-chain components on the substrate side of cyt c_1 become reduced, while cyt c_1 and the cytochromes to the oxygen side of it become fully oxidized. If a substrate is added (e.g., ascorbate and N,N,N',N'-tetramethylene-phenylenediamine, TMPD), which can reduce cyt c, oxygen consumption is restored and ATP is formed in the presence of ADP and P_i, thus demonstrating a phosphorylation site between cytochrome c and oxygen. That cyt c is the substance being reduced by the substrate can also be shown spectrophotometrically. If ascorbate and TMPD are added in the absence of ADP, then cyt c is reduced, while cyt $(a + a_3)$ remains oxidized. If ADP is now added, then cyt c becomes more oxidized and cyt $(a + a_3)$ more reduced. This indicates that the phosphorylation site is between cyt c and cyt $(a + a_3)$. This is a basic statement of the famous crossover theorem of Chance and Williams (1956) by which they identified the phosphorylation sites in the respiratory chain. A phosphorylation site was postulated to be one at which, in the absence of ADP (or P_i), the redox state of the respiratory carriers abruptly "crossed over" from reduced to oxidized. When ADP is added, the carrier to the substrate side of the "crossover" point becomes more oxidized, while the component to the oxygen side becomes more reduced. This explanation may become clearer when we discuss theories of the control of respiration by ADP and P_i. The crossover theorem also applies to other multienzyme systems [glycolysis, for example (Baierlein, 1967)].

Similar experiments show that oxidation of NADH is blocked by rotenone (with the loss of one phosphorylation site), while oxidation of succinate is unaffected. The same general approach was used to localize nhFe in the respiratory chain. Fragmentation of the chain by chemical means (D. E. Green and Baum, 1970) showed grossly the distribution of nhFe in the various segments. Precise location was made by Beinert and his colleagues (Beinert and Palmer, 1965) using inhibitors and electron spin resonance (esr) to detect the unpaired electron of oxidized nhFe. The g values for the three kinds of nonheme iron in the respiratory chain are shown in Eq. (2). (The abscissa in an esr spectrum is in multiples of the constant g, corresponding to values of the magnetic field at which the signal comes into resonance. The g values in esr spectroscopy are analogous to wavelength in conventional spectrophotometry.)

All the reactions of the respiratory chain [with the exception of cytochrome oxidase, i.e., the reaction of cyt $(a + a_3)$ with oxygen] and the associated phosphorylations are written as reversible. One of the most striking advances of recent years was the demonstration that these reactions are reversible and that the reverse reaction affords still another device by which the cell can regulate the redox state of the intracellular compartments (Klingenberg, 1964). For example, if the respiratory chain is blocked by cyanide, oxygen consumption stops and all respiratory carriers are reduced with the highest percent reduction at the substrate end of the chain. If to such a preparation ascorbate plus TMPD is added, nothing happens. However, if ATP is added, cyt c, cyt c_1, and the rest of the carriers on the substrate side of cyt c become more reduced. Such a preparation can be shown, for example, to reduce NAD in the presence of ascorbate plus TMPD, cyanide, and ATP. As expected, this reaction is blocked by antimycin, rotenone, etc. Similarly, succinate in the presence of ATP can be shown to reduce NAD if antimycin is present, and this reaction is also blocked by rotenone or by TTB.

The mechanism by which electron transport is coupled to ATP formation is the subject of that sort of vigorous controversy which bespeaks a rapidly evolving field or, less charitably, a highly confused one. Most simply stated, there are three types of mechanisms currently being discussed: the chemical hypothesis, which postulates energy transfer between nonphosphorylated and phosphorylated intermediates; the chemi-osmotic coupling hypothesis, which suggests that the energy for ATP formation is created by a proton and/or a membrane potential resulting from electron transport; and finally the conformational hypothesis, which holds that the energy ultimately appearing in ATP first appears in an "energized" conformation of the mitochondrial membrane. These hypotheses will be discussed in turn, and I shall attempt to show that they have many features in common. For a more detailed discussion the reader is referred to Racker (1970).

The chemical hypothesis, first proposed by Slater (1953), is shown in one of its many versions in Eq. (3a–d):

$$X + C \xrightleftharpoons[\text{transport}]{\text{electron}} X{\sim}C \tag{3a}$$

$$X{\sim}C + P_i \xrightleftharpoons{\hspace{1.5cm}} X{\sim}P + C \tag{3b}$$

$$X{\sim}P + ADP \xrightleftharpoons{\hspace{1.5cm}} ATP + X \tag{3c}$$

$$X{\sim}C \xrightarrow[\text{agents}]{\text{uncoupling}} X + C \tag{3d}$$

This scheme provides a framework in which to discuss some useful concepts as well as for outlining briefly the action of certain inhibitors which affect the coupling mechanism. The substance designated C in Eq. (3) is considered to be an electron carrier such as NAD or a cytochrome. Electrons passing through an energy-conserving site cause the hypothetical high-energy intermediate $X{\sim}C$ to be formed. It should be understood that "high-energy" in this context means a chemical bond energy equivalent to ATP [i.e., about 11 kcal/mole under physiological conditions (R. C. Phillips et al., 1969)].[*] In the second reaction $X{\sim}C$ undergoes phosphorolysis, yielding $X{\sim}P$ and releasing C. Finally, $X{\sim}P$ reacts with ADP to form ATP and release X. Equation (3d) describes the action of uncoupling agents, which, according to this hypothesis, act to cause the breakdown of $X{\sim}C$, regenerating free X and free C without the intervention of P_i and ADP and without the formation of ATP.

[*] The biochemist's idea of a "high-energy" bond has been frequently attacked, most recently by Banks (1969), who describes the concept of high-energy bonds to intermediates as odd, curious, confusing, and misleading. "The mythology leads to a way of thinking about molecules which is not only chemically nonsensical but may well impede progress in biology   " Rejoinders by Linus Pauling, D. Wilkie, and A. F. Huxley appeared in a later issue of the same journal (e.g., Pauling, 1970).

It is clear that the respiration rate (the rate of oxygen consumption, as most commonly measured) depends on the rate at which free X and free C are supplied. The regeneration of X and C from intermediate X$\sim$C depends on the presence of both P_i [to regenerate C via reaction (3b)] and ADP [to release X via reaction (3c)]. Thus, in the absence of P_i and ADP, respiration should proceed only until the supply of free X and C are exhausted and then drop to zero. This situation is never observed experimentally. There are always reactions which lead to breakdown of X$\sim$C, either discharging the bond energy as heat (via naturally occurring uncoupled agents) or utilizing the energy of the X$\sim$C bond directly (without the concomitant formation of ATP) to, for example, maintain ionic gradients across the mitochondrial membrane.

An obvious way to quantitatively compare the degree to which the respiration rates of various preparations depend on the presence of P_i and the phosphate "acceptor" ADP is to measure the rates of oxygen consumption of the preparation in the presence and in the absence of ADP and P_i. Experimentally, ADP is usually the component omitted, although equivalent results are obtained when phosphate is left out. The ratio of oxygen consumption rates obtained in the presence and absence of ADP,[*] for example, may range as high as 50. Quite arbitrarily a preparation with an RCR of 6 or above is considered to be "tightly coupled" (i.e., to have a respiratory rate strongly dependent on the presence of ADP and phosphate). On the other hand, "loosely coupled" preparations have an RCR of less than 6. An "uncoupled" preparation has a respiratory rate which is totally independent of P_i and ADP; no ATP is formed. The best known uncoupling agent is 2,4-dinitrophenol (DNP). However, the more recently discovered cyanocarbonylphenyl-hydrazones (the CCPs) are much more effective compounds (Heytler, 1963). Trifluoromethoxy-CCP completely uncouples phosphorylation in the mouse lymphoma cells at 70 pmoles of inhibitor per milligram, about 1000 times less than required with DNP.[†] More efficient still are the salicylanilides (R. L. Williamson and Metcalf, 1967), the most potent of which is effective at one tenth the concentration of trifluoromethoxy-CCP. Inorganic arsenate also is an uncoupling agent. It is believed to react in Eq. (3b) in place of phosphate but to form an unstable S$\sim$As compound which spontaneously hydrolyzes.

Compounds such as aurovertin and the antibiotic oligomycin are true inhibitors of oxidative phosphorylation, since their site of action is on

---

[*] Such ratios are described in the literature as respiratory control ratios (RCR), respiratory control indices (RCI), or acceptor control ratios (ACR).

[†] The abbreviations used throughout this chapter are nmole (nanomole) and pmole (picomole) for 10^{-9} and 10^{-12} mole, respectively.

the coupling mechanism. Other compounds may inhibit oxidative phosphorylation secondarily by inhibiting electron transport (e.g., rotenone) or the Krebs cycle (e.g., malonate). The mode of action of aurovertin and oligomycin, according to Eq. (3), is the inhibition of reaction (3b). Thus these compounds are observed experimentally to inhibit respiration and phosphorylation in a tightly coupled system. The formulation of Eq. (3) implies that there exist certain exchange reactions which should be inhibited by uncoupling agents or phosphorylation inhibitors such as oligomycin. These so-called "partial reactions" of the coupling mechanism include ATP-P_i exchange, ADP-ATP exchange, H_2O-P_i exchange, H_2O-ATP exchange, and the ATPase reaction stimulated by uncoupling agents. Although some puzzling features remain, such reactions do exist and have many of the expected properties. For example, in a nonphosphorylating preparation (made nonphosphorylating with respiratory inhibitors, deprivation of oxidizable substrate, etc.), $^{32}P_i$ is incorporated into ATP under circumstances in which there is no net change in the concentrations of either P_i or ATP. This exchange is considered to be the sum of reactions (3b) and (3c). As expected, the exchange is inhibited by uncoupling agents (which decompose X~C, a participant in the reaction) and by oligomycin, which inhibits reaction (3b). Similarly, the ADP-ATP exchange [Eq. (3c)] can be demonstrated by tracer methods. The ATPase stimulated by uncoupling agents and inhibited by oligomycin is made up of reactions (3c) and (3b) running in reverse (from right to left). The two oxygen exchanges require a more extensive formulation of the coupling mechanism than given in Eq. (3) and will not be discussed here (see Boyer, 1968).

The effect of some of the inhibitors discussed on the endogenous respiration of intact mouse lymphoma cells (L5178Y) is shown in Table I (Gregg et al., 1968). The amounts of the first four compounds required

TABLE I.

INHIBITION OF THE ENDOGENOUS RESPIRATION OF LYMPHOMA CELLS BY A VARIETY OF RESPIRATORY AND GLYCOLYTIC INHIBITORS

Inhibitor	Effective concentration (M)	
	50% Inhibition	100% Inhibition
Rotenone	0.85×10^{-8}	1.7×10^{-8}
Antimycin	1.4×10^{-8}	2.8×10^{-8}
Oligomycin	4.5×10^{-8}	—
Aurovertin	4.2×10^{-8}	—
Cyanide	0.25×10^{-4}	1.0×10^{-4}
Amytal	0.70×10^{-3}	1.4×10^{-3}

for complete inhibition (or maximum inhibition in the case of oligomycin) were of the order of 8–20 pmoles of inhibitor per milligram
of cellular protein. This is from one third to one tenth the amounts
required on a protein basis for corresponding levels of inhibition in
isolated rat liver mitochondria (Slater, 1967a), thus indicating that
these complex inhibitors effectively reach the mitochondrial level even
in intact cells. These data also suggest that there is no essential difference
between the mitochondria of these malignant cells in culture and the
mitochondria of normal rat liver. This conclusion is more strongly supported by subsequent studies on mitochondria isolated from the lymphoma cells (see Section II,B).

The popularity of this representation [Eq. (3)] has suffered in recent
years because, in spite of considerable effort on the part of highly skilled
investigators, none of the presumed high-energy intermediates have been
isolated. The search has concentrated on phosphorylated intermediates
because of the obvious utility of ^{32}P as a tracer and has thus far failed.
Hence biochemists are faced with the discouraging prospect of searching
for a nonphosphorylated intermediate with no known label. But that
such a nonphosphorylated high-energy intermediate does exist is central
to all three theories of the coupling mechanism, as will become apparent.
Evidence for the existence of such an intermediate rests on the experimental facts that both ion transport and energy-linked transhydrogenase
(Lee, et al., 1964) can be driven by the energy supplied by $X \sim C$ (or a
related compound; see below) when ATP formation is inhibited by
oligomycin. Studies on amino acid incorporation in the mouse lymphoma
cell suggest that valine incorporation into acid-insoluble materials may
also be driven in part by $X \sim C$ (Gregg et al., 1968).

As the chemical coupling hypothesis was celebrating its eighteenth
birthday it came under formidable assault by Peter Mitchell, father
of the chemiosmotic coupling hypothesis. Only the main features of
this ingenious proposal will be discussed here. For more detailed accounts the reader is referred to recent reviews (Mitchell, 1966; Slater,
1967b; Greville, 1969; Racker, 1970). The main features of the hypothesis
are briefly outlined by Deamer (1969). Mitchell rejects the existence
of a high-energy intermediate involving a respiratory carrier [$X \sim C$ in
Eq. (3)]; however, his theory does involve a high-energy intermediate
which can be designated $A \sim X$. In some versions of the chemical
coupling hypothesis there is an additional step:

$$X \sim C + A \leftrightarrow A \sim X + C$$

Thus there is no disagreement about the existence of a high-energy
intermediate. The argument centers about its mode of formation. According to the Mitchell hypothesis, the intermediate is formed via a pH

gradient and/or a membrane potential created by the translocation of protons during the respiratory process. Protons are removed from substrate on one side of the coupling membrane (specifically, the inner mitochondrial membrane) and translocated to the other side. A cyclic "proton current" is generated by driving the protons back through a reversible ATPase located in the membrane. The effect of the reversible ATPase is to remove the elements of water from ADP and P_i, forming the anhydride ATP and expelling H^+ on one side and OH^- on the opposite side of the membrane. The theory requires two proton-translocating systems in the coupling membrane, each consisting of a proton carrier and a H_2 carrier. One of the proton-translocating systems consists of three loops of the oxidation chain, while the other is the oligomycin-sensitive ATPase. It should be noted parenthetically that, of the many ATPases in the cell, only the one inhibited by oligomycin is relevant to the respiratory-chain coupling mechanism.

The chemiosmotic hypothesis requires a vesicular structure which allows an unambiguous recognition of the inside and outside of the coupling membrane. This structural requirement is more stringent than that associated with the chemical hypothesis, which simply regards the coupling membrane as a spatial "organizer" on which the proteins involved in the coupling mechanism are arranged.

The action of uncouplers is also viewed differently in the two theories. The explanation of uncoupling in the chemical hypothesis [Eq. (3d)] has always been somewhat unsatisfactory, since it is unclear how compounds of such diverse chemical properties could cause hydrolysis of the same intermediate. In the Mitchell hypothesis, uncouplers are those compounds which affect the permeability of the coupling membrane in such a way as to reduce the membrane potential below that required to form ATP (about 270 mV).

In the earliest form of the chemiosmotic hypothesis emphasis was on the proton gradient across the coupling membrane as the driving force for ATP synthesis. The pH gradients required were quickly shown to be excessively high, and more recently stress has been on the membrane potential as the driving force. However, Slater's calculations (1967b) indicate that the concentration gradients required to maintain the requisite membrane potential are also extremely high. Moreover, formulation of the respiratory chain in the chemiosmotic hypothesis requires alternating electron and hydrogen carriers. This causes two difficulties. The first is that Mitchell places UQ between cyt b and cyt c_1, while most of the evidence places it on the substrate side of cyt b. The second difficulty concerns the third phosphorylation site which is located in the cytochrome chain where no known hydrogen carriers exist. The chemiosmotic hypothesis also fails to encompass the various exchange reactions or the energy-linked transhydrogenase located in the coupling membrane, all of which are readily accommodated by the chemical hypothesis.

The challenge posed by the Mitchell hypothesis has led to the discovery of a proton-translocating reversible ATPase in mitochondria (and chloroplasts) and to a very detailed examination of the permeability properties of the mitochondrial membranes. This led to the discovery of the ion impermeability of the inner membrane as required by the chemiosmotic hypothesis. However, both Slater (1967b) and Racker (1970) point out that the existence of the proton-translocating ATPase as a proton or cation pump in parallel with the phosphorylation system would explain much of the data invoked in favor of the Mitchell hypothesis.

In summary, the proton-translocating reversible ATPase and the ion impermeable coupling membrane required by the chemiosmotic hypothesis do exist. However, there is no present evidence for the proton-translocating oxidation-reduction chain postulated by Mitchell nor for the exchange diffusion system required if ATP synthesis is to be driven by a membrane potential alone. The chemiosmotic hypothesis more nearly satisfactorily explains the structural requirements for coupled phosphorylation and the action of uncoupling agents than does the chemical hypothesis. A unifying theory combining features of both the chemical and chemiosmotic hypotheses has been proposed by Racker (1970).

The suggestion that the high-energy intermediate ($C{\sim}X$ or $A{\sim}X$) might be a conformational change in the mitochondrial inner membrane rather than the formation of a chemical species originated with Boyer (1965). Simple correlations between gross morphology and energetic state of the mitochondrion soon followed. Hackenbrock (1964), using rat liver mitochondria, and Green and his associates (Penniston et al., 1968), using beef heart mitochondria, showed that electron micrographs of sectioned mitochondria contained different structures which appeared to correlate with the energetic state of the mitochondria. More recently, however, Hackenbrock (1968) has found that one of the morphological changes ("condensed" to "orthodox") requires electron transport only and is independent of the energy-coupling mechanism. The reverse reaction remains sensitive to DNP and presumably involves the coupling enzymes. However, Sordahl et al. (1969) have attempted to reproduce both the results of Penniston et al. and of Hackenbrock and conclude that no direct relationship between inner membrane conformation and mitochondrial function could be demonstrated. Stoner and Sirak (1969) propose that the "energized-twisted" conformational state observed by

Penniston *et al.* (1968) is an artifact passively induced during the process of fixation of the mitochondria for electron microscopic examination. To summarize, the conformation hypothesis is an attempt to identify the high-energy intermediate as a structural rather than a chemical entity. The experimental evidence adduced to support the hypothesis is so far unconvincing.

Some 13 different candidates for the high-energy intermediate have been proposed with nearly as many detailed chemical mechanisms for their formation. These aspects are critically reviewed by Lardy and Ferguson (1969). Cross and Wang (1970) have recently presented indirect evidence for a phosphorylated intermediate in the coupling mechanism in line with the mechanism proposed by Wang (1970). Storey (1970) has presented a particularly interesting amplification of the chemical hypothesis, emphasizing the role of thiol groups in the production of high-energy compounds and in energy transduction (see also Painter and Hunter, 1970). Storey points out that, although the stoichiometry of coupling requires three coupling sites, thermodynamic considerations [specifically, the inadequate differences in midpoint potentials between cyt b and c and cyt c and $(a + a_3)$] require that the energy of oxidation of reduced cyt $(a + a_3)$ be conserved. Thus there must be a fourth energy conservation site (although not a phosphorylation site) at the terminal reaction of the respiratory chain. There is abundant experimental evidence, primarily from experiments on reversed electron transfer, to show that high-energy intermediates generated at one coupling site can be used to drive reactions at another, or, in other words, that there is a pool of high-energy intermediates interconnecting the phosphorylation sites. Storey's proposal is that the energy conserved in the oxidation of cyt $(a + a_3)$ enters this pool of high-energy intermediates, making thermodynamically possible the generation of ATP at sites II and III. Storey also offers a plausible synthesis of the two different modes of uncoupler action arising from the chemical and chemiosmotic hypotheses.

C. The Relationship of Energy Metabolism to Malignancy

For over 50 years biochemists have sought a metabolic difference between normal and malignant cells, hoping that such a difference, if found, would provide an entering wedge for rational chemotherapy. Ironically, one of the earliest metabolic differences discovered between normal and malignant cells, that of aerobic glycolysis, is the only one which can be considered as almost the hallmark of a malignant cell.

TABLE II.

RESPIRATORY AND GLYCOLYTIC ACTIVITY OF NORMAL AND TUMOR TISSUE

	Respiratory Activity, Q_{O_2}	Aerobic Glycolysis, $Q_{O_2}^{CO_2}$	Anaerobic Glycolysis, $Q_{N_2}^{CO_2}$
Animal tumors			
Mouse lymphoma cells (L5178Y)	8.5	27	39
Ehrlich ascites	7	30	70
Flexner-Jobling carcinoma	7	25	31
Human tumors			
Breast (95%)[a]	7	9	15
Large round-cell sarcoma (100%)[a]	7	6	12
Normal tissues (rat)			
Liver	12	1	3
Kidney	21	0	3
Intestinal mucous membrane	10	3	14
Embryonic liver (5th day)	16	1	14
Retina	33	45	88

[a] The number in parentheses is the percentage of the tissue tested which was tumor. (From Gregg et al., 1968, and Aisenberg, 1961.)

This is shown in Table II, where a variety of normal tissues are compared to human and animal tumors with respect to respiratory activities and both aerobic and anaerobic glycolysis. The data for mouse lymphoma cells were taken from Gregg et al. (1968), the remaining data were adapted from Aisenberg (1961). For most normal tissue, the rate of aerobic glycolysis (in the units used here) is a small fraction of the corresponding respiratory activity. For the relatively slowly growing human tumors, the two activities are nearly equal, while for the rapidly growing animal tumors, the situation is the converse of normal. Respiratory activity is only a small fraction of the aerobic glycolysis.

The point is that aerobic glycolysis is a comparative rarity among normal tissues, while it is routinely observed in malignant cells (the data for rat retina are included in the table to remind the reader to beware of generalizations). It may be noted that the cell strain WI38 [a human diploid cell carried at the Wistar Institute and extensively studied by Cristofalo and Kritchevsky (1966)] is considered nontumorigenic and, for this reason, has been suggested as a suitable substrate for cultivation of virus for vaccine production (Hayflick, 1968). Apparently this normal diploid cell in culture has a Q_{O_2} (μl O_2/mg of protein/hr) of 5.6 (Cristofalo and Kritchevsky, 1966), which is somewhat

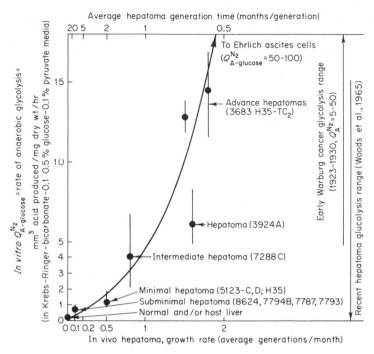

Fig. 2. Anaerobic glycolysis as a function of growth rate for a series of Morris hepatomas. (From Burk *et al.*, 1967.)

lower than any of the tumors listed in Table II and much lower than any of the normal tissue, while the aerobic glycolytic activity is 15.4, again much closer to tumor than to normal values. Thus, from the standpoint of these parameters, the cell should be considered quasi-malignant (but cf. Section III).

It is clear from Table II that the capacity for anaerobic glycolysis in rapidly growing tumors is also substantially greater than in normal tissue. Some investigators prefer to concentrate on anaerobic glycolysis because they find it more reproducible and more stable than aerobic glycolysis (Warburg, 1956).* A very striking demonstration of the relation between tumor growth rate and capacity for anaerobic glycolysis was provided by the data of Burk *et al.* (1967) on a series of Morris hepatomas. These data are shown in Fig. 2. The figure shows in a very convincing way the correlation between glycolytic rates and malig-

---

* With lymphoma cells there is no detectable difference in reproducibility or stability between the two reactions. However, long-term tests of stability (several hours) were never done.

nancy.* Since aerobic glycolysis is the reaction largely unique to malignant cells, we shall concentrate on it. That aerobic glycolysis is not only a function of growth rate is shown by the data in Table II for regenerating liver. However, intact embryos show very high rates of aerobic glycolysis even if the membranes surrounding them are undamaged (Neubert *et al.*, 1971). This recent work contradicts that of Warburg and his collaborators (Warburg, 1930b), about which experimental details are vague.

Having established a correlation between growth rate and glycolysis, the question is, why? What is there about the metabolism of malignant cells which leads to the high glycolytic rate? The question was asked and an answer proposed by the discoverer of the phenomenon, Otto Warburg (1930a, 1956). Warburg concluded that the respiration of the malignant cells was defective, either quantitatively or qualitatively. The result was presumed to be that the malignant cell had to depend on glycolysis, even under aerobic conditions, to supply sufficient energy for its rapid growth. Warburg modified his original theory somewhat as more information became available on mitochondrial properties generally and on the properties of tumor mitochondria specifically.† But his basic view (namely, that a defect in mitochondrial energy metabolism was responsible for the high aerobic glycolysis of tumor cells) remained unchanged for over 40 years. The Warburg hypothesis is no longer fashionable; however, this view stimulated a vast amount of work for nearly half a century on the energetic aspects of malignant and normal cells, and this great foundation of knowledge is the basis for present-day investigations. The Warburg hypothesis remains one of the most scientifically fruitful suggestions of the century.

The properties of tumor and malignant cell mitochondria will be discussed in more detail later (see Section II,B). Anticipating, it can be said that, although malignant cell mitochondria do differ somewhat from the standard to which mitochondrial preparations are usually compared (namely, rat liver mitochondria), the number of differences grows steadily less as more experience is gained in techniques for isolation of tumor mitochondria. It is the author's view that such differences as

---

* Creation of the series of transplantable hepatomas of varying growth rates was a milestone in tumor enzymology. For more detailed accounts of enzyme levels as a function of growth rate in these tumors, the reader is referred to Weber (1968) and Lea *et al.* (1969).

† Although Warburg (1913) had discovered that the respiration of liver cells resided in "grana" which could be sedimented from a lysate, it was not until 1949 that cellular respiration was shown to be unequivocally associated with mitochondria (Kennedy and Lehninger, 1949).

do exist are far from adequate as an explanation for the high aerobic glycolysis of malignant cells.

Since lactic dehydrogenase is the enzyme unique to aerobic glycolysis, the question, "Is aerobic glycolysis essential for the growth of tumor cells?" may be rephrased, "Is lactic dehydrogenase activity essential for the growth of tumor cells?" This question has been investigated using oxamic acid, a specific inhibitor of lactic dehydrogenase (LDH), and also through application of specific antibody to the enzyme.

Papaconstantinou and Colowick (1961a) showed that oxamate inhibited aerobic and anaerobic glycolysis and glucose uptake to the same extent. They further showed (1961b) that the growth of HeLa cells in monolayer culture under aerobic conditions was completely inhibited by oxamate at the concentrations required to inhibit aerobic glycolysis. These authors tentatively concluded that LDH activity was essential for growth but pointed out that this was not equivalent to saying that aerobic glycolysis was essential for growth, since the principal function of LDH might be the regeneration of NADP, as had been shown for liver by Navazio et al. (1957). Later work in Colowick's laboratory (Goldberg et al. 1965) established that oxamate had little effect on growth of HeLa cells when glucose was absent, and these authors concluded that glucose was itself toxic and inhibitory to growth when glycolysis was inhibited. The proposed basis of this toxicity was depletion of the pool of cellular ATP in the process of transforming glucose to fructose diphosphate, since the subsequent regeneration of ATP by the later reactions of glycolysis was prevented by the unavailability of NAD. Thus ATP depletion prevented growth.

Glycolysis and growth of intact mouse and rabbit malignant cells was similarly inhibited by specific antibody to LDH (Ng and Gregory, 1966), although the antibody had no effect on normal cells. These authors showed that the antibody did not enter normal cells but that malignant cells, with their drastically altered surface properties (see Section III), were permeable to the protein, in agreement with other data showing that intact proteins could pass through the membrane of malignant but not normal cells (see Ng and Gregory, 1966).

Since the rate of NADPH oxidation has an important effect on the rate of the oxidative reactions of the pentose phosphate cycle and Wenner (1959) had shown that pyruvate was an important electron acceptor for the oxidation of NADPH in ascites tumor cells, Gumaa and McLean (1969) investigated the effects of pyruvate and oxamate on the oxidative and nonoxidative portions of the pentose phosphate pathway. In their controls about one third of the total pentose phosphate was formed by the oxidative pathway (the oxidative decarboxylation of 6-phosphogluconic acid) and two thirds from the nonoxidative pathways (from glucose-6-phosphate via the transaldolase and transketolase reactions). Pyruvate increased the oxidative pathway somewhat (about 30%) but increased the nonoxidative pathway some 400%, while oxamate completely blocked the nonoxidative formation of pentose phosphate. These results indicate that the effects of pyruvate and oxamate on the pentose phosphate pathway were mediated largely through the corresponding changes in concentration of fructose-6-phosphate (increased 50% by pyruvate, decreased 60% by oxamate) rather than through changes in the redox state of NADP.

Coe and Strunk (1970) have tested the effect of oxamate on the levels of glycolytic intermediates for up to 30 minutes after addition of the inhibitor. Their results are in general agreement with those from Colowick's laboratory (Papaconstantinou and Colowick, 1961a,b). In summary, these results seem to suggest that aerobic lactate production by tumor cells is not a response to the cell's energy demands but is, at least in part, a device for producing NADP.

II. Studies of Energy Metabolism at Various Levels of Organization

A. Intact Cells: Mitochondrial-Cytoplasmic Interactions

1. THE PASTEUR AND CRABTREE EFFECTS

In 1876 Louis Pasteur observed that admitting air to an anaerobic suspension of yeast cells reduced their rate of glucose consumption. Subsequent work has shown that this phenomenon, christened the Pasteur effect by Warburg (1926), occurs in a variety of organisms, including mammalian cells in culture. The Pasteur effect on lactate formation for five lines of cultured cells and for the ascites form of the mouse lymphoma is shown in Table III (Gregg et al., 1968). Here the cell lines are arranged in order of increasing respiratory activity (Q_{O_2}); this is also the order of increasing anaerobic glycolysis (except for HeLa, which is discussed below). Thus, in spite of the difference in origin (the first three lines are from Syrian hamster, the lymphoma from mouse, and HeLa from human tissue), the differences in energy metabolism between the lines are primarily differences in overall metabolic activity. The ratio of total aerobic to total anaerobic activity is nearly constant for the first five cell types.

Comparing the C13 (BHK21/C13) and P183 lines is especially interesting, since the latter arises from the former upon treatment of the

TABLE III.

RESPIRATORY AND GLYCOLYTIC ACTIVITY OF CELLS IN CULTURE[a]

| Cell Line | Q_{O_2} (μl O_2/mg protein/hr) | Glycolysis (μmoles lactate/mg protein/hr) | |
		Aerobic	Anaerobic
C13	5.57 ± 0.61 (3)	0.850 ± 0.068 (2)	1.40 ± 0.094
CHO	6.80 ± 0.90 (14)	1.92 ± 0.30 (5)	1.97 ± 0.29
P183	8.08 ± 3.56 (5)	1.77 ± 0.17 (3)	2.14 ± 0.21
Lymphoma (ascites)	10.70 ± 0.17 (3)	1.50 ± 0.29 (3)	2.53 ± 0.40
Lymphoma (cultured)	14.10 ± 1.10 (4)	1.99 ± 0.28 (4)	2.85 ± 0.28
HeLa	12.70 ± 1.80 (3)	0.223 ± 0.032 (3)	0.434 ± 0.046

[a] Respiratory measurements on each batch of cells were carried out at least four times. The average value was then averaged with those of other experiments to give the values shown. The values in parentheses are the number of separate experiments; aerobic and anaerobic glycolyses were always measured together.

C13 cells with polyoma virus. This "malignant transformation" (Sanford, 1965) increases both aerobic and anaerobic glycolysis 50–100% without changing the respiratory activity, in confirmation of the previous report of Paul *et al.* (1966). A number of interesting points arise when the Pasteur effect is examined in these cell lines. The reader will note that all lines except CHO show a strong inhibition of lactate formation by air. The CHO is the only line we have examined which fails to show the Pasteur effect. Several years ago, the CHO line used in our laboratory had a doubling time of about 12 hours and exhibited a normal Pasteur effect. Subsequently the line altered. The altered line has a stable generation time of about 16 hours, but the Pasteur effect has disappeared.

Baierlein (1967) has reported that human leukocytes suspended in serum show a normal Pasteur effect but that, when cells are suspended in buffer, the Pasteur effect is lost. A number of possible explanations were eliminated, but the basis of the loss was not firmly established. Baierlein concluded that the most probable explanation was damage to the leukocyte mitochondria on suspension in buffer. Some properties of CHO mitochondria which may be relevant to this problem will be discussed later (see Section II,B).

Another unexpected result (Table III) was the high Pasteur effect found with HeLa cells. Racker (1965, p. 226) and his co-workers had previously reported that the strain of HeLa used in their laboratory showed no Pasteur effect on lactate production from exogenous glucose, and H. J. Phillips and McCarthy (1956) found a lower glycolysis anaerobically than aerobically. However, in contrast to the results reported in Table III, where the incubation time was 10 minutes at 10-mM glucose, the other workers used long incubations and very small amounts of substrate. Also, HeLa cells are frequently contaminated with mycoplasma, but the effect of such contamination on energy metabolism, although implicated in numerous other metabolic changes (Levine et al., 1968; Russell et al., 1968; Gill and Pan, 1970), has never been investigated.

A number of explanations for the Pasteur effect have been advanced. An early explanation postulated a toxic effect of oxygen on essential —SH groups of glycolytic enzymes. This view became untenable when it was found that the Pasteur effect was abolished by uncoupling agents which obviously had no effect on oxygen concentration in the medium. The currently favored hypothesis, although not completely satisfactory, hinges on the known kinetic properties of hexokinase and phosphofructokinase (Uyeda and Racker, 1965). The important parameters are inhibition of the former by its product glucose-6-phosphate, inhibition of the latter by ATP, and the fact that both inhibitions are reversed by P_i. Two other factors are invoked; these are intracellular compartmentation of nucleotides and the concept of a competition between glycolysis and terminal respiration for essential cofactors. Only a general discussion will be given here. For a more penetrating analysis, the reader is referred to Wenner (1967).

Under aerobic conditions, phosphofructokinase is kept in a largely inhibited state by high levels of cytoplasmic ATP, possibly assisted by citrate The inhibition of phosphofructokinase causes an accumulation of glucose-6-phosphate which inhibits hexokinase, the enzyme known to control glucose uptake in normal cells. The concentration of P_i in the cytoplasm is kept low by the demands of the respiratory chain and the triose phosphate dehydrogenase reaction. Small changes in the ATP/ADP ratio or the fructose-6-phosphate/citrate ratio could exert "fine tuning" without large changes in concentrations of ATP, ADP, or P_i (see Section I,B,1).

When oxygen is exhausted from the medium, the respiratory chain stops phosphorylating. The cytoplasmic ATP concentration falls, since mitochondrial ATP is no longer available, and the P_i level correspondingly rises. These effects act in concert to relieve the inhibition of hexokinase and phosphofructokinase, and the glycolytic rate increases. Under anaerobic conditions, the control point of glycolysis may shift to pyruvate kinase, which is inhibited by high concentrations of P_i (J. R. Williamson, 1965).

Evidence supporting the idea that the Pasteur effect results from a competition between glycolytic and respiratory chains comes from a variety of studies on reconstructed systems (Uyeda and Racker, 1965), homogenates (Racker, 1965, pp. 193–208 and 211–218), and intact cells (Gregg et al., 1968). This view suggests that the magnitude of the Pasteur effect* should depend on the relative activities of the glycolytic and respiratory systems. For two cell lines with nearly the same respiratory activity but widely different glycolytic activities, the Pasteur effect should be less the higher the glycolytic activity. This is confirmed by the data in Table III on the C13 and P183 lines. The Pasteur effect is 39% in the former and only 18% in the P183 cells in which glycolytic activity is approximately doubled. Similarly, the HeLa line used in these studies had a high respiratory activity but an unexpectedly low glycolytic activity. The Pasteur effect in HeLa was the highest of any line tested. However, this view obviously fails to explain the lack of a Pasteur effect in the CHO cells.

Wenner (1967) discusses the evidence against nucleotide compartmentation, and Nigam (1968) reviews the data which fail to support the idea of a general control of glycolysis by nucleotides (specifically, ATP). Sauer (1968) concluded that, in the Ehrlich-Lettre cells, control of hexokinase by glucose-6-phosphate alone was adequate to explain the changes in glucose utilization following glucose addition under aerobic conditions. I.-Y. Lee et al. (1967) find that their data on glycolysis in Ehrlich ascites cells are inconsistent with either control by ADP or P_i. One may conclude that, although considerable evidence points to phosphofructokinase and hexokinase as agents controlling glycolytic rates, much remains to be learned before a generalized explanation of the Pasteur effect can be given.

The Crabtree (1929), or "reverse Pasteur," effect is the name given to the inhibition of respiration by added glucose. Crabtree originally observed the effect only in tumors. We now know that the effect occurs in cells, such as malignant cells or retina, which have high glycolytic rates. Figure 3 shows an experimental demonstration of the Crabtree effect (Gregg et al., 1968) in the mouse lymphoma cells. The endogenous respiration rate (measured with the Clark oxygen electrode) decreases 50% on addition of 15-mM glucose. Inhibition is relieved by successive additions of the uncoupling agent F_3-CCP (final concentration, 7×10^{-7} M). The similar effects of uncoupling agents on the Pasteur and Crabtree effects suggest that the phenomena are closely related and that both involve a cytoplasmic-mitochondrial interaction. The extensive studies

---

* Quantitatively, the Pasteur effect is

$$\left(\frac{\Delta\text{lactate}_\text{anaerobic} - \Delta\text{lactate}_\text{aerobic}}{\Delta\text{lactate}_\text{anaerobic}} \right) \times 100$$

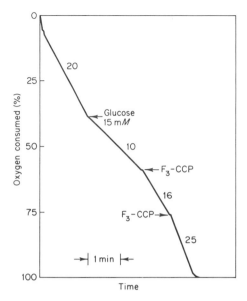

Fig. 3. The Crabtree effect and stimulation of respiration by an uncoupling agent in intact lymphoma cells. F_3-CCP (1 μl, 7×10^{-4} M) was added as indicated. The final concentration of F_3-CCP was 7×10^{-7} M. At zero time the solution contained 617 ng atoms O_2.

on the Crabtree effect, largely done on Ehrlich ascites cells, were reviewed by Ibsen (1961).

Like the Pasteur effect, the Crabtree effect has been demonstrated in reconstructed systems by Racker and his associates (Racker, 1965, pp. 204–205). Also, like the Pasteur effect, the reverse Pasteur effect is believed to represent a competition between the respiratory chain and glycolysis for cofactors, but, as Racker points out, in both effects it is quite possible that more complex regulatory mechanisms are superimposed on the simple regulation by competition.

It is frequently claimed that the Crabtree effect results from a drop in intracellular pH due to increased lactate formation. This phenomenon is more properly termed a quasi-Crabtree effect. In the experiment shown in Fig. 3, the rapidity of response to glucose and the equally rapid reversal by uncoupling agent make unlikely an effect of pH due to lactate accumulation.

2. Oxidation of Cytoplasmic NADH: Shuttle Systems

Many attempted explanations for the high aerobic glycolysis of tumor cells center around NAD (NADH) because of its obligatory involvement in glycolysis. Since early preparations of tumor mitochondria required added NAD for maximum activity, it was proposed that NAD leaked into the cytosol from tumor mitochondria *in vivo* and thus stimulated glycolysis. This view was abandoned when it was subsequently shown in many laboratories that tumor mitochondria could be isolated which were fully active in the absence of added NAD. A more recent idea, proposed by S. Green and Bodansky (1963) and McKee *et al.* (1968), is that NADase is abnormally low in malignant tissue and that the high cytoplasmic level of NAD thus resulting causes the high rate of tumor glycolysis. However, so far the low NADase activity has been demonstrated only in the Ehrlich ascites cell.

The question was raised earlier as to how cytoplasmic NAD could be regenerated, since NADH is generally not oxidized by intact mitochondria. Inevitably this fact was associated with the high aerobic glycolysis of malignant tissue. In the important paper of Boxer and Devlin (1961), they discussed systems (to which was given the name NADH shuttles) in which a cytoplasmic enzyme and a mitochondrial enzyme collaborate to effectively transfer reducing equivalents from cytoplasmic NADH to mitochondrial NAD or flavoprotein. A typical NADH shuttle is the α-glycerolphosphate cycle shown in Eq. (4).

$$
\text{Cytoplasmic:} \quad
\begin{array}{l}
\text{CH}_2\text{OH} \\
| \\
\text{C}{=}\text{O} \;+\; \text{NADH} \\
| \\
\text{CH}_2\text{OPO}_3\text{H}_2
\end{array}
\longrightarrow
\begin{array}{l}
\text{CH}_2\text{OH} \\
| \\
\text{CHOH} \;+\; \text{NAD} \\
| \\
\text{CH}_2\text{OPO}_3\text{H}
\end{array}
$$

$$
\text{DHAP} \qquad\qquad\qquad\qquad\qquad \alpha\text{-GP}
$$

$$
\text{H}_2\text{C(OH)C({=}O)CH}_2\text{OPO}_3\text{H}_2 \tag{4}
$$

$$
\text{Mitochondrial:} \quad \alpha\text{-GP} \;\xrightarrow[\text{flavoprotein}]{\text{mitochondrial}}\; \text{fp-H}_2 \;+\; \text{DHAP}
$$

$$
\text{fp-H}_2 \;\xrightarrow[\text{oxygen}]{\text{respiratory chain}}\; \text{fp} \;+\; \text{H}_2\text{O}
$$

In the cytoplasm the normal glycolytic intermediate dihydroxyacetone phosphate (DHAP) is reduced by the soluble α-glycerolphosphate dehydrogenase, in the presence of NADH, to form α-glycerolphosphate (α-GP) and NAD, thus effectively regenerating the NAD required for continuing glycolysis. α-Glycerolphosphate can, in turn, be oxidized by a flavin-linked mitochondrial dehydrogenase to regenerate DHAP (which diffuses back into the cytosol) and reduced flavoprotein. Reducing equivalents are transferred from the reduced flavoprotein to oxygen via the respiratory chain. Thus the overall reaction is

$$
\text{NADH} \;+\; \text{H}^+ \;+\; \tfrac{1}{2}\text{O}_2 \longrightarrow \text{NAD}^+ \;+\; \text{H}_2\text{O} \tag{5}
$$

Considerable data exist on the activity of the cytoplasmic dehydrogenase in a variety of tissues, but quantitative comparisons between different workers are made difficult by the variety of assays used. The work of Boxer and Shonk showed that lymphoid tissue has relatively low activity in the rat, mouse, and hamster, with liver and muscle having very high values. The enzyme activity in liver is reduced in rats fed carcinogenic dyes. It is lower still in Novikoff solid hepatoma and lowest of all in ascites hepatoma. The activity in ascites hepatoma is one four-hundredth that of normal liver. This is not due to increasing growth rate, since regenerating rat liver shows values comparable to normal liver (Boxer and Shonk, 1960).

Sacktor and Dick (1964) report an enzyme activity of 0.1 unit/mg of protein for mouse liver and less than one tenth of that for spleen. Using the same assay we find approximately 2×10^{-4} unit/mg of protein for the lymphoma cells in culture and only a slightly higher level for the ascites form (Hutson and Gregg, 1971). However, Letnansky and Klc (1969) report a 300-fold difference in α-glycerolphosphate dehydrogenase activity among four different strains of the Ehrlich ascites cell. Thus, although rapidly growing malignant cells such as the lymphoma or the Novikoff ascites have very low activities of the dehydrogenase and this enzyme activity declines dramatically in going from normal to malignant tissues derived therefrom, a similar range of activities is shown by different strains of a single tumor cell.

In all, three major shuttle systems were proposed: the α-glycerolphosphate cycle, the β-hydroxybutyric acid cycle, and the malate cycle (Wenner, 1967). Boxer and Devlin (1961) pointed out that, in a variety of tumor tissues, the various enzymes of the NADH shuttles were either very low or absent. From their data they concluded that the high aerobic glycolysis so typical of tumor metabolism was due to the requirement that tumor cells regenerate cytoplasmic NAD by reducing pyruvate, since they lacked the necessary shuttle enzymes to transfer reducing equivalents from cytoplasmic NADH to the respiratory chain. This paper focused the attention of biochemists on the cytoplasmic NADH of tumor and normal cells. The final result was to show that the hypothesis of Boxer and Devlin was of limited applicability. Only the α-glycerolphosphate shuttle could function under normal physiological conditions (Borst, 1962). Further work brought more devastating criticism. Wenner (1967) pointed out that, since the rate of anaerobic glycolysis of tumor cells (where mitochondrial regeneration of cytoplasmic NAD could not occur) was much higher than the aerobic rate, the latter could certainly not be limited by the mitochondrial contribution to the reoxidation of cytoplasmic NAD. Moreover, Terranova and his colleagues (1969;

Dionisi *et al.*, 1970) showed convincingly that in two strains of Ehrlich ascites cell which differed markedly in their content of shuttle enzymes there was no difference in the amount of lactate formed under aerobic conditions. Thus the presence or absence of shuttle enzymes had little effect on the rate of aerobic glycolysis.

Chance and his associates (Galeotti *et al.*, 1970) have pointed out that the high cytoplasmic NADH level of tumor cells is not necessarily a disability. The cells of rapidly growing tumors or of cells in culture with doubling times of 12–20 hours must drive the essentially reductive biosynthetic reactions at a rapid rate. Although the reductant in most biosynthetic sequences is NADPH, as mentioned before, enzyme systems (the transhydrogenases) exist for carrying out the reaction

$$NADH + NADP \rightarrow NAD + NADPH \qquad (6)$$

Chance's view implies that there should be a correlation between cytoplasmic NADH levels and growth rate. A suitable selection of normal and malignant tissue or of cells in culture covering a wide span of growth rates, combined with the highly sensitive radiometric enzymatic recycling assays for NADH developed by Serif and Butcher (1966), should quickly demonstrate whether the proposed relationship between cytoplasmic NADH levels and growth rates actually exists. Such a study, carefully done, would greatly enhance our understanding of the overall metabolism of tumor cells.

3. Response to Inhibitors by Mitochondria *in Vivo* and *in Vitro*

This poorly understood aspect of mitochondrial-cytoplasmic interaction has so far been investigated only in the author's laboratory with intriguing but poorly understood results. The first aspect is the enhanced sensitivity of lymphoma cell mitochondria to inhibitors when the mitochondria are in their normal intracellular environment, vis-à-vis their sensitivity to the same inhibitors when isolated mitochondria are tested. Relevant data are shown in Table IV (Machinist and Gregg, 1970). Here are compared the amounts (picomoles per milligram of protein) of three inhibitors required for maximal respiratory inhibition of whole lymphoma cells, isolated lymphoma mitochondria, and rat liver mitochondria.* It is clear that the endogenous respiration of whole cells is from 3 to 16 times more sensitive to these three very different kinds of respiratory inhibitors than are the corresponding mitochondria. This result was quite unexpected. We had doubts about the ability of the inhibitors to penetrate to the mitochondria at all in the intact cells and anticipated

---

* These data are taken from the work of Ernster *et al.* (1963).

TABLE IV.

AMOUNTS OF INHIBITORS REQUIRED FOR MAXIMAL RESPIRATORY INHIBITION OF
LYMPHOMA CELLS AND LYMPHOMA AND RAT LIVER MITOCHONDRIA

	Lymphoma				Rat Liver, Mitochondria (pmoles/ mg protein)
	Whole Cells		Mitochondria		
Inhibitor	$M \times 10^8$	pmoles/ mg protein	$M \times 10^8$	pmoles/ mg protein	
Rotenone	3.4	8	0.51	23	28
Oligomycin	2.8	7	2.7	55	50
Antimycin	6.7	18	5.8	288	230

that, in the presence of so much nonmitochondrial protein, nonspecific adsorption of the inhibitors might lead to a grossly reduced sensitivity of the whole cells. The agreement between the levels of inhibitor required for maximum respiratory inhibition in lymphoma mitochondria, compared to rat liver mitochondria, was heartening and gave further indication that the mitochondria of the malignant lymphoma cells in culture were not intrinsically different from normal mammalian mitochondria. Whether the difference in inhibitor sensitivity observed here is a true cytoplasmic-mitochondrial interaction is not clear. If it were, experiments on the sensitivity of homogenate respiration to inhibitors should give the same results as for whole cells, while addition of mitochondria-free supernatant fluid might be expected to increase the inhibitor sensitivity of isolated mitochondria.

A second aspect of inhibitior response is shown in Fig. 4 (Machinist and Gregg, 1970). Titration of the endogenous respiration of intact lymphoma cells with antimycin gives a linear reduction in respiratory activity from 0 to 100% inhibition (Gregg et al., 1968). This is in striking contrast to curve C, which is the titration curve obtained with antimycin and isolated lymphoma mitochondria. As with isolated rat liver mitochondria (Ernster et al., 1963), there is a large threshold below which addition of inhibitor has no effect on mitochondrial respiration. The obvious further step was taken of carrying out the inhibitior titration in the presence of small amounts of added mitochondria-free supernatant fluid from the original homogenate (curve B). This had the expected result of reducing the threshold. Unfortunately, this experiment, although done successfully more than once, was extremely difficult to repeat. But the implication is clear: There is some substance in the nonmitochondrial fraction of the cell which abolishes the threshold for antimycin

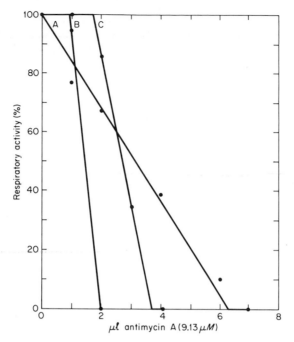

Fig. 4. Titration of respiratory activity with antimycin: (A) titration of the endogenous respiration of intact lymphoma cells (from Gregg *et al.*, 1968), (C) titration with isolated lymphoma mitochondria, and (B) same as (C) but with the addition of mitochondria-free supernatant [both (B) and (C) from Machinist and Gregg, 1970].

inhibition seen with isolated mitochondria. Or, conversely stated, the threshold phenomenon may be an artifact of mitochondrial isolation and thus does not represent a normal property of mitochondria *in vivo*.

4. Miscellaneous

Other aspects of whole-cell studies which are poorly understood are the influences of respiration and phosphorylation inhibitors and of anaerobiosis on cell growth, division, and the phenomenon of the development of malignancy. Currie and Gregg (1965) studied the effects of rotenone, antimycin, 2,4-dinitrophenol, and *m*-Cl-cyanocarbonyl-phenylhydrazone on cell growth of CHO cells synchronized with thymidine (Tobey *et al.*, 1966). A typical result with DNP is shown in Fig. 5. The effect of the uncoupler at 0.1 mM was to slow each phase of the cell cycle nearly equally, while a 10-fold higher concentration prevented cell division altogether. These data could be interpreted as showing that a constant supply of energy is required throughout the cell cycle, in contrast to the energy-store hypothesis of Bullough (1952). The more potent uncoupler m-Cl-CCP showed similar effects on randomly growing CHO cells (i.e., growth was slowed at 1×10^{-5} M and prevented at 10 times this amount). Rotenone at 0.33 μM, when tested on synchronized CHO cells, delayed the midpoint of mitosis (M phase) from 7.5 hours after release of thymidine inhibition to approximately 15 hours, and a second round of cell division did not occur. At a 10-fold higher concentration of inhibitor, the beginning of M was delayed some 10 hours after the control vessels, and cell division was incomplete, with the cell count reaching a peak at 22 hours and then dropping. Similar results were obtained with antimycin and oligomycin.

This work was continued by Tobey *et al.* (1969), who tested the effects of antimycin, rotenone, and oligomycin in mitotic CHO cells. In this system oligomycin, rotenone, and antimycin inhibited mitosis some 70% at concentrations of 6.9×10^{-4}, 1.2×10^{-5}, and 2.7×10^{-7} M, respectively. If these levels are compared on a protein basis with those required to completely inhibit cellular respiration in the mouse lymphoma cells (Table IV), it can be calculated that the concentrations used by Tobey *et al.* (1969) were from 4×10^{3} to 5×10^{6} higher. The same criticism applies to the earlier work of Currie and Gregg (1965).

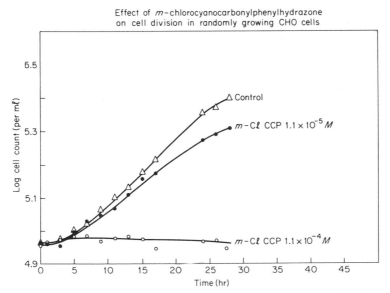

Fig. 5. The effect of m-chlorocyanocarbonylphenylhydrazone on cell division in randomly growing Chinese hamster cells. (From Currie and Gregg, 1966.)

All these experiments were done before development of oxygen electrode amplifiers sufficiently sensitive to measure changes in oxygen tension in suspensions of growing cells (Carr *et al.*, 1971b). Hence no direct measurement of effect of the inhibitors on respiration was made. In the absence of such data, preferably coupled with measurements of glycolytic rates (lactate formation), no conclusions can be drawn concerning the energy requirements for various portions of the cell cycle. Since such high concentrations were required, effects of the inhibitors on protein conformation must be considered as an alternative explanation of the results. Such effects have been shown to occur for high concentrations of antimycin (Rieske *et al.*, 1967) and rotenone (Butow, 1967).

Warburg has stoutly maintained that "Es gibt keine Ausnahme von der Regel, dass die Körperzellen des Menschen obligate Aerobier sind, die im Körper nur auf Kosten der Sauerstoffatmung wachsen konne, während die Krebszellen fakultative Anaerobier sind, die wie Hefezellen auf Kosten einer Gärung (die Milchsauregärung) wachsen konne" (Warburg, 1966a).* In support of this view, Warburg has cited the ingenious experiments of Malmgren and Flanigan (1955), who showed that tetanus spores (obligate anaerobes) would not germinate if injected into normal mice but that when they were injected into tumor-bearing mice the mice died of tetanus. This result was interpreted to mean that only in tumor-bearing mice were there anaerobic areas (in the region of the tumor) which permitted germination of the spores (Warburg, 1964). Although Warburg's statement seems to imply that tumor cells will grow indefinitely under anaerobic conditions, while normal cells will not, this is not what is meant. Warburg (1966b) makes it clear that tumor cells, like yeast (Pasteur, 1876), degenerate in the complete absence of oxygen, and he proposes the breaking off of cells from such a degenerating tumor as the basis for metastasis.

The work in a great many laboratories has suggested that cells in culture will not grow indefinitely in the absence of oxygen (e.g., Graff and Moser, 1962; Paul, 1965, p. 260; Sanford and Westfall, 1969). However, many of the reports (and those reaching a contrary conclusion) offer no evidence that oxygen was excluded, since no measurements of actual oxygen concentration were made and trace oxygen was not removed from the N_2 or N_2-CO_2 mixtures used to create "anaerobiosis." The possibility of leakage of air back into the vessels was also not eliminated.

---

* "There are no exceptions to the rule that cells of the human body are obligate aerobes which can grow in the body only at the expense of oxygen respiration, while cancer cells are facultative anaerobes which, like yeast, can grow at the expense of a (lactic acid) fermentation."

We have carried out such experiments (Hutson and Gregg, 1971) with the mouse lymphoma cells using an oxygen electrode (Carr *et al.*, 1971a) to ensure that oxygen was absent. While oxygen was removed by bubbling prepurified N_2 through the vessel, the pH control unit was turned off. By the time anaerobiosis was achieved (about 90 minutes), the pH had risen from 7.0 to 7.42 due to loss of CO_2. The pH (and CO_2 concentration of the medium) was returned to its initial value by brief gassing with CO_2 and the pH subsequently controlled at 7.0. The cells, which had an initial doubling time of 12.1 hours, grew anaerobically for 24 hours with a doubling time of 15.8 hours. During this period there was no increase in number of dead cells as measured by trypan blue exclusion. Following the period of slow growth, growth stopped and the number of dead cells rose rapidly. In some experiments anaerobiosis was maintained to allow a population resistant to anaerobiosis to develop. None did. All cells were dead after 96 hours. In other experiments an attempt was made to reverse the extinction of the culture by reaerating when only 29% of the cells were dead (after about 44 hours of anaerobiosis). These efforts failed. The damage done to the cells in anaerobiosis was irreversible. To show that the medium would still support normal growth after reaeration, a concentrated suspension of the ascites form of the mouse lymphoma was inoculated into the fermentor. After the usual lag (on transition from *in vivo* to *in vitro* growth), the cells grew normally.

The ability of cells (particularly tumor cells) to survive *brief* periods of anaerobiosis is another matter. Bickis and Quastel (1965), in a much-cited paper, showed that Ehrlich ascites cells could, according to their calculations, supply as much energy (ATP) anaerobically as aerobically through the greatly increased rate of anaerobic glycolysis. Glycine was incorporated into acid-insoluble material at the same rate aerobically and anaerobically as long as glucose was present. This result may seem at variance with the inability of cells to grow anaerobically. Even if ATP formation were the only consideration, the result is puzzling when it is considered that 30 moles of ATP is formed via respiratory-chain phosphorylation from 1 mole of glucose catabolized, while only 2 moles of ATP is formed glycolytically. With the additional 2 moles of ATP formed via substrate-level phosphorylation in the Krebs cycle, there would appear to be a requirement for a 16-fold increase in glycolysis to meet the energy requirement if growth were to continue at the same rate. However, the 16:1 ratio assumes that the 2 moles of NADH formed glycolytically are not effectively oxidized via the respiratory chain. If they are considered to be so oxidized, then the energy balance in favor of the respiratory chain is 19:1. However, aerobic ATP synthesis

is obviously not the only reaction in which oxygen is a necessary participant. A number of essential biosynthetic reactions in mammalian cells
require the participation of molecular oxygen (Bloch, 1962). Even
though the experiments of Bickis and Quastel (1965) were done with
the most favorable possible tissue (as we have seen, the Ehrlich ascites
cells have an enormous capacity for anaerobic glycolysis even relative
to other tumor cells; see Table II), it is likely that glycine incorporation
was limited in their experiments by some factor other than ATP formation and that "La vie sans l'air" is impossible for mammalian cells.

Even though studies on metabolic control systems must ultimately
be done with intact cells (if not intact organs or whole animals), cell
homogenates offer the experimentalist certain advantages. Most of the
normal cell constituents are (or can be) present, although somewhat
diluted. The cell membrane barrier is eliminated. This is of obvious
importance to the student of cellular energy metabolism because the
cell wall is ordinarily impermeable to some of the compounds in which
he is most interested (e.g., AMP, ADP, or ATP).* A further aspect
of removing the permeability barrier is that the concentrations of added
metabolites are precisely known. As shown by Racker (1965, p. 217),
a 20-fold increase in extracellular phosphate concentration is required
to give a 2-fold increase in intracellular phosphate in ascites tumor cells.
There is the additional advantage that the experimenter has the option
of testing the effect on, for example, the glycolytic rate of a homogenate,
of adding additional amounts of purified glycolytic enzymes or of subcellular fractions such as mitochondria or nuclei. These advantages of
homogenate preparations in the study of energy metabolism were most
vigorously exploited by Racker and his co-workers and are summarized
in his excellent book, *Mechanisms in Bioenergetics* (1965), which, in
addition to the details of homogenate studies on the Pasteur and Crabtree effects, offers a thorough discussion of oxidative phosphorylation,
the respiratory chain, and the general principles of energy metabolism
at the cellular level.

B. Studies on the Energy Metabolism of Isolated Mitochondria

The concept that the mitochondria of malignant cells are qualitatively
or quantitatively deficient has played a major role in attempting to explain the phenomenon of aerobic glycolysis as discussed earlier. In this
section we shall review recent studies on mitochondria of various cell

* The HeLa cells used by Racker and his co-workers (Racker, 1965, p. 227)
were somewhat permeable to AMP.

types: cultured, cultured malignant, and finally ascites tumor cells. Except where indicated otherwise, most of the work described in this section was done at Los Alamos (Machinist and Gregg, 1970).

Rat liver is the source of the most intensively studied mitochondria and has thus become the standard to which other mitochondrial preparations are compared. As with other standards, it has disadvantages. Since the liver is an easily homogenized tissue, mitochondria can be isolated with minimal damage. Also, because the liver is essentially a chemical factory, liver has an extraordinarily high concentration of mitochondria [30–35% of the liver protein (Schneider and Hogeboom, 1951)]. On this basis, the lymphoma cells do indeed appear to be depleted of mitochondria, since they contain only about 6% of the total protein as mitochondria. This figure is not grossly different from that found in ascites and solid tumors generally (see references in Sato and Hagihara, 1970). However, heart muscle, which is certainly a highly aerobic and hardworking tissue from the standpoint of energy requirement, contains only about 5% mitochondria on a protein basis (calculated from data in Hatefi and Fakouhi, 1968; Isaacs et al., 1969; LaNoue et al., 1970). Therefore, the relevance of mitochondrial content per se is not clear.

The question of mitochondrial quality is more difficult to answer. As in establishing the purity of a protein, determining mitochondrial quality requires evaluation of a number of parameters. First among these is the degree of respiratory control (as evidenced by the acceptor control ratio; see Section I,B,2), a characteristic highly correlated with mitochondrial integrity. Respiratory control is readily lost on aging of the mitochondria or by other deleterious treatments which have no discernible effect on the other parameters we shall discuss. The second most important factor for judging mitochondrial quality is the familiar P/O ratio. As pointed out before, this should theoretically be 3.0 for NAD-linked substrates and 2.0 for flavin-linked substrates such as succinate and α-glycerolphosphate. Values very close to theoretical are routinely obtained with a variety of mitochondrial preparations.

The principal function of mitochondria is usually considered to be the production of energy. For a particular substrate, the amount of energy abstracted from it is the product of the P/O ratio and the Q_{O_2} which yields moles of ATP per milligram of mitochondrial protein per hour. Thus the rate (Q_{O_2}) at which a particular substrate is oxidized is also important as an indication of mitochondrial quality. However, the situation is more complex here than with the other quality parameters. The highest values of Q_{O_2} for succinate oxidation, for example, are obtained with preparations of mitochondrial fragments which are nonphosphorylating. Hence, the requirement that the high respiratory

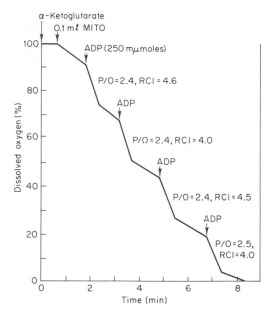

Fig. 6. Response of mouse lymphoma cell mitochondria to ADP in saline-phosphate buffer, pH 7.4. The substrate, α-ketoglutarate (20 mM), was added at zero time. (From Machinist and Gregg, 1970.)

rate be coupled to ATP formation must be kept in mind. Also, as mentioned earlier, the oxidation of NADH does not occur in many species of intact mitochondria. Heart mitochondria are exceptional in this respect. Some preparations of intact heart mitochondria with high respiratory control and P/O ratios vigorously oxidize added NADH (Blanchaer and Griffith, 1966). Another criterion for mitochondrial quality is the ability to function optimally in the absence of bovine serum albumin, which seems to stabilize some mitochondrial preparations (e.g., Weinbach *et al.*, 1967; Warshaw, 1969).

With these four criteria for mitochondrial quality in mind, let us examine the properties of isolated lymphoma mitochondria. Figure 6 shows an oxygen electrode trace of the response of lymphoma mitochondria to aliquots of ADP in the presence of P$_i$ and the substrate α-ketoglutarate. This common form for the presentation of such data deserves discussion in some detail. When the mitochondrial suspension is added in the absence of ADP, a slow respiration rate is established. This is the state 4 of Chance and Williams (1956) in which respiration is limited by the availability of ADP. On addition of 250 nmoles of ADP, a more rapid respiratory rate ensues (state 3). When all the ADP has been phosphorylated to ATP, a state 3 to 4 transition results. This cycle can be repeated until the oxygen in the vessel is exhausted. From such tracings many of the parameters of interest relative to mitochondrial quality can be calculated. The respiratory control ratio (or index) in this preparation varies from 4.0 to 4.5. The P/O ratio can be calculated from the amount of ADP added and the amount of oxygen consumed in state 3 during which the ADP is phosphorylated to ATP. In the case shown here, the P/O ratio is 2.4. The Q_{O_2} is determined from the amount of mitochondrial protein added, the oxygen consumed during one or more state 3 periods, and the time spent in state 3. The Q_{O_2} in this experiment was approximately 140 μl of O_2 per milligram of mitochondrial protein per hour.

Several interesting points emerge. The respiratory control ratios are somewhat lower than with rat liver mitochondria but not dramatically. The Q_{O_2} is as high as with rat liver, but the P/O ratios are definitely lower. The author has never seen a report of a mitochondrial preparation from either tumors or cells in culture which gave a P/O ratio with α-ketoglutarate higher than reported here. The reader will recall that, because of the substrate-level phosphorylation associated with α-ketoglutarate oxidation, the theoretical P/O ratio for this substrate should be 4.0. This experiment (Fig. 6) would be more convincing if done in the presence of malonate to inhibit the further oxidation of succinate; even in its present form, it suggests that the substrate-level phosphorylation step in the Krebs cycle is missing in tumor or cultured cell mitochondria. If this were true, then the maximum theoretical P/O ratio for complete oxidation of a Krebs cycle substrate would be 2.8.

In Table V some of the more important quality parameters are listed

TABLE V.

SOME QUALITY PARAMETERS OF LYMPHOMA MITOCHONDRIA[a]

Substrate	Respiratory Control Ratio	P/O Ratio	Q_{O_2} (μl O_2/mg protein/hr)
Succinate	5.56	1.80	299
Glutamate	4.69	2.60	192
α-Ketoglutarate	4.18	2.43	143
Malate	3.63	2.46	127

[a] The assay medium contained 0.25-M mannitol; 0.2-mM EDTA; 1.0-mM MgCl$_2$; 10-mM KCl; 10-mM potassium phosphate, pH 7.2; and 5–10-mM substrate. The final volume was 2.0, and the temperature 37°C.

for the lymphoma mitochondria with four different substrates. The Q_{O_2} values shown here are those expected with rat liver mitochondria; as usual, succinate is the most rapidly oxidized substrate. Both the P/O and respiratory control ratios are somewhat lower than observed with the best rat liver preparations.

The effect of 0.2% added bovine serum albumin on succinate oxidation is shown in Table VI for both lymphoma mitochondria and mitochondria isolated from the line of Chinese hamster cells studied extensively at Los Alamos. In contrast to the results shown here, rat liver mitochondria show no effect of the addition of serum albumin unless the mitochondria are damaged. With lymphoma mitochondria, it increases the respiratory activity, primarily by increasing the state 3 rate and, consequently, increases the respiratory control ratio. There is no effect on the P/O ratio. With the mitochondria from Chinese hamster cells, the effect of serum albumin is different. Here the effect is primarily to decrease the state 4 rate with consequent increase in both respiratory control and P/O ratio but no change in respiratory activity [a similar effect of serum albumin on fetal heart mitochondria was observed by Warshaw (1969)].

In summary, there are ways in which tumor and cultured cell mitochondria differ from the mitochondria of normal cells. The differences are small, however, and, in the author's opinion, insignificant.

The history of mitochondriology is replete with examples where mitochondria of a certain material (for example, insect flight muscle or animal skeletal muscle) were originally considered to be subnormal in one

TABLE VI.

EFFECT OF 0.2% BOVINE SERUM ALBUMIN (BSA) ON SUCCINATE OXIDATION BY MITOCHONDRIA FROM CELLS IN CULTURE[a]

		Q_{O_2} (μl O_2/mg protein/hr)			
	BSA	State 4	State 3	ACR	P/O
Lymphoma mitochondria					
	−	67.6	172	2.51	1.64
	+	74.0	297	4.04	1.75
Chinese hamster cell mitochondria					
	−	70.0	172	2.46	1.37
	+	43.0	171	4.20	1.64

[a] The medium was that described in Table V.

or more respects. Subsequent advances in methods of mitochondrial preparation made it possible to isolate mitochondria from previously refractory materials which had all the properties to be expected of the intact organelles. This history will probably be repeated with tumor mitochondria. One aspect which should be further explored is the addition of exogenous substrates to the isolation medium for cultured cell or tumor mitochondria. Enzymology abounds in examples where providing a substrate for an enzyme protects that enzyme from degradation during isolation. It is an additional unusual property of lymphoma mitochondria, for example, that they have a very low or nonexistent content of endogenous substrate. Tarjan and von Korff (1967) have shown that the presence of endogenous substrate enhances mitochondrial activity, respiratory control, and P/O ratios. It may be that the modest defects of the lymphoma mitochondria, particularly the serum albumin requirement, may be overcome by simply providing exogenous substrate during the mitochondrial isolation.

C. Nuclei

Betel and Klouwen (1964) have strongly stressed the role of nuclear respiration and accompanying phosphorylation in the energy metabolism of lymphoid cells including the L5178Y lymphoma (Klouwen and Appleman, 1967). Evaluation of these results is hampered by the enormous variability in response of their nuclear preparations to, for example, antimycin. Betel and Klouwen (1964) originally reported 63% inhibition of respiration by 400 pmoles/mg of protein, based on their figure of approximately 17 mg of nuclear protein per milligram of DNA-P for the rat thymus preparations. Recently (Betel, 1969) 100% respiratory inhibition was reported with 400 nmoles/mg of protein, a 1000-fold decrease in antimycin sensitivity, if the original factor is used to convert milligrams of DNA-P into milligrams of protein. However, in 1967 (Betel and Klouwen, 1967) they reported 75% inhibition (the maximum obtained) by 12 pmoles/mg of protein, a 30-fold higher sensitivity than in 1964 and 30,000-fold higher than in 1969.

In any event, the data of Table I suggest that, under the conditions of these experiments, nuclear respiration was not significant in mouse lymphoma cells where complete inhibition of respiration was obtained with 18 pmoles of antimycin per milligram of protein. The subject of nuclear respiration and phosphorylation was reviewed by Conover (1967).

III. A Hypothesis on the Relationship of Increased Glycolysis to Tumorigenicity

The relative uniqueness of aerobic glycolysis as a hallmark of the malignant cell has been stressed. Various explanations for the high glycolysis of malignant cells have been discussed and shown to be inadequate. To close this discussion on a more positive note, I should like here to offer a potential explanation of this relationship which appears sufficiently plausible to justify further investigation. It is well documented that when a cell undergoes a malignant transformation, there occur very early changes in the cell membrane. These changes have the following effects:

1. Loss of contact inhibition of locomotion and growth (see Wallach, 1969).
2. Loss of normal hormonal response [e.g., tumor cells are unresponsive to stilbesterol (Burk *et al.*, 1967)].
3. Loss of some surface antigens which may prevent normal cells from being invasive and acquisition of others, the well-known tumor antigens (Wallach, 1969).
4. Exposure of sites previously in an occult form in the cell membrane (Burger, 1969; Inbar and Sachs, 1969).
5. Finally and most importantly from the standpoint of aerobic glycolysis, the following postulated change: The cell, in the course of malignant transformation, loses much of its capacity to exclude exogenous glucose (Hatanaka *et al.*, 1970).

Loss of the last property has the following consequences. When the cell is exposed to high glucose concentrations [and the physiological level is 15–1000 times the K_m for glycolysis in tumor cells (Burk *et al.*, 1967; Hatanaka *et al.*, 1970)],* glucose enters the cell in potentially toxic concentrations, since the normal barrier to its entry is largely obliterated. The response of the cell is to synthesize greatly increased amounts of glycolytic enzymes so that glucose can be "detoxified" by conversion to lactate and excretion into the medium. The idea of the high glycolytic rate of tumor cells as resulting from a membrane defect was apparently first suggested by Bloch-Frankenthal and Weinhouse (1959), while "glucose toxicity" was demonstrated by Graff and his

---

* A substrate concentration of $10 \times K_m$ gives approximately $0.9V_{max}$. Thus, for malignant cells at physiological glucose concentrations (5.56 mM), glycolysis would proceed at nearly the maximum velocity of which the multienzyme complex is capable.

colleagues (1965). Graff *et al.* showed that Ehrlich ascites cells in culture would grow at low glucose levels with production of very little lactic acid. Good growth (and negligible lactate production) was achieved at 0.05-mM glucose in a chemostat. Moreover, Graff *et al.* (1965) showed that cells grown on a low glucose concentration for 12 days were killed when exposed to physiological concentrations of glucose. The overall thrust of these results is clear: Aerobic lactate formation is not required for growth of cells in culture, and glucose, even at physiological concentrations, is toxic to cells not adapted to it. No assays of glycolytic rates were done on high and low glucose cells. Presumably, the low glucose cells had a greatly reduced content of glycolytic enzymes and were unable to synthesize higher levels of the necessary enzymes quickly enough to overcome the toxicity of intracellular glucose when the cells were again exposed to physiological concentrations. The precise enzymatic basis of the glucose toxicity observed by Graff *et al.*, remains to be established (cf. Section I,C).

The postulated inability of tumorigenic cells to exclude glucose may have other consequences. The formation of glycogen by cells in culture grown in a glucose-rich medium is puzzling, but this could be a means of detoxifying intracellular glucose. In other words, control on the intracellular glucose concentration, exerted by normal cells at the level of the cell membrane, is exerted in tumor cells by transforming intracellular glucose into something else, either glycogen or lactate.

It is well known that in yeast high levels of medium glucose suppress the formation of functional mitochondria (e.g., Jayaraman *et al.*, 1966). Functional mitochondria are formed only when the glucose level is lowered. Such an effect may play a role in the low levels of mitochondria found in tumor tissue (see Section II,B). This question could be readily approached experimentally by simply comparing the Q_{O_2} of cells grown under high (20-mM) and low (0.1-mM) glucose. If there were a substantial increase in Q_{O_2} at low glucose concentration, the increase would almost certainly be due to increased numbers of mitochondria. This tentative conclusion could be confirmed by quantitative electron microscopy or by measuring the increased level in the low glucose cells of a convenient mitochondrial constituent such as cytochrome *c*.

According to the hypothesis offered here, the high glycolytic rates of tumor cells result from membrane changes accompanying the malignant transformation. Thus it is an effect of the conversion to the malignant state rather than a cause. This explanation is consistent with the striking correlations obtained by Burk *et al.* (1967) and others between glycolytic rates and growth rates of tumors. The fact that similar membrane architecture may exist in a few nonmalignant cell types (such as retina or leukocytes) is not surprising. The membrane alteration may be advantageous to cells which require very high glycolytic rates.

IV. Summary and Concluding Remarks

Some time was spent attempting to convince the reader of the virtue, if not the necessity, of carefully controlled growth conditions for cells used in studies on intermediary metabolism. The requirement for a population of malignant cells for studies on tumor cells was also stressed. This is an area in which much remains to be done. There are very few cell lines of known malignancy, and the number needs to be expanded greatly. Since large numbers of animals and long times are required, this sort of work is best done perhaps by establishments such as the National Cancer Institute. It would be a valuable public service and would greatly enhance the utility of cells in culture in studies on the biochemical basis of malignancy.

Although glycolysis was given short shrift, a large proportion of this chapter was spent on aspects of terminal oxidation. The reader, on finishing this section, may exclaim with MacDuff, "Confusion now hath made his masterpiece" (Shakespeare, ca. 1581). Some confusion is intrinsic to an unfinished area; the author apologizes for any additional contributions. It is generally true that tumor cells have high glycolytic rates which correlate with growth rates. Cells in culture are also highly glycolytic. A proposal that this results from a combination of glucose toxicity and increased cellular permeability to glucose was made. Other explanations, such as a role for pyruvate as electron acceptor for the regeneration of NADP, are also possible. The view that the high glycolytic rate arises from a deficiency in energy metabolism of tumor cells is probably incorrect, as is the explanation based on lack of NADH shuttle systems.

Criteria for mitochondrial quality were discussed and some properties of lymphoma cell mitochondria presented. The conclusion was drawn that the ways in which lymphoma mitochondria fail to meet the most exacting criteria for mitochondrial quality are of secondary importance and probably reflect our lack of knowledge of the best isolation procedure for these organelles from cells in culture. The nuclear respiration of lymphoma cells was considered to be insignificant.

By this point both the reader and author are in a position to judge to what extent the aims originally suggested for this chapter have been met. It is comforting to think that what is good will endure, while what is bad will be set right by the work of subsequent writers in this fascinating field in which they, too, may find Eternal Delight.

ACKNOWLEDGMENTS

I wish to express my deep appreciation to Mr. John Hanners for his devoted help in many things; to Dr. Judith Hutson for invaluable assistance both in the laboratory and in writing this chapter; to Dr. Paul Kraemer, Dr. William Currie, and Dr. Joseph Machinist for helpful discussions; and finally to Mrs. William (Betty) Sullivan for the truly extraordinary skill, cheerfulness, and patience she contributed to the preparation of this account.

REFERENCES

Aisenberg, A. C. (1961). "The Glycolysis and Respiration of Tumors." Academic Press, New York.

Baierlein, J. L. (1967). Mechanism of the Pasteur effect in human leukocytes. Ph.D. Dissertation, Boston University.

Banks, B. E. C. (1969). Thermodynamics and biology. *Chem. Brit.* **5**, 514–519.

Beinert, H., and Palmer, G. (1965). Contributions of EPR spectroscopy to our knowledge of oxidative enzymes. *Advan. Enzymol.* **27**, 105–198.

Betel, I. (1969). The endogenous substrate for nuclear oxidative phosphorylation. *Arch. Biochem. Biophys.* **134**, 271–274.

Betel, I., and Klouwen, H. M. (1964). Oxidative phosphorylation in isolated rat-thymus nuclei. *Biochim. Biophys. Acta* **85**, 348–350.

Betel, I., and Klouwen, H. M. (1967). Oxidative phosphorylation in nuclei isolated from rat thymus. *Biochim. Biophys. Acta* **131**, 453–467.

Bickis, I. J., and Quastel, F. R. S. (1965). Effects of metabolic inhibitors on energy metabolism of Ehrlich ascites carcinoma cells. *Nature (London)* **205**, 44–46.

Blanchaer, M. C., and Griffith, T. J. (1966). Control of reduced nicotinamide adenine dinucleotide oxidation by pigeon-heart mitochondria. *Can. J. Biochem.* **44**, 1527–1537.

Bloch, K. (1962). Oxygen and biosynthetic patterns. *Fed. Proc., Fed. Amer. Soc. Exp. Biol.* **21**, 964–975.

Bloch-Frankenthal, L., and Weinhouse, S. (1959). Metabolism of neoplastic tissue. XII. Effects of glucose concentration on respiration and glycolysis of ascites tumor cells. *Cancer Res.* **19**, 1082–1090.

Borst, P. (1962). The aerobic oxidation of reduced diphosphopyridine nucleotide formed by glycolysis in Ehrlich ascites-tumour cells. *Biochim. Biophys. Acta* **57**, 270–282.

Boxer, G. E., and Devlin, T. M. (1961). Pathways of intracellular hydrogen transport. *Science* **134**, 1495–1501.

Boxer, G. E., and Shonk, C. E. (1960). Low levels of soluble DPN-linked α-glycerophosphate dehydrogenase in tumors. *Cancer Res.* **20**, 85–91.

Boyer, P. D. (1965). Carboxyl activation as a possible common reaction in substrate-level and oxidative phosphorylation and in muscle contraction. *In* "Oxidases and Related Redox Systems" (T. E. King, H. S. Mason, and M. Morrison, eds.), pp. 994–1017. Wiley, New York.

Boyer, P. D. (1968). Oxidative phosphorylation. In "Biological Oxidations" (T. P. Singer, ed.), pp. 193–235. Wiley (Interscience), New York.

Bullough, W. S. (1952). The energy relations of mitotic activity. Biol. Rev. Cambridge Phil. Soc. 27, 133–168.

Burger, M. M. (1969). A difference in the architecture of the surface membrane of normal and virally transformed cells. Proc. Nat. Acad. Sci. U.S. 62, 994–1001.

Burk, D., Woods, M., and Hunter, J. (1967). On the significance of glucolysis for cancer growth, with special reference to Morris rat hepatomas. J. Nat. Cancer Inst. 38, 839–863.

Butow, R. A. (1967). The effect of rotenone on beef liver glutamic dehydrogenase. Biochemistry 6, 1088–1093.

Bygrave, F. L. (1966). The effect of calcium ions on the glycolytic activity of Ehrlich ascites-tumour cells. Biochem. J. 101, 480–487.

Carr, L. J., Hiebert, R. D., Currie, W. D., and Gregg, C. T. (1971a). A stable sensitive, and inexpensive amplifier for oxygen electrode studies. Anal. Biochem. 41, 492–502.

Carr, L. J., Larkins, J. H., and Gregg, C. T. (1971b). A multi-channel recording oxygen electrode amplifier for biochemical studies. Anal. Biochem. 41, 503–509.

Chance, B., and Williams, G. R. (1956). The respiratory chain and oxidative phosphorylation. Advan. Enzymol. 17, 65–134.

Chance, B., Ernster, L., Garland, P. B., Lee, C.-P., Light, P. A., Ohnishi, T., Ragan, C. I., and Wong, D. (1967). Flavoproteins of the mitochondrial respiratory chain. Proc. Nat. Acad. Sci. U.S. 57, 1498–1505.

Coe, E. L., and Strunk, R. C. (1970). The effect of oxamate on glycolysis in intact ascites tumor cells. I. Kinetic evidence for a dual glycolytic system. Biochim. Biophys. Acta 208, 189–202.

Conover, T. E. (1967). Respiration and adenosine triphosphate synthesis in nuclei. Curr. Top. Bioenerg. 2, 235–267.

Crabtree, H. G. (1929). Observations on the carbohydrate metabolism of tumours. Biochem. J. 23, 536–545.

Cristofalo, V. J., and Kritchevsky, D. (1966). Respiration and glycolysis in the human diploid cell strain WI-38. J. Cell. Physiol. 67, 125–132.

Cross, R. L., and Wang, J. H. (1970). Evidence of a phosphorylated intermediate in mitochondrial oxidative phosphorylation. Biochem. Biophys. Res. Commun. 38, 848–854.

Currie, W. D., and Gregg, C. T. (1965). Inhibition of the respiration of cultured mammalian cells by oligomycin. Biochem. Biophys. Res. Commun. 21, 9–15.

Currie, W. D., and Gregg, C. T. (1966). Unpublished data.

Deamer, D. W. (1969). ATP synthesis: The current controversy. J. Chem. Educ. 46, 198–206.

De Duve, C., Wattiaux, R., and Baudhuin, P. (1962). Distribution of enzymes between subcellular fractions in animal tissues. Advan. Enzymol. 24, 291–358.

Del Monte, U. (1967). Changes in oxygen tension in Yoshida ascites hepatoma during growth. Proc. Soc. Exp. Biol. Med. 125, 853–856.

Dionisi, O., Cittadini, A., Gelmuzzi, G., Galeotti, T., and Terranova, T. (1970). The role of the α-glycerophosphate shuttle in the reoxidation of cytosolic NADH in Ehrlich ascites tumour cells. Biochim. Biophys. Acta 216, 71–79.

Ernster, L., Dallner, G., and Azzone, C. F. (1963). Differential effects of rotenone and amytal on mitochondrial electron and energy transfer. J. Biol. Chem. 238, 1124–1131.

Estabrook, R. W., Shigematsu, A., and Schenkman, J. B. (1970). The contribution of the microsomal electron transport pathway to the oxidative metabolism of liver. *Advan. Enzyme Regul.* **8,** 121–130.
Galeotti, T., Azzi, A., and Chance, B. (1969). The flavoproteins of ascites tumor cells. *Arch. Biochem. Biophys.* **131,** 306–309.
Galeotti, T., Azzi, A., and Chance, B. (1970). The reoxidation of cytoplasmic reducing equivalents in Ehrlich ascites tumor cells. *Biochim. Biophys. Acta* **197,** 11–24.
Garland, P. B., Shepherd, D., Nicholls, D. G., and Ontko, J. A. (1967). Energy-dependent control of the tricarboxylic acid cycle by fatty acid oxidation in rat liver mitochondria. *Advan. Enzyme Regul.* **6,** 3–30.
Gibson, Q. H. (1968). Cytochromes. *In* "Biological Oxidations" (T. P. Singer, ed.), pp. 379–413. Wiley (Interscience), New York.
Gill, P., and Pan, J. (1970). Inhibition of cell division in L5178Y cells by arginine-degrading mycoplasmas: The role of arginine deiminase. *Can. J. Microbiol.* **16,** 415–419.
Goldberg, E. B., Nitowsky, H. M., and Colowick, S. P. (1965). The role of glycolysis in the growth of tumor cells. IV. The basis of glucose toxicity in oxamate-treated, cultured cells. *J. Biol. Chem.* **240,** 2791–2796.
Graff, S., and Moser, H. (1962). New interpretations in cancer biology. *J. Hosp. Joint Dis.* **23,** 59–79.
Graff, S., Moser, H., Kastner, O., Graff, A. M., and Tannenbaum, M. (1965). The significance of glycolysis. *J. Nat. Cancer Inst.* **34,** 511–519.
Green, D. E., and Baum, H. (1970). "Energy and the Mitochondrion." Academic Press, New York.
Green, D. E., Murer, E., Hultin, H. O., Richardson, S. H., Salmon, B., Brierley, G. P., and Baum, H. (1965). Association of integrated metabolic pathways with membranes. I. Glycolytic enzymes of the red blood corpuscle and yeast. *Arch. Biochem. Biophys.* **112,** 635–647.
Green, S., and Bodansky, O. (1963). Quantitative aspects of the relationship between nicotinamide adenine dinucleotide and the enzyme nicotinamide adenine dinucleotide glycohydrolase in Ehrlich ascites cells. *J. Biol. Chem.* **238,** 2119–2122.
Gregg, C. T. (1970). Unpublished experiments.
Gregg, C. T., Machinist, J. M., and Currie, W. D. (1968). Glycolytic and respiratory properties of intact mammalian cells. Inhibitor studies. *Arch. Biochem. Biophys.* **127,** 101–111.
Greville, G. D. (1969). A scrutiny of Mitchell's chemiosmotic hypothesis of respiratory chain and photosynthetic phosphorylation. *Curr. Top. Bioenerg.* **3,** 1–78.
Gumaa, K. A., and McLean, P. (1969). Effect of oxamate, pyruvate, nicotinamide and streptozotocin on the pentose phosphate pathway intermediates in ascites tumour cells. *Biochem. Biophys. Res. Commun.* **35,** 86–93.
Günther, T., and Goecke, C. (1966). Änderung der Permeabilität bei der Isolierung von Zellen. *Z. Naturforsch. B* **21,** 1171–1174.
Hackenbrock, C. R. (1964). Ultrastructural bases for metabolically linked mechanical activity in mitochondria. I. Reversible ultrastructural changes with change in metabolic steady state in isolated liver mitochondria. *J. Cell Biol.* **30,** 269–297.
Hackenbrock, C. R. (1968). Ultrastructural bases for metabolically linked mechanical activity in mitochondria. II. Electron transport-linked ultrastructural transformations in mitochondria. *J. Cell Biol.* **37,** 345–369.
Hatanaka, M., Todaro, G. J., and Gilden, R. V. (1970). Altered glucose transport

kinetics in murine sarcoma virus-transformed BALB/3T3 clones. *Int. J. Cancer* **5**, 224–228.

Hatefi, Y., and Fakouhi, T. (1968). Control of β-hydroxybutyrate and acetoacetate oxidation by inorganic phosphate and adenosine-5'-diphosphate in heart mitochondria. *Arch. Biochem. Biophys.* **125**, 114–125.

Hawtrey, A. O. (1958). Fractionation of cell types in Ehrlich ascites fluid by sucrose gradient centrifugation. *Proc. Soc. Exp. Biol. Med.* **99**, 28–30.

Hayflick, L. (1968). Cell substrates for human virus vaccine preparation—General comments. *Nat. Cancer Inst., Monogr.* **29**, 83–91.

Hewitt, H. B. (1958). Studies of the dissemination and quantitative transplantation of a lymphocytic leukaemia of CBA mice. *Brit. J. Cancer* **12**, 378–401.

Heytler, P. G. (1963). Uncoupling of oxidative phosphorylation by carbonyl cyanide phenylhydrazones. I. Some characteristics of *m*-Cl-CCP action on mitochondria and chloroplasts. *Biochemistry* **2**, 357–361.

Hutson, J., and Gregg, C. T. (1971). Unpublished experiments.

Ibsen, K. H. (1961). The Crabtree effect: A review. *Cancer Res.* **21**, 829–841.

Inbar, M., and Sachs, L. (1969). Structural difference in sites on the surface membrane of normal and transformed cells. *Nature (London)* **223**, 710–712.

Isaacs, G. H., Sacktor, B., and Murphy, T. A. (1969). The role of the α-glycerophosphate cycle in the control of carbohydrate oxidation in heart and in the mechanism of action of thyroid hormone. *Biochim. Biophys. Acta* **177**, 196–203.

Jarrett, O., and Macpherson, I. (1968). The basis of the tumorigenicity of BHK21 cells. *Int. J. Cancer* **3**, 654–662.

Jayaraman, J., Cotman, C., and Mahler, H. R. (1966). Biochemical correlates of respiratory deficiency. VII. Glucose repression. *Arch. Biochem. Biophys.* **116**, 224–251.

Kennedy, E. P., and Lehninger, A. L. (1949). Oxidation of fatty acids and tricarboxylic acid cycle intermediates by isolated rat liver mitochondria. *J. Biol. Chem.* **179**, 957–972.

Kilburn, D. G., Lilly, M. D., Self, D. A., and Webb, F. C. (1969). The effect of dissolved oxygen partial pressure on the growth and carbohydrate metabolism of mouse LS cells. *J. Cell Sci.* **4**, 25–37.

Klein, G. (1951). Comparative studies of mouse tumors with respect to their capacity for growth as "ascites tumors" and their average nucleic acid content per cell. *Exp. Cell Res.* **2**, 518–573.

Klingenberg, M. (1964). Reversibility of energy transformations in the respiratory chain. *Angew. Chem., Int. Ed. Engl.* **3**, 54–61.

Klingenberg, M. (1968). The respiratory chain. *In* "Biological Oxidations" (T. P. Singer, ed.), pp. 3–54. Wiley (Interscience), New York.

Klouwen, H. M., and Appelman, A. W. M. (1967). Synthesis of adenosine triphosphate in isolated nuclei and intact cells. *Biochem. J.* **102**, 878–884.

Koprowska, I. (1956). Exfoliative cytology in the study of ascites tumors. *Ann. N.Y. Acad Sci.* **63**, 738–747.

LaBauve, P. M., and Gregg, C. T. (1970). Unpublished experiments.

LaNoue, K., Nicklas, W. T., and Williamson, J. R. (1970). Control of citric acid cycle activity in rat heart mitochondria. *J. Biol. Chem.* **245**, 102–111.

Lardy, H. A., and Ferguson, S. M. (1969). Oxidative phosphorylation in mitochondria. *Ann. Rev. Biochem.* **38**, 991–1034.

Lea, M. A., Sasovetz, D., and Morris, H. P. (1969). Some factors affecting carbohydrate metabolism in hepatic tissues of different growth rates. *Int. J. Cancer* **4**, 487–494.

Lee, C.-P., Azzone, G. F., and Ernster, L. (1964). Evidence for energy-coupling in non-phosphorylating electron transport particles from beef-heart mitochondria. *Nature (London)* **201**, 152–155.

Lee, I.-Y., Strunk, R. C., and Coe, E. L. (1967). Coordination among rate-limiting steps of glycolysis and respiration in intact ascites tumor cells. *J. Biol. Chem.* **242**, 2021–2028.

Letnansky, K., and Klc, G. M. (1969). Glycerophosphate oxidoreductases and the glycerophosphate cycle in Ehrlich ascites tumor cells. *Arch. Biochem. Biophys.* **130**, 218–226.

Levine, E. M., Thomas, L., McGregor, D., Hayflick, L., and Eagle, H. (1968). Altered nucleic acid metabolism in human cell cultures infected with mycoplasma. *Proc. Nat. Acad. Sci. U.S.* **60**, 583–589.

Lindahl, P. E., and Klein, G. (1955). Separation of Ehrlich ascites tumour cells from other cellular elements. *Nature (London)* **176**, 401–4012.

Loewenthal, H., and Jahn, G. (1932). Übertragungsversuche mit carcinomatöser Mäuse-Ascitesflüssigkeit und ihr Verhalten gegen physikalische und chemische Einwirkungen. *Z. Krebsforsch.* **37**, 339–447.

Luthardt, T., and Fischer, W. (1964). Atmung, glykolyse und Überlebensdauer von monolayer-gewebekultur-zellen nach verbringen in suspension. *Z. Naturforsch. B.* **19**, 916–922.

McKee, R. W., Dickey, A., and Parks, M. E. (1968). NADase activity and glycolysis in Ehrlich-Lettre carcinoma and liver cell particulates. *Arch. Biochem. Biophys.* **126**, 760–763.

Machinist, J. M., and Gregg, C. T. (1970). Unpublished results.

Machinist, J. M., Hutson, J., and Gregg, C. T. (1971). Unpublished results.

Malmgren, R. A., and Flanigan, C. C. (1955). Location of the vegetative form of *Clostridium tetani* in mouse tumors following intravenous spore administration. *Cancer Res.* **15**, 473–478.

Mitchell, P. (1966). Chemiosmotic Coupling in Oxidative and Photosynthetic Phosphorylation. Glynn Research Laboratory, Bodwin, England. 192 pp.

Nakamura, W., and Hosoda, S. (1968). The absence of glucose in Ehrlich ascites tumor cells and fluid. *Biochim. Biophys. Acta* **158**, 212–218.

Navazio, F., Ernster, B. B., and Ernster, L. (1957). Studies on TPN-linked oxidations. II. The quantitative significance of liver lactic dehydrogenase as a catalyzer of TPNH-oxidation. *Biochim. Biophys. Acta* **26**, 416–421.

Neubert, D., Peters, H., Teske, S., Köhler, E. and Barrach, H.-J. (1971). Studies on the problem of "aerobic glycolysis" occuring in mammalian embyros. *Naunyn-Schmiedebergs Arch. Pharmak.* **268**, 235–241.

Ng, C. W., and Gregory, K. F. (1966). Antibody to lactate dehydrogenase. II. Inhibition of glycolysis and growth of tumor cells. *Biochim. Biophys. Acta* **130**, 477–485.

Nigam, V. N. (1968). ATP level and control of glycolysis in Novikoff ascites-hepatoma cells. *Enzymologia* **36**, 257–268.

Oesper, P. (1968). Error and trial: The story of the oxidative reaction of glycolysis. *J. Chem. Educ.* **45**, 607–610.

Painter, A. A., and Hunter, E., Jr. (1970). Phosphorylation coupled to oxidation of thiol groups (GSH) by cytochrome c with disulfide (GSSG) as an essential catalyst. I. Demonstration of ADP formation from AMP and HPO_4^{2-}. *Biochem. Biophys. Res. Commun.* **40**, 360–368.

Papaconstantinou, J., and Colowick, W. P. (1961a). The role of glycolysis in the growth of tumor cells. I. Effects of oxamic acid on the metabolism of Ehrlich ascites tumor cells *in vitro. J. Biol. Chem.* **236**, 278–284.

Papaconstantinou, J., and Colowick, S. P. (1961b). The role of glycolysis in the growth of tumor cells. II. The effect of oxamic acid on the growth of HeLa cells in tissue culture. *J. Biol. Chem.* **236**, 285–288.

Pasteur, L. (1876). "Etudes sur la biere." Gauthier-Villars, Paris.

Paul, J. (1965). Carbohydrate and energy metabolism. *In* "Cells and Tissues in Culture" (E. N. Willmer, ed.), Vol. 1, pp. 239–276 Academic Press, New York.

Paul, J., Broadfoot, M. M., and Walker, P. (1966). Increased glycolytic capacity and associated enzyme changes in BHK21 cells transformed with polyoma virus. *Int. J. Cancer* **1**, 207–218.

Pauling, L. (1970). Structure of high-energy molecules. *Chem. Brit.* **6**, 468–472.

Penniston, J. T., Harris, R. A., Asai, J., and Green, D. E. (1968). The conformational basis of energy transformations in membrane systems. I. Conformational changes in mitochondria. *Proc. Nat. Acad. Sci. U.S.* **59**, 624–631.

Phillips, H. J., and McCarthy, H. L. (1956). Oxygen uptake and lactate formation of HeLa cells. *Proc. Soc. Exp. Biol. Med.* **93**, 602–605.

Phillips, R. C., George, P., and Rutman, R. J. (1969). Thermodynamic data for the hydrolysis of adenosine triphosphate as a function of pH, Mg^{2+} ion concentration, and ionic strength. *J. Biol. Chem.* **224**, 3330–3342.

Racker, E. (1965). "Mechanisms in Bioenergetics," p. 228. Academic Press, New York.

Racker, E. (1970). Function and structure of the inner membrane of mitochondria and chloroplasts. *In* "Membranes of Mitochondria and Chloroplasts" (E. Racker, ed.), pp. 127–171. Van Nostrand-Reinhold, New York.

Rieske, J. S., Baum, H., Stoner, C. D., and Lipton, S. H. (1967). On the antimycin-sensitive cleavage of complex III of the mitochondrial respiratory chain. *J. Biol. Chem.* **242**, 4854–4866.

Rose, I. A., and Rose, Z. A. (1969). Glycolysis: Regulation and mechanisms of the enzymes. *Compr. Biochem.* **17**, 93–161.

Rothblat, G., and Cristofalo, V. (1971). Personal communication.

Russell, W. C., Niven, J. S. F., and Berman, L. D. (1968). Studies on the biology of the mycoplasma-induced "stimulation" of BHK21-C13 cells. *Int. J. Cancer* **3**, 191–202.

Sacktor, B., and Dick, A. R. (1964). Oxidation of extramitochondrial diphosphopyridine nucleotide by various tissues of the mouse. *Science* **145**, 606–607.

Sanford, K. K. (1965). Malignant transformation of cells *in vitro. Int. Rev. Cytol.* **18**, 249–311.

Sanford, K. K., and Westfall, B. B. (1969). Growth and glucose metabolism of high and low tumor-producing clones under aerobic and anaerobic conditions *in vitro. J. Nat. Cancer Inst.* **42**, 953–959.

Sanford, K. K., Woods, M. W., Scott, D. B. M., and Kerr, H. A. (1969). Stability *in vitro* of diverse capacities for tumor production and glucose metabolism in three murine clones derived from the same cell. *J. Nat. Cancer Inst.* **42**, 945–952.

Sato, N., and Hagihara, B. (1970). Spectrophotometric analyses of cytochromes in ascites hepatomas of rats and mice. *Cancer Res.* **30**, 2061–2068.

Sauer, L. A. (1968). Hexokinase inhibition after the addition of glucose to Ehrlich-Lettre ascites tumor cells under aerobiosis. *J. Biol. Chem.* **243**, 2429–2436.

Schneider, W. C., and Hogeboom, G. H. (1951). Cytochemical studies of mammalian tissues: The isolation of cell components by differential centrifugation: A review. *Cancer Res.* **11**, 1–22.

Self, D. A., Kilburn, D. G., and Lilly, M. D. (1968). The influence of dissolved oxygen partial pressure on the level of various enzymes in mouse LS cells. *Biotechnol. Bioeng* **10**, 815–828.

Serif, G. S., and Butcher, F. R. (1966). Pyridine nucleotides. I. Radiometric determination of minute quantities of diphosphopyridine nucleotides. *Anal. Biochem.* **15**, 278–286.

Shakespeare, W. (ca. 1581). "Macbeth," Act II, Scene III.

Slater, E. C. (1953). Mechanism of phosphorylation in the respiratory chain. *Nature (London)* **172**, 975–978.

Slater, E. C. (1966). Oxidative phosphorylation. *Compr. Biochem.* **14**, 334.

Slater, E. C. (1967a). Application of inhibitors and uncouplers for a study of oxidative phosphorylation. *Methods Enzymol.* **10**, 48–57.

Slater, E. C. (1967b). An evaluation of the Mitchell hypothesis of chemiosmotic coupling in oxidative and photosynthetic phosphorylation. *Eur. J. Biochem.* **1**, 317–326.

Sordahl, L. A., Blailock, Z. R., Kraft, G. H., and Schwartz, A. (1969). The possible relationship between ultrastructure and biochemical state of heart mitochondria. *Arch. Biochem. Biophys.* **132**, 404–415.

Stoner, C. D., and Sirak, H. D. (1969). Passive induction of the "energized-twisted" conformational state in bovine heart mitochondria. *Biochem. Biophys. Res. Commun.* **35**, 59–66.

Storey, B. T. (1970). Chemical hypothesis for energy conservation in the mitochondrial respiratory chain. *J. Theor. Biol.* **28**, 233–259.

Suzuki, R., Sato, K., and Tsuiku, S. (1968). Pyruvate inhibition of glycolysis in Ehrlich ascites tumor cells. *Arch. Biochem. Biophys.* **124**, 1–3.

Tarjan, E. M., and Von Korff, R. W. (1967). Factors affecting the respiratory control ratio of rabbit heart mitochondria. *J. Biol. Chem.* **242**, 318–324.

Terranova, T., Baldi, S., and Dionisi, O. (1969). Further observations on the glucose-induced respiration in Ehrlich ascites cells treated with rotenone. *Arch. Biochem. Biophys.* **130**, 594–603.

Tobey, R. A., Petersen, D. F., Anderson, E. C., and Puck, T. T. (1966). Life cycle analysis of mammalian cells. III. The inhibition of division in Chinese hamster cells by puromycin and actinomycin. *Biophys. J.* **6**, 567–581.

Tobey, R. A., Petersen, D. F., and Anderson, E. C. (1969). Energy requirements for mitosis in Chinese hamster cells. In "Biochemistry of Cell Division" (R. Baserga, ed.), pp. 39–56. Thomas, Springfield, Illinois.

Uyeda, K., and Racker, E. (1965). Regulatory mechanisms in carbohydrate metabolism. VII. Hexokinase and phosphofructokinase. *J. Biol. Chem.* **240**, 4682–4693.

Wallach, D. F. H. (1969). Cellular membrane alterations in neoplasia: A review and a unifying hypothesis. *Curr. Top. Microbiol. Immunol.* **47**, 152–176.

Wang, J. H. (1970). Oxidative and photosynthetic phosphorylation mechanisms. *Science* **167**, 25–30.

Warburg, O. (1913). Über sauerstoffatmende Körnchen aus Lebercellen und über saurerstoffatmung in Berkfeld-filtraten wässriger leberextract. *Pfluegers Arch. Gesamte Physiol. Menschen Tiere* **154**, 599–617.

Warburg, O. (1926). Über die wirking von Blausäueräthylester (Äthylcarbylamin) auf die Pasteursche reaktion. *Biochem Z.* **172**, 432–441.

Warburg, O. (1930a). "The Metabolism of Tumours." Constable, London.

Warburg, O. (1930b). "The Metabolism of Tumours," p. 236. Constable, London.

Warburg, O. (1956). On the origin of cancer cells. *Science* **123**, 309–314.

Warburg, O. (1964). Prefatory chapter. *Ann. Rev. Biochem.* **33**, 1–14.

Warburg, O. (1966a). Tatigkeitsbericht der Max-Planck-Gesellschaft; Max-Planck-Institut fur Zellphysiologie, Berlin-Dahlem. *Z. Angew. Chem.* **53**, 697.

Warburg, O. (1966b). Oxygen, the creator of differentiation. *In* "Current Aspects of Biochemical Energetics" (N. O. Kaplan and E. Kennedy, eds.), pp. 10–109. Academic Press, New York.

Warshaw, J. B. (1969). Cellular energy metabolism during fetal development. *J. Cell Biol.* **41**, 651–657.

Weber, G. (1968). Carbohydrate metabolism in cancer cells and the molecular correlation concept. *Naturwissenschaften* **55**, 418–429.

Weinbach, E. C., Garbus, T., and Sheffield, H. G. (1967). Morphology of mitochondria in the coupled, uncoupled and recoupled states. *Exp. Cell Res.* **46**, 129–143.

Wenner, C. E. (1959). Oxidation of reduced triphosphopyridine nucleotide by ascites tumor cells. *J. Biol. Chem.* **234**, 2472–2479.

Wenner, C. E. (1967). Progress in tumor enzymology. *Advan. Enzymol.* **29**, 321–390.

Williamson, J. R. (1965). Metabolic control in the perfused rat heart. *In* "Control of Energy Metabolism" (B. Chance, R. W. Estabrook, and J. R. Williamson, eds.), pp. 333–346. Academic Press, New York.

Williamson, R. L., and Metcalf, R. L. (1967). Salicylanilides: A new group of active uncouplers of oxidative phosphorylation. *Science* **158**, 1694–1695.

THE GASEOUS ENVIRONMENT OF THE

MAMMALIAN CELL IN CULTURE

William F. McLimans

> . . . it seems a crime that such a gas
> as CO_2 . . . be essential for respiration
> and growth.
> HANKS, *1955*

I. Introduction

The gaseous environment of the mammalian cell in culture is an important experimental parameter for those interested in the kinetics of *in vitro* cell growth and function, spontaneous transformation, the isolation of new types of viral agents, the biosynthesis or bioconversion of physio-logically active substances, etc. Illustrative examples include (1) the altered reactivity of viral infected cells or animals held under various gas environments as reported for influenza (Magill and Francis, 1936; Kalter and Teppermann, 1952; Berry et al., 1955; Kalter et al., 1955; Sawicki et al., 1961), hog cholera (Segre, 1964), mengovirus (Giron et al., 1967), and Theiler's GDVII virus (Pearson, 1950); (2) the effects of altered gas environment on interferon production in mice (Huang and Gordon, 1968) or of growth hormone elaboration in cultures of human anterior pituitary cells (Gailani et al., 1970); and, finally, (3) its possible role in the selection of tumorigenic clones of cells (Gotlieb-Stematsky et al., 1966) or, indeed, in the etiology of the malignant state itself (Mottram, 1927; Loomis, 1959a; Kieler and Bicz, 1962).

It is not the purpose of this chapter to detail possible applications of altered cell reactivity as catalyzed by changes in the gaseous environ-ment of the cell in culture, but rather to attempt to explore the physio-logical basis for the effects of the gaseous overlay on the growing cell in vitro. One is concerned here not only with the concentrations of the components of the gas phase in cell culture but also with the equi-librium and stability of this phase during continuous culture.

The stability of the culture milieu has been somewhat neglected in studies of the primary cell in culture. This is not surprising when one considers the complexity of the problem of maintaining, in the culture flask, a constant cellular environment, as well as the dramatic progress, particularly in the field of virology, that has been made through the use of the available uncontrolled culture systems. Justifiably, little con-cern was warranted over the excessive and ever-changing transient states in the culture environment, nor could the "sameness" or lack of correla-tion of the cell in culture to its types or characteristic in the tissue of origin serve as a valid basis to delay the more exciting search for new viruses. Yet, from the beginning, there have been hopes that obser-vations of the cell in culture would yield information applicable to spe-cific interpretation of physiological events as they occur in the tissue of origin—the pathology of disease, stress, aging, etc.—a hope not yet realized. In fact, it may be suggested that the realization of the latter goals may well result in medical advances equal to or surpassing the already impressive accomplishments achieved through application of tis-sue culture techniques.

Conventional cell culture techniques offer a system in which many of the cells present in the tissue of origin are traumatized or destroyed during the process of tissue dissociation (Anderson, 1953; Kaltenbach, 1954; Laws and Strickland, 1956), freshly isolated cells appear to be "metabolically smothered" through loss of key enzymes (Laws and Strickland, 1956; Kalant and Young, 1957; Lata and Reinerston, 1957; Zimmerman *et al.*, 1960), wide and uncontrolled variations in the physical and chemical characteristics of the culture medium occur (Westfall *et al.*, 1955; Schwartz, 1960), and growth and metabolic regulating factors, so important in the body, are little understood or applied in the culture system. It is not surprising, therefore, that relatively few of the cells found in the tissue of origin find these cell culture systems compatible with growth and function (Zaroff *et al.*, 1961; Pious *et al.*, 1963), a point previously commented on by Levintow and Eagle (1961) and Morgan (1958).

Interestingly enough, the few cells that do survive the traumatic experience of dissociation and uncontrolled culture environment appear to be characterized by a high degree of "sameness." Eagle has noted, "The surprising aspect . . . has been that the nutritional requirements and metabolic activities associated with the growth of mammalian cells in culture have been so basically similar, regardless of the species or the tissue from which those cells had been derived" (Eagle, 1960).

It thus appears that success in the majority of conventional cell culture techniques is based on the ability and apparent "ruggedness" of a few selected cells in the tissue of origin to withstand the dissociation procedure and grow *in vitro,* in spite of the rather large variations in the *milieu interieur.* Whether or not these uncontrolled conditions appear as carcinogenic triggers remains to be proved.

II. Physiological Parameters

Over the past few years, we have attempted to study some of the physiological parameters involved in the culture of primary mammalian cells of liver and anterior pituitary tissues. The objective has been to devise a methodology which would expose the cell to the minimum possible physiological stress in its isolation from the tissue and its transfer to a controlled environment system, which, in turn, would minimize the transient states of the culture milieu. In brief these studies included the following observations.

A. Dissociation of the Cell from Its Tissue of Origin

In the instance of liver tissue, a number of conventional procedures commonly employed for tissue cell dissociation were evaluated and found to be unsatisfactory for our purposes. A detailed study of the role of osmolarity during cell dissociation suggested that hyperosmolar solutions could be used to dissociate liver cells in an acceptable manner (McLimans, 1969). However, recent observations in our laboratory (Marcelo et al., 1971) confirm the important observations of Barnard et al. (1969) regarding the efficacy of purified trypsin (lyophilized Tryptar, Armour Pharmaceutical Co.) as an agent for dissociation of mammalian cells from their tissue of origin. Superior results have been consistently achieved employing 6250 N.F. units/ml as diluted in an appropriate serum-free media (McCoys) adjusted with $NaHCO_3$ to yield a pH of 7.4 under 2% CO_2 with the osmolarity held at 300 mOsm. The tissue fragments are washed and placed in thin films of the Tryptar-medium solution and incubated at 37°C under flowing 2% CO_2 for a sufficient time to permit complete penetration of the tissue, i.e., liver, 3 hours; anterior pituitary, 20–24 hours. Gentle and minimal pipetting assures a high yield with even dispersion and cell viabilities, as judged on trypan blue exclusion, of 80–100%. Preliminary experiments suggest that low concentrations of trypsin under the above conditions appear as efficacious for dissociation of mammalian cells as Tryptar.

B. Culture Surface

Growth of differentiated hepatic cells has been consistently observed only on collagen surfaces. Only minimum growth has been observed on glass or plastic surfaces (McLimans et al., 1968b).

C. Controlled Environment Culture System

Prior reports (McLimans et al., 1966; McLimans, 1969) have detailed the controlled environment systems that we have developed. Experience with these prototypes has indicated that the most reliable system appears to be one operated on a proportional medium feed to a culture, as based on time (1–6 hours) or as sensed by a parameter such as pH. Such a modified system has been developed incorporating a thin-film culture unit with a collagen surface continually exposed to a constant gaseous environment. This unit has been in operation in our laboratories for over a year with reasonable reliability of continuous and flexible operation (Figs. 1 and 2).

The essential components of the system include a medium reservoir of a volume sufficient for a feeding schedule of 1 or 2 days. An appro-

Fig. 1. The described constant environment culture system. Note three independent incubators, contained "rocker racks," and gas "feed" control panel.

priate vaccine-stoppered port is available to replenish the reservoir medium on a daily basis. Upon demand, air pressure is applied to the reservoir, forcing the medium to flow gently over the bottom surface of the upper manifold; the pressure is released and the medium drains back to the reservoir. In the medium transit over this surface, approximately 0.4 ml is deposited and retained in five glass tubing traps, equally spaced along the bottom of the manifold. The effluent end of the trap extends to the bottom of each of five culture flasks. A drain tube from each flask passes over an inexpensive pneumatic valve constructed of

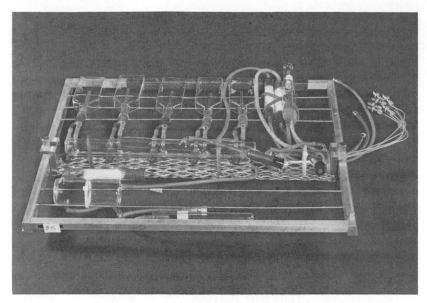

Fig. 2. Details of rocker rack containing two manifolds, five culture flasks, and balloon valves on drain side as operative through lower manifold. Note vaccine-stoppered media reservoir as connected to upper manifold and effluent collection reservoir mounted in front of the tray.

glass with a rubber balloon insert. When the balloon is collapsed, the drain is open; slight air pressure expands the balloon, sealing the drain. Sequential operation of these valves controls the pressure to the medium reservoir and permits gassing of one flask for a finite time during each cycle. When the drain is open, gas forces medium from the trap into the flask just prior to the rock cycle. A mixing with the culture medium in the flask occurs, and on the next rock excess medium is drained off to the lower manifold and carried via air pressure to a collecting reservoir. The collecting reservoir may be tapped at daily intervals for analysis as desired. The entire unit is synchronously programmed and electronically controlled by an ATC timer. It is thus feasible to replenish about 10% of the culture medium at any desired time interval down to 1–2-hour cycles. This procedure, when coupled with substrate analysis of effluent media, assures a reasonably constant level of nutrients in the culture flask at all times as well as constant removal of cell products, i.e., lactate, hormones, etc. It should be noted that some protection of thermolabile components is achieved since maximum exposure of the culture medium to 37.5°C is limited to 1–2 days. The cultures are gassed for 1 minute of each 3-minute cycle with the objective of achieving a better control of evaporation and thus of osmolarity.

D. Oxygen and Gas Diffusion

The role of oxygen and gas diffusion in culture systems has been extensively detailed in prior reports from this laboratory (McLimans *et al.*, 1968a,b). It seems sufficient to mention that mathematical models have been constructed which relate the depth of the culture fluid over-lay to the oxygen available to mammalian cells cultured under static conditions (Figs. 3 and 4). These models suggest that the maintenance of a given rate of oxygen utilization by some culture systems may be critically dependent on this fluid depth and on the solubility and rate of diffusion of oxygen in the culture fluid. The importance of these concepts as applied to the isolation and growth of normal cells representative of the tissue of origin is noted (Figs. 3 and 4).

Further, it has been demonstrated experimentally that the thickness of fluid overlay in conventional tissue culture systems limits the oxygen

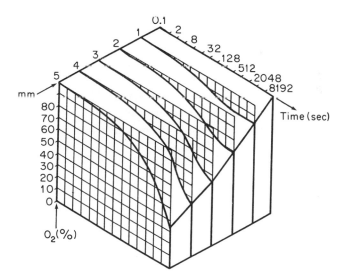

Fig. 3. Kinetics of gas diffusion as related to mammalian cells in culture demonstrating the relationship of culture fluid depth (x) vs. time in culture (t) vs. the percentage of oxygen remaining at the plane $x = 0$. Thus, for example, at 2-mm solution depth 93% of the original equilibrated concentration of oxygen will be present after 128 seconds of respiration of the monolayer.

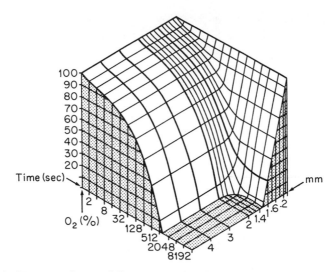

Fig. 4. Kinetics of gas diffusion as related to the normal liver cell *in vivo*, demonstrating the relationship of culture fluid depth (x) vs. time in culture (t) vs. percentage of oxygen remaining at the plane $x = 0$. Thus, for example, with a solution depth of 1.6 mm the total initial equilibrated concentration of oxygen will be exhausted in 1280 seconds.

available to mammalian cells growing as a submerged monolayer. A rocker culture system has been described which circumvents critical problems associated with thin-film culture while permitting nearly unlimited access of oxygen to the cell monolayer (McLimans *et al.*, 1968a; Tunnah *et al.*, 1968). Good growth of primary hepatic cells as isolated sheets has been obtained (Fig. 5) (McLimans, 1969).

Experimental studies employing the described system led us to the conclusion that the gaseous environment is indeed one of the vital factors involved in the growth of anterior pituitary and liver cells *in vitro*, and also that the most interesting component of the gaseous environment of the cell in culture appears to be CO_2.

III. Carbon Dioxide

Powers suggested in 1927 that ". . . pCO_2 is the most troublesome of all biologic factors to measure and any variable that is hard to measure is likely to be neglected" (Powers, 1927). While this state of affairs has changed over the years, it still seems that in many fields relatively

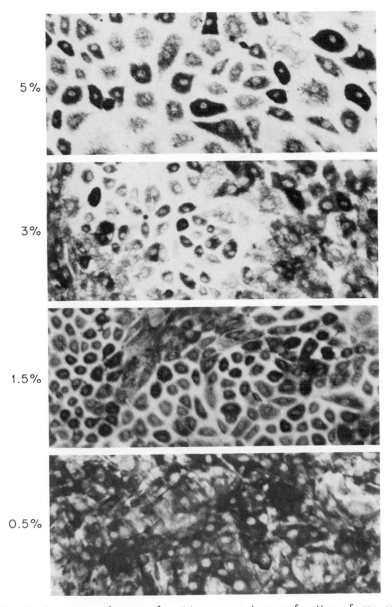

5%

3%

1.5%

0.5%

Fig. 5. Increasing glycogen deposition as an inverse function of gas overlay concentration of CO_2. Thin film, rocker culture of fetal mouse liver cells, 14 days in culture—NCTC 109, 20% fetal calf sera. Cultures were maintained under steady-state conditions with pH 7.28–7.4, with 5, 3, 1.5, and 0.5% CO_2. pH achieved by varying the bicarbonate level. Gomori's hexamine silver histochemical stain. 400×.

little investigational effort has been directed toward the role of CO_2 as influencing the *in vitro* growth and function of mammalian cells. It is hoped that the following brief account, as based on selected works from diverse areas and disciplines, illustrates the importance of pCO_2 in the *in vitro* growth and metabolic regulation of cells in culture over and above that of the more obvious role of pH control of the culture milieu.

A. Background

About the time that Sörenson (1909) published his classic paper on enzymatic activity as influenced by hydrogen ion concentration and introduced the term pH, Henderson (1908) delineated the relationship between CO_2 and pH. Physiological evidence of its significance was presented about 10 years later by Van Slyke via his extensive studies on the normal and abnormal variations in the acid-base balance of the blood (1921), as well as by Shock and Hastings (1935). Even at that early date it was clear that the regulation of CO_2 appeared equally important to the pH control of the body's economy and/or intermediary metabolism.

When CO_2 is present in the atmosphere over a cell culture, a partial pressure of CO_2 exists which, in time, equilibrates with the aqueous media to form H_2CO_3 and CO_2. The amount of free CO_2 under these conditions is far greater than the concentration of H_2CO_3 (Bull, 1943). According to Buytendyk *et al.* (1927), the ". . . greater part of the physically dissolved CO_2 at room temperature does remain as the anhydride, only a small portion changing to H_2CO_3. The convention by which the total hydrated plus anhydrous CO_2 is at the same time called 'physically dissolved' and formulated as H_2CO_3 is inconsistent and inexact. The convention is, however, convenient, and causes no errors if the H_2CO_3 symbol is used consistently with the same meaning." The H_2CO_3 ionizes to yield the negatively charged carbonate ion and H^+. Thus, while a gradient exists, it is apparent that the concentration of H_2CO_3 is proportional to the concentration of free pCO_2.

The dissociation equilibrium relating the interacting constituents is

$$CO_2 + H_2O \rightleftharpoons H_2CO_3 \rightleftharpoons H^+ + HCO_3^-$$

Employing the equilibrium constant, one obtains

$$K_{[H_2CO_3]} = \frac{[H^+][HCO_3^-]}{[H_2CO_3]}$$

Solving for [H+] yields the Henderson equation*:

$$[\text{H}^+] = K_{[\text{H}_2\text{CO}_3]} \frac{[\text{H}_2\text{CO}_3]}{[\text{HCO}_3^-]}$$

Since pH $= \log 1/[\text{H}^+]$ and pK $= \log 1/K$, Hasselbalch could therefore transform the Henderson equation to a logarithmic form, enabling him to solve for pH rather than [H+]:

$$\text{pH} = \text{p}K + \log \frac{[\text{HCO}_3^-]}{[\text{H}_2\text{CO}_3]} \qquad \text{Henderson-Hasselbalch equation}$$

The Henderson-Hasselbalch equation is fundamental to our understanding of any quantitative treatment of acid-base equilibrium when applied to the bicarbonate system.

The equation as modified by Loomis clearly illustrates the relationships which exist between pCO_2, pH, and concentration of bicarbonate (Loomis, 1959b). Since, according to Henry's law,

$$[\text{CO}_2] = \alpha\text{-pCO}_2$$

where $[\text{CO}_2]$ is concentration, pCO_2 its pressure, and α the solubility coefficient of CO_2, the alternative equation may therefore be represented as

$$\text{pH} = \frac{\text{p}K + \log[\text{HCO}_3^-]}{\alpha - \text{pCO}_2} \qquad \text{Loomis equation}$$

In any cell culture system, the growing cells serve both as an oxygen sink—oxygen is being removed—and as a CO_2 pump—CO_2 is being added to the medium. Thus, while the CO_2 is attempting to establish a physical equilibrium between the $\text{pCO}_2(\text{air})$-$\text{pCO}_2(\text{culture media})$, the cells are pumping CO_2 into the media, tending to distort the equilibrium. The net result, from a practical viewpoint, is the establishment of a pCO_2 gradient which gradually decreases as it reaches the pCO_2

---

* Current knowledge (Waddell and Bates, 1969) suggests that the practical pH value of a solution is a measure of neither hydrogen ion concentration nor activity. In reality, pH is defined in terms of the following equation.

$$\text{pH}(x) = \text{pH}(s) + \frac{F}{2.3026RT} (Ex - Es)$$

where $\text{pH}(s)$, pH assigned to standard solution; F, Faraday; R, gas constant; T, temperature in degrees Kelvin; Ex, unknown electrode; Es, reference electrode; and E, electromotive force of pH cell in which an electrode reversible to H^+ is coupled with a reference electrode.

of the air overlay. The magnitude of this pCO_2 gradient appears dependent on the concentration of cells per square millimeter on the flask surface, the type and metabolic rates of respiration of these cells, the rate of diffusion of the gas, and the thickness of the media through which the gas must diffuse. Obviously, the final results are going to be markedly different depending on whether the flask is stoppered or whether an open—cotton plug—system is utilized. In the stoppered cell culture flask, the pCO_2 (both air and media) will continue to increase with a concomitant rise in H_2CO_3, lactate, etc., and a drop in pH until a level of acidity is reached which terminates the cell respiration and production of CO_2, with realization of the final true equilibrium—an undesirable state of affairs to those interested in cell culture. Periodic (2–4-day) removal of the stopper during media replacement reduces the pCO_2 to lower levels with the result that this ever-changing transient state becomes less dramatic, yet continues as cyclic over the life span of the culture. In shaker or roller cultures the pCO_2 internal gradient becomes almost negligible, yet the necessity of stoppering the tubes still results in an ever-changing pCO_2 unless the roller apparatus is placed in a CO_2 incubator with flowing CO_2 and cotton tube stoppers. Under the described conditions of cyclic variations in the CO_2 levels between media changes, the cells are no doubt stressed, yet the effects of this stress on cell type or function remain to be appreciated.

In cell culture flasks which have been closed with a gas-impermeable seal, i.e., a stopper, the course of events dramatically illustrates the rapidity of the transient states which develop during incubation. The important data of Searcy et al. (1964) have been summarized by the author and are presented in Table I.

TABLE 1.

EFFECT OF INCUBATION TIME ON pCO_2 pH, AND ACTUAL BICARBONATE LEVELS OF MONKEY HEART CULTURES IN STOPPERED FLASKS

Incubation time (hr)	pCO_2 (mm Hg)	pH	Bicarbonate (mEq/liter)
0	33.6[a]	7.31[a]	16.6[a]
0.25	37.3	7.26	16.6
2	41.6	7.23	17.0
60	56.0	6.99	13.3

[a] Author's average of data from three flasks, as presented by Searcy et al. (1964).

Searcy *et al.* point out that *in vitro* culture of cells by similar procedures subjects those cells, within 2 hours, to acid-base changes analogous to *in vivo* development of metabolic as well as respiratory acidosis. Thus, for example, between 2 and 60 hours of incubation, the average decrease rate in pH is equivalent to 0.004 unit/hour—a rate which at 60 hours yields a pH level not well tolerated *in vivo*.

From the above, it is apparent that a controlled environmental system for the *in vitro* culture of mammalian cells appears mandatory if one is interested in the study of the morphology and function of the normal cell, let alone the precise role of pCO_2 as a potential regulator of cell metabolism. This may explain why relatively little work has been done with the view of delineating the role of CO_2 in metabolism, growth, and development of the mammalian cell in culture. Great technical difficulties are encountered when one attempts to maintain cultures under precisely controlled environmental conditions wherein, for example, pH and substrates are held constant while pCO_2 is varied. On the other hand, sufficient work has been done to demonstrate the exciting possibilities inherent in studies of pCO_2 as a regulator of cell metabolism.

It should be emphasized that the reports of the vital role of CO_2 per se may be a semantic oversimplification since intracellular pH as affected by the environmental pCO_2 may be as important as the actual fixation of CO_2. Calculations of intracellular pH have been made from measurements of the distribution of CO_2, 5,5-dimethyl-2,4-oxazolidinedione, ammonia, bromphenol blue, 2,4-dinitrophenol, rates of pH-dependent reactions, etc. (Waddell and Bates, 1969). It seems clear that it is hazardous to predict the intracellular pH from a knowledge of the extracellular pH alone. Further, the weight of evidence suggests that the intracellular pH of most animal cells is approximately 7.0 under normal conditions *in vivo*. The intracellular water has been reported to be very sensitive to acidification from increasing environmental pCO_2, while the addition of highly ionized acids or bases—including HCl and $NaHCO_3$—to the extracellular water causes only a negligible effect on the pH of the intracellular water (Waddell and Bates, 1969). In the authors' view, actual intracellular pH may not be best expressed as being poised at a particular level but rather as representing an average pH which masks the finite pH gradients which may exist throughout the cell—gradients which would, in the final expression, be zones of altered charge reactivities. Differential staining of cell structures by acid and basic dyes suggests, perhaps, the existence of such gradients. Such a system, if true, would indeed be highly selective and responsive to changes in the cells' environmental pCO_2.

B. CO_2 and the Microbial Cell

Carbon dioxide has been shown to influence morphogenesis in microbes, higher plants, and animals. The effects of CO_2 on cellular growth and/or function range from sporangeal differentiation in *Blastocladiella* (Cantino, 1961), mold yeast dimorphism (Drouhet and Mariat, 1952; Bartnicki-Garcia and Nickerson, 1962), spherulation in *Coccidiodes immitis* (Lones and Peacock, 1960), sporulation by bacteria (Powell and Hunter, 1955) and fungi (Chin and Knight, 1957; Niederpruem, 1963), root development by pea (Geisler, 1963), sexual differentiation in *Hydra* (Loomis, 1957), its role as a teratogenic agent in mammals (Haring, 1960) to CO_2 fixation as an important factor in energy metabolism for occyte maturation, and early zygote development (Biggers *et al.*, 1967; Wales *et al.*, 1969).

In 1927, Valley and Rettger proved that bacteria would not grow in the total absence of CO_2. Gladstone *et al.* (1935) and Rahn (1941) confirmed and extended these findings. The inhibition of tubercle bacilli growth in the absence of CO_2 (Wherry and Ervin, 1918), the role of CO_2 in successful growth of *meningococci* (Cohen and Fleming, 1918), and the facilitation of growth of *gonococci* by CO_2 (Chapin, 1918) all contributed to our early understanding of the biological importance of CO_2.

There are numerous examples of the apparent regulatory role of CO_2 in the growth and function of microbial cells. It is known that the growth of many bacterial cultures are inhibited below a minimum concentration of CO_2 (Werkman, 1951). Osterlind (1948) recorded over 60% inhibition of growth of *Scenedesmus quadricauda* in the presence of 15 mmoles/liter of CO_2. Growing yeast and bacteria fix 5–10 times the CO_2 rate that resting cells do (Liener and Buchanan, 1951; McLean *et al.*, 1951). CO_2 has been observed to increase the respiratory rate even for resting cells (J. O. Harris, 1954). Generally, high concentrations of CO_2 inhibit the growth and formation of by-products by molds (Foster, 1949). Yet exceptions occur, as in the instance of acceleration of growth of some molds by air overlay concentrations of up to 20%. In aerated liquid cultures CO_2 has been demonstrated to influence the rate of spore germination (Nyiri, 1967), the development of *hyphae* (Nyiri and Lengyel, 1968), and the rate of carbohydrate uptake and penicillin biosynthesis (Nyiri and Lengyel, 1965) of *Penicillium chrysogenum*. In the biosynthesis of polysaccharides by *Streptococcus bovis*, no other culture condition appears so important as the enrichment of the air overlay with 5% CO_2 (Dain *et al.*, 1956). In this instance Prescott *et al.* (1965) demonstrated the utilization of CO_2 and acetate in amino acid synthesis. Inhibition of photosynthesis in dense algal cultures by CO_2 levels of 3% has been reported (Ammann and Lynch, 1967). Similar effects have been noted at lower CO_2 levels, with more dilute algal concentrations (Osterlind, 1948; Steemann, 1955). Inhibitory effects of high levels of CO_2 in several reactions associated with the Krebs cycle have also been noted (Walker and Brown, 1957; Ranson et al., 1960).

Subsequent studies on the role of CO_2 and bacterial growth, based on the inhibition of growth of many bacterial species in the absence or at minimal levels of CO_2, suggested that CO_2 was required for growth of all living cells (Sorokin, 1962). While this is no doubt correct, the interesting question arises as to which stage of cell growth or development is responsive to the presence of CO_2 in the culture milieu. Sorokin uncoupled cell division from growth and observed not only that CO_2 was unnecessary in the process of cell division of Chlorella but also that it acted as a powerful inhibitor (Sorokin, 1962). This inhibitory action appeared to depend on the presence and concentrations of other ions, particularly the molar concentration of the bicarbonate. However, Sorokin noted that above a critical level of CO_2 concentration (10^{-2} M) the inhibitory action of CO_2 on cell division could not be suppressed by bicarbonate and CO_2 acquired a major role in regulating cell division. Since CO_2 may reach high levels of concentration in multicellular organisms, Sorokin suggested that a careful consideration of its role in normal and neoplastic growth might be rewarding. Mer and Causton (1965), while supporting the vital role of CO_2, seriously questioned the data of Sorokin since a concomitant pH effect may be related to the role of bicarbonate.

The demonstration of fixation of CO_2 by heterotrophic organisms was first made via the classic studies of Wood and Werkman (1935). Since that time about a dozen distinct reactions and enzymes have been found to contribute to the enzymatic basis of CO_2 fixation (Wood and Utter, 1965). Wood and Utter (1965) suggested that "CO_2 fixation reactions play an important role in a variety of metabolic areas and that considering the variety and essential nature of these processes it may be stated that most if not all heterotrophic cells are dependent on CO_2 fixation." They further reviewed the mechanisms of CO_2 fixation and noted that they encompass fixation by reduction, by phosphoenolpyruvate carboxykinase, and by biotin-containing enzymes as well as fixation to noncarbon atoms such as in the formation of carbamyl phosphate. The importance and scope were magnified when it was recognized that CO_2 fixation plays an important role in the synthesis of glucose and glycogen by liver (Krebs, 1954; Utter, 1959), as well as in fatty acid synthesis by other cells (Wakil, 1962).

Lachica (1968) reviewed the role of CO_2 in carboxylation reactions, in assimilation of CO_2 and its subsequent incorporation into the carbon skeleton of the amino acid molecules, in replenishing the Krebs cycle intermediates which are drained away during growth, in regulation of synthesis of Krebs cycle intermediates and aspartic acid, and in the important *de novo* synthesis of fatty acids via malonyl-CoA from CO_2 and acetyl-CoA. The effects of CO_2 on protein structure by transforming uncharged amino groups into negatively charged carbamate groups (Hastings and Longmore, 1965) have also been noted.

C. CO_2 as a Regulator of Mammalian Cell Activities

If a multipurpose environmental "master switch"—to borrow Scholander's term as first expressed during his early studies of the physiology of CO_2 in mammals (Scholander, 1963)—existed for the metabolic control of the mammalian cell, it could be postulated that, among other attributes, it would have to possess the following characteristics: (1) The substance would have to be capable of very rapid diffusion through extracellular fluids as well as into and out of the cells themselves. Generally, the efficiency of any feedback control system is proportional to the rate or speed of the response and adjustment to the sensor signals. (2) An equilibrium concentration of the substance would have to be present and be of such a nature that shifts from optimum levels of reactivity would bring about a stimulation or inhibition in the desired manner. In this instance reference to a single entity may be marked oversimplification as the ultimate control may be via competitive inhibitors or fluctuating gradients, as previously noted in the discussion of intracellular pH. (3) The proposed regulator of the mammalian cell should be demonstrated to be not only essential to a variety of key metabolic interactions but quantitatively reversible within physiological limits without inherent toxicity. (4) Finally, the interrelationship of basic biological systems from microbe to man suggests that the theoretical regulator should be demonstrated to be operative in a wide range of organisms.

The most promising candidate, that which fulfills most of the requirements as an environmental regulator of prime importance to cell growth and function, appears to be CO_2.

The first suggestion of CO_2 as a metabolic regulator was perhaps that of Lillie (1909), who theorized that CO_2 was the reaction product whose rate of removal from the cell determined the velocity of the chemical processes concerned in stimulation and contraction. Some years later, Bauer (1925) observed that CO_2 appeared to depress cell division in a fibroblast culture with apparent inhibition during prophase but little or no effect during telophase. Haywood and Root (1930) were led to state that "In two respects CO_2 claims a place of peculiar interest for the physiologist. First as a metabolic product, it is present wherever there are active cells. Second, as many workers have observed, it has the effect of very promptly depressing certain cell activities." They suggested that CO_2 was a naturally produced physiological inhibitor of cell division in marine eggs, causing an inhibition that was reversible and thus nontoxic in its kinetics of inhibition. This confirmed and amplified the observations of Smith and Clowes (1924), who noted that the rate of cell division in marine eggs was reversibly and linearly arrested by elevating the pCO_2. CO_2 has been shown to affect embryonic differentiation in *Fundulus* (Trinkaus and Drake, 1956), *Rana pipiens* (Flickinger, 1958), and chick embryo (Spratt, 1949). High levels of CO_2 are known to inhibit metabolism and respiration of sperm (Salisbury *et al.*, 1960) and other tissues (Danes and Kieler, 1958; Loomis, 1959a; Kieler and Bicz, 1962).

Warburg *et al.* (1924) were the first to demonstrate a stimulatory effect of CO_2/HCO_3^- on anaerobic glycolysis in slices of Flexner-Jobling rat carcinoma. Fleeger *et al.* (1967) concluded, from their studies of testis tissue, that a minimal level of CO_2 was essential in order to maintain maximum glycolysis and respiration but that high levels inhibited these metabolic systems. Craig and Beecher (1943), under conditions of constant pH, noted a twofold increase in the rate of glycolysis and respiration of rat retina when the CO_2 of the gas phase was raised from 1 to 5%. A further increase in CO_2, up to 20%, did not affect the rate of glycolysis, although the rate of oxygen consumption was decreased. Danes and Kieler (1958), employing L strain mouse fibroblasts and Yoshida rat ascites tumor cells, demonstrated maximal respiration in the range 0.5–2% CO_2 in the gas overlay. The maximal respiration increase, as compared to room air (0.03% CO_2), was 32% for L cell and 60% in the Yoshida cell system. On the other hand, Bicz (1960) could not determine any differences in the response of normal and malignant leukocytes to variations in CO_2 concentrations, although the optimal CO_2 concentration, judged on respiratory rate, appeared to be in the range 1–2%, as opposed to the observation of Seelich *et al.* (1957) of a stimulatory effect of CO_2 on normal but not on leukemic leukocytes.

Nazario and Reissig (1964) suggested that CO_2 was involved in the biosynthesis of arginine and the pyrimidines as a precursor of carbamyl phosphate, which, in turn, donated carbamyl residues in reactions catalyzed, respectively, by ornithine transcarbamylase and aspartic transcarbamylase. Some responses to CO_2 observed in *Neurospora* mutants deficient in the synthesis of arginine or pyrimidines (Charles, 1962; Reissig and Nazario, 1962) were not explainable in terms of the role of CO_2 as a carbamyl phosphate precursor.

Mayfield *et al.* (1967) suggested that bicarbonate was incorporated into carbamylaspartate and into the C-2 of uracil nucleotides in Novikoff ascites tumor cell and that the incorporation was increased by both glutamine and ammonia.

T. G. Cooper *et al.* (1968) demonstrated that the active species in carboxykinase and carboxytransphosphorylase reactions was CO_2 and not bicarbonate. Bicarbonate appeared to be the active species in pyruvate carboxylase reactions.

Evans and Slotin (1940) reported that in pigeon liver mince, CO_2 was fixed via pyruvate to form α-ketoglutarate. Solomon *et al.* (1941) demonstrated that CO_2 was fixed in liver glycogen to the extent of about 11%. Studies conducted under carefully controlled experimental conditions led Hastings and Dowdle (1960) to conclude that ". . . CO_2 molecules are just as important to the metabolism by the liver as are other types of molecules—and that homeostasis of the CO_2 tension is just as important as pH—perhaps even more so." In fact, in their studies they were surprised to find such a great effect of CO_2 on glycogen and fatty acid synthesis when the CO_2 variations were held within the physiological range. Thus they noted that increasing the HCO_3^- concentration from 10 to 40 mM at pH 7.4 produced as large an increase in glycogen synthesis from glucose as did a corresponding increase in pH from 7.1 to 7.7 at constant HCO_3^- concentrations.

Longmore (1966), a colleague of Hastings, summarized their important studies as being centered around the effect of CO_2 concentration on rates of metabolism *in vitro* of rat liver slices. The conversion of glycerol to glycogen was profoundly affected by the level of CO_2/HCO_3^- and pH, the stimulatory effect of increased CO_2 levels in isolated rat liver mitochondrial ATPase activity suggesting that CO_2 effects on glycogen synthesis were related in part to an increase in energy available for that purpose.

Further studies (Hastings and Longmore, 1965) suggested that the CO_2 effect might operate at the phosphorylation step of glucose, fructose, and glycerol. Raising the CO_2 concentration at pH 7.4 increased the phosphorylation of glucose 70%. Whether this was due to an effect of CO_2 on the activity of hexokinase or glucokinase or to an increase in ATP activity was not resolved. That ATP turnover may be involved was suggested by Fanestil *et al.* (1963), who, under carefully controlled conditions, noted that the presence of HCO_3^- (0.06 M) led to a marked increase in the rate of ATPase activity. This occurred when CO_2-N_2 or CO_2-O_2 gas mixtures were used. No difference in ATPase activity was noted between oxygen and air. Racker (1962) had previously noted that bicarbonate stimulates the ATPase activity of a protein preparation from beef heart mitochondria.

D. CO_2 and the Culture of Mammalian Cells

In any discussion concerning the role of CO_2 in cell culture systems, emphasis should again be placed on the difficulty of delineating whether the observed effect is due to pCO_2, HCO_3^-, and/or intracellular pH. Further, it must be brought out that the known existence of transient states in conventional cultures, as opposed to physiological steady-state systems, intensifies the problem of discerning the true role of CO_2.

Conventional cell culture methodology usually employs one of a number of apparent buffer systems for the control of pH of the culture medium. In many instances the buffer chosen is the carbon dioxide-carbonate system (CO_2/HCO_3^-). Some investigators consider the Co_2/HCO_3^- system to be unstable and thus inconvenient as a buffer system. While this view is certainly correct in instances involving cell or media manipulation as in preparation of dilutions or plating of cells, it appears less than realistic when one considers two aspects of this problem. The apparent instability of the CO_2/HCO_3^- system may be an advantage in that under steady-state conditions it permits rapid directional pH control and adjustment via variation of either the gas or its substrate. Perhaps more important than the role of CO_2/HCO_3^- as a buffer is its role in influencing the metabolic equilibrium of the cell. The use of other buffer systems may not be free of hazards. It has been recorded that tris buffer inhibits glycogen synthesis from glucose by liver slices *in vitro* and that the presence of CO_2 is required for the conversion of glucose to glycogen (Hastings and Longmore, 1965). Further, Ashford and Holmes (1931) demonstrated that the oxygen consumption of rabbit brain was 50% higher in bicarbonate as compared to a phosphate buffer, while Laser (1935) noted similar results employing rat retina.

The CO_2/HCO_3^- system has been considered essential for the propagation of mammalian cells (Carter *et al.*, 1967; Holland, 1967; De Wulf and Hers, 1968). Yet Gwatkin and Siminovitch (1960) observed that oxalacetic acid substituted for CO_2 when established cell lines were employed. However, oxalacetic acid could not substitute when primary cultures (monkey kidney) were employed. Thus some of the controversy regarding the CO_2/HCO_3^- system may have arisen as a result of the nature of the cultures employed for experimental observations.

About 45 years ago, Bauer (1925), and, later, Gomirato (1934), noted that the application of CO_2 appeared to inhibit the movement of mito-chondria in chick fibroblasts, particularly during cell division. As early as 1927, Mottram (1927, 1928) demonstrated complete inhibition of cel-lular outgrowth of cells from tissue explants of Jensen sarcoma and rat kidney when the pCO_2 of the gas overlay in the cultures exceeded 400 mm. Conversely, rat kidney proliferation was inhibited if CO_2 was removed from the gas overlay. Excellent outgrowth was obtained at intermediate levels of pCO_2.

Brues and Naranjo (1948) noted that ^{14}C from labeled bicarbonate was rapidly taken up by the cell in culture and retained in an organic form. Brues and Buchanan (1948) demonstrated that proliferating cells may fix a large amount of carbon from this source.

M. Harris (1952) suggested that bicarbonate appeared essential for outgrowth of chick heart fibroblasts as determined using dialyzed media. In a later publication, M. Harris (1954) extended these observations and noted that chick heart fibroblasts failed to proliferate in unsupple-mented bicarbonate-free media and did not develop uniformly below concentrations of 0.05%. At higher levels, the increase in outgrowth paral-leled the rise in bicarbonate concentration. He was perhaps the first to suggest that the primary role of bicarbonate, quite apart from its buffering action of extracellular fluids, is maintenance of intracellular pH values.

These observations are not restricted to primary cultures of the explant type but also apply to well-established cell lines (HeLa, conjunctival, Chang), as has been reported by Geyer and Neimark (1958). The latter investigators demonstrated that within 24 hours of removal of CO_2 from the culture environment, net multiplication ceased—the cause was as-sumed to be interference with nucleic acid synthesis—while the con-version of glucose and acetate to CO_2 continued. Of great interest was their observation that the addition of normal cell extract restored the net multiplication of HeLa and Chang, while the extracts of CO_2-de-ficient cells had an adverse effect on the deficient cells.

Swim and Parker (1958) concluded that CO_2 was essential for growth and maintenance of six permanent strains of fibroblasts as established in their laboratories. They demonstrated that in stoppered flasks the mammalian cells are capable of producing sufficient metabolic CO_2 to satisfy their nutritional requirement, as indicated by the fact that omis-sion of CO_2 from the culture milieu had no effect on the growth of the cells in stoppered flasks. These observations explain why Eagle (1956) could not demonstrate a CO_2 requirement for HeLa and L cells in stoppered flasks.

Mackenzie *et al.* (1961) established a simple procedure for controlling the pH of media employing a bicarbonate-carbonic acid system. Various pH levels of the culture media were achieved by varying the H_2CO_3 and equilibrating with 0.5–40% CO_2 in air. Periods of observation of growth rates of HeLa and Chang were very short (3 days) and made precise interpretation more difficult. However, they observed maximum growth over a pH range 7.38–7.87. Effects of pH on growth appeared independent of CO_2 concentration within limits studied. The accumulation of lipid-rich particles in acidosis also appeared to be independent of CO_2 concentration and growth rate. (For further discussion, see Chapter 8, Volume 1.)

An important contribution to the role of CO_2 as a potential regulator of the metabolism of the cell in culture was reported from Chang's laboratories (Geyer and Chang, 1958; Chang, 1959; Chang *et al.*, 1961) where it was demonstrated that over 90% of ^{14}C of $NaH^{14}CO_3$ fixed by HeLa or conjunctival cells was recovered in the acid-soluble and nucleic acid fractions. Most of the activity was associated with the isolated purines and pyrimidines of the cells in culture. Further, it was noted that a combination of ribonucleosides and oxalacetic acid provided conditions as good as or better than CO_2 for multiplication of CO_2-depleted conjunctival cells. The conclusion reached was that the chief function of CO_2 in human cells was to provide specific precursors for synthesis of purines, pyrimidines, and oxalacetate. It is unfortunate that these authors did not incorporate a primary culture in their experimental protocols. One recalls the failure of oxalacetate to substitute for CO_2 in primary monkey kidney (Gwatkin and Siminovitch, 1960).

In experiments seriously limited by the short range of incubation (2 days) Kieler and Gromek (1967) noted that the stimulating effect of CO_2 on cell growth could not be attributed exclusively to its effect on glycolysis and respiration as an energetic process in the cells. They suggested that enhancement of growth and development of tumor cells by CO_2 might be a result of the effect of CO_2 on NADP production in the course of fatty acid synthesis and, indirectly, on glucose metabolism. The optimal level of CO_2 appeared to be 5%, judged on the basis of fatty acid synthesis employing Ehrlich ascites carcinoma cells, spleen cells of mouse origin, mouse lung cells, embryonic mouse cells, L929 fibroblasts, and tetraploidal Ehrlich ascites carcinoma cells.

Gala and Reece (1963) employed rat anterior pituitary cultures (3-day) to demonstrate that maximum lactogen production was obtained in the presence of 5% CO_2, as compared to air or pure O_2, and, further (Gala and Reece, 1964), that an increase of $NaHCO_3$ resulted in a 19% increase in lactogen production. In our laboratories we have noted that a smaller amount of human growth hormone appeared to be elaborated from fetal anterior pituitary cultures held under 5%, as opposed to 2%, CO_2 overlay (Gailani *et al.*, 1970).

The usual cell cultures are incubated under about 5% CO_2 with appropriate $NaHCO_3$ concentration to achieve a desired pH (Parker, 1950). Five percent CO_2 apparently has been selected on the basis of its being the usual concentration found in the alveolar spaces of the lung. From the data herein presented, it appears that this may not be the optimal level. In fact, it is suggested that it may be much lower, perhaps in the 0.5–2.0% range. In any event, precise determination of the optimal level for various types of cultures and tissues certainly appears warranted.

The use of more than optimum levels of CO_2 as a gas overlay in cell culture systems may not be without its hazards, particularly in attempts to study the growth and function of the normal cell *in vitro*. Haring (1960) demonstrated that CO_2, unrelated to hypoxia, was a teratogenic agent capable of arresting or retarding growth. She exposed pregnant rats to 6% CO_2 for a single 24-hour period on the fifth day of pregnancy. Her experiments clearly demonstrated that these conditions produced offspring with a high incidence of cardiac malformations. There was evidence of temporary stimulation of differentiation and growth processes, i.e., localized tissue overgrowths found in lungs, thymus, and heart.

E. CO_2 and the Malignant Cell

This chapter is primarily concerned with the role of the gaseous environment of the cell in culture, stressing primary culture of normal cells. As such, the theoretical role of CO_2 in the etiology of the malignant state is outside its scope. On the other hand, a very brief review of this important area seems warranted since, in the author's opinion, the uncontrolled pCO_2 in the culture gas overlay may be a prime factor in many instances of so-called "spontaneous transformation" of cells in culture.

Mottram (1927) was one of the first investigators interested in the role of the gaseous environment and the malignant cell. He employed rat kidney cells, rat fibroblasts, and Jensen's sarcoma cell to demonstrate lack of growth in the absence of oxygen and a selective response to high and low levels of CO_2. The potential role of CO_2 and the malignant cell was advanced in a later paper (Mottram, 1928).

It has been suggested that neoplastic cells *in vitro* do not require a high atmospheric content of oxygen (Fischer and Andersen, 1926; Wright, 1928; Warburg, 1956a,b). Rovin (1962) has demonstrated survival of organ cultures of neoplastic tissue in atmospheres with as little as 1% oxygen. Conversely, a 5% concentration of CO_2 appeared essential for survival of some neoplasms *in vitro*.

While the role of CO_2 and its effects on various biological systems has been known for many years, it remained for W. F. Loomis to integrate these observations with his own investigations relating to the effects of pCO_2 on differentiation of *Hydra* (Loomis, 1957) and to propose a theory relating to the origin of the cancer cell (Loomis, 1959b). He suggested that free CO_2 was a physiologically produced, nontoxic, reversible inhibitor of cell growth; that normal adult cells limited their own growth by producing inhibitory levels of free CO_2 consequent to their own respiration; that, in essence, cytostasis was achieved by "damping the cells' furnace with a blanket of CO_2" (Loomis, 1959a); that, conversely, embryonic and malignant cells achieved the ability to grow continuously by using an ATP-generating system—glycolysis—which did not result in the production of CO_2; and that in the normal cell, cytostasis resulted from an insufficient supply of phosphate bond energy as a result of *in vivo* inhibition of respiration by the CO_2 produced by that cell. Thus it is known that cell survival appears to require about one fifth of the energy necessary for maximum cell growth (Warburg, 1956a). Cytostasis results from the 6–15% atmospheres of pCO_2 present in some adult mammalian tissue. It is on this basis that Loomis suggested (1959a) that four-fifths inhibition of respiration would generate enough ATP for maintenance but not a sufficient amount for growth. Essentially, he suggested that since CO_2 may act as a physiological regulator of cell multiplication, cancer may represent the *de novo* appearance of a cell which possesses metabolic reactions less susceptible to CO_2 inhibition than those of normal cells.

The brilliant studies from Kieler's laboratory (Kieler and Bicz, 1962), based on the relationship between pCO_2 and cellular energy metabolism, supported the critical role of CO_2 as visualized by Loomis. Kieler's observations led him to support Loomis' concept that CO_2 may act as a physiological regulator of energy metabolism and cell multiplication and, therefore, be important to their view of the malignant state. It was his view that the high glycolytic capacity of tumor cells presented a metabolic mechanism which caused tumor cells to be less susceptible to CO_2 inhibition than normal cells. Kieler pointed out that if the glycolytic capacity were sufficiently high, CO_2 might even have a stimulatory effect on total energy production and suggested that stroma cells might stimulate the growth of tumor cells by their production of CO_2

and a postulated factor which would enable tumor cells to utilize CO_2 in support of their glycolytic metabolism. (For further discussion, see Chapter 4, Volume 1.)

IV. Carbon Monoxide

Carbon monoxide (CO) has been of little concern to those working with cell culture systems. Accordingly, we were surprised to note that research-grade bottled CO_2, as commonly employed in cell culture laboratories, is certified to contain not more than 50 ppm CO. Conversely, one may assume that such bottled CO_2 may contain up to 50 ppm CO. Table II indicates an analysis typical of that made by a highly reputable manufacturer and the calculated level of CO that would be found in the gas overlay of a culture gassed with 5% CO_2 and 20% O_2.

It may be disputed that a level of up to 2.5 ppm CO or up to 0.8 ppm hydrocarbons (determined as methane) in the instance of 20% O_2 is of no significance. Yet caution is indicated when one considers that the maximum allowable atmospheric concentration or threshold value for occupational exposure of man in industry in the United States is 50 ppm CO for 8 hours and the recommended CO threshold level for workers at 5–8000 feet altitude is 25 ppm (Goldsmith and Landaw, 1968). Under these standards, recognition should be made of the protective barrier of hemoglobin which avidly binds CO and thus protects the *in vivo* cell and its contained cytochrome from the direct effects of exposure to large amounts of CO. In typical cell culture, no such protective barrier exists and CO is thus free to bind to the vital cellular elements, i.e., cytochrome (P-450). The potential net result in this in-

TABLE 2.

GAS CONTAMINANTS IN COMMERCIAL BOTTLED GAS

	Cylinders, research grade (ppm)		Estimated level in gas overlay of cell culture (ppm)
	O_2	CO_2	
Carbon monoxide		<50	5% CO_2 < 2.5 ppm
Nitrous oxide	<1		20% O_2 < 0.2 ppm
Hydrocarbons (as methane)	16		20% O_2 < 3.2 ppm

stance would perhaps be a depression in the rate of oxygen utilization or respiration of the naked cell. Further, the threat of N_2O and hydrocarbons as carcinogenic factors remains even if all CO is removed. For the above reasons a brief statement of the role of CO would seem a relevant point of discussion in this chapter.

It has been suggested (Goldsmith and Landaw, 1968) that a true CO cycle exists in nature since CO may be produced by plants and many lower animal species (Simpson et al., 1960; Wittenberg, 1960; Westlake et al., 1961; Loewus and Delwiche, 1963), may be utilized for metabolic purposes in certain bacteria and plants (Krall and Tolbert, 1957; Yagi, 1958), and may be oxidized to CO_2 in animals and man (Clark, 1950; Luomanmaki, 1966), as well as endogenously produced as a result of heme catabolism, etc.

The endogenous production of CO was first demonstrated by analysis of expired air of normal men (Sjöstrand, 1949). Sjöstrand (1949) further showed that CO was produced by the catabolism of heme in a fixed ratio. Ludwig et al. (1957a,b) fixed the site of origin of CO as arising directly from the α-methane bridge carbon of heme. Confirmation of endogenous production with introduction of a more sensitive assay was realized by Landaw and Winchell (1966). However, Coburn et al. (1963, 1964), on the basis of quantitative studies of endogenous production of CO, suggested that about 25% of the normal amount of CO produced by man (0.42/ml/hour) cannot be accounted for on the basis of catabolism of hemoglobin.

The concentration of CO usually encountered in the blood of normal nonsmokers, as determined by gas chromatography, is 0.096 ± 0.056 ml/100 ml (Ayres et al., 1966).

At the cellular level, oxidation and utilization of CO by bacteria (chemolithotropic) has been reported (Hasemann, 1927; Kistner, 1953, 1954; Hirsch, 1965). Growth of such bacteria was enhanced if the culture was maintained under continuous illumination, an effect suggested to be due to photoreversal binding of CO to cytochrome c oxidase (Hirsch, 1968). The photodissociation of cytochrome a $\frac{2+}{3}$ · CO compound formed the basis of Warburg's determination (Warburg and Negelein, 1929) of the photochemical action spectrum of the respiratory enzymes.

Conney et al. (1957) indicated that CO inhibited the oxidative metabolism of amino azo dyes by liver microsomes, which suggested the presence of a CO-sensitive step in microsomal oxidation reactions. D. Y. Cooper et al. (1965) presented evidence that the CO-binding pigment (cytochrome, P-450) functions as an oxygen-activating enzyme for several drug oxidations. Thus liver microsome oxidation of drugs was inhibited by CO and this inhibition was reversed by monochromatic light of a 450-mμ wavelength. Stannard (1940) suggested that the oxidation of CO, in some cases, was superimposed on inhibition of the cytochrome-cytochrome oxidase system (frog muscle). Fenn and Cobb (1932a) demonstrated stimulation *in vitro* of muscle by exposure to O_2-CO mixtures. It was established later that this was due to oxidation of CO-CO_2 (Fenn and Cobb, 1932b). Griffin and Hollocher (1967) suggested that soluble succinic dehydrogenase from beef heart bound CO and that it might be recognized by its competitive action on O_2. CO inhibits the metabolism of testosterone to hydrotestosterone (Conney *et al.*, 1968), as well as the activity of liver tryptophan oxygenase (Warburg *et al.*, 1967). Myoglobin binds CO (Bartlett, 1968). Treatment of Ehrlich ascites cells *in vivo* with aqueous solution (2%) of CO resulted (9 days after transplant) in a marked increase in nucleic acid content (Jeney and Medve, 1967) similar to the results of experiments with yeast (Jeney *et al.*, 1961).

A metabolic role of CO at the cellular level may be feasible; yet, because of the paucity of data, its delineation remains unclear. Certainly, its potential role as a contaminant in the gas overlay of a cell culture should be established.

ACKNOWLEDGMENT

The author expresses his appreciation for the assistance given by Miss Suzanne Zajac in the preparation of this manuscript.

The studies were conducted under U.S.P.H. Grant ES 00030-06.

REFERENCES

Ammann, E. C. B., and Lynch, V. H. (1967). Gas exchange of algae. III. Relation between the concentration of carbon dioxide in the nutrient medium and the oxygen production of *Chlorella pyrenoidosa*. *Appl. Microbiol.* **15**, 487–491.

Anderson, N. (1953). The mass isolation of whole cells from rat liver. *Science* **117**, 627–628.

Ashford, C. A., and Holmes, E. G. (1931). Further observations on the oxidation of lactic acid by brain tissue. *Biochem. J.* **25**, 2028–2049.

Ayres, S. M., Criscitiello, A., and Giannelli, S., Jr. (1966). Determination of blood carbon monoxide content by gas chromatography. *J. Appl. Physiol.* **21**, 1368–1370.

Barnard, P. J., Weiss, L., and Ratcliffe, T. (1969). Changes in the surface properties of embryonic chick neural retinal cells after dissociation. *Exp. Cell Res.* **54**, 293–301.

Bartlett, D., Jr. (1968). Pathophysiology of exposure to low concentrations of carbon monoxide. *Arch. Environ. Health* **16**, 719–727.

Bartnicki-Garcia, S., and Nickerson, W. J. (1962). Induction of yeastlike development in *Mucor* by carbon dioxide. *J. Bacteriol.* **84**, 829–840.

Bauer, J. T. (1925). The effect of carbon dioxide on cells in tissue cultures. *Bull. Johns Hopkins Hosp.* **37**, 420–427.

Berry, L. J., Mitchell, R. B., and Rubenstein, D. (1955). Effect of acclimatization to altitude on susceptibility of mice to influenza A virus infection. *Proc. Soc. Exp. Biol. Med.* **88**, 543–548.

Bicz, W. (1960). The influence of carbon dioxide tension on the respiration of normal and leukemic human leukocytes. I. Influence on endogenous respiration. *Cancer Res.* **20**, 184–190.

Biggers, J. D., Whittingham, D. G., and Donohue, R. P. (1967). The pattern of energy metabolism in the mouse oocyte and zygote. *Proc. Nat. Acad. Sci. U.S.* **58**, 560–567.

Brues, A. M., and Buchanan, D. L. (1948). Studies of the overall CO_2 metabolism of tissues and total organisms. *Cold Spring Harbor Symp. Quant. Biol.* **13**, 52–62.

Brues, A. M., and Naranjo, A. (1948). Preliminary studies on C^{14} metabolism of tissue cultures. *Anat. Rec.* **100**, 644–645.

Bull, H. B. (1943). "Physical Biochemistry." Wiley, New York.

Buytendyk, F. J. J., Brinkman, R., and Mook, H. W. (1927). LXXXI. A study of the system carbonic acid, carbon dioxide and water. I. Determination of the true dissociation constant of carbonic acid. *Biochem. J.* **21**, 576–584.

Cantino, E. C. (1961). The relationship between biochemical and morphological differentiation in non-filamentous aquatic fungi. In "Microbial Reaction to Environment" (G. G. Meynell and H. Gooder, eds.), pp. 243–271. Cambridge Univ. Press, London and New York.

Carter, N. W., Rector, F. C., Jr., Campion, D. S., and Seldin, D. W. (1967). Measurement of intracellular pH of skeletal muscle with pH-sensitive glass microelectrodes. *J. Clin. Invest.* **46**, 920–933.

Chang, R. S. (1959). Participation of bicarbonates in RNA and protein synthesis as indicated by virus propagation in human cells. *J. Exp. Med.* **109**, 229–238.

Chang, R. S., Liepins, H., and Margolish, M. (1961). Carbon dioxide requirement and nucleic acid metabolism of HeLa and conjunctival cells. *Proc. Soc. Exp. Biol. Med.* **106**, 149–152.

Chapin, C. W. (1918). Carbon dioxid in the primary cultivation of the gonococcus. *J. Infec. Dis.* **23**, 342.

Charles, H. P. (1962). Response of *Neurospora* to carbon dioxide. *Nature (London)* **195**, 359–360.

Chin, B., and Knight, S. G. (1957). Growth of *Trichophyton mentagrophytes* and *Trichophyton rubrum* in increased carbon dioxide tensions. *J. Gen. Microbiol.* **16**, 642–646.

Clark, R. T., Jr. (1950). Evidence for conversion of carbon monoxide to carbon dioxide by the intact animal. *Amer. J. Physiol.* **162**, 560–564.

Coburn, R. F., Blakemore, W. S., and Forster, R. E. (1963). Endogenous carbon monoxide production in man. *J. Clin. Invest.* **42**, 1172–1178.

Coburn, R. F., Williams, W. J., and Forster, R. E. (1964). Effect of erythrocyte destruction on carbon monoxide production in man. *J. Clin. Invest.* **43**, 1098–1103.

Cohen, M. B., and Fleming, J. S. (1918). The diagnosis of epidemic meningitis and the control of its treatment by rapid bacteriologic and serologic methods. *J. Infec. Dis.* **23**, 337–341.

Conney, A. H., Brown, R. R., Miller, J. A., and Miller, E. C. (1957). The metabolism

164   William F. McLimans

of methylated aminoazo dyes. VI. Intracellular distribution and properties of the demethylase system. *Cancer Res.* **17,** 628–633.

Conney, A. H., Levin, W., Ikeka, M., Kuntzman, R., Cooper, D. Y., and Rosenthal, O. (1968). Inhibitory effect of carbon monoxide on the hydroxylation of testosterone by rat liver microsomes. *J. Biol. Chem.* **243,** 3912–3915.

Cooper, D. Y., Levin, S., Narasimhulu, S., Rosenthal, O., and Estabrook, R. W. (1965). Photochemical action spectrum of the terminal oxidase of mixed function oxidase systems. *Science* **147,** 400–402.

Cooper, T. G., Tchen, T. T., Wood, H. G., and Benedict, C. R. (1968). The carboxylation of phosphoenolpyruvate and pyruvate. I. The active species of "CO_2" utilized by phosphoenolpyruvate carboxykinase, carboxytransphosphorylase, and pyruvate carboxylase. *J. Biol. Chem.* **243,** 3857–3863.

Craig, F. N., and Beecher, H. K. (1943). Effect of carbon dioxide tension on tissue metabolism (retina). *J. Gen. Physiol.* **26,** 473–478.

Dain, J. A., Neal, A. L., and Seeley, H. W. (1956). The effect of carbon dioxide on polysaccharide production by *Streptococcus bovis*. *J. Bacteriol.* **72,** 209–213.

Danes, B. S., and Kieler, J. (1958). The influence of CO_2 tension on cellular respiration studied by the Cartesian Diver Technique. *C. R. Trav. Lab. Carlsberg, Ser. Chim.* **31,** 61–75.

De Wulf, H., and Hers, H. G. (1968). The interconversion of liver glycogen synthetase a and b *in vitro*. *Eur. J. Biochem.* **6,** 552–557.

Drouhet, E., and Mariat, F. (1952). Etude des facteurs déterminant le développement de la phase levure de *Sporotrichum schencki*. *Ann. Inst. Pasteur, Paris* **83,** 506–514.

Eagle, H. (1956). The salt requirements of mammalian cells in tissue culture. *Arch. Biochem. Biophys.* **61,** 356–366.

Eagle, H. (1960). Metabolic studies with normal and malignant human cells in culture. *Harvey Lect.* **54,** 156–175.

Evans, E. A., Jr., and Slotin, L. (1940). The utilization of carbon dioxide in the synthesis of α-ketoglutaric acid. *J. Biol. Chem.* **136,** 301–302.

Fanestil, D. D., Hastings, A. B., and Mahowald, T. A. (1963). Environmental CO_2 stimulation of mitochondrial adenosine triphosphate activity. *J. Biol. Chem.* **238,** 836–842.

Fenn, W. O., and Cobb, D. M. (1932a). Stimulation of muscle respiration by carbon monoxide. *Amer. J. Physiol.* **102,** 379–392.

Fenn, W. O., and Cobb, D. M. (1932b). Burning of carbon monoxide by heart and skeletal muscle. *Amer. J. Physiol.* **102,** 393–401.

Fischer, A., and Andersen, E. B. (1926). Über das wachstrum von normalen und bösartigen gewebezellen unter erhöhtem sauerstoffdruck. *Z. Krebsforsch.* **23,** 12–27.

Fleeger, J. L., Van Demark, N. L., and Johnson, A. D. (1967). The effect of carbon dioxide level and temperature on the *in vitro* metabolism of testis tissue. *Comp. Biochem. Physiol.* **23,** 475–482.

Flickinger, R. A. (1958). Induction of neural tissue in ventral explants from frog gastrulae by carbon dioxide shock. *Science* **127,** 145–146.

Foster, J. W. (1949). "Chemical Activities of Fungi." Academic Press, New York.

Gailani, S. D., Nussbaum, A., McDougall, W. J., and McLimans, W. F. (1970). Studies on hormone production by human fetal pituitary cell cultures. *Proc. Soc. Exp. Biol. Med.* **134,** 27–32.

Gala, R. R., and Reece, R. P. (1963). Influence of type and level of gas environment on anterior pituitary lactogen production *in vitro. Proc. Soc. Exp. Biol. Med.* **114,** 422–426.

Gala, R. R., and Reece, R. P. (1964). Influence of sodium bicarbonate level and culture temperature on anterior pituitary lactogen production *in vitro. Proc. Soc. Exp. Biol. Med.* **115,** 232–235.

Geisler, G. (1963). Morphogenetic influence of ($CO_2 + HCO_3^-$) on roots. *Plant Physiol.* **38,** 77–80.

Geyer, R. P., and Chang, R. S. (1958). Bicarbonate as an essential for human cells *in vitro. Arch. Biochem. Biophys.* **73,** 500–506.

Geyer, R. P., and Neimark, J. M. (1958). Response of CO_2-deficient human cells *in vitro* to normal cell extracts. *Proc. Soc. Exp. Biol. Med.* **99,** 599–601.

Giron, D. J., Pindak, F. F., and Schmidt, J. P. (1967). Effect of a space cabin environment on viral infection. *Aerosp. Med.* **38,** 832–834.

Gladstone, G. P., Fildes, P., and Richardson, G. M. (1935). Carbon dioxide as an essential factor in the growth of bacteria. *Brit. J. Exp. Pathol.* **16,** 335–348.

Goldsmith, J. R., and Landaw, S. A. (1968). Carbon monoxide and human health. *Science* **162,** 1352–1359.

Gomirato, G. (1934). Die Wirkung der Köhlensäure und des Stickstoffes auf die *in vitro* zezüchteten Zellen. *Arch. Exp. Zellforsch. Besonders Gewebezuecht.* **15,** 186–189.

Gotlieb-Stematsky, T., Yaniv, A., and Gazith, A. (1966). Spontaneous malignant transformation of hamster embryo cells *in vitro. J. Nat. Cancer Inst.* **36,** 477–482.

Griffin, J. B., and Hollocher, T. C. (1967). Evidence for the binding of oxygen and carbon monoxide by succinic dehydrogenase. *Biochem. Biophys. Res. Commun.* **26,** 405–410.

Gwatkin, R. B. L., and Siminovitch, L. (1960). Multiplication of single mammalian cells in a nonbicarbonate medium. *Proc. Soc. Exp. Biol. Med.* **103,** 718–721.

Hanks, J. H. (1955). Balanced solutions, inorganic requirements and pH control, *In* "An Introduction to Cell and Tissue Culture" (W. F. Scherer *et al.,* eds), pp. 5–8. Burgess, Minneapolis, Minnesota.

Haring, O. M. (1960). Cardiac malformations in rats induced by exposure of the mother to carbon dioxide during pregnancy. *Circ. Res.* **8,** 1218–1227.

Harris, J. O. (1954). The influence of carbon dioxide on oxygen uptake by "resting cells" of bacteria. *J. Bacteriol.* **67,** 476–479.

Harris, M. (1952). The use of dialyzed media for studies in cell nutrition. *J. Cell. Comp. Physiol.* **40,** 279–302.

Harris, M. (1954). The role of bicarbonate for outgrowth of chick heart fibroblasts *in vitro. J. Exp. Zool.* **125,** 85–98.

Hasemann, W. (1927). Zersetzung von Leuchtgas und Kohlenoxyd durch Bakterien. *Biochem. Z.* **184,** 147–171.

Hastings, A. B., and Dowdle, E. B. (1960). Effect of CO_2 tension on synthesis of liver glycogen *in vitro. Trans. Ass. Amer. Physicians* **73,** 240–246.

Hastings, A. B., and Longmore, W. J. (1965). Carbon dioxide and pH as regulatory factors in metabolism. *Advan. Enzyme Regul.* **3,** 147–159.

Haywood, C., and Root, W. S. (1930). A quantitative study of the effect of carbon dioxide upon the cleavage rate of the arbacia egg. *Biol. Bull.* **59,** 63–70.

Henderson, L. J. (1908). The theory of neutrality regulation in the animal organism. *Amer. J. Physiol.* **21,** 427–448.

Hirsch, P. (1965). Bacterial oxidation and utilization of carbon monoxide. *Bacteriol. Proc.* p. 90 (abstr.).

Hirsch, P. (1968). Photosynthetic bacterium growing under carbon monoxide. *Nature (London)* **217**, 555–556.

Holland, R. A. B. (1967). Kinetics of combination of O_2 and CO with human hemoglobin F in cells and in solution. *Resp. Physiol.* **3**, 307–317.

Huang, K., and Gordon, F. B. (1968). Production of interferon in mice: Effect of altered gaseous environments. *Appl. Microbiol.* **16**, 1551–1556.

Jeney, E., and Medve, F. (1967). Effect of CO, KCN and NaN_3 on the nucleic acid content of Ehrlich ascites tumor cells. *Biochem. Pharmacol.* **16**, 1899–1902.

Jeney, E., Szendrey, A., and Vig, E. (1961). Détermination des acides nucléiques de *Saccharomyces italicus* (Castelli) traités par l'oxyde de carbone (CO), le cyanure de potassium (CNK) ou l'azide de sodium (N_3Na). *Biochem. Pharmacol.* **7**, 23–30.

Kalant, H., and Young, F. (1957). Metabolic behavior of isolated liver and kidney cells. *Nature (London)* **179**, 816–817.

Kaltenbach, J. P. (1954). The preparation and utilization of whole cell suspensions obtained from solid mammalian tissues. *Exp. Cell Res.* **7**, 568–571.

Kalter, S. S., and Teppermann, J. (1952). Influenza virus proliferation in hypoxic mice. *Science* **115**, 621–622.

Kalter, S. S., Prier, J. E., and Zaman, H. (1955). Virus proliferation in hypoxic mice and chick embryo. *J. Exp. Med.* **102**, 475–488.

Kieler, J., and Bicz, W. (1962). The function of CO_2 in cell metabolism and growth. *Biol. Interactions Norm. Neoplastic Growth, Contrib. Host-Tumor Probl., Symp., 1961*, pp. 89–100.

Kieler, J., and Gromek, A. (1967). Effect of carbon dioxide on the production of NADPH in the synthesis of fatty acids in normal and tumor cells. *Arch. Immunol. Ther. Exp.* **15**, 109–111.

Kistner, A. (1953). On a bacterium oxidizing carbon monoxide. *Proc., Kon. Ned. Akad. Wetensch., Ser. C* **56**, 443–450.

Kistner, A. (1954). Conditions determining the oxidation of carbon monoxide and of hydrogen by *Hydrogenomonas carboxydovorans*. *Proc., Kon. Ned. Akad. Wetensch., Ser. C* **57**, 186–195.

Krall, A. R., and Tolbert, N. E. (1957). A comparison of the light dependent metabolism of carbon monoxide by barley leaves with that of formaldehyde, formate and carbon dioxide. *Plant Physiol.* **32**, 321–326.

Krebs, H. A. (1954). Considerations concerning the pathways of syntheses in living matter. *Bull. Johns Hopkins Hosp.* **95**, 19–33.

Lachica, R. V. F. (1968). Carbon dioxide fixation. *Enzymologia* **34**, 81–98.

Landaw, S. A., and Winchell, H. S. (1966). Endogenous production of carbon-14 labeled carbon monoxide: An *in vivo* technique for the study of heme catabolism. *J. Nucl. Med.* **7**, 696–707.

Laser, H. (1935). Metabolism of retina. *Nature (London)* **136**, 184.

Lata, G., and Reinerston, J. (1957). Cholesterol synthesis by isolated rat liver cells. *Nature (London)* **178**, 47–48.

Laws, J., and Strickland, L. (1956). Metabolism of isolated liver cells. *Nature (London)* **178**, 309–310.

Levintow, L., and Eagle, H. (1961). Biochemistry of cultured mammalian cells. *Annu. Rev. Biochem.* **30**, 605–640.

Liener, I. E., and Buchanan, D. L. (1951). The fixation of carbon dioxide by growing yeast and nongrowing yeast. *J. Bacteriol.* **61**, 527–534.

Lillie, R. S. (1909). On the connection between changes of permeability and stimulation and on the significance of changes in permeability to carbon dioxide. *Amer. J. Physiol.* **24**, 14–44.

Loewus, M. L., and Delwiche, C. C. (1963). Carbon monoxide production by algae. *Plant Physiol.* **38**, 371–374.

Lones, G. W., and Peacock, C. L. (1960). Role of carbon dioxide in the dimorphism of *Coccidiodes immitis*. *J. Bacteriol.* **79**, 308–309.

Longmore, W. J. (1966). Effect of physiological variations in pH and CO_2 concentrations on metabolism *in vitro*. *Fed. Proc., Fed. Amer. Soc. Exp. Biol.* **25**, 887–892.

Loomis, W. F. (1957). Sexual differentiation in *Hydra*. *Science* **126**, 735–739.

Loomis, W. F. (1959a). pCO_2 inhibition of normal and malignant growth. *J. Nat. Cancer Inst.* **22**, 207–217.

Loomis, W. F. (1959b). Feedback control of growth and differentiation by carbon dioxide tension and related metabolic variables. In "Cell, Organism and Milieu" (D. Rudnick, ed.), pp. 253–294. Ronald Press, New York.

Ludwig, G. D., Blakemore, W. S., and Drabkin, D. L. (1957a). Production of carbon monoxide and bile pigment by haemin oxidation. *Biochem. J.* **66**, 38P (abstr.).

Ludwig, G. D., Blakemore, W. S., and Drabkin, D. L. (1957a). Production of carbon monoxide by hemin oxidation. *J. Clin. Invest.* **36**, 912 (abstr.).

Luomanmaki, K. (1966). The metabolism of carbon monoxide. *Ann. Med. Exp. Biol. Fenn.* **44**, Suppl., 1–55.

Mackenzie, C. G., Mackenzie, J. B., and Beck, P. (1961). The effect of pH on growth, protein synthesis, and lipid-rich particles of cultured mammalian cells. *J. Biophys. Biochem. Cytol.* **9**, 141–156.

McLean, D. J., Robinson, N. H., and Purdie, E. F. (1951). The influence of the metabolic state and of the medium on carbon dioxide fixation by *Serratia marcescens*. *J. Bacteriol.* **61**, 617–626.

McLimans, W. F. (1969). Physiology of the cultured mammalian cell. In "Axenic Mammalian Cell Reactions" (G. L. Tritsch, ed.), pp. 307–367. Dekker, New York.

McLimans, W. F., Mount, D. T., Bogitch, S., Crouse, E. J., Harris, G., and Moore, G. E. (1966). A controlled environment system for study of mammalian cell physiology. *Ann. N.Y. Acad. Sci.* **139**, 190–213.

McLimans, W. F., Crouse, E. J., Tunnah, K. V., and Moore, G. E. (1968a). Kinetics of gas diffusion in mammalian cell culture systems. I. Experimental. *Biotechnol. Bioeng.* **10**, 725–740.

McLimans, W. F., Blumenson, L. E., Tunnah, K. V., and Moore, G. E. (1968b). Kinetics of gas diffusion in mammalian cell culture systems. II. Theory. *Biotechnol. Bioeng.* **10**, 741–763.

Magill, T. P., and Francis, T., Jr. (1936). Studies with human influenza virus cultivated in artificial medium. *J. Exp. Med.* **63**, 803–811.

Marcelo, C. L., Kurland, J., Styles, B. D., Kwasniewski, B. A., and McLimans, W. F. (1971). Manuscript in preparation.

Mayfield, E. D., Jr., Lyman, K., and Bresnick, E. (1967). Incorporation of bicarbonate-^{14}C into pyrimidines and into ribonucleic acid of the Novikoff ascites tumor cell. *Cancer Res.* **27**, 476–481.

Mer, C. L., and Causton, D. R. (1965). Carbon dioxide and cell division. *Nature (London)* **206**, 34–35.

Morgan, J. F. (1958). Tissue culture nutrition. *Bacteriol. Rev.* **22**, 20.

Mottram, J. C. (1927). The role of carbon dioxide in the growth of normal and tumour cells. *Lancet* **2**, 1232–1234.

Mottram, J. C. (1928). Carbon dioxide tension in tissues in relation to cancerous cells. *Nature* (*London*) **121**, 420–421.

Nazario, M., and Reissig, J. L. (1964). Induction of aspartic transcarbamylase by carbon dioxide. *Biochem. Biophys. Res. Commun.* **16**, 42–46.

Niederpruem, D. J. (1963). Role of carbon dioxide in the control of fruiting of *Schizophyllum commune*. *J. Bacteriol.* **85**, 1300–1308.

Nyiri, L. (1967). Effect of CO_2 on the germination of *Penicillin chrysogenum* spores. *Z. Allg. Mickrobiol.* **7**, 107–111.

Nyiri, L., and Lengyel, Z. L. (1965). Studies on automatically aerated biosynthetic processes. I. The effect of agitation and CO_2 on penicillin formation in automatically aerated liquid cultures. *Biotechnol. Bioeng.* **7**, 343–354.

Nyiri, L., and Lengyel, Z. L. (1968). Studies on ventilation of culture broths. I. Behavior of CO_2 in model systems. *Biotechnol. Bioeng.* **10**, 133–150.

Osterlind, S. (1948). The retarding effect of high concentrations of carbon dioxide and carbonate ions on the growth of a green alga. *Physiol. Plant.* **1**, 170–175.

Parker, R. C. (1950). "Methods of Tissue Culture." Harper (Hoeber), New York.

Pearson, H. E. (1950). Factors affecting the propagation of Theiler's GDVII mouse encephalomyelitis virus in tissue cultures. *J. Immunol.* **64**, 447–454.

Pious, D. A., Hamburger, R. N., and Mills, S. E. (1963). Clonal analysis of primary culture. *Fed. Proc., Fed. Amer. Soc. Exp. Biol.* **22**, 382 (abstr.).

Powell, J. F., and Hunter, J. R. (1955). The sporulation of *Bacillus sphaericus* stimulated by association with other bacteria: An effect of carbon dioxide. *J. Gen. Microbiol.* **13**, 54–58.

Powers, E. B. (1927). A simple colorimetric method for field determination of the carbon dioxide tension and free carbon dioxide, bicarbonates and carbonates in solution in natural waters. I. A theoretical discussion. *Ecology* **8**, 333–338.

Prescott, J. M., Ragland, R. S., and Harley, R. J. (1965). Utilization of CO_2 and acetate in amino acid synthesis by *Streptococcus bovis*. *Proc. Soc. Exp. Biol. Med.* **119**, 1097–1102.

Racker, E. (1962). Adenosine triphosphate and oxidative phosphorylation. *Fed. Proc., Fed. Amer. Soc. Exp. Biol.* **21**, 54 (abstr.).

Rahn, O. (1941). Notes on the CO_2-requirement of bacteria. *Growth* **5**, 113–118.

Ranson, S. L., Walker, D. A., and Clarke, I. D. (1960). Effects of carbon dioxide on mitochondrial enzymes from *Ricinus*. *Biochem. J.* **76**, 216–221.

Reissig, J. L., and Nazario, M. (1962). Regulación genética de la síntesis de arginina y uridina. *In* "Resúmen de Communicaciones," p. 76. II. Sesiones Científicas de Biología, Córdoba.

Rovin, S. (1962). The influence of carbon dioxide on the cultivation of human neoplastic explants *in vitro*. *Cancer Res.* **22**, 384–387.

Salisbury, G. W., Van Demark, N. L., Lodge, J. R., and Cragle, R. G. (1960). Inhibition of spermatozoan metabolism by pCO_2, pH, K ion and antibacterial compounds. *Amer. J. Physiol.* **198**, 659–664.

Sawicki, L., Baron, S., and Isaacs, A. (1961). Influence of increased oxygenation on influenza virus infection in mice. *Lancet* **2**, 680–682.

Scholander, P. F. (1963). The master switch of life. *Sci. Amer.* **209**, 92–106.

Schwartz, B. (1960). A critical analysis of the closed system technique for lens culture. *AMA Arch. Ophthalmol.* **63**, 593–606.

Searcy, R. L., Giddings, J. A., and Gordon, G. F. (1964). Instrumental approach for monitoring carbon dioxide homeostasis during tissue cultivation. *Amer. J. Med. Electron.* **3**, 119–122.

Seelich, F., Letnansky, K., Frisch, W., and Schneck, O. (1957). Zur Frage des Stoffwechsels normaler und pathologischer menschlicher Leukocyten. *Z. Krebsforsch.* **62**, 1–8.

Segre, D. (1964). Cytopathic effects of normally non-cytopathogenic viruses in tissue cultures held under increased oxygen pressure. *Proc. Soc. Exp. Biol. Med.* **117**, 567–569.

Shock, N. W., and Hastings, A. B. (1935). Studies of the acid-base balance of the blood. IV. Characterization and interpretation of displacement of the acid-base balance. *J. Biol. Chem.* **112**, 239–262.

Simpson, F. J., Talbot, G., and Westlake, D. W. S. (1960). Production of carbon monoxide in the enzymatic degradation of rutin. *Biochem. Biophys. Res. Commun.* **2**, 15–18.

Sjöstrand, T. (1949). Endogenous production of carbon monoxide in man under normal and pathological conditions. *Scand. J. Clin. Lab. Invest.* **1**, 201–214.

Smith, H. W., and Clowes, G. H. A. (1924). The influence of carbon dioxide on the velocity of division of marine eggs. *Amer. J. Physiol.* **68**, 183–202.

Solomon, A. K., Vennesland, B., Klemperer, F. W., Buchanan, J. M., and Hastings, A. B. (1941). The participation of carbon dioxide in the carbohydrate cycle. *J. Biol. Chem.* **140**, 171–182.

Sörensen, S. P. L. (1909). Etudes enzymatiques. II. Sur la mesure et l'importance de la concentration des ions hydrogène dans les réactions enzymatiques. *C. R. Trav. Lab. Carlsberg* **8**, 1–168.

Sorokin, C. (1962). Carbon dioxide and bicarbonate in cell division. *Arch. Mikrobiol.* **44**, 219–227.

Spratt, N. T. (1949). Carbon dioxide requirements of the early chick embryo. *Anat. Rec.* **105**, 583–584 (abstr.).

Stannard, J. N. (1940). An analysis of the effect of carbon monoxide on the respiration of frog skeletal muscle. *Amer. J. Physiol.* **129**, 195–213.

Steemann, N. E. (1955). Carbon dioxide as a carbon source and narcotic in photosynthesis and growth of *Chlorella pyrenoidosa*. *Physiol. Plant.* **8**, 317–335.

Swim, H. E., and Parker, R. F. (1958). The role of carbon dioxide as an essential nutrient for six permanent strains of fibroblasts. *J. Biophys. Biochem. Cytol.* **4**, 525–528.

Trinkaus, J. P., and Drake, J. W. (1956). Exogenous control of morphogenesis in ioslated *Fundulus* blastoderms by nutrient chemical factors. *J. Exp. Zool.* **132**, 311–342.

Tunnah, K. V., McLimans, W. F., and Moore, G. E. (1968). Rocker cell culture incubator. *Biotechnol. Bioeng.* **10**, 698–706.

Utter, M. F. (1959). The role of CO_2 fixation in carbohydrate utilization and synthesis. *Ann. N.Y. Acad. Sci.* **72**, 451–461.

Valley, G., and Rettger, L. F. (1927). Influence of carbon dioxide on bacteria. *J. Bacteriol.* **14**, 101–137.

Van Slyke, D. D. (1921). Studies of acidosis. XVII. The normal and abnormal variations in the acid-base balance of the blood. *J. Biol. Chem.* **48**, 153–176.

Waddell, W. J., and Bates, R. G. (1969). Intracellular pH. *Physiol. Rev.* **49**, 285–329.

Wakil, S. J. (1962). Enzymatic synthesis of fatty acids. *Comp. Biochem. Physiol.* **4**, 123–158.

Wales, R. G., Quinn, P., and Murdoch, R. N. (1969). The fixation of carbon dioxide by the eight-cell mouse embryo. *J. Reprod. Fert.* **20**, 541–543.

Walker, D. A., and Brown, J. M. A. (1957). Effects of carbon dioxide concentration on phosphoenolpyruvic carboxylase activity. *Biochem. J.* **67**, 79–83.

Warburg, O. (1956a). On the origin of cancer cells. *Science* **123**, 309–314.

Warburg, O. (1956b). On respiratory impairment in cancer cells. *Science* **124**, 269–270.

Warburg, O., and Negelein, E. (1929). Über das Absorptionsspektrum des Atmungsferments. *Biochem. Z.* **214**, 64–100.

Warburg, O., Posener, K., and Negelein, E. (1924). Ueber den Stoffwechsel der Carcinomzelle. *Biochem. Z.* **152**, 309–344.

Warburg, O., Geissler, A. W., and Lorenz, S. (1967). Bemerkung über die Tryptophan-Oxygenase. *Hoppe-Seyler's Z. Physiol. Chem.* **348**, 899–901.

Werkman, C. H. (1951). Assimilation of carbon dioxide by heterotropic bacteria. *In* "Bacterial Physiology" (C. H. Werkman and P. W. Wilson, eds.), pp. 407–427. Academic Press, New York.

Westfall, B. B., Peppers, E. V., and Earle, W. R. (1955). The change in concentration of certain constituents of the medium during growth of the strain HeLa cells. *Amer. J. Hyg.* **61**, 326–333.

Westlake, D. W. S., Roxburgh, J. M., and Talbot, G. (1961). Microbial production of carbon monoxide from flavonoids. *Nature* (*London*) **189**, 510–511.

Wherry, W. B., and Ervin, D. M. (1918). The necessity of carbon dioxid for the growth of *B. tuberculosis. J. Infec. Dis.* **22**, 194–197.

Wittenberg, J. W. (1960). The source of carbon monoxide in the float of the Portuguese man-of-war, *Physalia physalis L. J. Exp. Biol.* **37**, 698–705.

Wood, H. G., and Utter, M. F. (1965). The role of CO_2 fixation in metabolism. *Essays Biochem.* **1**, 1–27.

Wood, H. G., and Werkman, C. H. (1935). The utilization of CO_2 by the propionic acid bacteria in the dissimilation of glycerol. *J. Bacteriol.* **30**, 332 (abstr.).

Wright, G. P. (1928). The oxygen tension necessary for the mitosis of certain embryonic and neoplastic cells. *J. Pathol. Bacteriol.* **31**, 735–752.

Yagi, T. (1958). Enzymic oxidation of carbon monoxide. *Biochim. Biophys. Acta* **30**, 194–195.

Zaroff, L., Sato, G., and Mills, S. E. (1961). Single cell platings from freshly isolated mammalian tissues. *Exp. Cell Res.* **23**, 565–575.

Zimmerman, M., Delvin, T., and Pruss, M. (1960). Anaerobic glycolysis of dispersed cell suspensions from normal and malignant tissues. *Nature* (*London*) **185**, 315–316.

6

UPTAKE AND UTILIZATION OF AMINO ACIDS BY CELL CULTURES

M. K. Patterson, Jr.

I. Introduction

Amino acids play a diverse role in the metabolism of cells. Their direct participation in the synthesis of structural proteins and enzymes related to all aspects of metabolism should designate them as "limiting factors." Thus their transport, catabolism, anabolism, and ultimate utilization potentially have an effect at all levels of cell growth and metabolism.

Cell and tissue culture techniques have provided an excellent tool for the study of the interrelationship of amino acids and the metabolic "machinery" of the cell. Evaluation of the research conducted in this area, however, must be made while bearing in mind the limitations of the system, although the limitations may reflect a lack of knowledge of what is physiological and what is not. Although it is not within the scope of this chapter to discuss the overall nutrition of cells (see Chapters 2 and 3, Volume 1), evaluation of the amino acid uptake and metabolism of cells must include a consideration of the availability of their nutrients. The amino acid composition of culture medium has been for the most part based on the levels found in biological fluids. Alterations of the levels were subsequently made by observing the growth or survival of cell populations. It appears, however, that some form of adaptation by the cell has been required for continued survival in these media, whether they are chemically defined or contain biological materials such as serum. The criteria for adequate amino acid nutrition generally has been growth, not function. Proliferating cells, whose primary goal appears to be growth, may differ markedly in their amino acid requirements from a metabolic standpoint when compared to differentiated tissues such as liver or endocrine cells, whose ultimate goal is functional (Green and Todaro, 1967). Until recently no qualitative differences, other than those assigned to organ specificity (Eagle et al., 1966a), had been observed between malignant and normal cells' requirements for amino acids. Neuman and McCoy (1956) found that the Walker 256 carcinoma had an apparent requirement for L-asparagine. Subsequently other tumors, notably certain leukemias, were found to have a similar requirement (Haley et al., 1961). A new concept in cancer chemotherapy, "amino acid depletion therapy," has developed because of these findings and now is being used clinically. The metabolic defect in certain malignant cells requiring L-asparagine appears to be an insufficient capacity to synthesize this amino acid (Patterson and Orr, 1967, 1968; Horowitz et al., 1968; Broome, 1968). Recent reports concerning glutamine (Roberts et al., 1970), serine (Regan et al., 1969), phenylalanine and tyrosine (Demopoulos, 1966), cystine (G. E. Foley et al., 1969), and isoleucine (Ley and Tobey, 1970) suggest that amino acid depletions, or in the case of tryptophan an amino acid excess (Gold, 1970), may be effective in the clinical treatment of metabolic diseases.

Numerous other factors, possibly more appropriately called pitfalls, must be considered in evaluating amino acid metabolism of cells in culture. A now classic example is the series of early studies on L-arginine metabolism which was shown to be invalid by the observation that mycoplasma, a not too uncommon contaminant of cells in culture, metabolized the amino acid by pathways different from that of the cells in culture (Schimke and Barile, 1963). Another factor, that of cell density, may alter metabolic pathways of amino acids. Eagle et al. (1966a) showed that certain cells lose their requirement for cystine, glycine, serine, glutamine, and asparagine at high population densities; later studies questioned the inclusion of asparagine (Patterson *et al.*, 1969a) and cystine (G. E. Foley *et al.*, 1969). Leakage of amino acids from the cell into the medium, or of intermediates and cofactors in their synthesis, or of essential metabolites derived from amino acids are suggested reasons for the paradoxical requirements (Eagle, 1965). Low-density cell cultures, such as those used in cloning studies, often have unique nutritional requirements, e.g., addition of certain α-keto acid analogues of amino acids (Neuman and McCoy, 1958). Why these particular amino acids or their metabolites are lost and not others has not been elucidated.

In considering the uptake and utilization of amino acids by cells in culture, it must, therefore, be reemphasized that the technique itself imposes certain limitations on the interpretation of such studies.

II. Uptake of Amino Acids

The uptake of amino acids by cells in culture has been determined by analysis of several parameters, each of which appears to be interacting. Thus a measure of the transport of an amino acid across the cell membrane is a function of the composition and concentration of the culture medium and of the intracellular fluids. These parameters will be discussed as separate entities.

A. *Transport by Plasma Membranes*

1. MECHANISM OF TRANSPORT

The basic concepts of the movement of amino acids across mammalian cell membranes and the kinetics involved have been developed primarily by Christensen (1969) and co-workers using Ehrlich ascites tumor cells. Cells are capable of concentrating, intracellularly, pools of amino acids against a concentration gradient by a stereospecific mechanism requiring energy. Passive or facilitated diffusions are biologically significant, however, under certain conditions and in certain systems (Stein, 1968). While heterogeneity exists for the active transport of amino acids, a series of mediated systems showing classic Michaelis-Menten kinetics, and thereby saturable, have been defined (Christensen, 1969) as *A*, which is sodium-ion-dependent and whose reactivity centers around alanine and glycine; *ASC*, which is also sodium-ion-dependent and whose reactivity centers around alanine, serine, and cysteine; L, a sodium-independent system whose reactivity centers around leucine and phenylalanine; and Ly^+, a system defined for cationic amino acid transport of an individual amino acid. Thus, while system L is ascribed to the transport of phenylalanine, blockage of this system with an analogue of the amino acid does not completely inhibit the transport of the amino acid. Some transport of phenylalanine occurs via the A system. Moreover, almost every neutral amino acid examined is inhibitory to the uptake of another neutral amino acid. Thus, leucine (5 mM) effectively inhibits phenylalanine (1 mM) uptake.

The exodus of amino acids appears to occur through the reversal of some or all of the processes for mediated entry, differing only in the value of K_m, which is much higher on the inside of the cell than on the outside (Christensen and Handlogten, 1968). For example, the exodus of phenylalanine can be accelerated by the addition of leucine or lysine to an amino-acid-free medium (Christensen, 1969).

The specificity of the amino acid transport system resides primarily in the reactive α-amino and carboxyl sites, because blockage of these groups inhibits transport. Enzymes that metabolize amino acids, however, are most stringent in their specificity toward the side chain. Such "mirror-image" specificity has led Berlin (1970) to propose that "the selectivity of cells for exogenous compounds is enhanced by sequential action of membrane carriers and enzymes." Such a proposed mechanism is of interest because of the effects of amino acids on enzymes (see Section III,C).

Until recently most "transport" studies have been conducted using slices or explants. As pointed out by Guidotti et al. (1969), however, "isolated" cells in culture have certain advantages over these systems: (1) a more rapid and complete contact of the cell population with compounds under study, thus decreasing reaction times and increasing metabolic responses (which, however, imposes certain limitations on the values obtained for amino acid accumulation due to concomitant utilization); (2) a less complicated system involving only two compartments (incubation medium and intracellular space) and eliminating a third, the extracellular space; kinetic analysis is thereby simplified; (3) a more uniform composition of cells, consisting of randomly mixed cells or a defined population, thus minimizing biological variability within the experiments.

The transport of amino acids by cells in culture has been studied under a variety of conditions in efforts to elucidate the effects of culture media, enzyme treatment, cell differentiation, hormones, and malignancy or other metabolic diseases.

2. Transport in Metabolic Disorders

Malignancy and other metabolic diseases manifest themselves in such a way as to suggest membrane alterations; altered electrophoretic mobility, so-called "contact inhibition," and membrane-associated "tumor-specific" antigens are but a few of the changes reported for malignant cells. Reports by Eagle *et al.* (1961a), using HeLa, KB, normal intestinal epithelium, and conjunctiva cells, and Hare (1967), using normal and polyoma-transformed hamster (HTC-3049-91TC) cell lines, concluded that only small differences in amino acid uptake existed between normal and malignant cells. A similar conclusion was drawn by Foster and Pardee (1969) using 3T3 and Py3T3 cells (polyoma-virus-transformed 3T3 cells).

Cell cultures derived from normal tissues and from patients with "metabolic" disorders of amino acids provide a simple test for hypotheses that mutations of certain specialized cells express the biochemical lesion or that all somatic cells are affected (Hsia, 1970) and also a test to differentiate those cells with impaired transport (Platter and Martin, 1966) or metabolism at the enzymatic level of amino acids (Uhlendorf and Mudd, 1968; Scriver, 1969). Although cystinuria (Knox, 1966), Hartnup disease (Jepson, 1966), and certain forms of hyperglycinuria (Wyngaarden and Seagle, 1966) were reported to involve cellular amino acid transport, they have not been tested in a cell culture system.

The rate of transport of α-amino-1-^{14}C-isobutyric acid, a nonmetabolized amino acid, was shown to decrease with age in isolated chick cardiac cells (Guidotti *et al.*, 1969). Whether this decrease reflected a decreased efficiency of the amino acid transport system or reflected a change in composition of the cell population resulting during ontogeny of the heart was not elucidated. Addition of insulin to the system resulted in an increased uptake of glycine, serine, and valine and slightly increased the uptake of proline, threonine, leucine, lysine, and phenylalanine; the uptake of histidine was unaffected, and methionine uptake decreased in the presence of the hormone. The uptake of glycine by HeLa cells, however, was suppressed by cortisone; 19-nortestosterone, an anabolic steroid, had little effect (Mohri, 1967).

3. Effect of Exogenous Factors on Transport

The validity of extrapolating studies of amino acid transport by cells in a simplified salt solution to a media containing the nutrients required for growth of cells in culture has been challenged in studies by Kuchler (1967) and Hayes and Kuchler (1970). They showed that sodium ion/potassium ion ratios effectively regulate the accumulation of amino acids in LM strain mouse fibroblasts and should be rigorously controlled in transport studies. Kuchler (1967) found that whole but not dialyzed serum inhibited leucine uptake in these cells and suggested that the free amino acids of the whole serum suppressed transport. However, addition of nondialyzed horse serum had no effect on the transport of α-aminoisobutyric acid by L strain mouse fibroblasts (Kuchler and Marlow-Kuchler, 1965), nor did the addition of fresh calf serum to confluent 3T3 cells (Cunningham and Pardee, 1969) affect the transport rates of cycloleucine-^{14}C or a mixture of 15 "naturally" occurring L-amino acids-^{3}H. In contrast, studies with WI38 cells allowed to reach a stationary phase were stimulated to reenter DNA synthesis and cell division by addition of serum. This suggested that serum may exert an effect on the cell membrane and alter amino acid transport (Wiebel and Baserga, 1969).

The efficacy of pronase and trypsin in dispersing cells has been evaluated in some detail (J. F. Foley and Aftonomos, 1970) but its effects on amino acid transport have not been thoroughly investigated. A release from density-dependent growth inhibition of confluent chick embryo cultures occurred after the addition of trypsin, pronase, or ficin at concentrations too low to promote cell dispersion. It was suggested that this resulted from alterations in membrane (Sefton and Rubin, 1970). It would be of interest to study amino acid transport in cells treated in a similar manner. Treatment of HeLa cells with neuraminidase decreased α-aminoisobutyric-^{14}C transport without altering the rate of efflux from preloaded cells (Brown and Michael, 1969), but ribonuclease was ineffective.

These results suggest that cellular metabolic disorders (other than those of a heritable nature or reflecting hormone imbalance) do not manifest themselves by changes in the capacity to transport amino acids but by changes in the transport "rate." The latter in turn is controlled by extracellular and intracellular levels of amino acids and cofactors for activation of membrane "transport sites."

B. Uptake as Measured by Changes in the Culture Medium

Although many investigators have related changes in medium composition during growth of cells in culture to utilization of these components, a more precise demonstration of utilization is preferred. Although a significant decrease in the concentration of a component of the culture medium is probably reflected by an increase in the intracellular level of the component, the resultant intracellular level may, in fact, cause a decrease in the utilization of that component. Earlier studies reported changes of amino acid concentrations in the culture medium during a specified time. More recently this "change" has been related to changes in cell population or efficiency of incorporation into protein. While, indeed, the latter studies may reflect "utilization," they have been included in this section for clarity in organization.

1. QUALITATIVE CHANGES

Changes in the amino acid concentration of medium by cells in culture have been studied using a variety of cell lines cultured under a variety of conditions (Lucy, 1960). Table I shows some of the more recent studies. Table II is an extraction from these studies. Earlier studies (Pasieka et al., 1958a,b; Lucy and Rinaldini, 1959) show similar characteristic patterns of cells maintained under conditions supporting survival rather than proliferation. The increase or decrease of essential or nonessential amino acids in the culture medium reflect differences in culture media, cell type, or proliferative phase. For example, a change in medium composition by addition of serum shifts the production of tyrosine and glutamine into media to one of uptake (Table II, column

TABLE I.

STUDIES ON THE CHANGES OF THE AMINO ACIDS OF MEDIUM BY CELLS IN CULTURE

Column in Table II	Reference	Cell line	Medium
A	Kagawa et al. (1960)	Lp1	0.4% lactalbumin hydrolysate DM-11 (chemically defined) DM-11 plus 5% serum
B	Pasieka et al. (1960)	L, HeLa, J96, J111, J128	M-150 (chemically defined)
C	McCarty (1962)	HeLa, KB, Hep 2, intestine, pituitary	MEM
D	Mohberg and Johnson (1963)	L929	MB752/1
E	Chung et al. (1966)	16 lines	199 plus 20% serum
F	Kruse et al. (1967)	Jensen sarcoma	7a
G	Griffiths and Pirt (1967)	LS	Medium B
H	Griffiths (1970a)	WI38	MEM and BME

TABLE II.

QUALITATIVE UPTAKE OR PRODUCTION OF AMINO ACIDS BY CELLS IN CULTURE[a]

Amino acid	Column							
	A[b]	B[h]	C[o]	D[r]	E[v]	F[bb]	G[gg]	H[hh]
Arginine	U	U[i]	U	U	U[w]	U	U	U
Cystine/cysteine	P[c]	U	P[p]	U	U	U	U	U
Glutamine	P[d]	U	U	U	—	U	U	U
Histidine	U	U[i]	U	U	U[x]	U	U	U
Isoleucine	U	U	U	U	U	U	U	U
Leucine	U	U	U	U	U	U	U	U
Lysine	U	U	U	U	P[y]	U	U	U
Methionine	O[e]	U[j]	U	U	P[y]	U	U	U
Phenylalanine	U	U	U	U	U	U	U	U
Threonine	U[f]	U[k]	U	U	U	U[cc]	U	U
Tryptophan	U	U[i]	U	—	—	—	U	U
Tyrosine	P[e]	U[i]	U	U	O[x]	U[cc]	U	U
Valine	U	U	U	U	P	U	U	U
Alanine	U	U[k]	P	P[s]	U[x]	P	—	—
Asparagine	—	—	—	—	—	U	—	—
Aspartic acid	U[f]	U[l]	P	U	U	P[dd]	P	—
Glutamic acid	—	U[m]	P	P	P	P	P	—
Glycine	O[g]	O[n]	P[q]	U[s]	U[z]	P[ee]	—	—
Proline	U[f]	U[i]	P[q]	U[t]	U[x]	U[ff]	P	—
Serine	U[f]	U	P[q]	P[u]	U[aa]	U	—	—

[a] U, uptake; P, production; O, no change.
[b] DM-11 (amino acid medium); other media the same except where noted.
[c] Lactalbumin hydrolysate (peptide medium), U.
[d] Includes glutamic acid; DM-11 plus 5% serum (protein medium), U.
[e] Protein medium, U; peptide medium, U.
[f] Peptide medium, P.
[g] Protein medium, U; peptide medium, P.
[h] HeLa; other cells (J series) the same except where noted.
[i] J96, O.
[j] J96, O; J128, P.
[k] J96, O; J128, O.
[l] J96, O; J111, O.
[m] J96; O; J111, P; J-128, P.
[n] J96, P; J111, P.
[o] HeLa; other cells the same except where noted.
[p] Pituitary, U; KB, U; Hep 2, U.
[q] Pituitary, U.
[r] MB752/1 medium; other media the same except where noted.
[s] Medium A (see text), P; medium C, P.
[t] Medium A, P.
[u] Medium A, U.
[v] HeLa-S; other cells (HeLa-Q, HeLa-F, and HeLa-S3) the same except where noted.
[w] Arginine depleted in HeLa-Q and HeLa-S3.
[x] HeLa-S3, P.
[y] HeLa-Q, U; HeLa-F, U; HeLa-S3, U.
[z] HeLa-Q, O; HeLa-F, P; HeLa-S3, P.
[aa] HeLa-S3, P.
[bb] During rapid proliferation (days 1–3 on perfusion); others the same except where noted.
[cc] Slow proliferation (days 6–8), P.
[dd] Slow proliferation (days 6–8), U.
[ee] Rapid proliferation (days 0–2), U.
[ff] Slow proliferation (days 3–5 and 6–8), P.
[gg] Batch culture; chemostat culture qualitatively the same.
[hh] Minimum essential medium; basal medium Eagles, qualitatively the same.

A). Similarly, different cell lines showed variability in production or uptake (Table II, columns B, C, and E). Nonessential amino acids are used when supplied in the media (Table II, columns A, B, E, and F) and produced when omitted from the media (Table II, columns C and G). A reduction of essential amino acids, omission of cysteine and glycine, plus the addition of serine (Table II, medium A, column D) result in the production of proline and glycine and an uptake of serine. Moreover, the Jensen sarcomas cease to use threonine, tyrosine, and proline and begin producing them when proliferation slows; the converse is true for aspartic acid (Table II, column F). Doubling the amino acid concentration of the medium does not alter their uptake (Table II, column H).

2. QUANTITATIVE CHANGES

The quantitative uptake of individual amino acids by cells in culture has been used to establish their "growth-limiting" concentrations, and on this basis new culture media have been developed. Relationships between amino acid uptake, cellular protein composition, and maximum growth yields have been established.

Glutamine is the growth-limiting amino acid in most tissue culture media, since it is the first to be exhausted. Griffiths and Pirt (1967) showed that 98% of the glutamine but only 20–80% of the other amino acids disappeared from the media in 120-hour batch cultures of mouse LS cells. Such apparent utilization, however, reflected not only its central role in metabolic processes but its instability in culture media. These investigators, as well as numerous others, have shown glutamine to be unstable in tissue culture media, resulting in the nonenzymatic production of pyrrolidone carboxylic acid and ammonia or degradation by cellular glutaminase to produce glutamic acid and ammonia. Certain cells in culture can synthesize glutamine in amounts sufficient for growth, provided there is a sufficiently high population density and provided that glutamine synthetase activity is sufficiently high to utilize endogenously synthesized glutamic acid (Paul and Fottrell, 1963).

On the basis of percentage uptake, isoleucine and leucine would also be rate-limiting amino acids (Kagawa et al., 1960; Mohberg and Johnson, 1963; Griffiths and Pirt, 1967; Griffiths, 1970a). An increase of 1.5-fold of these two amino acids plus arginine in MEM gave a 20% increase in the cell yield of WI38 cells cultured for 168 hours (Griffiths, 1970a). Daily media changes of MEM, however, resulted in a 140% increase. Elimination of the problem of amino acid depletion, it would seem, can best be resolved by perfusion culture systems (Kruse and Miedema,

1965; see also Chapter 2, Volume 2) or chemostat cultures (Griffiths and Pirt, 1967).

Studies on the rates of amino acid uptake showed that glutamine was the most rapidly used, followed by lysine, leucine, and isoleucine (Kagawa et al., 1960; McCarty, 1962; Kruse et al., 1967; Griffiths and Pirt, 1967). Comparing the relative rates of uptake with the composition of cellular protein of Jensen sarcoma cells, Kruse et al. (1967) found the ratios to be disproportionate. Isoleucine utilization rates, for example, were in excess of that percentage of the amino acid found in protein. Modification of the media to contain valine, isoleucine, and leucine in the same relative concentrations as those found in the cellular protein resulted in a disappearance of the "excessive" uptake of isoleucine. This study, as well as those of Mohberg and Johnson (1963), Ling et al. (1968), and Stoner (1970), suggests that media may best be formulated on the basis of cellular demand (Kruse et al., 1970).

The efficiency of conversion of amino acids to cell material has been evaluated in LS cells grown in batch and chemostat cultures (Griffiths and Pirt, 1967). When the cells were adapted to grow in media containing glutamic acid in place of glutamine, conversion of amino acid nitrogen to cell nitrogen was 100% efficient when optimum growth rate was achieved in the chemostat cultures; in contrast only 33% efficiency was achieved in batch cultures with glutamine medium. A recent study evaluating the effects of a chemically defined medium, which included inorganic phosphate, iron, and zinc, however, showed a two- to threefold better utilization of amino acids on a per-cell basis (Higuchi, 1970) than did the studies of Griffiths and Pirt (1967).

The uptake of amino acids by cells whose movement has been "contact-inhibited" (see Abercrombie, 1970, for a recent review) or by cells whose division has been "density-dependent"-inhibited (Castor, 1968) has been the object of recent studies. Although the latter phenomenon has been attributed to cell-to-cell contact (Stoker et al., 1966), several observations suggest that it is influenced by properties of the growth medium. Thus, with continuous perfusion cultures (Kruse and Miedema, 1965; see also Chapter 2, Volume 2) or frequent medium changes (Rhode and Ellem, 1968), the effects of "contact inhibition of growth" were reduced. However, Griffiths (1970a), when comparing the growth yield of WI38 cells cultured in BME and MEM, found that whereas cell yield in the BME was far less than that in the MEM, no amino acid was found to be growth-limiting. This investigator suggested, therefore, that the inhibition of growth in crowded cultures depends more on the uptake of nutrients, which in turn was a reflection of a reduced cell surface area, than on a nutrient depletion of the medium. In pursuit of this theory, the effect of insulin on the uptake of amino acids by confluent WI38 cells was studied (Griffiths, 1970b). It was found that insulin lowered amino acid uptake despite a higher cell protein synthesis and content. Although the phenomenon of "contact inhibition" was not altered, a more efficient utilization of amino acids did occur with an increase in metabolic activity and cell size. Glucose uptake was stimulated by insulin, however, and has implications in the more efficient utilization of the intracellular amino acids.

C. Intracellular Pools

It is generally assumed that intracellular amino acid pools, derived from exogenous sources or by *de novo* synthesis, exist as single entities. Several investigators, however, have suggested that intracellular "compartmentation" of amino acids may exist, with the result that certain endogenously and exogenously supplied amino acids are metabolized differentially (Kipnis *et al.*, 1961; Eagle *et al.*, 1966a; Broome, 1968). No direct evidence exists for this event, but double-labeling experiments should aid in answering the problem.

The degree of concentration by cells to form intracellular amino acid pools varies with the exogenous level (Eagle *et al.*, 1961a; also Table III). At approximate physiological levels (serum concentrations) most of the essential amino acids were concentrated 2- to 10-fold (Table IV). With some of the amino acids, however, e.g., valine and lysine, the degree of concentration varied within wide limits and seemed independent of exogenous levels; but with others, e.g., threonine, the degree of concentration increased with lower levels of exogenous supply (Table IV, columns F and G). In general, nonessential amino acids were more highly concentrated intracellularly than essential amino acids when nonessential amino acids were included in the medium (Table IV; columns A and F did not contain the nonessentials). This suggests a greater capacity for pooling of nonessential amino acids. When the nonessentials were omitted, however, they were even more effectively concentrated (Table IV, columns A and F), although their absolute concentration (Table III, columns A and F) was lower. Eliminating nonessential amino acids also resulted in an increased intracellular level of certain essential amino acids, e.g., glutamine and threonine (Table III, columns E and F) and a concomitant increase in the distribution ratio of all the essential amino acids (Table IV, columns E and F). Doubling the concentration of the essential amino acids of MEM failed to increase the rate of protein synthesis in L cells in culture (Kuchler, 1964) but did increase the

TABLE III.

INTRACELLULAR CONCENTRATION OF AMINO ACIDS (mM)

Amino acid	Column and Cell						
	A^a HeLa	B^b HeLa	C^b JTC4	D^b L	E^c L	F^c L	G^c L
Essential amino acids							
Arginine	(0.03)	—	—	—	0.84	0.28	—
Cystine	<0.05	—	—	—	Trace	Trace	Trace
Glutamine	8.1	—	—	—	Trace	3.30	4.57
Histidine	0.26	0.74	0.86	3.1	Trace	Trace	—
Isoleucine	1.00	2.36	2.68	9.2	0.33	0.54	0.99
Leucine	0.73	5.50	5.60	21.9	0.66	0.29	0.56
Lysine	0.29	2.67	4.29	16.3	0.48	0.29	—
Methionine	0.19	—	—	—	·0.30	0.36	0.36
Phenylalanine	0.52	1.69	1.64	7.2	0.87	0.85	0.68
Threonine	0.96	3.21	4.15	22.7	1.74	3.10	5.01
Tryptophan	<0.10	—	—	—	—	—	—
Tyrosine	0.81	1.60	1.55	6.5	0.58	0.52	0.59
Valine	0.79	2.85	2.93	12.4	0.54	0.83	1.08
Nonessential amino acids							
Alanine	1.43	6.34	7.48	46.0	3.18	3.19	7.52
Asparagine	0.15	—	—	—	—	—	—
Aspartic acid	1.27	3.61	2.95	22.1	0.63	0.64	1.10
Glutamic acid	10.8	4.27	7.55	103	3.87	7.88	9.84
Glycine	0.79	5.13	3.90	70.4	9.99	0.38	0.48
Proline	0.80	2.23	2.55	18.6	2.74	0.81	1.33
Serine	(0.03)	4.85	7.25	45.0	2.13	0.30	—

[a] Piez and Eagle (1958). Media consisted of 13 essential amino acids, salts, vitamins, and glucose with 5% dialyzed human serum. Values in parenthesis are estimates.

[b] Mohri (1967). HeLa (column B) and JTC4 rat heart fibroblast (column C) grown on 80% Hanks balanced salt solution, 20% bovine serum, and 0.4% lactalbumin hydrolysate; L cells (column D) grown in 95% Hanks balanced salt solution, 5% bovine serum, and 0.4% lactalbumin hydrolysate, until 24 hours prior to harvest, when the cells were cultured in the same medium as HeLa and JTC4.

[c] Kuchler (1964). L cells: column E, grown in medium 199 supplemented with 0.5% bactopeptone; column F, MEM plus 5% horse serum; column G, MEM containing a double concentration of each essential amino acid plus 5% horse serum.

level of the intracellular essential and nonessential amino acids (Table III, columns F and G). The intracellular level, however, was not doubled as one might expect if the pool level were a constant and independent function of each amino acid.

TABLE IV.

DISTRIBUTION RATIOS (INTRACELLULAR AND EXTRACELLULAR) OF AMINO ACIDS

	Column						
Amino acid	A[a]	B[b]	C[b]	D[b]	E[c]	F[c]	G[c]
Essential amino acids							
Arginine	2	—	—	—	0.9	2.0	—
Cystine	<2	—	—	—	—	—	Trace
Glutamine	11	—	—	—	—	11.0	7.4
Histidine	8	2.8	2.7	11	—	—	—
Isoleucine	8	2.6	2.3	11	1.0	3.6	2.8
Leucine	8	2.5	2.2	10	1.0	4.8	2.7
Lysine	3	1.2	2.1	8.9	0.6	2.0	—
Methionine	5	—	—	—	1.3	6.0	3.0
Phenylalanine	8	3.4	2.6	13	1.9	6.1	2.6
Threonine	5	4.2	4.2	32	3.2	15.0	9.2
Tryptophan	<8	—	—	—	—	—	—
Tyrosine	9	3.5	2.3	13	1.6	7.4	4.2
Valine	7	2.2	2.2	13	1.0	3.9	2.0
Nonessential amino acids							
Alanine	18	4.0	3.7	34	3.9	15.7	18.3
Asparagine	30	—	—	—	—	—	—
Aspartic acid	106	9.7	4.9	52	1.1	6.9	12.2
Glutamic acid	14	4.4	5.7	106	5.1	20.0	20.4
Glycine	158	13	6.3	109	11.6	19.0	12.0
Proline	>160	2.3	1.7	21	9.1	20.0	22.1
Serine	10	4.3	4.3	43	2.6	3.8	—

[a,b,c] Footnotes are the same as those for Table III.

Ling et al. (1968) found that the intracellular concentration of amino acids in MBIII cells was 200% greater when chemically defined medium containing 54 mM amino acid nitrogen was used, 170% greater at 81 mM, and 100% greater at 162 mM when compared to medium containing 13.5 mM amino acids. Two other parameters were studied: cell number and average cell protein content. These also were greater at the 54 mM amino acid nitrogen level.

The preceding emphasizes that if the utilization of amino acids is regulated by factors such as feedback mechanisms, product inhibition, repression and derepression, and enzyme stabilization (see Schimke and Doyle, 1970, for critique of terminology), both the intracellular and extracellular levels of amino acids must be carefully defined.

III. Utilization of Amino Acids

In Section II it was pointed out that some of the studies might be referred to as "uptake" where in fact they reflected the "utilization" of the amino acids. It is obvious that an overlap will occur when arbitrary divisions of subject matter are made. The studies to be discussed in this section were selected because they represent attempts to define the intracellular utilization of amino acids in biosynthetic pathways and regulatory processes.

A. Macromolecular Synthesis

1. PROTEIN SYNTHESIS

The most direct precursory role of amino acids is that involving protein synthesis. Definitive studies in this area by Eagle et al. (1961a) showed that a lower critical threshold of intracellular amino acid concentration from 0.01 to 0.05 mM had to be exceeded to initiate cell growth and protein synthesis. It is not yet clear, however, which of the reaction sequences of protein synthesis is the rate-limiting step at these low concentrations of amino acids (Eagle and Levintow (1965); see Section III,C).

The relationship of individual amino acids to the synthesis of proteins has been studied by using labeled precursors. Table V lists some examples of specific (secretory and structural) proteins whose synthesis by cells in culture has been studied by use of labeled amino acid incorporation. For example, collagen synthesis by dispersed rat bone cells has been measured by the addition of proline-U-^{14}C to the culture medium (Birge and Peck, 1966). Since collagen hydroxyproline is derived from free proline, both collagen and noncollagen protein could be estimated because the ratio of proline to hydroxyproline in collagen is nearly 1. The rate of collagen synthesis increased as the cell density rose and cell proliferation slowed. Similar results were obtained in other connective tissues (Green and Goldberg, 1964; Prockop et al., 1964) and suggested possible differential amino acid utilization for protein synthesis in functional and proliferating cells (Goldberg and Green, 1967). Similarly, incorporation of labeled amino acids into other proteins listed in Table V shows specificity that relates to cell types, e.g., liver versus spleen in immunoglobulin production, or to cell cycle phases. The use of immunochemical techniques wherein specific proteins, e.g.,

TABLE V.

Sᴘᴇᴄɪꜰɪᴄ Pʀᴏᴛᴇɪɴ Sʏɴᴛʜᴇsɪs ʙʏ Cᴇʟʟs ɪɴ Cᴜʟᴛᴜʀᴇ

Protein synthesized	Cells	Reference
Collagen	Bone	Birge and Peck (1966)
	Fibroblasts	Goldberg and Green (1967)
Immunoglobulins	Human	van Furth *et al.* (1966)
Serum	Primate and rodent	Stecher and Thorbecke (1967)
Membrane	L	Gerner *et al.* (1970)
Histones	Chinese hamster	Gurley and Hardin (1968)
	HeLa	Robbins and Borun (1967)
Glycoprotein	L5178Y	Kessel and Bosmann (1970)
Milk	Mammary gland	Schingoethe *et al.* (1967)
Enzymes	Hepatoma, etc.	Schimke (1964a)

enzymes [Schimke (1964a) is an example], are isolated is discussed further in Section III,C.

2. Oᴛʜᴇʀ Mᴀᴄʀᴏᴍᴏʟᴇᴄᴜʟᴇs

The secondary role of amino acids in the biosynthesis of macromolecules involves either the incorporation of their metabolic products or their regulation of protein synthesis. A discussion of these facets follows.

B. Conversion to Metabolites

For the most part, cells in culture utilize amino acids through established metabolic pathways (Meister, 1965). For instance, most are active amino donors in transaminase reactions (Barban and Schulze, 1959). A particular advantage of the system has been to establish the conversion of an amino acid to a specific product via a tissue-specific reaction and to study the effects of exogenous factors on these reactions.

1. Gʟᴜᴛᴀᴍɪɴᴇ ᴀɴᴅ Gʟᴜᴛᴀᴍɪᴄ Aᴄɪᴅ

The diverse precursory role of glutamine (Meister, 1968) should not be unexpected because of its apparent paradoxical requirement by cells in culture and decomposition in culture medium. Decomposition results from two separate reactions: a nonenzymatic reaction catalyzed by phosphate and bicarbonate ions, resulting in the production of pyrrolidone carboxylic acid and ammonia, and an enzymatic breakdown, catalyzed by glutaminase, giving rise to glutamic acid and ammonia. Pyrrolidone carboxylic acid appears to be metabolically inert. Glutamic acid through transamination and entry into the Krebs cycle reportedly contributes carbon atoms to proline, alanine, lactic acid, and CO_2, but not to aspartic acid in NCTC clone 929 mouse cells growing in a synthetic medium (Kitos and Waymouth, 1966). In contrast, HeLa cells metabolize glutamic acid to yield aspartic acid and proline (Levintow et al., 1957). In a similar cell system the carbons of glutamine-^{14}C and glutamic acid-^{14}C were incorporated effectively into pyrimidines but only "slightly" incorporated into the nucleic acid purines (Salzman et al., 1958). The amide nitrogen was a direct precursor of two nitrogen atoms of the purine ring and of the guanine amino group, as well as one nitrogen atom of the pyrimidine ring and of the cytosine amino group. A similar labeling of the pyrimidines by aspartic acid-^{14}C in Jensen sarcoma cultures (McCoy, 1961) supports a conclusion that glutamine carbons enter via its conversion to aspartic acid. In contrast, an insignificant incorporation of asparagine carbon and amide nitrogen into the nucleic acids of mouse leukemic cells has been reported (Ubuka and Meister, 1971). The amide nitrogen of glutamine is also the donor of the amide nitrogen of asparagine (Levintow, 1957; Patterson and Orr, 1967, 1968).

2. ARGININE

The validity of earlier studies on the metabolism of arginine by cells in culture has been challenged because some pleuropneumonia-like organisms (PPLO), an occasional contaminant of cell culture systems, utilize the amino acids by pathways not found in mammalian tissues (Schimke and Barile, 1963). It was shown that the conversion of arginine to ornithine in HeLa-S3, KB, or L cell cultures did not occur except when contaminated with PPLO. Subsequent studies, however, showed that arginase (conversion of arginine to urea and ornithine) was inducible by higher levels of arginine and that addition of manganese to the medium stabilized the arginase molecule (Schimke, 1964a). Such a conversion would explain the observation by Kruse (1961) that protein hydrolysates of Walker 256 carcinoma and Jensen sarcoma cells cultured in the presence of arginine-U-^{14}C contained label only in arginine and proline, a product of ornithine metabolism.

3. METHIONINE AND CYSTINE

Methionine is utilized as a methyl donor and as a sulfur donor toward cystine synthesis. Its role as a methyl donor is discussed in Section

III,C. The following reaction sequence has been demonstrated in liver tissues:

methionine-^{35}S $\rightarrow$ homocysteine ($+$ serine) $\rightarrow$ cystathionine
$$\rightarrow \text{cysteine}^{35}\text{S} + \text{homoserine}$$

Although earlier studies suggested that cystine was required by cells in culture, subsequent studies showed that at sufficiently high population densities this requirement could no longer be demonstrated (Eagle *et al.*, 1961b). At intermediate population densities, cell growth could be supported by the intermediates depicted in the reaction sequence. Although further studies showed this to be correct for heteroploid cells, human diploid cell lines still required cystine, irrespective of population density. All diploid cell lines studied were blocked in the conversion of cystathionine to cystine (Eagle *et al.*, 1966b). In recent studies, cultures of both diploid and heteroploid human leukemic cells, with only one exception, exhibited an absolute requirement for cystine even at high population densities (G. E. Foley *et al.*, 1969). Whereas the earlier studies were done with monolayer cultures, the later studies of G. E. Foley *et al.* (1969) used suspension cultures—and "perhaps of more importance, the cell lines considered herein are of human lymphocytic origin."

4. Tyrosine and Phenylalanine

The utilization of tyrosine for the production of melanin has been demonstrated in a pigmented hamster melanoma cell line (MM 2, RPMI 3460) by Ulrich *et al.* (1968). A 20-fold increase in tyrosine uptake concomitant with an increased melanin synthesis over nonpigmented cells was shown. Phenylalanine was not converted to tyrosine in these studies, but Eagle *et al.* (1957) have reported a significant conversion (12–25%) in a HeLa variant cell population when tyrosine was omitted from the culture medium.

5. Serine and Glycine

The interconversion of serine and glycine and utilization of the latter for *de novo* synthesis of purines have been studied in KB cell cultures (Lembach and Charalampous, 1967). Although a myoinositol deficiency had no effect on the level of serine hydroxymethylase activity, the size of the intracellular glycine pool, or the rate of interconversion of serine and glycine, the rate of incorporation of glycine into purine nucleotides was increased when serine or glycine was added to the medium. When the cells were grown in a medium without the added amino acids and depended on their own biosynthetic capabilities, the inositol deficiency impaired the ability of the cell to maintain normal amino acid concentration gradients. Such an impairment lowered the intracellular amino acid level to such an extent that nucleotide synthesis decreased (Charalampous, 1969). These studies may explain earlier reports that serine or glycine was "stimulatory" or essential since the medium used contained low levels of inositol (McCoy, 1960; Neuman and Tytell, 1960).

6. TRYPTOPHAN

The metabolism of tryptophan in mammalian tissues occurs mainly by two pathways, both involving its oxidation, with products of nicotinic acid or melatonin.

The conversion of tryptophan to melatonin, a hormone that is found in pineal glands and that produces lightening of the skin, has been extensively studied in cell culture systems. The reaction proceeds through the following sequence:

$$\text{tryptophan} \rightarrow \text{5-hydroxytryptophan} \rightarrow \text{serotonin} \rightarrow \text{melatonin}$$

Rat pineal glands cultured for 48 hours in the presence of tryptophan-^{14}C showed conversion of the amino acid to intermediates and product of the above reaction (Wurtman et al., 1968, 1969). Noradrenaline stimulated melatonin, suggesting either an increase in the formation of new melatonin-forming enzyme, an inhibition of the metabolism of one of the intermediates of the reaction, or an increase in the transport of tryptophan. Support for the latter mechanism was provided by the observation that noradrenaline, as well as other catecholamines, stimulated the incorporation of tryptophan-^{14}C into protein of pineal gland cultures (Axelrod et al., 1969). Recent evidence, however, showing that dibutyryl cyclic AMP stimulates the production of labeled melatonin from labeled tryptophan (Klein et al., 1970) suggests that melatonin synthesis is modulated by control at the enzymatic level (Klein and Weller, 1970).

7. HISTIDINE AND PROLINE

Some amino acids, once incorporated into peptide linkage, undergo a "secondary" alteration in structure. Thus, primary cultures of rat leg muscle cells are capable of converting histidine-^{14}C as well as methionine-methyl-^{14}C into protein-bound 3-methylhistidine-^{14}C, a component of contractile protein (Actin) (Reporter, 1969). Moreover, the methylation of histidine increased as the cells completed proliferation and entered the major phase of differentiation. The hydroxylation of proline (Rosenbloom *et al.*, 1967) and sequential hydroxylation and glycosylation of lysine (Rosenbloom *et al.*, 1968) in completed peptides in collagen biosynthesis have been shown in unossified tibiae cartilage cultures.

C. Regulation and Metabolic Events

The role of amino acids in the regulation of cellular metabolic events has proved diverse. They regulate not only the enzymes involved in their own metabolism, but also apparently unrelated enzymes, which in turn affect a number of cellular parameters such as ultrastructure and the cell cycle. Some enzymes of amino acid metabolism whose regulation has been studied are listed in Table VI.

1. Specific Enzyme Regulation

As suggested in Section II, the intracellular levels of amino acids regulate their own biosynthesis by several proposed mechanisms. Examples of enzyme "repression" in cell culture systems have been reported and include asparagine synthetase (Prager and Bachynsky, 1968; Patterson, 1970a,1971), argininosuccinate synthetase and argininosuccinase (Schimke, 1964a,b), glutamine synthetase (DeMars, 1958), feedback control of the enzyme systems for serine (Pizer, 1964), and proline biosynthesis (Rickenberg, 1961). Eagle *et al.* (1965) observed, however, that there was no uniformity in the susceptibility of amino acid biosynthesis to these mechanisms. Thus, growth of several human cell lines for 2–10 generations in a medium containing either none of the end-product amino acid or 1–10 mM exogenous amino acid yielded varying results. Prior growth of cells in glycine or serine reduced their subsequent biosynthesis from glucose, but glycine synthesis from serine was unaffected; alanine synthesis from glucose and proline from glutamine or ornithine were unaffected. Moreover, the addition of relatively high concentrations of preformed amino acids inhibited glycine synthesis from glucose or serine. There was no effect on the synthesis of serine from glucose; ornithine from arginine; or homocysteine, cystathionine, and cystine from methionine, homocysteine, and cystathionine, respectively; but paradoxically the synthesis of alanine from glucose and proline from ornithine was increased. Failure of aspartic acid or glutamic acid to affect their continued synthesis from glutamine was attributed mainly

TABLE VI.

Enzymes of Amino Acid Metabolism Studied in Cell Culture Systems

Enzyme	Cell Line	Culture conditions	Enzyme activity
Alanine transaminase	Hepatoma	Glucocorticoid supplement	Increased[a]
Arginase	HeLa, Chang liver	High arginine or MnCl$_2$	Increased[b,c]
Argininosuccinate synthetase	HeLa, HeLa-S3, KB, L	Low arginine	Increased[c]
	Human fibroblast	Low arginine	No change[d]
Argininosuccinase	HeLa, HeLa-S3, KB, L	Low arginine	Increased[c]
Asparagine synthetase	Jensen sarcoma	No asparagine	Increased[e]
Glutaminase	Kidney cortex	Glutamine	Increased[f]
Glutamic-oxalacetate transaminase	L929	Low aspartate	No change[g]
Glutamic-pyruvate transaminase	L929	Low aspartate	No change[g]
Glutamine synthetase	HeLa, L	Glutamine present	Decreased[h]
		Glutamic minus glutamine	Increased[h]
	Chick retina	Hormone present	Increased[i]
Histidase	Hepatoma	—	Present[j]
Ornithine δ-aminotransferase	Hepatoma	—	Present[k]
	Chang liver	—	Present[l]
Phenylalanine hydroxylase	Mouse	—	Present[m]
Proline oxidase	Hepatoma	—	Present[n]
Threonine dehydrase	Hepatoma	—	Present[n]
Transaminases	HeLa, L	—	Present[o]
Tyrosine transaminase	Hepatoma	Cortisone supplement	Increased[p]
	HTC	Serum factor	Increased[q]
	H41E	Cortisone supplement	Increased[i,r]
	Normal liver	Dexamethasone	Increased[s]
Alkaline phosphatase	HeLa	Low phenylalanine or tryptophan concentration	Inhibited increase caused by prednesolone[t]
	BSC1	Cyst(e)ine	Decreased[u]

[a] Lee and Kenney (1970).
[b] Paul and Fottrell (1963); Eliasson (1967a).
[c] Schimke (1964a,b).
[d] Tedesco and Mellman (1967).
[e] Patterson (1970a).
[f] Majumdar (1969).
[g] Kitos and Waymouth (1966).
[h] DeMars (1958); Paul and Fottrell (1963).
[i] Moscona et al. (1968).
[j] Pitot et al. (1964).
[k] Pitot et al. (1964); Pitot and Jost (1967).
[l] Strecker and Eliasson (1966).
[m] Tourian et al. (1969).
[n] Pitot et al. (1964).
[o] Barban and Schulze (1959).
[p] Thompson et al. (1966); Martin et al. (1969).
[q] Gelehrter and Tomkins (1969).
[r] Pitot and Jost (1967).
[s] Gerschenson et al. (1970).
[t] Griffin and Cox (1966).
[u] Griffin and Cox (1967).

to their failure to equilibrate freely with the cellular pool. While the possibility existed that the cells studied were already maximally repressed, these investigators concluded that in physiological concentrations "amino acids do not usually limit the rate of their own biosynthesis." Although these studies suggest a lack of regulation of enzyme content, the levels of the individual enzyme activities were not determined.

Variations in arginase activity in cell cultures related to variations in the composition of the medium have been reported by several investigators (Schimke, 1964a; Paul et al., 1964; Eliasson, 1967a; Strecker et al., 1970). Schimke (1964a) showed that the level of arginase activity in HeLa-S3 cells was directly related to the concentration of arginine in the medium. Addition of manganese to the medium further increased the arginase activity of the cells. The mechanism controlling the increased enzyme activity in the presence of arginine was shown by immunochemical studies using antiarginase antibody to be related to an increase in the rate of de novo synthesis of the enzyme as well as a decrease in its rate of turnover. The addition of manganese apparently stabilized the enzyme, resulting in an accumulation by decreasing enzyme breakdown. Subsequent studies by Eliasson (1967a,b) using Chang liver cells grown in suspension cultures have shown that in addition to the arginine effect, a temporary omission of a single essential amino acid (glutamine or phenylalanine) from the medium resulted in a rapid increase in arginase activity of these cells following restitution of the normal cell environment. The enzyme activity was also repressed by a metabolic product of the arginase-initiated reaction sequence from arginine to proline. Thus, by increasing the levels of leucine, isoleucine, valine, ornithine (Eliasson and Strecker, 1966), or norvaline (Strecker et al., 1970), which are known to inhibit enzymes of the reaction sequence, the arginase activity was increased (Eliasson, 1967a). From studies using actinomycin D and puromycin, these investigators concluded that the synthesis of arginase occurred in response to a preformed stable messenger RNA and that the repression of this synthesis occurred at the level of translation of this preformed RNA. Studies on the enzyme ornithine δ-aminotransferase in Chang liver cells previously led Strecker and Eliasson (1966) to the same conclusion on the control mechanism of this enzyme. Subsequently Strecker et al. (1970) showed that excess methionine when added to the culture media of Chang liver cells resulted in inhibition of cell proliferation, accumulation of cellular protein, and a 65% increase in the specific activity of ornithine δ-aminotransferase. The toxic effects were not elicited in the presence of excess glycine, serine, histidine, tyrosine, or tryptophan, nor was the enzyme activity increased. However, homocysteine, a potential metabolic product of methionine, produced the same effects as did methionine but at one tenth the concentration. Arginine in excess produced none of the toxic effects but significantly elevated the enzyme activity. The cellular accumulation of both protein and ornithine δ-aminotransferase in these cultures in the presence of excess methionine gave a total accumulation of these constituents, not much less than the corresponding total accumulation in the control cultures in spite of the decreased cell proliferation. These investigators suggest that these results might be considered as an example of homeostatic control involved in regulating the size and function of an organ, even though suspension cultures were used.

A relation between cellular organization and enzyme induction has been suggested by the finding that induction of glutamine synthetase by hydrocortisone in embryonic chick retina explants or aggregates was lost when carried as a monolayer culture (Morris and Moscona, 1970). This loss of inducibility was complete 24 hours after dispersion, but inducibility could be preserved throughout 24 hours by maintaining the freshly dispersed cells at 4°C. Moreover, in dispersed HeLa (DeMars, 1958) and L cells (Paul and Fottrell, 1963), this enzyme was "repressed" by glutamine, whereas both glutamate and glutamine were effective only in partially depressing the enzyme activity in the explanted retina (Kirk and Moscona, 1963). In the L cell (Paul and Fottrell, 1963) and the explanted retina systems (Alescio and Moscona, 1969), protein synthesis was required for the increase in glutamine synthetase whether resulting by "derepression" of the former system or hydrocortisone treatment in the latter. In the chick retina system, RNA synthesis was required for induction of glutamine synthetase by hydrocortisone, but sustained RNA synthesis was not (Moscona et al., 1968). This suggested that long-lived transcripts for the enzyme accumulated in the cells. Four hours after induction, low doses of actinomycin D (0.05 μg/ml) blocked further increases in both enzyme activity and amount, whereas high doses (10 μg/ml) failed to block continued induction (Alescio et al., 1970). The low dose of actinomycin D was proposed to inhibit the formation of a derepressor, which counteracted a translational repressor, whose formation was blocked by high levels of actinomycin D. Induction of the synthetase, similar to that seen with hydrocortisone, was also accomplished by a pulse treatment of the explants with cycloheximide followed by inhibition of transcription by actinomycin D (Moscona et al., 1970). This "precocious" induction of glutamine synthetase was accompanied by premature histological differentiation of the retina (Piddington and Moscona, 1965). A similar enhancing effect of glutamine on differentiation of germinal cells in culture has been reported (Steinberger and Steinberger, 1966).

The asparagine synthetase activity of the Jensen sarcoma cells increases when cultured in media devoid of asparagine or containing L-asparaginase (Patterson et al., 1969a). Although the initial change was the result of cell selection (Patterson et al., 1969b) the asparagine synthetase activity of the surviving cell population responded to asparagine depletion by further adaptive increases in activity (Patterson, 1970a, 1971). These investigations have suggested a "repression"-type control by L-asparagine of enzyme synthesis (Patterson, 1970b) similar to that observed in rat liver (Patterson and Orr, 1969; Patterson, 1971).

The active transaminase systems for the formation of glutamic and aspartic acids, phenylalanine, glycine, tyrosine, and alanine from the corresponding α-keto acids was present in cell-free extracts of HeLa and L cells (Barban and Schulze, 1959). To determine whether the glucocorticoid induction of several specific transaminases of liver resulted from direct steroid action on the hepatic cells, investigators have utilized cell culture systems. Examples of this were reported by Pitot et al. (1964) using the H4-II-E cell line (derived by cloning the Reuber hepatoma H-35), Tomkins et al. (1969) using the HTC hepatoma cell line, Gerschenson et al. (1970) using normal rat liver hepatocyte cultures, and Lee and Kenney (1970) in Reuber (H-35) hepatoma cells.

2. Mechanism of Enzyme Control

One of the more definitive studies conducted on an amino-acid-metabolizing enzyme and its regulation by exogenous factors has been by Tomkins and his co-workers (1969) on the tyrosine aminotransferase induction in HTC hepatoma cell cultures by adrenal steroids. From these studies a mechanism of control of enzymes by hormones has been proposed. Their work showed that the steroid inducer stimulated the rate of tyrosine aminotransferase synthesis and caused an accumulation of enzyme-specific messenger RNA. The constant presence of the steroid was required to maintain the induced rate of enzyme synthesis, but if RNA synthesis was blocked after induction, the enzyme synthesis became "constitutive." Thus, synthesis of RNA was required for the enzyme induction, but continued RNA synthesis was not required to maintain the enzyme synthesis either at the basal or induced rates. Studies with synchronized cells showed that the enzyme was noninducible during the postsynthetic (G_2), mitosis (M), or the early pre-DNA synthesis (G_1) periods and that during this period a repressor, acting at the post-transcriptional stage, was formed. These investigators proposed that steroids "have only a single action, that is, to antagonize a post-transcriptional repressor which both inhibits messenger translation and promotes messenger degradation." They proposed a model involving two genes: a structural gene for the induced enzyme and a regulatory gene for the repressor. The repressor was assumed to inhibit reversibly the translation of the messenger into tyrosine aminotransferase.

On balance, protein degradation is as important as protein synthesis in the regulation of enzyme concentration (Schimke, 1969). This is especially pertinent due to the observations that inhibitors of protein synthesis inhibit enzyme degradation (Kenney, 1967; Grossman and Mavrides, 1967; Auricchio et al., 1969). Actinomycin D, for example, inhibits degradation of tyrosine aminotransferase in HTC and Reuber hepatoma cells in culture (Reel and Kenney, 1968).

3. Amino Acids and Hormones as Regulators

The relationship of hormones and amino acids in regulating enzymes involved in the synthesis or metabolism of amino acids is not clear. Analysis of the kinetics and immunochemical data of increases in rat liver tryptophan pyrrolase after injections of corticosteroids and tryptophan has suggested that the former controlled the rate of enzyme formation and the latter the rate of enzyme degradation (Schimke, 1969). An apparent correlation between tyrosine aminotransferase and serine dehydratase induction by hydrocortisone and amino acid accumulation in liver and hepatoma tissues in vivo would suggest that the transport of the amino acid played a role in the induction (Baril et al., 1969). Hydrocortisone, however, failed to alter the influx of α-aminoisobutyric acid in Reuber hepatoma cells grown in monolayers (Krawitt et al., 1970). Insulin, which also induced the enzyme, increased the transport of the nonmetabolizable amino acid, but only after the tyrosine amino transferase activity had reached maximum levels. The increased transport required the presence of natural amino acids in the incubation medium, although the transport was 6–10 times greater when the cells were incubated in Hank's solution alone. This competition by the natural amino acids suggests that "a different protocol for α-aminoisobutyric acid experiments may be required to give results in this system comparable to those obtained in vivo."

4. Regulation of RNA and DNA Synthesis

The control of macromolecular synthesis by amino acids has been studied primarily by depletion techniques. Deleting asparagine from the medium, either by the addition of asparaginase or by formulation of the medium, resulted in a precipitous decline in protein synthesis by 6C3HED lymphoma cells (Ellem *et al.*, 1970). The initial decrease in protein synthesis was followed by a gradual decline concomitant with an exponential decline in DNA synthesis. Later, RNA synthesis declined in a sequence of rRNA > DNA like RNA > tRNA. Both the transcription and maturation of rRNA were inhibited. Earlier, Kubinski and Koch (1966) had reported that RNA synthesis was altered when human amnion cells (Fernandez strain) were cultured in "nutritionally depleted" medium. Under these conditions a DNA-like RNA, presumed to be mRNA, was predominantly synthesized with a parallel inhibition of the synthesis of ribosomal RNA precursor. It was proposed that inhibition of RNA synthesis, after such a nutritional "stepdown," was mediated by an increased level of deacylated tRNA, which inhibited rRNA synthesis without appreciably affecting tRNA or mRNA synthesis (Farnham and Dubin, 1965). Similar effects were found for a variety of heteroploid mammalian cells (Studzinsky and Ellem, 1968) and HeLa cells (Smulson and Thomas, 1969). In the latter study, the amino acylation level of tRNA was reduced 24% when the cells were incubated in medium lacking tyrosine; concomitantly RNA synthesis decreased. It had been suggested that tRNA (aminoacylated or non-acylated) reacted directly with RNA polymerase in bacterial systems. Later studies, however, indicated that there was essentially no reduction in the activity of nuclear RNA polymerase from HeLa cells incubated in Eagle's medium minus amino acids, although RNA synthesis was significantly reduced (Smulson, 1970). A possible explanation for the inhibition of rRNA synthesis during nutritional depletion may be found in the studies of Vaughan *et al.* (1967) on HeLa cells starved for methionine. Under their conditions, production of complete ribosomes was inhibited, but the 45 S and 32 S ribosomal precursors of RNA continued to be formed. Synthesis of ribosomal proteins continued as a result of preexisting cell protein turnover (Maden and Vaughan, 1968). By contrast, cells deprived of valine, which is essential for growth but unlike methionine is without special function in RNA methylation, continued to produce ribosomes at a reduced rate (Maden *et al.*, 1969). This resulted from a decreased rate of 45 S RNA synthesis and conversion to 32 S RNA and suggested that some particular protein may regulate these rates.

Depletion of methionine, but not valine, reduced the rate of synthesis of pre-tRNA (a class of small RNA molecules) and tRNA and the rate of apparent conversion of the former to the latter in HeLa cells (Bernhardt and Darnell, 1969). The pre-tRNA apparently required methylation in its conversion to tRNA. Recent evidence suggests that aminoacyl-tRNA may act as a repressor of protein synthesis (Gallo *et al.*, 1970).

Profiles of isoaccepting tRNAs of mouse leukemic cells showed quantitative and qualitative differences between L-asparaginase-resistant and -sensitive cells. The latter contained a "peak 4" asparaginyl-tRNA not found in the former. Small but reproducible differences have also been reported in the leucyl-, seryl-, threonyl-, and prolyl-tRNAs and a pronounced difference in the tyrosyl-tRNA of normal human lymphoblasts (cell line NC37) and leukemic human lymphoblasts (cell line F152) (Gallo and Pestka, 1970). The binding of a particular aminoacylated tRNA to polysomes may thus repress protein synthesis not unlike the N-formylmethionyl-tRNA's association at a chain initiation codon on mRNA.

5. REGULATION OF CELL CYCLE EVENTS

Protein synthesis is a continuing process throughout the reproductive cell cycle, although its rate during mitosis may be reduced. The pattern of synthesis, however, changes from one stage of the cycle to another (Kolodny and Gross, 1969). Histones, for example, are synthesized coincident with the S phase, and the elevation of certain enzyme activities has been associated with specific phases of the cycle. Since the amino acid composition of enzymes and proteins involved varies, omission of a single amino acid might effectively regulate a cell cycle event. Omission of asparagine from the medium of 6C3HED cultures resulted in a slower rate of DNA synthesis and a prolongation of the S phase (Ellem et al., 1970). In Jensen sarcoma cells, deletion of asparagine inhibited all phases of the cycle except mitosis (Patterson and Maxwell, 1970). However, when Shah (1963) added 5-methyltryptophan to cultures of hamster cells (Al) and HeLa cells, proteins synthesis was inhibited 75% but histone and DNA synthesis continued.

Effects of specific amino acid deprivations were observed in certain cell lines. EB_3 cell line of Burkitt lymphoblasts when cultured in "arginine-free" medium showed marked inhibition of protein and RNA synthesis, and the G_2 phase of the cell cycle was no longer detectable (Weinberg and Becker, 1970). DNA synthesis, however, was reduced only 50% under the experimental conditions. This response differed from that observed following the deletion of other essential amino acids. For example, cystine, valine, histidine, serine, phenylalanine, or glycine deprivation resulted in an inhibition of the synthesis of protein, RNA, and DNA and the disappearance of the S and G_2 phases. Threonine, tryptophan, leucine, methionine, isoleucine, or lysine deprivation caused a decrease in the macromolecular synthesis but did not markedly affect the S and G_2 phases.

Suspension cultures of Chinese hamster cells (line CHO), "arrested in G_1" by growing to high density, initiated DNA synthesis and divided in synchrony after addition of isoleucine and glutamine (Ley and Tobey, 1970). Deficiencies in other amino acids resulted in a random distribution of cells throughout the cell cycle. Subsequent studies with Syrian hamster (BHK21/C13) and mouse L929 cells showed that the accumulation of cells in G_1 resulted from a specific deletion of isoleucine from the culture medium (Tobey and Ley, 1971). These investigators proposed that isoleucine may be an essential constituent of a specific nucleoprotein involved in "genome" replication. Of particular interest in these studies is that the isoleucine effect was "difficult" to demonstrate in cells infected with PPLO.

6. ULTRASTRUCTURE ALTERATIONS

The effects of amino acid deprivation on protein, RNA, and DNA ultimately reflect alterations in cell ultrastructure. The most common observation following deletion of amino acids from the medium is a disappearance of polysomes, a phenomenon not unexpected because of its effect on the maturation of ribosomes. Restoration of complete medium, however, after 10 hours to cells such as the L cell (Chen et al., 1968) resulted in a rapid reestablishment of the initial concentration of polysomes. This suggests a rather stable cohort of mRNA and rRNA. A similar phenomenon is seen during metaphase, when the rate of protein synthesis is decreased (Steward et al., 1968). Reassembly of the polysomes and increased protein synthesis after mitosis (early G_1) occurred independently of de novo RNA synthesis, suggesting that RNA is passed from mother to daughter cells to direct protein synthesis in the early stages of the ensuing cell cycle.

Deprivation of glutamine in Chang liver cultures for 24 hours caused a reduction in the number of free ribosomes, polysomes, rough endoplasmic reticular profiles, and mitochondria (Kochhar, 1968). Restoration of glutamine caused an increase within 2 hours in the rough endoplasmic reticulum profile, polysomes, and free ribosomes, and within 5 hours had nearly regained the appearance of normal cells but with two well-defined zones of the particulates. Within 24 hours they had the structural features of cells growing exponentially. Contradictory results, showing the only "striking" change in Chang liver cells cultivated 20 hours in glutamine-deficient medium to be in mitochondrial morphology, have been reported (Jagendorf-Elfvin and Eliasson, 1969).

Chinese hamster cells (CH/3b) cultured in medium lacking any one of the essential amino acids increased the frequency of chromosome aberrations (Freed and Schatz, 1969). The aberrant metaphases observed were breaks and exchanges in chromosomes and chromatid and partial or complete "pulverization." The predominant effect was chromatid exchange, suggesting that breakage was more likely to have occurred after replication. They dismissed the possibility that the aberrations were directly correlated with histone synthesis by noting that omission of arginine and lysine, constituents of histone, caused aberrations similar to those seen with tryptophan and cystine, which are not normally found in the protein. Similar exchanges were reported in variant cells derived from Jensen sarcoma cells cultured in medium devoid of asparagine (Hsu and Manna, 1959).

7. ANALOGUES OF AMINO ACIDS

The use of amino acid analogues in studying the transport of amino acids across plasma membranes has already been discussed. Cell culture techniques have also provided a system for testing the efficacy of these analogues as therapeutic or cytotoxic agents and to aid in the elucidation of cellular mechanisms. For example, L-canavanine, and analogue of arginine that incorporates into protein (Kruse et al., 1959), reversibly inhibited DNA synthesis in polyoma-virus-transformed hamster cells (Hare, 1969). The inhibition did not involve a block in the synthesis of either RNA, histone, or nonhistone nuclear proteins, although subsequent studies suggested the involvement of abnormal canavanyl proteins critical to the process of DNA replication (Hare, 1970).

Studies by Sisken and Iwasaki (1969) showed differential effects of several amino acid analogues on protein synthesis, mitosis, and the cell cycle of the Fernandez line of human amnion cells. Whereas analogues of alanine, tryptophan, and methionine affected the duration of mitosis only, phenylalanine analogues, especially p-fluorophenylalanine, prolonged the duration of metaphase with little or relatively minor effects on the duration of the rest of the mitotic cycle. The effects of 2-β-thienylalanine and p-chlorophenylalanine confirmed earlier studies with p-fluorophenylalanine (Sisken and Wilkes, 1967) and indicated that a mitosis-related protein was synthesized prior to mitosis and was conserved and reutilized by daughter cells for their subsequent divisions.

IV. Conclusion

Cell culture techniques have been used to elucidate diverse biological problems; similarly, diversity has been extended to the selection of the literature cited in this survey. The studies discussed represent examples of recent applications of the techniques to the problems of amino acid uptake and utilization.

The classic studies of Christensen and associates on the transport of amino acids and its treatment as a problem of kinetics of mixed catalytic systems (Christensen, 1969) provided clues to the relationship of extracellular and intracellular levels of amino acids. The heterogeneity of the system is not unexpected because of the problems of not only transporting nutrients into the cell but also being responsible for the exodus of metabolites from the cell. Thus, as with any catalytic system the rate of a reaction is a function of substrate and product (the extracellular or intracellular component, whichever the case may be) as well as cofactors (Na^+ and K^+) and cosubstrates, e.g., other amino acids or related compounds showing an affinity for the transport site. Owing to these considerations, it is not surprising that variable results, dependent on the balance of nutrients (and not just amino acids) in culture medium, have been obtained. Moreover, this balance could conceivably determine whether a cell undergoes proliferation, retains or assumes a differentiated functional state, or is maintained as a nonfunctional entity.

The extent of plasma membrane alteration and its effects on amino acid transport is not clear. In viral-transformed cells or confluent populations, for example, where membrane alterations occur, transport appears normal. Treatment of cells with enzymes such as trypsin, pronase, or neuraminidase and their effects on transport have not been sufficiently studied, although evidence suggestive of an effect is available.

Hormonal effects on transport appear variable and may relate more to the synthesis than to the activation of "transport sites." Studies using "target tissues" of hormones, such as those described for norepinephrine and tryptophan transport in pineal glands may clarify the mechanisms involved. Other exogenous factors such as myoinositol and trace elements affect transport and emphasize their importance in formulating culture media.

The extensive number of studies on the uptake of exogenous amino acids by a variety of cells in culture have revealed a remarkably similar pattern. Usually glutamine, arginine, leucine, and isoleucine are used in greatest concentrations followed by lysine, valine, and phenylalanine. However, each cell line appears to have a characteristic pattern which reflects subsequent utilization of the amino acids for metabolic pathways. In rapidly proliferating cells the rate of uptake of essential amino acids on a per-cell basis increases proportionately and decreases with decreasing rate of proliferation. On the basis of cell protein content, however, the uptake correlates with the relative proportions of the amino acids in the cell protein. Future studies with a reduction in the variables imposed in culture systems and a clear definition of parameters (expression of data) should ultimately lead to better formulation of culture media.

The major portion of the amino acids utilized are directed toward protein synthesis. Through this process they indirectly control cellular processes. The specific role that they play in regulating their own incorporation into proteins has not been defined, although their function in the synthesis and maturation of the protein-synthesizing components are now under study.

The regulation of amino-acid-metabolizing enzymes, either by amino acids or hormones, suggests that in most cases the mechanisms are retained by cells in culture. The extent to which these mechanisms can be controlled by exogenous factors such as amino acids should lead to a better understanding of their role in metabolic disorders.

ACKNOWLEDGMENT

The author thanks Dr. D. E. Kizer and Dr. R. F. Kampschmidt for their critical appraisal of this manuscript and Mrs. Gwen Taft for her assistance in its final editing and typing.

REFERENCES

Abercrombie, M. (1970). Contact inhibition in tissue culture. In Vitro 6, 128–142.

Alescio, T., and Moscona, A. A. (1969). Immunochemical evidence for enzyme synthesis in the hormonal induction of glutamine synthetase in embryonic retina in culture. Biochem. Biophys. Res. Commun. 34, 176–182.

Alescio, T., Moscona, M. H., and Moscona, A. A. (1970). Induction of glutamine synthetase in embroynic retina. Effects of partial and complete inhibition of RNA synthesis on enzyme accumulation. Exp. Cell Res. 61, 342–346.

Auricchio, F., Martin, D., Jr., and Tomkins, G. M. (1969). Control of degradation and synthesis of induced tyrosine aminotransferase studied in hepatoma cells in culture. Nature (London) 224, 806–808.

Axelrod, J., Shein, H. M., and Wurtman, R. J. (1969). Stimulation of C^{14}-melatonin synthesis from C^{14}-tryptophan by noradrenaline in rat pineal in organ culture. Proc. Nat. Acad. Sci. U.S. 62, 544–549.

Barban, S., and Schulze, H. O. (1959). Transamination reactions of mammalian cells in tissue culture. J. Biol. Chem. 234, 829–831.

Baril, E. F., Potter, V. R., and Morris, H. P. (1969). Amino acid transport in rat liver and Morris hepatomas: Effect of protein diet and hormones on the uptake of α-aminoisobutyric acid-^{14}C. Cancer Res. 29, 2101–2115.

Berlin, R. D. (1970). Specificities of transport systems and enzymes. Science 168, 1539–1545.

Bernhardt, D., and Darnell, J. E., Jr. (1969). tRNA synthesis in HeLa cells: A precursor to tRNA and the effects of methionine starvation on tRNA synthesis. *J. Mol. Biol.* **42**, 43–56.

Birge, S. J., Jr., and Peck, W. A. (1966). Collagen synthesis by isolated bone cells. *Biochem. Biophys. Res. Commun.* **22**, 532–539.

Broome, J. D. (1968). L-Asparaginase: The evolution of a new tumor inhibitory agent. *Trans. N.Y. Acad. Sci.* [2] **30**, 690–704.

Brown, D. M., and Michael, A. F. (1969). Effect of neuraminidase on the accumulation of alpha-aminoisobutyric acid in HeLa cells. *Proc. Soc. Exp. Biol. Med.* **131**, 568–570.

Castor, L. N. (1968). Contact regulation of cell division in an epithelial-like cell line. *J. Cell. Physiol.* **72**, 161–172.

Charalampous, F. C. (1969). Metabolic functions of myo-inositol. VII. Role of inositol in the transport of α-aminoisobutyric acid in KB cells. *J. Biol. Chem.* **244**, 1705–1710.

Chen, H. W., Hersh, R. T., and Kitos, P. A. (1968). Environmental effects on the polysome content of artificially cultured mouse cells. *Exp. Cell Res.* **52**, 490–498.

Christensen, H. N. (1969). Some special kinetic problems of transport. *Advan. Enzymol.* **32**, 1–20.

Christensen, H. N., and Handlogten, M. E. (1968). Modes of mediated exodus of amino acids from the Ehrlich ascites tumor cell. *J. Biol. Chem.* **243**, 5428–5438.

Chung, R. A., Brown, R. W., Beacham, L. H., and Huang, I. (1966). Amino acid metabolism of different mammalian cell lines. *Can. J. Biochem.* **44**, 1145–1157.

Cunningham, D. D., and Pardee, A. B. (1969). Transport changes rapidly initiated by serum addition to "contact inhibited" 3T3 cells. *Proc. Nat. Acad. Sci. U.S.* **64**, 1049–1056.

DeMars, R. (1958). The inhibition by glutamine of glutamyl transferase formation in cultures of human cells. *Biochim. Biophys. Acta* **27**, 435–436.

Demopoulos, H. B (1966). Effects of low phenylalanine-tyrosine diets on S91 mouse melanomas. *J. Nat. Cancer Inst.* **37**, 185–190.

Eagle, H. (1965). Metabolic controls in cultured mammalian cells. *Science* **148**, 42–51.

Eagle, H., and Levintow, L. (1965). Amino acid and protein metabolism. I. The metabolic characteristics of serially propagated cells. *In* "Cells and Tissues in Culture" (E. N. Wilmer, ed.), Vol. 1, pp. 277–296. Academic Press, New York.

Eagle, H., Piez, K. A., and Fleischman, R. (1957). The utilization of phenylalanine and tyrosine for protein synthesis by human cells in tissue culture. *J. Biol. Chem.* **228**, 847–861.

Eagle, H., Piez, K. A., and Levy, M. (1961a). The intracellular amino acid concentrations required for protein synthesis in cultured human cells. *J. Biol. Chem.* **236**, 2039–2042.

Eagle, H., Piez, K. A., and Oyama, V. I. (1961b). The biosynthesis of cystine in human cell cultures. *J. Biol. Chem.* **236**, 1425–1428.

Eagle, H., Washington, C. L., and Levy, M. (1965). End product control of amino acid synthesis by cultured human cells. *J. Biol. Chem.* **240**, 3944–3950.

Eagle, H., Washington, C. L., Levy, M., and Cohen, L. (1966a). The population-dependent requirement by cultured mammalian cells for metabolites which they can synthesize. II. Glutamic acid and glutamine; aspartic acid and asparagine. *J. Biol. Chem.* **241**, 4994–4999.

Eagle, H., Washington, C. L., and Friedman, S. M. (1966b). The synthesis of homocystine, cystathionine, and cystine by cultured diploid and heteroploid human cells. *Proc. Nat. Acad. Sci. U.S.* **56**, 156–163.

Eliasson, E. E. (1967a). Repression of arginase synthesis in Chang liver cells. *Exp. Cell Res.* **48**, 1–17.

Eliasson, E. E. (1967b). Regulation of arginase activity in Chang liver cells in the absence of net protein synthesis. *Biochem. Biophys. Res. Commun.* **27**, 661–667.

Eliasson, E. E., and Streckcr, H. J. (1966). Arginase activity during the growth cycle of Chang's liver cells. *J. Biol. Chem.* **241**, 5757–5763.

Ellem, K. A. O., Fabrizio, A. M., and Jackson, L. (1970). The dependence of DNA and RNA synthesis on protein synthesis in asparaginase-treated lymphoma cells. *Cancer Res.* **30**, 515–527.

Farnham, A. E., and Dubin, D. T. (1965). Effect of puromycin aminonucleoside on RNA synthesis in L cells. *J. Mol. Biol.* **14**, 55–62.

Foley, G. E., Barell, E. F., Adams, R. A., and Lazarus, H. (1969). Nutritional requirements of human leukemic cells. Cystine requirements of diploid cell lines and their heteroploid variants. *Exp. Cell Res.* **57**, 129–133.

Foley, J. F., and Aftonomos, B. (1970). The use of pronase in tissue culture: A comparison with trypsin. *J. Cell. Physiol.* **75**, 159–161.

Foster, D. O., and Pardee, A. B. (1969). Transport of amino acids by confluent and nonconfluent 3T3 and polyoma virus-transformed 3T3 cells growing on glass cover slips. *J. Biol. Chem.* **244**, 2675–2681.

Freed, J. J., and Schatz, S. A. (1969). Chromosome aberrations in cultured cells deprived of single essential amino acids. *Exp. Cell Res.* **55**, 393–409.

Gallo, R. C., and Pestka, S. (1970). Transfer RNA species in normal and leukemic human lymphoblasts, *J. Mol. Biol.* **52**, 195–219.

Gallo, R. C., Longmore, J. L., and Adamson, R. H. (1970). Asparaginyl-tRNA and resistance of murine leukemias to L-asparaginase. *Nature (London)* **227**, 1134–1136.

Gelehrter, T. D., and Tomkins, G. M. (1969). Control of tyrosine aminotransferase synthesis in tissue culture by a factor in serum. *Proc. Nat. Acad. Sci. U.S.* **64**, 723–730.

Gerner, E. W., Glick, M. C., and Warren, L. (1970). Membranes of animal cells. V. Biosynthesis of the surface membrane during the cell cycle. *J. Cell. Physiol.* **75**, 275–280.

Gerschenson, L. E., Andersson, M., Molson, J., and Okigaki, T. (1970). Tyrosine transaminase induction by dexamethasone in a new rat liver cell line. *Science* **170**, 859–861.

Gold, J. (1970). Inhibition of Walker 256 intramuscular carcinoma in rats by administration of L-tryptophan. *Oncology* **24**, 291–303.

Goldberg, B., and Green, H. (1967). Collagen synthesis on polyribosomes of cultured mammalian fibroblasts. *J. Mol. Biol.* **26**, 1–18.

Green, H., and Goldberg, B. (1964). Collagen and cell protein synthesis by an established mammalian fibroblast line. *Nature (London)* **204**, 347–349.

Green, II., and Todaro, G. J. (1967). The mammalian cell as differentiated microorganism. *Annu. Rev. Microbiol.* **21**, 573–600.

Griffin, M. J., and Cox, R. P. (1966). Studies on the mechanism of hormone induction of alkaline phosphatase in human cell cultures. II. Rate of enzyme synthesis and properties of base level and induced enzymes. *Proc. Nat. Acad. Sci. U.S.* **56**, 946–953.

Griffin, M. J., and Cox, R. P. (1967). Studies on the mechanism of substrate induction and L-cyst(e)ine repression of alkaline phosphatase in mammalian cell cultures. *J. Cell Sci.* **2**, 545–555.

Griffiths, J. B. (1970a). The quantitative utilization of amino acids and glucose and contact inhibition of growth in cultures of the human diploid cell, WI-38. *J. Cell Sci.* **6**, 739–749.

Griffiths, J. B. (1970b). The effects of insulin on the growth and metabolism of the human diploid cell, WI-38. *J. Cell Sci.* **7**, 575–585.

Griffiths, J. B., and Pirt, S. J. (1967). The uptake of amino acids by mouse cells (Strain LS) during growth in batch culture and chemostat culture: The influence of cell growth rate. *Proc. Roy. Soc., Ser. B.* **168**, 421–438.

Grossman, A., and Mavrides, C. (1967). Studies on the regulation of tyrosine aminotransferase in rats. *J. Biol. Chem.* **242**, 1398–1405.

Guidotti, G. G., Lüneburg, B., and Borghetti, A. F. (1969). Amino acid uptake in isolated chick embryo heart cells. Effect of insulin. *Biochem. J.* **114**, 97–105.

Gurley, L. R., and Hardin, J. M. (1968). The metabolism of histone fractions. I. Synthesis of histone fractions during the life cycle of mammalian cells. *Arch. Biochem. Biophys.* **128**, 285–292.

Haley, E. E., Fischer, G. A., and Welch, A. D. (1961). The requirement for L-asparagine of mouse leukemia cells L5178Y in culture. *Cancer Res.* **21**, 532–536.

Hare, J. D. (1967). Location and characteristics of the phenylalanine transport mechanism in normal and polyoma-transformed hamster cells. *Cancer Res.* **27**, 2357–2363.

Hare, J. D. (1969). Reversible inhibition of DNA synthesis by the arginine analogue canavanine in hamster and mouse cells *in vitro*. *Exp. Cell Res.* **58**, 170–174.

Hare, J. D. (1970). Nuclear alterations in mammalian cells induced by L-canavanine. *J. Cell. Physiol.* **75**, 129–131.

Hayes, E. C., and Kuchler, R. J. (1970). The influence of Na⁺ and K⁺ on amino acid accumulation in L strain fibroblasts. *Fed. Proc., Fed. Amer. Soc. Exp. Biol.* **29**, 539 (abstr. 1639).

Higuchi, K. (1970). An improved chemically defined culture medium for strain L mouse cells based on growth responses to graded levels of nutrients including iron and zinc ions. *J. Cell. Physiol.* **75**, 65–72.

Horowitz, B., Madras, B. K., Meister, A., Old, L. J., Boyse, E. A., and Stockert, E. (1968). Asparagine synthetase activity of mouse leukemias. *Science* **160**, 533–535.

Hsia, D. Y. (1970). Study of hereditary metabolic diseases using *in vitro* techniques. *Metab., Clin. Exp.* **19**, 309–339.

Hsu, T. C., and Manna, G. K. (1959). High frequency of chromatid breaks in two *in vitro* cell populations. *Amer. Natur.* **93**, 207–208.

Jagendorf-Elfvin, M., and Eliasson, E. E. (1969). Reversible changes in the ultrastructure of Chang liver cell mitochondria following incubation of the cells in a glutamine-deficient medium. *J. Cell Biol.* **41**, 905–909.

Jepson, J. B. (1966). Hartnup disease. *In* "The Metabolic Basis of Inherited Disease" (J. B. Stanbury, J. B. Wyngaarden, and D. S. Fredrickson, eds.), 2nd ed., p. 1283. McGraw-Hill, New York.

Kagawa, Y., Kaneko, K., Takaoka, T., and Katsuta, H. (1960). Amino acid consumption by strain L cells (mouse fibroblasts) in protein-free media. *Jap. J. Exp. Med.* **30**, 95–113.

Kenney, F. T. (1967). Turnover of rat liver tyrosine transaminase: Stabilization after inhibition of protein synthesis. *Science* **156**, 525–528.

Kessel, D., and Bosmann, H. B. (1970). Effects of L-asparaginase on protein and glycoprotein synthesis. *FEBS Lett.* **10**, 85–88.

Kipnis, D. M., Reiss, E., and Helmreich, E. (1961). Functional heterogeneity of the intracellular amino acid pool in mammalian cells. *Biochim. Biophys. Acta* **51**, 519–524.

Kirk, D. L., and Moscona, A. A. (1963). Synthesis of experimentally induced glutamine synthetase (glutamotransferase activity) in embryonic chick retina *in vitro*. *Develop. Biol.* **8**, 341–357.

Kitos, P. A., and Waymouth, C. (1966). The metabolism of L-glutamate and L-5-carboxypyrrolidone by mouse cells (NCTC Clone 929) under conditions of defined nutrition. *J. Cell. Physiol.* **67**, 383–398.

Klein, D. C., and Weller, J. (1970). Input and output signals in a model neural system: The regulation of melatonin production in the pineal gland. *In Vitro* **6**, 197–204.

Klein, D. C., Berg, G. R., Weller, J., and Glinsmann, W. (1970). Pineal gland: Dibutyryl cyclic adenosine monophosphate stimulation of labeled melatonin production. *Science* **167**, 1738–1740.

Knox, W. E. (1966). Cystinuria *In* "The Metabolic Basis of Inherited Disease" (J. B. Stanbury, J. B. Wyngaarden, and D. S. Fredrickson, eds.), 2nd ed., p. 1262. McGraw-Hill, New York.

Kochhar, O. S.(1968). Effect of glutamine starvation on the ultrastructural organization of Chang's cultured cells. *Exp. Cell Res.* **49**, 598–611.

Kolodny, G. M., and Gross, P. R. (1969). Changes in patterns of protein synthesis during the mammalian cell cycle. *Exp. Cell Res.* **56**, 117–121.

Krawitt, E. L., Baril, E. F., Becker, J. E., and Potter, V. R. (1970). Amino acid transport in hepatoma cell cultures during tyrosine aminotransferase induction. *Science* **169**, 294–296.

Kruse, P. F., Jr. (1961). Arginine metabolism in cell cultures of the Walker carcinosarcoma 256 and Jensen sarcoma. *Pathol. Biol.* **9**, 576–578.

Kruse, P. F., Jr., and Miedema, E. (1965). Production and characterization of multiple-layered populations of animal cells. *J. Cell Biol.* **27**, 273–279.

Kruse, P. F., Jr., White, P. B., Carter, H. A., and McCoy, T. A. (1959). Incorporation of canavanine into protein of Walker carcinosarcoma 256 cells cultured *in vitro*. *Cancer Res.* **19**, 122–125.

Kruse, P. F., Jr., Miedema, E., and Carter, H. A. (1967). Amino acid utilizations and protein synthesis at various proliferation rates, population densities, and protein contents of perfused animal cell and tissue cultures. *Biochemistry* **6**, 949–955.

Kruse, P. F., Jr., Keen, L. N., and Whittle, W. L. (1970). Some distinctive characteristics of high density perfusion cultures of diverse cell types. *In Vitro* **6**, 75–88.

Kubinski, H., and Koch, G. (1966). Regulation of the synthesis of various ribonucleic acids in animal cells. *Biochem. Biophys. Res. Commun.* **22**, 346–351.

Kuchler, R. J. (1964). The flexible nature of the amino acid pool in L strain fibroblasts. *Proc. Soc. Exp. Biol. Med.* **116**, 20–25.

Kuchler, R. J. (1967). The role of sodium and potassium in regulating amino acid accumulation and protein synthesis in LM-strain mouse fibroblasts. *Biochim. Biophys. Acta* **136**, 473–483.

Kuchler, R. J., and Marlowe-Kuchler, M. (1965). The transport and accumulation of α-aminoisobutyric acid into L-strain mouse fibroblasts. *Biochim. Biophys. Acta* **102**, 226–234.

Lee, K., and Kenney, F. T. (1970). Induction of alanine transaminase by adrenal steroids in cultured hepatoma cells. *Biochem. Biophys. Res. Commun.* **40**, 469–475.

Lembach, K., and Charalampous, F. C. (1967). Metabolic functions of myoinositol. V. Utilization of glycine and serine in nucleotide and nucleic acid biosynthesis by inositol-deficient KB cells. *J. Biol. Chem.* **242**, 2599–2605.

Levintow, L. (1957). Evidence that glutamine is a precursor of asparagine in a human cell in tissue culture. *Science* **126**, 611–612.

Levintow, L., Eagle, H., and Piez, K. A. (1957). The role of glutamine in protein biosynthesis in tissue culture. *J. Biol. Chem.* **227**, 929–941.

Ley, K. D., and Tobey, R. A. (1970). Regulation of initiation of DNA synthesis in Chinese hamster cells. II. Induction of DNA synthesis and cell division by isoleucine and glutamine in G_1-arrested cells in suspension culture. *J. Cell Biol.* **47**, 453–459.

Ling, C. T., Gey, G. O., and Richters, V. (1968). Chemically characterized concentrated corodies for continuous cell culture (the 7 C's culture media). *Exp. Cell Res.* **52**, 469–489.

Lucy, J. A. (1960). The amino acid and protein metabolism of tissues cultivated *in vitro. Biol. Rev.* **35**, 533–571.

Lucy, J. A., and Rinaldini, L. M. (1959). The amino acid metabolism of differentiating skeletal myoblasts *in vitro. Exp. Cell Res.* **17**, 385–398.

McCarty, K. (1962). Selective utilization of amino acids by mammalian cell cultures. *Exp. Cell Res.* **27**, 230–240.

McCoy, T. A. (1960). Neoplasia and nutrition. *World Rev. Nutr. Diet.* **1**, 181–203.

McCoy, T. A. (1961). Some aspects of asparagine metabolism in the Jensen and JA sarcomas *in vitro. Pathol. Biol.* **0**, 574–575.

Maden, B. E. H., and Vaughan, M. H. Jr. (1968). Synthesis of ribosomal proteins in the absence of ribosome maturation in methionine-deficient HeLa cells. *J. Mol. Biol.* **38**, 431–435.

Maden, B. E. H., Vaughan, M. H., Warner, J. R., and Darnell, J. E. Jr. (1969). Effects of valine deprivation on ribosome formation in HeLa cells. *J. Mol. Biol.* **45**, 265–275.

Majumdar, A. (1969). Behavior of glutaminase in normal cells cultured *in vitro. Indian J. Cancer* **6**, 44–48.

Martin, D., Jr., Tomkins, G. M., and Granner, D. (1969). Synthesis and induction of tyrosine aminotransferase in synchronized hepatoma cells in culture. *Proc. Nat. Acad. Sci. U.S.* **62**, 248–255.

Meister, A. (1965). "Biochemistry of the Amino Acids," 2nd ed., Vol. 2, pp. 593–1020. Academic Press, New York.

Meister, A. (1968). On the synthesis and utilization of glutamine. *Harvey Lect.* **63**, 139–178.

Mohberg, J., and Johnson, M. J. (1963). Amino acid utilization by 929-L fibroblasts in chemically defined media. *J. Nat. Cancer Inst.* **31**, 611–625.

Mohri, T. (1967). Effects of cortisol and 19-nortestosterone on the free amino acid pools and amino acid uptake of cultured cells. *Endocrinology* **81**, 454–460.

Morris, J. E., and Moscona, A. A. (1970). Induction of glutamine synthetase in embryonic retina: Its dependence on cell interactions. *Science* 167, 1736–1738.

Moscona, A. A., Moscona, M. H., and Saenz, N. (1968). Enzyme induction in embryonic retina: The role of transcription and translation. *Proc. Nat. Acad. Sci. U.S.* 61, 160–167.

Moscona, A. A., Moscona, M. H., and Jones, R. E. (1970). Induction of glutamine synthetase in embryonic neural retina *in vitro* by inhibitors of macromolecular synthesis. *Biochem. Biophys. Res. Commun.* 39, 943–949.

Neuman, R. E., and McCoy, T. A. (1956). Dual requirement of Walker carcinosarcoma 256 *in vitro* for asparagine and glutamine. *Science* 124, 124–125.

Neuman, R. E., and McCoy, T. A. (1958). Growth-promoting properties of pyruvate, oxalacetate, and α-ketoglutarate for isolated Walker carcinosarcoma 256 cells. *Proc. Soc. Exp. Biol. Med.* 98, 303–306.

Neuman, R. E., and Tytell, A. A. (1960). Stimulatory effects of glycine, L-serine, folic acid and related compounds on growth of cell cultures. *Proc. Soc. Exp. Biol. Med.* 103, 762–767.

Pasieka, A. E., Morton, H. J., and Morgan, J. F. (1958a). The metabolism of animal tissues cultivated *in vitro*. II. Amino acid metabolism of chick embryonic kidney, chick embryonic liver, and monkey kidney cortex cultures. *Can. J. Biochem. Physiol.* 36, 171–184.

Pasieka, A. E., Morton, H. J., and Morgan, J. F. (1958b). The metabolism of animal tissues cultivated *in vitro*. III. Amino acid metabolism of strain L cells in completely synthetic media. *Can. J. Biochem. Physiol.* 36, 771–782.

Pasieka, A. E., Morton, H. J., and Morgan, J. F. (1960). The metabolism of animal tissues cultivated *in vitro*. IV. Comparative studies on human malignant cells. *Cancer Res.* 20, 362–367.

Patterson, M. K., Jr. (1970a). The role of cell culture in the development of L-asparaginase as a therapeutic agent. *Ann. Okla. Acad. Sci.* 1, 1–11.

Patterson, M. K., Jr. (1970b). Studies on the control of asparagine biosynthesis in mammalian tissues. *Recent Results Cancer Res.* 33, 22–30.

Patterson, M. K., Jr. (1971). Effects of L-asparaginase on asparagine synthetase levels of normal and malignant tissues. *Colloq. Intern. Centre Natl. Rech. Scien* (*Paris*) 197 (to be published).

Patterson, M. K., Jr., and Maxwell, M. D. (1970). Effects of L-asparagine deprivation on the cell cycle of the Jensen sarcoma. *Cancer Res.* 30, 1064–1067.

Patterson, M. K., Jr., and Orr, G. R. (1967). L-asparagine biosynthesis by nutritional variants of the Jensen sarcoma. *Biochem. Biophys. Res. Commun.* 26, 228–233.

Patterson, M. K., Jr., and Orr, G. R. (1968). Asparagine biosynthesis by the Novikoff hepatoma. Isolation, purification, property, and mechanism studies of the enzyme system. *J. Biol. Chem.* 243, 376–380.

Patterson, M. K., Jr., and Orr, G. R. (1969). Regeneration, tumor, dietary, and L-asparaginase effects on asparagine biosynthesis in rat liver. *Cancer Res.* 29, 1179–1183.

Patterson, M. K., Jr., Orr, G. R., and Conway, E. (1969a). Studies on the aspartic acid "sparing effect" on the nutritional requirement of L-asparagine for tumors *in vitro*. *Proc. Soc. Exp. Biol. Med.* 131, 131–134.

Patterson, M. K., Jr., Maxwell, M. D., and Conway, E. (1969b). Studies on the asparagine requirement of the Jensen sarcoma and the derivation of its nutritional variant. *Cancer Res.* 29, 296–300.

Paul, J., and Fottrell, P. F. (1963). Mechanism of D-glutamyltransferase repression in mammalian cells. *Biochim. Biophys. Acta* **67**, 334–336.

Paul, J., Fottrell, P. F., Freshney, I., Jondorf, W. R., and Struthers, M. G. (1964). Regulation of enzyme synthesis in cultured cells. *Nat. Cancer Inst., Monogr.* **13**, 219–228.

Piddington, R., and Moscona, A. A. (1965). Correspondence between glutamine synthetase activity and differentiation in the embryonic retina *in situ* and in culture. *J. Cell Biol.* **27**, 247–252.

Piez, K. A., and Eagle, H. (1958). The free amino acid pool of cultured human cells. *J. Biol. Chem.* **231**, 533–545.

Pitot, H. C., and Jost, J. (1967). Control of biochemical expression in morphologically related cells *in vivo* and *in vitro*. *Nat. Cancer Inst., Monogr.* **26**, 145–166.

Pitot, H. C., Peraino, C., Morse, P. A., Jr., and Potter, V. R. (1964). Hepatomas in tissue culture compared with adapting liver *in vivo*. *Nat. Cancer Inst., Monogr.* **13**, 229–246.

Pizer, L. I. (1964). Enzymology and regulation of serine biosynthesis in cultured human cells. *J. Biol. Chem.* **239**, 4219–4226.

Platter, H., and Martin, G. M. (1966). Tryptophane transport in cultures of human fibroblasts. *Proc. Soc. Exp. Biol. Med.* **123**, 140–143.

Prager, M. D., and Bachynsky, N. (1968). Asparagine synthetase in normal and malignant tissues: Correlation with tumor sensitivity to asparaginase. *Arch. Biochem. Biophys.* **127**, 645–654.

Prockop, D. J., Pettengill, O., and Holtzer, H. (1964). Incorporation of sulfate and the synthesis of collagen by cultures of embryonic chrondrocytes. *Biochim. Biophys. Acta* **83**, 189–196.

Reel, J. R., and Kenney, F. T. (1968). "Superinduction" of tyrosine transaminase in hepatoma cell cultures: Differential inhibition of synthesis and turnover by actinomycin D. *Proc. Nat. Acad. Sci. U.S.* **61**, 200–206.

Regan, J. D., Vodopick, H., Takeda, S., Lee, W. H., and Faulcon, F. M. (1969). Serine requirements in leukemic and normal blood cells. *Science* **163**, 1452–1453.

Reporter, M. (1969). 3-methylhistidine metabolism in proteins from cultured mammalian muscle cells. *Biochemistry* **8**, 3489–3496.

Rhode, S. L., III, and Ellem, K. A. O. (1968). Control of nucleic acid synthesis in human diploid cells undergoing contact inhibition. *Exp. Cell Res.* **53**, 184–204.

Rickenberg, H. V. (1961). Discussion. *Cold Spring Harbor Symp. Quant. Biol.* **26**, 366.

Robbins, E., and Borun, T. W. (1967). The cytoplasmic synthesis of histones in HeLa cells and its temporal relationship to DNA replication. *Proc. Nat. Acad. Sci. U.S.* **57**, 409–416.

Roberts, J., Holcenberg, J. S., and Dolowy, W. C. (1970). Antineoplastic activity of highly purified bacterial glutaminases. *Nature (London)* **227**, 1136–1137.

Rosenbloom, J., Bhatnagar, R. S., and Prockop, D. J. (1967). Hydroxylation of proline after the release of proline-rich polypeptides from ribosomal complexes during uninhibited collagen biosynthesis. *Biochim. Biophys. Acta* **149**, 259–272.

Rosenbloom, J., Blumenkrantz, N., and Prockop, D. J. (1968). Sequential hydroxylation of lysine and glycosylation of hydroxylysine during the biosynthesis of collagen in isolated cartilage. *Biochem. Biophys. Res. Commun.* **31**, 792–797.

Salzman, N. P., Eagle, H., and Sebring, E. D. (1958). The utilization of glutamine, glutamic acid and ammonia for the biosynthesis of nucleic acid bases in mammalian cell cultures. *J. Biol. Chem.* **230**, 1001–1012.

Schimke, R. T. (1964a). Enzymes of arginine metabolism in cell culture: Studies on enzyme induction and repression. *Nat. Cancer Inst., Monogr.* **13**, 197–218.

Schimke, R. T. (1964b). Enzymes of arginine metabolism in mammalian cell culture. I. Repression of arginosuccinate synthetase and argininosuccinase. *J. Biol. Chem.* **239**, 136–145.

Schimke, R. T. (1969). On the role of synthesis and degradation in regulation of enzyme levels in mammalian tissues. *In* "Current Topics in Cellular Regulation" (B. L. Horecker and E. R. Stadtman, eds.), Vol. 1, pp. 77–124. Academic Press, New York.

Schimke, R. T., and Barile, M. F. (1963). Arginine breakdown in mammalian cell culture contaminated with pleuropneumonia-like organisms (PPLO). *Exp. Cell Res.* **30**, 593–596.

Schimke, R. T., and Doyle, D. (1970). Control of enzyme levels in animal tissues. *Annu. Rev. Biochem.* **39**, 929–976.

Schingoethe, D. J., Hageman, E. C., and Larson, B. L. (1967). Essential amino acids for milk protein synthesis in the *in vitro* secretory cell and stimulation by elevated levels. *Biochim. Biophys. Acta* **148**, 469–474.

Scriver, C. R. (1969). Inborn errors of amino acid metabolism. *Brit. Med. Bull.* **25**, 35–41.

Sefton, B. M., and Rubin, H. (1970). Release from density dependent growth inhibition by proteolytic enzymes. *Nature (London)* **227**, 843–845.

Shah, V. C. (1963). Autoradiographic studies of the effects of antibiotics, amino acid analogs, and nucleases on the synthesis of DNA in cultured mammalian cells. *Cancer Res.* **23**, 1137–1147.

Sisken, J. E., and Iwasaki, T. (1969). The effects of some amino acid analogs on mitosis and the cell cycle. *Exp. Cell Res.* **55**, 161–167.

Sisken, J. E., and Wilkes, E. (1967). The time of synthesis and the conservation of mitosis-related proteins in cultured human amnion cells. *J. Cell Biol.* **34**, 97–110.

Smulson, M. E. (1970). Amino acid deprivation of human cells: Effects on RNA synthesis, RNA polymerase, and ribonucleoside phosphorylation. *Biochim. Biophys. Acta* **199**, 537–540.

Smulson, M. E., and Thomas, J. (1969). Ribonucleic acid biosynthesis of human cells during amino acid deprivation. *J. Biol. Chem.* **244**, 5309–5312.

Stecher, V. J., and Thorbecke, G. J. (1967). Sites of synthesis of serum proteins. I. Serum proteins produced by macrophages *in vitro. J. Immunol.* **99**, 643–652.

Stein, W. D. (1968). The transport of sugars. *Brit. Med. Bull.* **24**, 146–149.

Steinberger, A., and Steinberger, E. (1966). Stimulatory effect of vitamins and glutamine on the differentiation of germ cells in rat testes organ culture grown in chemically defined media. *Exp. Cell. Res.* **44**, 429–435.

Steward, D. L., Shaeffer, J. R., and Humphrey, R. M. (1968). Breakdown and assembly of polyribosomes in synchronized Chinese hamster cells. *Science* **161**, 791–793.

Stoker, M. G. P., Shearer, M., and O'Neill, C. (1966). Growth inhibition of polyoma-transformed cells by contact with static normal fibroblasts. *J. Cell Sci.* **1**, 297–310.

Stoner, G. D. (1970). Amino acid utilization by L-M strain mouse cells in a chemically defined medium. Ph.D. Dissertation, University of Michigan.

Strecker, H. J., and Eliasson, E. E. (1966). Ornithine δ-transaminase activity during the growth cycle of Chang's liver cells. *J. Biol. Chem.* **241**, 5750–5756.

Strecker, H. J., Hammar, U. B., and Volpe, P. (1970). Methionine toxicity and ornithine δ-aminotransferase in Chang's liver cells. *J. Biol. Chem.* **245**, 3328–3334.

Studzinski, G. P., and Ellem, K. A. O. (1968). Differences between diploid and heteroploid cultured mammalian cells in their response to puromycin aminonucleoside. *Cancer Res.* **28**, 1773–1782.

Tedesco, T. A., and Mellman, W. J. (1967). Argininosuccinate synthetase activity and citrulline metabolism in cells cultured from a citrullinemic subject. *Proc. Nat. Acad. Sci. U.S.* **57**, 829–834.

Thompson, E. B., Tomkins, G. M., and Curran, J. F. (1966). Induction of tyrosine α-ketoglutarate transaminase by steroid hormones in a newly established tissue culture cell line. *Proc. Nat. Acad. Sci. U.S.* **56**, 296–303.

Tobey, R. A., and Ley, K. D. (1971). Isoleucine-mediated regulation of genome replication in various mammalian cell lines. *Cancer Res.* **31**, 46–51.

Tomkins, G. M., Gelehrter, T. D., Granner, D., Martin, D., Jr., Samuels, H. H., and Thompson, E. B. (1969). Control of specific gene expression in higher organisms. *Science* **166**, 1474–1480.

Tourian, A., Goddard, J., and Puck, T. T. (1969). Phenylalanine hydroxylase activity in mammalian cells. *J. Cell. Physiol.* **73**, 159–170.

Ubuka, T., and Meister, A. (1971). Incorporation of C^{14}-amide-^{15}N-L-asparagine and amide-^{15}N-L-glutamine by mouse leukemic cells. *Colloq. Intern. Centre Natl. Rech. Scien. (Paris)* **197** (to be published).

Uhlendorf, B. W., and Mudd, S. H. (1968). Cystathionine synthase in tissue culture derived from human skin: Enzyme defect in homocystinuria. *Science* **160**, 1007–1009.

Ulrich, K., Tritsch, G. L., and Moore, G. E. (1968). Tyrosine utilization by pigmented hamster melanoma cells cultured *in vitro*. *Int. J. Cancer* **3**, 446–453.

van Furth, R., Schuit, H. R. E., and Hijmans, W. (1966). The formation of immunoglobulins by human tissues *in vitro*. III. Spleen, lymph nodes, bone marrow and thymus. *Immunology* **11**, 19–27.

Vaughan, M. H., Jr., Soeiro, R., Warner, J. R., and Darnell, J. E., Jr. (1967). The effects of methionine deprivation on ribosome synthesis in HeLa cells. *Proc. Nat. Acad. Sci. U.S.* **58**, 1527–1534.

Weinberg, A., and Becker, Y. (1970). Effect of arginine deprivation on macromolecular processes in Burkitt's lymphoblasts. *Exp. Cell Res.* **60**, 470–474.

Wiebel, F., and Baserga, R. (1969). Early alterations in amino acid pools and protein synthesis of diploid fibroblasts stimulated to synthesize DNA by addition of serum. *J. Cell. Physiol.* **74**, 191–202.

Wurtman, R. J., Larin, F., Axelrod, J., Shein, H. M., and Rosasco, K. (1968). Formation of melatonin and 5-hydroxyindole acetic acid from ^{14}C-tryptophan by rat pineal glands in organ culture. *Nature (London)* **217**, 953–954.

Wurtman, R. J., Shein, H. M., Axelrod, J., and Larin, F. (1969). Incorporation of ^{14}C-tryptophan into ^{14}C-protein by cultured rat pineals: Stimulation by L-norepinephrine. *Proc. Nat. Acad. Sci. U.S.* **62**, 749–755.

Wyngaarden, J. B., and Seagle, S. (1966). The hyperglycinurias In "The Metabolic Basis of Inherited Disease" (J. B. Stanbury, J. B. Wyngaarden, and D. S. Fredrickson, eds.), 2nd ed., p. 341. McGraw-Hill, New York.

7

PURINE AND PYRIMIDINE METABOLISM

OF CELLS IN CULTURE

William N. Kelley

Glossary

(Abbreviations used in text, figures, and tables)

A	Adenine	ATP	Adenosine 5′-triphosphate
AR	Adenosine	dAR	Deoxyadenosine
AMP	Adenosine 5′-monophosphate	dAMP	Deoxyadenosine 5′mono-
ADP	Adenosine 5′-diphosphate		phosphate

dADP	Deoxyadenosine 5'-diphosphate	dGTP	Deoxyguanosine 5'-triphosphate
dATP	Deoxyadenosine 5'-triphosphate	H	Hypoxanthine
AlCAR	5'-Aminoimidazole-4-carboxamide ribonucleotide	HR	Inosine
		IMP	Inosine 5'-monophosphate
		NAD	Nicotinamide adenine dinucleotide
AMP-S	Adenylosuccinic acid	NADP	Nicotinamide adenine dinucleotide phosphate
CMP	Cytidine 5'-monophosphate		
CDP	Cytidine 5'-diphosphate	OMP	Orotidine 5'-monophosphate
CTP	Cytidine 5'-triphosphate	PP-ribose-P	5-Phosphoribosyl-1-pyrophosphate
dCR	Deoxycytidine		
dCMP	Deoxycytidine 5'-monophosphate	PRA	Phosphoribosylamine
		SAlCAR	5-Aminoimidazole-4-(N-succinylcarboxamide) ribonucleotide
dCDP	Deoxycytidine 5'-diphosphate		
dCTP	Deoxycytidine 5'-triphosphate	dTR	Thymidine
		dTMP	Deoxythymidine 5'-monophosphate
CAP	Carbamyl phosphate		
DHFA	Dihydrofolic acid	dTDP	Deoxythymidine 5'-diphosphate
DON	6-Diazo-5-oxo-L-norlencine		
FGAR	Formylglycinamide ribonucleotide	dTTP	Deoxythymidine 5'-triphosphate
FGAM	Formylglycinamidine ribonucleotide	THFA	Tetrahydrofolic acid
		UMP	Uridine 5'-monophosphate
GAR	Glycinamide ribonucleotide	UDP	Uridine 5'-diphosphate
G	Guanine	UTP	Uridine 5'-triphosphate
GR	Guanosine	dUR	Deoxyuridine
GMP	Guanosine 5'monophosphate	dUMP	Deoxyuridine 5'-monophosphate
GDP	Guanosine 5'-diphosphate	dUDP	Deoxyuridine 5'-diphosphate
GTP	Guanosine 5'-triphosphate	dUTP	Deoxyuridine 5'-triphosphate
dGR	Deoxyguanosine		
dGMP	Deoxyguanosine 5'-monophosphate	X	Xanthine
		XR	Xanthosine
dGDP	Deoxyguanosine 5'-diphosphate	XMP	Xanthosine 5-monophosphate

I. Introduction

Purine and pyrimidine nucleotides are essential components of DNA and RNA as well as many of the coenzymes involved in bioenergetic processes and group transfer. Regulation of the pathways leading to the synthesis and degradation of these nucleotides has been studied intensively in many biological systems including rapidly growing cells in culture and appears to be a logical, orderly, and intricate process. In addition to the obvious importance of these pathways in cells in culture, elucidation of certain genetic alterations in these pathways coincident with the development of techniques of mammalian cell hybridization and expanding concepts of genetic regulation have provided important new tools for attacking many problems of mammalian biology.

In this chapter I wish to review (1) the actual pathways involved in purine and pyrimidine metabolism; (2) the nature of the regulatory controls, both genetic and molecular, insofar as they are known; (3) the known genetic defects involving these pathways; (4) the nutritional requirements of normal and mutant cells in culture; (5) the nature, specificity, and use of selective media; (6) the establishment of certain hybrid cell strains based on the presence of genetic markers in these pathways; and (7) the use of such hybrids for studying the function and organization of nuclei, nucleoli, chromosomes, and genes. Whenever possible, information obtained in cells in culture will be considered in preference to data obtained only in tissues or tissue extracts; human cells will be considered in preference to nonhuman cells; nonmalignant cells will be considered in preference to malignant cells; and diploid cells will be considered in preference to heteroploid cells. The reader is referred to several excellent recent reviews which consider certain of these topics in greater detail (Hartman, 1970; Kit, 1970; Murray et al., 1970; Blakely and Vitols, 1968; Wyngaarden and Kelley, 1972; Migeon and Childs, 1970; Harris, 1970; Krooth, 1970).

Owing to limitations of space, several broad areas relevant to metabolism of purines and pyrimidines in cell culture have not been considered. Although numerous important contributions are based on the incorporation of radioactive purine and pyrimidine bases and nucleosides into nucleic acids, these have not been considered in the present review unless specific enzymes were also investigated or the findings were particularly relevant to understanding the control of purine or pyrimidine metabolism. In addition, a number of enzymes which use purine or pyrimidine nucleotides but which are not primarily involved in the synthesis, interconversion, or degradation of these compounds (e.g., ATPase, adenyl cyclase, phosphodiesterase, DNA polymerase) are not discussed. I have also made no attempt to review membrane transport or the mechanism of action of the numerous nonphysiological inhibitors despite their obvious relevance to the metabolism of purines and pyrimidines under certain circumstances.

II. Purine Metabolism

A. Pathway and Regulation

Most of the early studies which helped to elucidate the pathways involved in the synthesis of purine nucleotides were performed in microorganisms (Hartman *et al.*, 1956; Warren *et al.*, 1957; Lukens and Buchanan, 1957; Flaks *et al.*, 1957). However, the general concepts based on these studies have proved to be generally relevant in mammalian cells. Although both genetic and molecular regulatory mechanisms have been amply demonstrated to have a role in the control of purine metabolism in bacterial cells, only the latter regulatory controls have been adequately studied in mammalian cells. Although it seems likely at this time that the general properties governing molecular regulation of this pathway in mammalian cells is similar to that observed in lower organisms, relatively minor differences probably do exist.

1. Purine Biosynthesis *De Novo*

The pathway leading to the synthesis of purine nucleotides *de novo* is illustrated in Fig. 1.

The first step leading to the synthesis of purine ribonucleotides *de novo* involves the formation of PP-ribose-P from ATP and ribose-5-phosphate (reaction 1-1). The enzyme catalyzing this reaction, PP-ribose-P synthetase, is regulated in mammalian cells as in bacteria by the "energy charge" of the cell as well as by a number of purine and pyrimidine nucleoside monophosphates, diphosphates, and triphosphates (Atkinson and Fall, 1967; Wong and Murray, 1969; Hershko *et al.*, 1969; Fox and Kelley, 1971, 1972).

The initial reaction specific for purine biosynthesis *de novo* (reaction 1-2) involves the formation of 5-phosphoribosyl-1-amine (PRA) from PP-ribose-P and glutamine. This step, which is catalyzed by PP-ribose-P amidotransferase, is thought to be rate-limiting for the entire pathway. As expected, there are several different types of molecular controls which appear to operate at this step in mammalian cells.

There is substantial evidence in human cells in culture (Kelley *et al.*, 1970a) as well in man *in vivo* (Kelley *et al.*, 1970b) that the intracellular concentration of PP-ribose-P is limiting. Thus, factors which alter the intracellular levels of PP-ribose-P lead to similar alterations in the rate of purine biosynthesis *de novo*. Although Henderson and Khoo concluded from their studies in Ehrlich ascites tumor cells that depletion of PP-ribose-P was unlikely to be important as long as an adequate supply of glucose was available in the cell (Henderson and Khoo, 1965), several examples have been described which are inconsistent with this hypothesis. For example, orotic acid clearly depletes intracellular PP-ribose-P and inhibits purine biosynthesis *de novo* in the presence of high concentrations of glucose (Kelley *et al.*, 1970a).

The importance of the other usual substrate for this enzyme, glu-

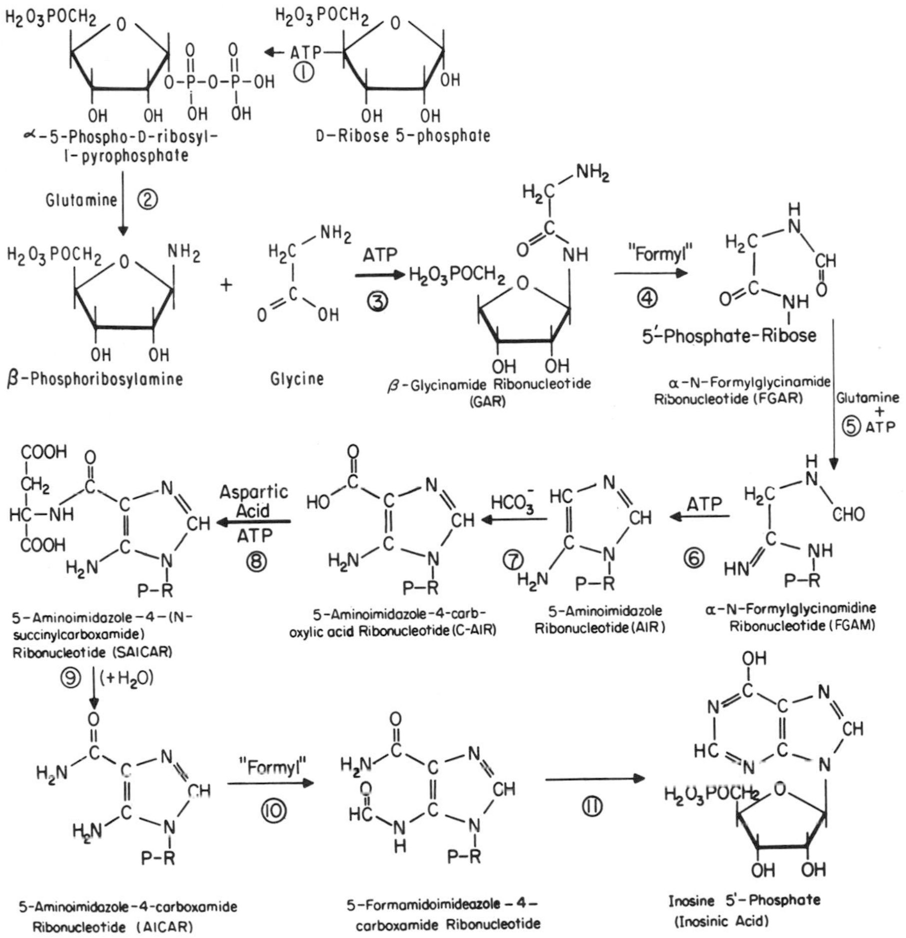

Fig. 1. Purine biosynthesis *de novo*. (1) PP-ribose-P synthetase, (2) PP-ribose-P amidotransferase, (3) phosphoribosyl-glycinamide synthetase, (4) phosphoribosyl-glycinamide formyltransferase, (5) FGAR amidotransferase, (6) AIR synthetase, (7) AIR carboxylase, (8) phosphoribosylaminoimidazolesuccinocarboxamide synthetase, (9) adenylosuccinase (catalyzed by the same enzyme that catalyzes reaction 2-8), (10) phosphoribosylaminoimidazolecarboxamide formyltransferase, and (11) inosinicase. (From Wyngaarden and Kelley, 1972.)

tamine, remains unclear at this time. There is increasing evidence in extracts prepared from Ehrlich ascites tumor cells (Henderson, 1963) and human leukemia cells (Reem and Friend, 1967) that ammonia can substitute for glutamine as a substrate for this enzyme. This would suggest, at least in these cell types, that in contrast to PP-ribose-P, the intracellular concentration of glutamine is not crucial for the regulation of purine biosynthesis *de novo.*

The activity of PP-ribose-P amidotransferase is also regulated by the purine nucleotide end products in the pathway. Numerous studies of the enzyme derived from avian liver (Wyngaarden and Ashton, 1959; Caskey *et al.,* 1964) and bacteria (Nierlich and Magasanik, 1965a) have demonstrated that the enzyme is inhibited in a cooperative or synergistic manner by 6-aminopurine nucleotides (e.g., AMP) and 6-hydroxypurine nucleotides (e.g., GMP). However, until very recently it has not been possible to assay PP-ribose-P amidotransferase directly in mammalian cells. The indirect assay of this enzyme which has gained the widest acceptance involves the incorporation of radioactive formate into formylglycinamide ribonucleotide (FGAR) in the presence of azaserine to block the further metabolism of FGAR (Fig. 1) (Henderson, 1962; Henderson and Khoo, 1965). Although this assay is actually a measure of the rate of the first three reactions in the pathway, since the initial step catalyzed by PP-ribose-P amidotransferase is limiting, it represents an indirect assay of the rate of this enzyme. Despite the relatively wide use of this assay it is subject to many limitations. The assay works only with intact cells, and therefore it is impossible to directly control the concentration of PP-ribose-P or possible nucleotide inhibitors. In addition, factors which alter the accumulation of FGAR-^{14}C may reflect a change in the further metabolism of FGAR rather than a change in activity of PP-ribose-P amidotransferase. As a result of these limitations it has not been possible to characterize the enzyme from mammalian cells in a very definitive way. The studies which have been performed using this technique in tumor cells (Henderson, 1962; Henderson and Khoo, 1965) and in human fibroblasts (Rosenbloom *et al.,* 1968a) suggested that the enzyme is subject to end-product inhibition. Although it has not been possible using this technique to demonstrate synergistic inhibition by 6-amino- and 6-hydroxypurine ribonucleotides, investigation of certain mutant strains strongly suggests that separate binding sites exist on the mammalian enzyme for these two types of purine nucleotides (Henderson *et al.,* 1967).

Recently, Hill and Bennett have assayed the PP-ribose-P amidotransferase directly in a mouse tumor, adenocarcinara 755, maintained in cell culture. This was accomplished with an assay based on the conversion of glutamine-^{14}C to glutamate-^{14}C. The enzyme, purified approximately 20-fold, was found to be very labile and to exhibit sigmoidal kinetics with PP-ribose-P as the variable substrate. In addition it was inhibited by most purine nucleoside mono-, di-, and triphosphates, although sensitivity to these inhibitors was not as great as with the pigeon liver enzyme (Hill and Bennett, 1969). PP-Ribose-P amidotransferase from Ehrlich ascites tumor cells was also found to be inhibited by thiopurine ribonucleotides using a similar assay (Tay et al., 1969). The use of this assay in extracts of diploid human fibroblasts has not been successful so far presumably because of the extremely low activity of the enzyme and the very high activity of glutaminase (McDonald and Kelley, 1971a).

Preliminary studies in Ehrlich ascites tumor cells suggested that PRA can also be synthesized from ribose-5-PO_4 and NH_3. This reaction appeared to be enzymatic and actually appeared to occur at a faster rate than observed with PP-ribose-P and glutamine. It was also subject to inhibition by purine nucleotides (Herscovics and Johnstone, 1964). In bacteria, PRA was formed from ribose-5-PO_4 and NH_3 but this reaction appeared to be nonenzymatic (Nierlich and Magasanik, 1965b).

Phosphoribosylamine (PRA), which is the product of this initial reaction, reacts with glycine to form glycinamide ribonucleotide (GAR) in a reaction requiring ATP (reaction 1-3). Glycinamide ribonucleotide receives a one-carbon "formyl" unit from N^5, N^{10}-methenyltetrahydrofolic acid to form formylglycinamide ribonucleotide (FGAR) (reaction 1-4). The amide group of glutamine is transferred to FGAR leading to the synthesis of formylglycinamidine ribonucleotide (FGAM) (reaction 1-5) in a reaction requiring ATP. Several studies in bacterial (Mizoburchi and Buchanan, 1968) as well as mammalian (Howard and Appel, 1968) tissue suggest that this enzyme is also subject to inhibition by purine nucleotides such as AMP and GMP. This might not be expected to be an important site of regulation since this is not a branch point in the pathway. Indeed, the high concentration of nucleotides required to effect inhibition suggests that under normal conditions this potential site of control is less important than inhibition of PP-ribose-P amidotransferase. This reaction is also inhibited by azaserine. This compound is used to retard the further metabolism of FGAR in the indirect assay of PP-ribose-P amidotransferase described above.

In the presence of Mg^{2+}, ATP, and the appropriate enzyme, ring closure occurs with the formation of an imidazole (reaction 1-6). The resulting intermediate, 5-aminoimidazole ribonucleotide (AIR), receives a carboxyl group at C-4 by a CO_2 fixation reaction (reaction 1-7). The carboxyl serves as a point of condensation of this intermediate with aspartic acid through an amide linkage involving another ATP as the source of energy (reaction 1-8). Hydrolysis of this intermediate yields 5-amino-4-imidazolecarboxamide ribonucleotide (AICAR) (reaction 1-9). AICAR receives a second formyl group from N^{10}-formyltetrahydrofolic acid in the presence of K^+ leading to the formation of 5-formamidoimidazole-4-carboxamide ribonucleotide (reaction 1-10). Ring closure completes the biosynthesis of the purine structure by forming inosine 5'-monophosphate (IMP) (reaction 1-11). Although none of the last steps (reactions 1-6 to 1-11) are thought to be regulated in mammalian cells, this possibility has not been subjected to rigorous examination.

2. PURINE NUCLEOTIDE INTERCONVERSIONS

IMP is a common intermediate in the synthesis of guanosine 5'-monophosphate (GMP) and adenosine 5'-monophosphate (AMP), the major purine nucleotide components of nucleic acids. The interconversions and catabolism of these purine nucleotides is summarized in Fig. 2.

GMP is synthesized from IMP by two steps involving the intermediate formation of XMP. This conversion of IMP to XMP, catalyzed by IMP dehydrogenase in the presence of DPN and K^+, occurs subsequent to a branch point in the pathway (reaction 2-1). Therefore, IMP dehydrogenase is the first enzyme uniquely involved in the biosynthesis of guanine nucleotides. As expected, the extent of conversion of IMP to GMP in lower organisms is controlled by the level of GMP in the cell

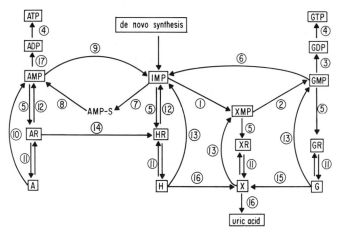

Fig. 2. Purine reutilization, interconversion, and catabolic pathways. (1) IMP dehydrogenase, (2) XMP aminase, (3) GMP kinase (same as reaction 4-3), (4) nucleoside diphosphate kinase (same as reactions 3-8 and 4-2), (5) 5'-nucleotidase, (6) GMP reductase, (7) adenylosuccinic acid synthetase, (8) adenylosuccinase, (9) AMP deaminase, (10) adenine phosphoribosyltransferase, (11) purine nucleoside phosphorylase, (12) adenosine kinase, (13) hypoxanthine-guanine phosphoribosyltransferase, (14) adenosine deaminase, (15) guanase, (16) xanthine oxidase, and (17) AMP kinase (same as reaction 4-4).

(Magasanik and Karihian, 1960). In sarcoma 180 cells, as in bacteria (J. H. Anderson and Sartorelli, 1968, 1969), GMP inhibits inosinic dehydrogenase in a manner competitive with respect to IMP.

XMP is converted to GMP in the presence of glutamine, ATP, Mg^{2+}, and the enzyme XMP aminase (reaction 2-2). There is some evidence in animal as well as bacterial cells that this enzyme can use NH_3 as well as glutamine (Lagerkvist, 1958a,b). However, the high concentration of NH_3 necessary suggests that glutamine is the preferred substrate.

Once synthesized, GMP can be utilized by the cell in several ways. It can be (1) phosphorylated to GDP and GTP by a GMP kinase and a nucleoside diphosphokinase, respectively, in the presence of ATP and Mg^{2+} (reactions 2-3 and 2-4) (Miech and Parks, 1965; Meich et al., 1969; Mourad and Parks, 1966; Sugino et al., 1966); (2) dephosphorylated to guanosine in the presence of 5'-nucleotidase (reaction 2-5); and (3) deaminated to IMP in the presence of GMP reductase (reaction 2-6). The first two reactions are not specific for GMP but appear to utilize other purine nucleotides; these will be discussed later. The last enzyme, GMP reductase, is thought to play a relatively minor role in mammalian cells, although its presence has been demonstrated (Hershko et al., 1963; Cook and Vibert, 1966).

The other branch point in the pathway after the formation of IMP involves the formation of adenylosuccinic acid which is catalyzed by adenylosuccinic acid (AMP-S) synthetase (reaction 2-7). This enzyme, which requires GTP and Mg^{2+}, is the first enzyme in the pathway uniquely involved in the biosynthesis of adenine nucleotides. As anticipated, AMP inhibits this enzyme obtained from bacteria (Wyngaarden and Greenland, 1963; Rudolph and Fromm, 1969). However, comparable studies on its regulation in mammalian cells have not been done.

AMP is formed by cleavage of AMP-S with the release of fumaric acid (reaction 2-8). The enzyme catalyzing the latter reaction also catalyzes the cleavage of SAICAR in the de novo pathway (reaction 1-9).

AMP can be converted to its di- and triphosphates (reactions 2-17 and 2-4), back to IMP by adenylic acid deaminase (reaction 2-9), or catabolized to adenosine by the 5'-nucleotidase (reaction 2-5).

The enzyme catalyzing the deamination of AMP to IMP has been studied intensively in mammalian tissue such as rabbit muscle (Lee, 1957a,b,c), calf brain (Setlow and Lowenstein, 1967), and erythrocytes (Askari and Franklin, 1965). It appears to be relatively specific for AMP and dAMP (Carter, 1951; Setlow and Lowenstein, 1967; Askari and Franklin, 1965). The mammalian enzyme exhibits sigmoidal kinetics with respect to an increasing concentration of AMP. ATP inhibits in a manner competitive with respect to AMP and GDP and GTP antagonizes the inhibitory effect of ATP (Setlow and Lowenstein, 1967, 1968; Muntz, 1953; Lee and Wang, 1968). The relatively high K_m for AMP (0.4–4 mM) suggests that the AMP concentration in the cell may be limiting (Murray et al., 1970). The catabolism of AMP to adenosine represents a purely catabolic pathway, which will be discussed later.

3. PURINE REUTILIZATION

In addition to the de novo pathway, the purine nucleotides, AMP, IMP, and GMP, can be synthesized directly from the free purine bases. This can potentially occur by two different mechanisms. Adenine is converted directly to AMP in the presence of PP-ribose-P in a reaction catalyzed by adenine phosphoribosyltransferase (reaction 2-10). Adenine could also be converted to AMP by the sequential actions of adenosine phosphorylase (reaction 2-11) and adenosine kinase (reaction 2-12) with the intermediate formation of adenosine. Both potential pathways appear to be operative in mammalian cells.

Adenine phosphoribosyltransferase has been found to be present in a variety of mammalian tissues (Hori and Henderson, 1966a; Murray, 1966a; Hori et al., 1967; Hori and Henderson, 1966b; Krenitsky et al., 1969). Although it is subject to product inhibiton by AMP, the importance of this as a control mechanism in the intact cell has not been established. Very few studies have been directed toward examining control of synthesis of adenine phosphoribosyltransferase; however, the observation that the level of activity is increased after partial hepatectomy (Murray, 1966b) and in embryonic tissue (Epstein, 1970) suggests that it may be important in cell replication. This is supported by the finding that the levels of adenine phosphoribosyltransferase activity in human fibroblasts in culture are 5–10 times higher than activity observed in other human tissues (Kelley, 1971a).

Adenine is converted to adenosine by a purine nucleoside phosphorylase, which has been studied extensively in a variety of mammalian tissues. However, since this enzyme appears to be primarily degradative in most tissues and utilizes other purine nucleosides more readily than adenosine it will be discussed in detail with the degradative enzymes. Adenosine kinase, which catalyzes the synthesis of adenosine to AMP, has also been studied quite well from many mammalian tissues (Lindberg et al., 1967; Holmsen and Rozenberg, 1968; Rozenberg and Holmsen, 1968; Ho et al., 1968) including Ehrlich ascites tumor cells (Pierre et al., 1967; Murray, 1968) and human tumor cells (Schnebli et al., 1966). This enzyme appears to be of particular significance in relation to resistance of certain cells in culture to purine analogues such as 6-methylmercaptopurine ribonucleoside and 2-flouroadenosine (Bennet *et al.*, 1966). Although it seems likely that the pathway in which adenine is converted directly to AMP is of relatively greater significance under usual conditions in mammalian cells, the latter pathway involving the intermediate formation of adenosine may be important in certain specialized circumstances.

Guanine, hypoxanthine, and, to a lesser extent, xanthine are converted directly to GMP, IMP, and XMP, respectively, in the presence of PP-ribose-P in a reaction catalyzed by hypoxanthine-guanine phosphoribosyltransferase (reaction 2-13) (Murray, 1966b, 1967; Kelley *et al.*, 1967a; Hill, 1970; Craft *et al.*, 1970). This enzyme is also subject to product inhibition by GMP and IMP, but, as is the case for adenine phosphoribosyltransferase, the importance of this as a control mechanism remains to be established (Henderson *et al.*, 1968a; Krenitsky and Papaioannow, 1969). Both hypoxanthine-guanine phosphoribosyltransferase and adenine phosphoribosyltransferase increase in activity in rabbit and mouse embryos in association with blastocyst formation and an acceleration in the rate of RNA synthesis (Epstein, 1970). The resistance of cells in culture to a variety of purine analogues such as 6-mercaptopurine, 6-thioguanine, and azathioprine is frequently due to the deficiency of this enzyme, which is necessary for the conversion of these analogues to their nucleotide derivatives (Brockman, 1963).

The conversion of guanine to its ribonucleotide derivative by the alternative pathway involving the intermediate formation of guanosine does not appear to be operative in human fibroblasts in culture (Friedman *et al.*, 1969; Kelley and Meade, 1971). The human purine nucleoside phosphorylase (reaction 2-11) is relatively active with guanine as well as xanthine and hypoxanthine as substrates (Krenitsky *et al.*, 1968). However, there is an apparent absence of guanosine kinase, which catalyzes the conversion of guanosine to GMP. Although several studies in mammalian tissues including Ehrlich ascites tumor cells indicate that inosine can be converted to IMP is a reaction apparently catalyzed by adenosine kinase (reaction 2-12) (Pierre *et al.*, 1967; Pierre and LePage, 1968; Tarr, 1964), this does not appear to be the case in human skin fibroblasts (Friedman *et al.*, 1969) and has not yet been demonstrated with a partially purified adenosine kinase from human tumor cells (Schnebli *et al.*, 1966).

4. PURINE NUCLEOTIDE CATABOLISM

The purine nucleotides are catabolized in most mammalian cells in culture to hypoxanthine and xanthine. Further catabolism to uric acid probably does not occur in cells in culture with the exception of those of hepatic origin since xanthine oxidase is absent. Catabolism to the free bases occurs by a somewhat different pathway for each of the purine nucleotides. Each purine nucleotide, AMP, IMP, XMP, and GMP, is dephosphorylated to its respective ribonucleoside derivative, adenosine, inosine, xanthosine, and guanosine, by a relatively specific 5'-nucleotidase (reaction 2-5) and probably by a nonspecific alkaline phosphatase (Fritzson, 1967; Sulkowski et al., 1963; Center and Behal, 1966; Segal and Brenner, 1960; Paterson and Hori, 1963; Itoh et al., 1968). This enzyme may also be important in the catabolism of 5'-deoxynucleotides in animal cells. Most recent studies have indicated that the enzyme is localized to the outside of the cell surface or in structures contiguous with the outside cell surface (Song and Bodansky, 1967; Sinha and Ghosh, 1964; Emmelot and Bos, 1965; Song et al., 1967; Baer and Drummond, 1968; Essner et al., 1965; Burger and Lowenstein, 1970), suggesting that it may play a role in the catabolism of nucleotides to nucleosides in preparation for transport into the cell. In sheep brain and rat heart, as well as in Ehrlich ascites tumor cells, the 5'-nucleotidase is inhibited by ATP as well as by a variety of other nucleoside triphosphates (Ipata, 1967, 1968; Murray and Friedrichs, 1969; Edwards and Maguire, 1970). This may be important in the regulation of purine catabolism. The finding of a heat-stable inhibitor of this enzyme in rat liver microsomes (Segal and Brenner, 1960) and in Ehrlich ascites tumor cells (Murray and Friedrichs, 1969) suggests that additional control mechanisms may be important in intact cells.

The purine nucleosides can be further catabolized to the free bases by a purine nucleoside phosphorylase (reaction 2-11). This enzyme has been studied intensively from a variety of animal sources including Ehrlich ascites tumor cells (Gotto et al., 1964; Paterson, 1965a,b,c; Pinto and Touster, 1966). A number of naturally occurring purine ribonucleosides and deoxyribonucleosides including guanosine, deoxyguanosine, inosine, deoxyinosine, and xanthosine are substrates for the enzyme, although adenosine appears to be a relatively poor substrate (Paterson, 1965a; Kim et al., 1968). Enzyme activity is also present in many animal cells which catalyzes the transfer of the ribose moiety from a purine ribonucleoside to a purine base (e.g., guanine + inosine → guanosine + hypoxanthine) by a mechanism which does not involve the intermediate formation of ribose-1-P (Abrams et al., 1965). Although this enzyme was found to copurify with purine nucleoside phosphorylase (Abrams et al., 1965), a separation of these two activities has been reported using sucrose gradients (Pinto and Touster, 1966). Another degradative enzyme, nucleoside hydrolase (inosine $+$ H_2O $\rightarrow$ hypoxanthine $+$ D-ribose), has not been demonstrated in mammalian tissues.

Adenosine and deoxyadenosine are readily deaminated to inosine and deoxyinosine, respectively, in a reaction catalyzed by adenosine deaminase (Zittle, 1946) (reaction 2-14). This enzyme, which has also been studied well in several animal tissues, does not appear to have any specialized regulatory function (Pfrogner, 1967; Baer et al., 1966), although it is inhibited by certain adenosine analogues (Baer et al., 1966) and its specific activity increases in mouse liver following 70% hepatectomy (Rothman et al., 1971). Two different molecular forms of adenosine deaminase exist in human tissues; the smaller form could be converted to the larger form in the presence of a nondialyzable, heat-labile fraction from lung homogenate (Akedo et al., 1970). The physiological importance of the multiple molecular forms or converting activity is unclear.

Guanine, which is formed from the catabolism of GMP and guanosine, is deaminated to xanthine by guanase (reaction 2-15). This enzyme is especially high in mammalian brain tissue but it is also present in other tissues such as liver (Kalckar, 1947; Currie et al., 1967; Kumar et al., 1965). Despite the high activity assayed in brain tissue in vitro, the functional activity of guanase in this tissue in vivo may be much less. An inhibitor of guanase, which is a protein, has been demonstrated in brain tissue; this protein begins to appear, at least in developing rat brain, after 15 days of age (Kumar, 1969). GTP may be important in regulation of the enzyme since in its presence the enzyme exhibits hyperbolic rather than sigmoidal kinetics (Josan and Krishman, 1968). The relative importance of this enzyme in cells in culture has not been established.

Hypoxanthine and xanthine can be oxidized to uric acid in a reaction catalyzed by xanthine oxidase (reaction 2-16). This enzyme has been extensively studied in the past. It is absent in most animal tissues other than liver and gastrointestinal mucosa. It is also absent in most cells in culture (Kelley and Wyngaarden, 1970) except possibly those of hepatic origin.

From the preceding discussion, it can be seen that purine nucleotides can be synthesized by several different pathways within the cell. In addition, these compounds are interconverted or catabolized by one of several pathways depending on cellular needs. The multiple sites where ATP can function as an inhibitor or an activator of critical enzymes in these pathways suggest that the intracellular concentration of this compound may be an important determinant of which metabolic pathway is preferred (Overgaard-Hansen, 1965).

B. Genetic Defects

1. PP-RIBOSE-P AMIDOTRANSFERASE

Henderson and associates (1968b) found that the early steps of purine biosynthesis *de novo* in fibroblasts cultured from two patients with extraordinary overexcretion and overproduction of uric acid appeared to be abnormally resistant to feedback inhibition after the addition of either 6-amino- and 6-hydroxypurine bases. Since the enzymes necessary for the formation of purine ribonucleotides, hypoxanthine-guanine phosphoribosyltransferase and adenine phosphoribosyltransferase, were present in these cells and purine nucleotide catabolism was not increased, this was regarded as suggestive evidence that the PP-ribose-P amidotransferase, at least in these two cells strains, had altered regulatory properties. Similar mutants have been described in bacterial cells as well as heteroploid mammalian cells. Since the regulatory sites of the amidotransferase, as discussed earlier, are distinct from its substrate site, selective loss of regulatory control is quite feasible. However, the mechanism of this altered responsiveness to feedback inhibitors can be resolved only by kinetics or inhibitor-binding studies of a partially purified enzyme. The technical requirements allowing such a study have not yet been solved in normal human tissue.

2. HYPOXANTHINE-GUANINE PHOSPHORIBOSYLTRANSFERASE

A functionally complete deficiency of hypoxanthine-guanine phosphoribosyltransferase in erythrocytes is associated with uric acid overproduction and a devastating neurological disorder characterized by self-mutilation, choreoathetosis, spasticity, and mental retardation (the Lesch-Nyhan syndrome) (Seegmiller *et al.*, 1967; Kelley, 1968; McDonald and Kelley, 1971b). A partial deficiency of this enzyme in erythrocytes is associated with a specific subtype of gout characterized by a striking increase in uric acid production (Kelley *et al.*, 1967b).

Skin fibroblasts derived from patients with these disorders also exhibit a deficiency of the hypoxanthine-guanine phosphoribosyltransferase enzyme. Fibroblasts derived from patients with the complete defect in erythrocytes have been studied most extensively. These mutant cells exhibit levels of hypoxanthine-guanine phosphoribosyltransferase activity which range from 0.1 to 10% of normal (Fujimoto and Seegmiller, 1970; Kelley and Meade, 1971). Study of the mutant enzyme from 11 of these cell strains indicates that the enzyme defect is the result of a mutation on the structural gene coding for this enzyme and that there is substantial heterogeneity in the mutations leading to this defect (Kelley and Meade, 1971). Cells from these patients exhibit an increased synthesis of FGAR, which represents an increased rate of the early steps of purine biosynthesis *de novo* similar to that which is observed *in vivo* (Rosenbloom *et al.*, 1968a,b). This is due in part to increased levels of PP-ribose-P which have been demonstrated in these cells (Rosenbloom *et al.*, 1968a).

The hypoxanthine-guanine phosphoribosyltransferase enzyme is coded by DNA in the X chromosome. Many studies have documented the presence of mosaicism in cells cultured from heterozygotes for this particular disorder (Rosenbloom *et al.*, 1967; Migeon *et al.*, 1968; J. Salzman *et al.*, 1968). The presence of two populations of cells, those with the normal and those with the mutant enzyme, has been demonstrated by radioautography of relatively disperse cultures after exposure to hypoxanthine-^{3}H (Rosenbloom *et al.*, 1967), by cloning techniques (Migeon *et al.*, 1968; J. Salzman *et al.*, 1968), and by the presence of both azaguanine-sensitive and -resistant cells (Migeon, 1970).

If cells from heterozygotes or mixtures of normal and mutant cells are allowed to grow to confluence, one finds that most of the cells appear to have hypoxanthine-guanine phosphoribosyltransferase activity. This phenomenon has been termed metabolic cooperation (Subak-Sharpe *et al.*, 1969; Friedman *et al.*, 1968) and appears to be due to the transfer of nucleotide end product from the normal to mutant cell upon contact of these cell types (Cox *et al.*, 1970). The recognition of this phenomenon becomes especially important if selective medium is being used in a culture of cells which are mixed with respect to this enzyme.

Mutations at this locus have been readily produced in Chinese hamster cells (Chu, 1971) and diploid human cells (Albertini and DeMars, 1970) by X-ray.

3. ADENINE PHOSPHORIBOSYLTRANSFERASE

A partial deficiency of adenine phosphoribosyltransferase has been described in two families (Kelley *et al.*, 1968, 1970c). In one family the defect was demonstrable in fibroblasts in culture, whereas in the other family adenine phosphoribosyltransferase activity in fibroblasts in culture was essentially normal (Kelley, 1971b). The enzyme is coded by an autosome and affected individuals are heterozygous for the defect. Cells completely deficient in this enzyme would be valuable in that they would be resistant to analogues such as azaadenine and would be ideal for hybridization studies. Although no human cell strains completely deficient in this enzyme have been discovered, such cells have been derived from established cell lines and have been used for chromosomal mapping (Kusano et al., 1971).

Adenine phosphoribosyltransferase activity is elevated in erythrocytes and possibly some other tissues obtained from patients genetically deficient in hypoxanthine-guanine phosphoribosyltransferase (Seegmiller et al., 1967; Kelley, 1968). This has been attributed to stabilization of the adenine phosphoribosyltransferase enzyme by the high PP-ribose-P levels present in the cell (Rubin et al., 1969; Greene et al., 1970). Fibroblasts cultured from these patients consistently have normal levels of the adenine phosphoribosyltransferase enzyme (Kelley, 1971a).

In addition to homonuclear diploid human cells which are derived from individuals with a specific enzymatic defect in vivo, a number of heteronuclear cell lines have been derived that also exhibit certain genetically determined alterations in purine metabolism. In general, these lines are usually isolated by application of a selective system to a mass culture (as will be discussed later). These cell lines have been discussed in detail by Gartler and Pious (1966) and will not be described in detail here. However, they primarily involve cell lines deficient in hypoxanthine-guanine posphoribosyltransferase (Szybalski and Szybalski, 1962; Littlefield, 1963, 1964a; J. D. Davidson et al., 1962; Morrow, 1970) and adenine phosphoribosyltransferase (Kusano et al., 1971).

III. Pyrimidine Metabolism

A. Pathway and Regulation

The sequence of events leading to the synthesis of pyrimidine nucleotides like in the purine pathway was initially elucidated in nonmammalian organisms. Subsequently, the validity of this pathway has been largely confirmed in mammalian cells including man. The pyrimidine nucleotides can also be synthesized de novo or by the utilization of preformed pyrimidine bases or ribonucleosides [Figs. 3(a) and 3(b)].

1. PYRIMIDINE BIOSYNTHESIS de novo

The first reaction important to pyrimidine biosynthesis in mammalian cells involves the synthesis of carbamyl phosphate (CAP), which can potentially serve not only as a substrate for the synthesis of pyrimidines but also for the synthesis of citrulline as part of the urea cycle. The forma-

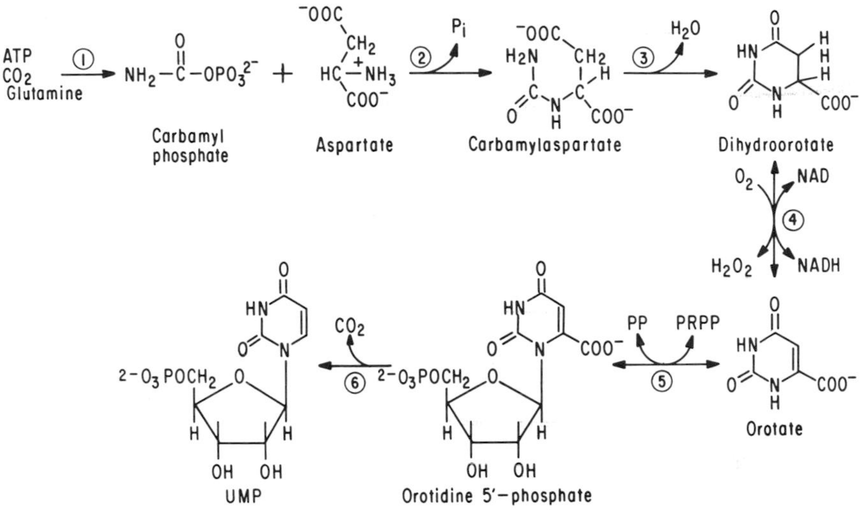

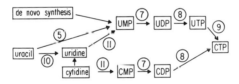

Fig. 3(a). Pyrimidine biosynthesis *de novo*. (1) Carbamyl phosphate synthetase, (2) aspartate transcarbamylase, (3) dihydroorotase, (4) dihydroorotic acid dehydrogenase, (5) orotate phosphoribosyltransferase, and (6) orotidine 5'-monophosphate decarboxylase. (Modified from Hartman, 1970.)

Fig. 3(b). Pyrimidine nucleotide interconversions. (7) CMP kinase (same as reaction 4-5), (8) nucleoside diphosphate kinase (same as reactions 2-4 and 4-2), (9) CTP synthetase, (10) uridine phosphorylase, and (11) uridine kinase.

tion of carbamyl phosphate is catalyzed by carbamyl phosphate synthetase and requires ATP, carbonate, and ammonia or glutamine (reaction 3-1). Several studies suggest that the glutamine-dependent enzyme is more important in mammalian cells than the ammonia-dependent enzyme. The evidence to support this statement may be summarized as follows: (1) Glutamine stimulates pyrimidine synthesis in animal cells in culture more than NH_3 (Hager and Jones, 1965; Ito et al., 1970); (2) incorporation of ^{15}N into nucleic acid uracil is greater from glutamine than NH_3 (N. P. Salzman et al., 1958); (3) several glutamine analogues such as DON and O-carbonylserine inhibit pyrimidine formation in Ehrlich ascites tumor cells (Levenberg, 1962; Hager and Jones,

1967a); (4) most of the enzymes involved in the synthesis of pyrimidines as well as glutamine are present in the cytoplasm (Hager and Jones, 1965; Jones and Hager, 1966; Ito *et al.*, 1970); and (5) the glutamine-dependent enzyme in bacteria is inhibited by pyrimidines (Pierard *et al.*, 1965; Lacroute *et al.*, 1965; P. M. Anderson and Meister, 1966). In liver tissue which must synthesize urea as well as pyrimidines the enzyme using NH_3 in preference to glutamine as a nitrogen donor may be restricted to mitochondria (Tatibana and Kazuhiko, 1969). However, the cytoplasmic glutamine-utilizing enzyme (Hager and Jones, 1967b) is probably capable of using NH_3 as a nitrogen donor to some extent (Tatibana and Kazuhiko, 1969; Ito *et al.*, 1970). It has recently been suggested that loss of the NH_3-requiring enzyme in malignant liver cells in culture may be related in some way to malignancy (Ito *et al.*, 1970).

Carbamyl phosphate (CAP) synthetase in mammalian tissues exhibits sigmoidal kinetics with ATP and is inhibited by UTP and CTP, suggesting that this is an important site for regulation of pyrimidine synthesis in animal cells. The highest levels of CAP synthetase are found in animal tissues with rapid cellular proliferation such as gastrointestinal tract, spleen, thymus, and testis, suggesting that it may be important in the regulation of cell growth (Tatibana and Kazuhiko, 1969; Inagaki and Tatibana, 1970).

Aspartate transcarbamylase catalyzes the irreversible carbamylation of L-aspartate by carbamyl phosphate to form carbamylaspartate (reaction 3-2). In lower organisms where this enzyme represents the first step unique to pyrimidine synthesis, it is regulated by pyrimidine nucleotides. In most animal cells the initial step of pyrimidine synthesis is CAP synthetase rather than aspartate transcarbamylase. Although early studies in mammalian cells suggested that this enzyme might be regulated by pyrimidine nucleotides (Bresnick, 1964), more recent studies indicate that it is not sensitive to inhibition by the ultimate end products of this pathway (Tatibana and Ito, 1967; Krakow and Vennesland, 1961; Miller and Massey, 1965a,b; Miller and Kerr, 1966). However, with the exception of the brain, which has high activity, the level of activity in various tissues appears to correlate with mitotic activity and it has been suggested that the level of activity may be an indicator of RNA synthesis *de novo* (Young *et al.*, 1967).

Ring closure to form dihydroorotic acid is catalyzed by dihydroorotase (reaction 3-3). Although this step does not represent a site of branching in the pathway or a recognized critical limiting step, this enzyme from animal cells appears to be inhibited by a variety of pyrimidines including CMP, dCMP, and thymidine (Bresnick and Blatchford, 1964; Bresnick and Hitchings, 1961; Bresnick, 1964). However, the actual importance of regulation at this step in the pathway has not been established.

Dihydroorotic acid is reversibly oxidized by dihydroorotic acid dehydrogenase to form orotic acid (reaction 3-4). Orotic acid is then converted to orotidine 5′-monophosphate (OMP) in a reaction requiring Mg^{2+} and PP-ribose-P which is catalyzed by orotate phosphoribosyltransferase (reaction 3-5). This enzyme, which has been purified from calf thymus and beef erythrocytes (Kasbekar et al., 1964; Hatfield and Wyngaarden, 1964a), is specific for 2,6-diketopyrimidines. This enzyme has also been found to increase in activity in concert with increased RNA synthesis (Bresnick, 1965). Although the enzyme is subject to product inhibition by OMP and pyrophosphate, it is not likely, because of the rapid metabolism of OMP, that this has any regulatory importance within the cell.

OMP is decarboxylated to form uridine 5′-monophosphate in a reaction catalyzed by orotidylic decarboxylase (reaction 3-6). The observation that this enzyme from several animal tissues copurifies with orotate phosphoribosyltransferase (Hatfield and Wyngaarden, 1964a; Creasy and Handschumacher, 1961; Appel, 1968) suggests that the two enzymatic activities reside in a single protein or a complex. This enzyme is also inhibited by high concentrations of nucleotides, although this is probably of little physiological importance under normal conditions (Appel, 1968).

2. Pyrimidine Nucleotide Interconversions

The pyrimidine ribonucleoside monophosphates as well as the deoxyribonucleoside derivatives can be phosphorylated to the di- and triphosphates by several different enzymes (reactions 3-7 and 3-8) [Fig. 3(b)]. These are discussed in detail later in the section on metabolism of the deoxyribonucleotides.

UTP can be aminated in the presence of glutamine or NH_3, ATP, Mg^{2+}, and CTP synthetase to yield CTP (reaction 3-9). In animal tissues including Novikoff hepatoma cells, glutamine appears to be a more important substrate for this enzyme than ammonia (Hurlbert and Kammen, 1960; N. P. Salzman et al., 1958; Eidinoff et al., 1958).

3. Pyrimidine Reutilization

In addition to the de novo pathway, UMP and CMP as well as TMP can be synthesized directly from performed pyrimidines. Theoretically this can occur by one of two mechanisms. Uracil and some of its ana-

logues can be converted directly to their nucleotide derivatives in a
reaction catalyzed by the orotate phosphoribosyltransferase enzyme (reaction 3-5) (Hatfield and Wyngaarden, 1964b; Reyes, 1969). Alternatively, the base can be converted to its ribonucleoside and this converted
to a nucleotide derivative by a nucleoside kinase. As mentioned earlier
the former mechanism appears to be of major importance in purine
metabolism. However, the latter sequence of events, at least in animal
cells, seems to be more important in pyrimidine metabolism (Reichard
and Skold, 1957; Canellakis, 1957b).

In animal tissues ribose is transferred to the pyrimidine base from
ribose-1-P with the formation of the ribonucleoside derivative and the
release of P_i (reaction 3-10):

$$\text{pyrimidine base} + \text{ribose-1-P} \rightarrow \text{pyrimidine ribonucleoside} + P_i$$

This enzyme is also active for the 5-halogenated derivatives and
5-methyluracil but is not active for cytosine, orotic acid, thymine, or
the purine bases (Paege and Schlenk, 1952; Canellakis, 1957a; Krenitsky
et al., 1964). It is inhibited by hypoxanthine and 6-mercaptopurine
(Gallo and Breitman, 1968). Recent studies in Ehrlich ascites tumor
cells have suggested that the availability of ribose-1-P may be a critical
factor regulating the incorporation of uracil into nucleic acids (Gotto
et al., 1969). A separate enzyme catalyzes the formation of pyrimidine
deoxyribonucleosides from the free bases, as will be discussed later.

Uridine is phosphorylated to form UMP in the presence of uridine
kinase and ATP (reaction 3-11). This enzyme, which has been extensively studied in Ehrlich ascites tumor cells (Skold, 1960), can also
use cytidine, 5-fluorouridine, 5-fluorocytidine, and 6-azauridine but not
5-methyluridine, deoxyuridine, deoxythymidine, or purine nucleosides.
The enzyme in tumor cells and human lymphocytes is inhibited by
CTP and UTP (E. P. Anderson and Brockman, 1964; Lucas, 1967).
An alteration in this enzyme appears to account, at least in some cases,
for resistance to 5-fluorouracil (Reichard et al., 1959; Skold, 1963) and
possibly 6-azauridine (Korbecki and Plagemann, 1969). Uridine kinase
activity increases 20-fold in lymphocytes during transformation. This
increase in activity is associated with an increase in RNA synthesis and
requires de novo synthesis of enzyme (Lucas, 1967).

It has been suggested that the transport of uridine into the cell
(Novikoff hepatoma) is limiting for its incorporation into the intracellular pools of phosphorylated intermediates (Plagemann and Roth, 1969;
Plagemann and Shea, 1971). Clearly, transport into the cell must be
evaluated when considering factors effecting incorporation of nucleosides
into nucleic acids (Stock et al., 1969; Nakata and Bader, 1969).

B. Genetic Defects

1. OROTIC ACIDURIA

Cells cultured from patients with orotic aciduria exhibit a marked deficiency of orotate phosphoribosyltransferase (OPRT) and orotidylic decarborylase (ODC) (Smith *et al.*, 1966). Pinsky and Krooth have found that the addition of azuridine, azaorotic acid, and barbituric acid leads to an increase in the level of these two enzymes in both normal and mutant cells (Pinsky and Krooth, 1967a,b; Krooth, 1970). The activity in the mutant cells after addition of these compounds actually approaches that found in normal cells under control conditions. However, it is not clear from these experiments whether the increase in level of activity represents *de novo* synthesis of the enzyme(s), its (their) stabilization or even enzyme "activation."

IV. Deoxyribonucleotide Metabolism

A. Pathway and Regulation

The formation and interconversion of deoxyribonucleotides are summarized in Fig. 4.

1. RIBONUCLEOTIDE REDUCTASE

The reduction of nucleoside diphosphates to their deoxy forms appears to be a crucial and probably the exclusive biosynthetic pathway for the synthesis of deoxyribonucleotides in mammalian cells (reaction 4-1). This reduction appears to be similar in bacterial and mammalian cells in its requirement for reduced lipoate or thioredoxin and thioredoxin reductase (Moore and Reichard, 1964; Larsson and Reichard, 1966; Moore and Hulbert, 1966; Larsson, 1963; Moore, 1967):

$$\text{thioredoxin—S}_2 + \text{NADPH} + \text{H}^+ \xrightarrow{\text{thioredoxin reductase}} \text{thioredoxin(SH)}_2 + \text{NADP}^+ \quad (1)$$

$$\text{nucleoside diphosphate} + \text{thioredoxin(SH)}_2 \xrightarrow[\text{Mg}^{2+}]{\text{ATP}} \text{deoxynucleoside diphosphate} + \text{thioredoxin—S}_2. \quad (2)$$

Regulation appears to be complex although logical in Novikoff cells, as it is in lower organisms (Moore and Hurlbert, 1966). The reduction of CDP and UDP is activated by ATP and inhibited by dTTP, dUTP,

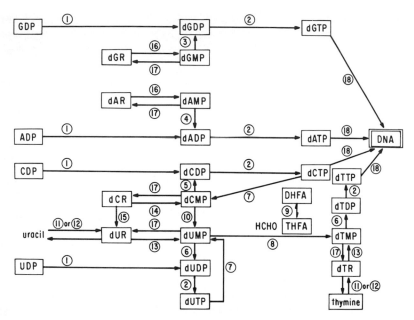

Fig. 4. Biosynthesis and interconversions of deoxyribonucleotides. (1) Ribonucleotide reductase, (2) nucleoside diphosphate kinase (same as reactions 2-4 and 3-8), (3) GMP kinase (same as reaction 2-3), (4) AMP kinase (same as reaction 2-17), (5) CMP kinase (same as reaction 3-7), (6) thymidylate kinase, (7) deoxyuridinetriphosphatase, (8) thymidylate synthetase, (9) dihydrofolate reductase, (10) deoxycytidylate deaminase, (11) deoxythymidine phosphorylase, (12) deoxythymidine synthetase, (13) thymidine kinase, (14) deoxycytidine kinase, (15) pyrimidine nucleoside deaminase, (16) purine deoxyribonucleoside kinase, (17) 5′-nucleotidase (same as reaction 2-5), and (18) DNA polymerase. (Adapted from Kit, 1970.)

dGTP, and dATP. The reduction of GDP is activated by dTTP and ATP and inhibited by dGTP and dATP as well as an excess of dATP. Most of these effects are produced at relatively physiological concentrations. Therefore, in this system, dTTP activates the reduction of GDP; the product of this reaction, dGDP, activates the reduction of ADP. The final product, dATP, inhibits all four reductions. Similar results were observed with adenocarcinoma 755 cells (Kimball *et al.*, 1967).

Ribonucleotide reductase activity increases in association with initiation of DNA synthesis in regenerating rat liver apparently due to *de novo* enzyme synthesis (King and Van Lancker, 1969). Polyoma virus infection of contact-inhibited monolayer cultures of kidney cells also leads to an increase in ribonucleotide reductase which is associated with the occurrence of DNA synthesis (Kara and Weil, 1967).

Inhibition of ribonucleotide reductase by hydroxyurea has been suggested as the mechanism by which this agent inhibits DNA synthesis and cell proliferation (Elford, 1968; Krakoff et al., 1968; Moore, 1969). Hydroxyurea inhibits the increased DNA synthesis observed in regenerating liver as well as after polyoma virus infection (Nordenskjold and Krakoff, 1968). Deoxyadenosine may inhibit DNA synthesis in Ehrlich ascites tumor cells by conversion to dATP with subsequent inhibition of the reductase (Klenow, 1962; Munch-Peterson, 1960).

2. INTERCONVERSIONS

The diphosphate derivatives are converted to the triphosphates in a reaction requiring another nucleoside triphosphate and nucleoside diphosphokinase. This enzyme has been studied well from several different mammalian sources (reaction 4-2) (Nakamura and Sugino, 1966a; Mourad and Parks, 1966). Mammalian nucleoside diphosphate kinase can act equally well on purine and pyrimidine ribonucleotide and deoxyribonucleotide diphosphates. The high activity of this enzyme in most tissues coupled with its relative nonspecificity make the enzyme an unlikely site for control of nucleotide synthesis (Mourad and Parks, 1966). Deoxynucleoside monophosphates as well as the nucleoside monophosphates are converted to the diphosphate derivatives by a series of four distinct enzymes in mammalian tissue (Sugino et al., 1966). The substrate specificity is for (1) GMP, dGMP (reaction 4-3); (2) AMP, dAMP (reaction 4-4); (3) CMP, UMP, dCMP (reaction 4-5); and (4) dTMP and dUMP (reaction 4-6), respectively. ATP or dATP can function as the phosphate donor.

The last enzyme, thymidylate kinase (reaction 4-6), is higher in growing tissues, although it tends to be lower than the other kinases listed above in animal cells which are not rapidly proliferating (Nakamura and Sugino, 1966a,b; Hiatt and Bojarski, 1960; Canellakis et al., 1959; Ives, 1965). In HeLa cells the increase in level of this enzyme occurs at the time of DNA synthesis. SV40 or polyoma infection leads to a two- to fivefold increase in activity in mouse kidney cultures, although vaccinia or herpes simplex infection produces no increase in activity (Kit et al., 1966a,b).

dUTP and dCTP can be catabolized to the deoxymonophosphates (reaction 4-7). The presence of this enzyme may account for the absence of uracil in natural DNA (Greenberg and Somerville, 1962; Sugino et al., 1964; Bertani et al., 1963). Although this enzyme may be important in the mechanism by which bacteriophage T4 turns off host cell DNA synthesis, relatively few studies have been carried out in mammalian cells.

dUMP is converted to dTMP in a reaction catalyzed by thymidylate synthetase and requiring tetrahydrofolate and magnesium (reaction 4-8):

$$\text{dUMP} + \text{5,10-methylene-5,6,7,8-tetrahydrofolate} \xrightarrow{\text{Mg}^{2+}} \text{dTMP} + \text{7,8-dihydrofolate}$$

This enzyme is not regulated in mammalian cells by the level of dTTP within the cell (Frearson et al., 1965) but appears to be controlled by the availability of folic acid derivatives. Aminopterin blocks this step as a result of inhibition of dihydrofolate reductase (reaction 4-9) necessary for the formation of the tetrahydrofolate derivative. The activity of thymidylate synthetase in animal cells is ordinarily low but rises rapidly in association with increased DNA synthesis (Ensinck et al., 1964; F. Maley and Maley, 1960, 1961; Silber et al., 1963).

The deamination of dCMP to form dUMP occurs in a reaction catalyzed by deoxycytidylate deaminase (reaction 4-10). The enzyme is activated by dCTP and inhibited by dTTP and dUTP (Scarano et al., 1960, 1962, 1967; G. F. Maley and Maley, 1964). Activation of this enzyme by dCTP and inhibition by dTTP appears to be related to changes in state of aggregation (G. F. Maley and Maley, 1968a,b). This type of regulation appears to be important since this enzyme is involved in the synthesis of dTTP from dCMP and dCTP.

A large number of studies have documented an increase in activity of deoxycytidylate deaminase in rapidly proliferating tissues (Kit, 1970; Blakely and Vitols, 1968). A microsomal protein which inhibits dCMP deaminase may account for the relatively low enzyme activity in non-proliferating tissues and high activity in rapidly growing cells and tissues (Fiala and Fiala, 1965).

3. Salvage Pathway

Deoxythymidine and deoxyuridine can be synthesized in several ways (Fig. 4). Deoxythymidine phosphorylase catalyzes the conversion of uracil or thymine to the deoxyribonucleoside derivative in the presence of deoxyribose-1-P (reaction 4-11) (Friedkin and Roberts, 1954; Zimmerman and Seidenberg, 1964; Zimmerman, 1964):

$$\text{thymine} + \text{deoxyribose-1-P} \rightarrow \text{deoxythymidine} + \text{P}_i$$

This reaction is inhibited by a variety of purine bases (Gallow and Breitman, 1968).

The second reaction involves the conversion of thymine to deoxythymidine in the presence of deoxyuridine and deoxythymidine synthetase (reaction 4-12). In contrast to the phosphorylase, this enzyme is only slightly inhibited by purine bases:

$$\text{thymine} + \text{deoxyuridine} \rightarrow \text{deoxythymidine} + \text{uridine}$$

Thymidine kinase catalyzes the formation of thymidine deoxyribonucleoside to its nucleotide derivative in the presence of ATP or dATP (Weissman et al., 1960) (reaction 4-13). The enzyme is also active with uridine and deoxyuridine as well as its fluorinated derivatives. The enzyme is inhibited by dTTP (Bresnick and Thompson, 1965; Bresnick et al., 1966; Kit et al., 1965) and beryllium (Mainigi and Bresnick, 1969). The activity of this enzyme is increased in a variety of rapidly growing cells such as hepatoma cells (Sneider et al., 1969), human leukemia cells (Roberts and Hall, 1969), regenerating rat liver (Beltz, 1962), and mammalian cells in culture (Brent et al., 1965) during exponential growth (Eker, 1965), after treatment with amethopterin, 5-fluorouridine or cytosine arabinoside (Eker, 1966), after infection with several DNA viruses (Dulbecco et al., 1965; M. Green et al., 1964), and after immunization (Raska and Cohen, 1967).

Deoxycytidine is converted to its deoxyribonucleotide derivative by an enzyme which is separate from thymidine kinase (reaction 4-14) (Kit et al., 1963). This enzyme also increases three- to fourfold after polyoma infection in mouse kidney cultures (Kara and Weil, 1967).

Deoxycytidine or cytidine can be deaminated to the respective uridine derivatives by a pyrimidine nucleoside deaminase which is present in a variety of animal tissues but apparently particularly high in human liver (reaction 4-15) (Zicha and Buric, 1969; Creasey, 1963; Tomchick et al., 1968). Most of the halogenated derivatives are also substrates for the enzyme. Its specific activity in liver increases following partial hepatectomy in the mouse (Rothman et al., 1971).

Deoxyguanosine and deoxyadenosine are converted to their deoxyribonucleotide derivatives by a relatively specific kinase (reaction 4-16). The deoxymononucleotides can be dephosphorylated to their respective deoxynucleosides by a 5'-nucleotidase (reaction 4-17; probably the same as reaction 2-5). DNA polymerase (reaction 4-18) will not be discussed.

B. Genetic Defects

No naturally occurring genetic defects involving this pathway have been demonstrated in man. However, as will be discussed later, cell strains lacking thymidine kinase have been derived from heteroploid somatic cells (Gartler and Pious, 1966) and have played a critical role in hydridization studies to be described later.

V. Nutritional Requirements

Basic artificial media used for the culture of mammalian cells contain no preformed purines and pyrimidines since these cells in culture are ordinarily capable of synthesizing these compounds *de novo* (Eagle, 1965). However, for these biosynthetic pathways to function, certain precursors are required; these include folic acid, L-glutamine, L-aspartic acid, glycine and bicarbonate, which are components of most minimal media used for cell culture. Although minimal media in combination with serum are adequate to maintain most cell types in culture, many cell lines and strains require additional supplementation to support maximal proliferation (Sanford, 1963). Very often it has been found the supplementation of purine and pyrimidine bases, ribonucleosides, and ribonucleotides improves growth rate and survival and lessens sensitivity of cells to trauma and manipulation. This is related in part to the relatively large expenditure of energy in terms of ATP consumption required to synthesize purine and pyrimidine ribonucleotides *de novo* as compared to the salvage pathways. In addition, certain specific cell strains have definite requirements based on variation in enzyme activity for preformed purines or pyrimidines. Good examples of this include the requirement of cells deficient in hypoxanthine-guanine phosphoribosyltransferase for adenine (although such a requirement can be avoided with high concentrations of folic acid) (Felix and DeMars, 1969) and the requirement of cells deficient in orotate phosphoribosyltransferase and orotidylic decarboxylase for preformed pyrimidines (Krooth, 1970).

In lieu of the previous studies on the salvage pathways for purines and pyrimidines, it can be anticipated that, in general, the free bases would be more effective than the nucleoside derivatives as a source of preformed purines, whereas the nucleoside derivatives would be more effective than the free bases as a source of preformed pyrimidines. It is not clear how purine or pyrimidine ribonucleotide derivatives would be superior to the bases or nucleosides since the former group of compounds are not transported effectively across cell membranes and probably must be catabolized to the nucleosides or bases prior to entering the cell.

The purine and pyrimidine supplements used in certain defined artificial media are listed in Table I.

TABLE I.

PURINE AND PYRIMIDINE SUPPLEMENTS IN DEFINED MEDIA (MG/LITER)

| | Media[a] | | | | | | | | |
	CMRL	A₂ + APG	199	N.C.I.	NCTC	F10, F12	Scherer's	Way-mouth	Orotic Aciduria[b]
Bases									
Adenine	—	—	10.0	—	—	—	2.5	—	15.0
Guanine	—	1.0	0.30	—	—	—	1.0	—	0.2
Hypoxanthine	—	1.0	0.30	25.0	—	4.0	—	25.0	15.0
Xanthine	—	—	0.30	—	—	—	1.0	—	0.2
Uracil	—	—	—	—	—	—	1.0	—	15.0
Cytosine	—	—	—	—	—	—	0.50	—	0.2
Thymine	—	—	0.30	—	—	—	0.50	—	—
Nucleosides									
Adenosine	10.0	10.0	—	—	—	—	—	—	—
Deoxyadenosine	—	—	—	—	10.0	—	—	—	—
Guanosine	10.0	10.0	—	—	—	—	—	—	—
Deoxyguanosine	—	—	—	—	10.0	—	—	—	—
Cytidine	10.0	10.0	—	—	10.0	—	—	—	—
Deoxycytidine	0.1	10.0	—	—	—	—	—	—	—
5-Methylcytidine	10.0	—	—	—	0.10	—	—	—	—
Thymidine	10.0	10.0	—	—	10.0	0.7	—	—	15.0
Uridine	—	—	—	—	—	—	—	—	15.0
Nucleotides									
5'-AMP	—	0.02	0.20	—	—	—	—	—	—
ATP	—	1.08	10.0	—	—	—	—	—	—
UTP	1.0	1.0	—	—	1.0	—	—	—	—
5-Methyl-CMP	—	0.2	—	—	—	—	—	—	—
Sugar precursors									
d-Ribose	—	0.2	0.5	—	—	—	—	—	—
2-Deoxyribose	—	0.2	0.50	—	—	—	—	—	—

[a] All collected from *Price and References Manual*, Grand Island Biological Company, Grand Island, New York 1970.

[b] Krooth (1964).

VI. Selection Systems

The use of media designed to select for or against the growth of certain cellular phenotypes has proved valuable if not indispensible for a variety of studies such as cell hybridization and determination of somatic cell mutation rate. Several general approaches have been commonly used. One method involves the addition to the culture medium of an analogue of a metabolic precursor which inhibits growth only after it has been acted on by a cellular enzyme. Cells which grow in the presence of this analogue may then be found to lack catalytic activity for the enzyme. An example of this type of selection system includes cells lacking hypoxanthine-guanine phosphoribosyltransferase. These cells grow selectively in the presence of azaguanine or 6-mercaptopurine, which kills cells having the enzyme. Another example is the use of 5-bromodeoxyuridine, which is toxic for most cells but allows the selective growth of cells lacking thymidine kinase (Kit *et al.,* 1963).

Another general method used for selection of certain cellular phenotypes involves the use of inhibitors of *de novo* purine or pyrimidine biosynthesis. In this situation cells are able to grow only if they can use preformed bases via the salvage pathways. For example, in the presence of amethopterin, which inhibits purine biosynthesis *de novo,* cells will grow only if they have the enzyme hypoxanthine-guanine phosphoribosyltransferase and therefore can synthesize adequate IMP from hypoxanthine despite the failure of the *de novo* pathway. Growth in the presence of amethopterin also requires the presence of thymidine kinase so that the cell can synthesize TMP from thymidine rather than from dUMP, which is inhibited by the folic acid analogue (DeMars and Hooper, 1960).

Although other examples could be cited, the specific selection systems described above have been of the most general value in recent years. Littlefield (1964b) was the first to combine the above selective systems in an ingenious and profitable manner. He was attempting to select hybrid cells which differed from each parent cell. Thus one parental line was selected which lacked hypoxanthine-guanine phosphoribosyltransferase (HGPRT⁻), whereas the other parental line lacked thymidine kinase (TK⁻). Selective media contained aminopterin to inhibit purine biosynthesis *de novo* and thymidylate synthetase. Hypoxanthine and thymidine were added to the media to ensure that adequate preformed purines and pyrimidines were available. This medium has been referred to as HAT for the hypoxanthine, aminopterin, and thymidine present, or THAG when glycine is also added. It will be recognized that neither the

HGPRT⁻ nor the TK⁻ cells could survive in the presence of HAT media since each cell strain was unable to synthesize IMP or TMP, respectively. Thus the only cells which survived in this sytem were the hybrids which contained adequate quantities of both enzymes. It was this method of selection which permitted isolation of the first hybrids between cells of different species and which has provided the means of selecting most of the other hybrids isolated since.

R. L. Davidson and B. Ephrussi (1965) adapted the HAT selective system to isolate hybrids when one parental line was heteroploid and sensitive to the presence of aminopterin while the other was a diploid cell carrying no selective marker. With this "half-selective" system these investigators isolated the first hybrids derived from diploid cells. The mutant parental cells could not use nucleotides from the medium and so, in the presence of aminopterin, could not grow. The same medium does not interfere with the growth of the diploid cells, but since they are normal cells, they grow slowly and form a monolayer background for the hybrid cells which form discrete multilayered colonies.

VII. Hybridization

The mating or hybridization of somatic cells from individuals of diverse genetic qualities, even of different species, has provided an extremely important new tool in biology and genetics. The relevance of this subject to a chapter involved with purine and pyrimidine metabolism derives from the statement above that nearly all the hybrid cells developed in the past have utilized mutants with enzymatic alterations in the purine pathway, the pyrimidine pathway, or both. This subject has been extensively and thoroughly reviewed in the recent past and will not be considered in detail here (Migeon and Childs, 1970; Harris, 1970). The discovery in the past decade of a number of human mutants demonstrable in cell culture which can be selected for or against by the techniques described above has allowed hybridization to be applied to human diploid cells.

Hybrid cells have already been used to answer a number of important questions in human biology. Detailed study of enzyme recombination in hybrid cells can be used to determine if the enzyme in question has a multiple subunit structure. Another important use of hybrid cells has been in the study of the chromosomal locus for human genes. Clearly, this approach is just beginning to yield the type of information which pottentially can be obtained from this system (Migeon, 1968;

Migeon *et al.*, 1969). Hybrid cells can be used to determine if more than one locus is involved in the development of an auxotroph or a genetically determined phenotype in man (Kao *et al.*, 1969; Nadler *et al.*, 1970). Control of gene, chromosomal, nucleolar, and nuclear functions can also be studied in human cells using this technique (Harris, 1970; H. Green *et al.*, 1966; Littlefield, 1969). Hybrids have also been used to determine the genetic nature of malignancy as well as susceptibility and response to viral transformation (R. L. Davidson *et al.*, 1966; R. L. Davidson and Yamamoto, 1968; Defendi *et al.*, 1967). There is little question that we are just beginning to see the potential value of studying hybrid cells in culture.

VIII. Future Uses of Human Cells in Culture with Genetically Determined Alterations in Purine and Pyrimidine Metabolism

It is easy to visualize a number of important developments in the future based on information currently available. Several examples might be cited. Selective systems have been developed for several well-defined enzymatic defects in purine and pyrimidine metabolism. Indeed, additional selective systems will be readily apparent as new genetic defects are defined. This should allow one to establish the mutation rate at specific loci in somatic cells. This would be particularly useful in male cells for an enzyme which is X-linked or in cells from either sex in which there is already heterozygosity at the locus in question. Once the spontaneous mutation rate can be established at various loci in these human cells, then the influence of various environmental factors and drugs, for example, can be evaluated.

Another possibility is that cells grown in culture could be used as a source of tissue for transplantation as an approach to enzyme replacement therapy. This tissue would be potentially more useful than solid tissues such as kidney in that relatively large quantities of cells could be provided at will. The Lesch-Nyhan syndrome due to a functional deficiency of hypoxanthine-guanine phosphoribosyltransferase is a good example of a disease that might be approached in this manner. Previous studies have shown that a very small quantity of functional enzyme appears to prevent the development of the bizarre neurological features characteristic of this disease (Kelley *et al.*, 1969). In addition, it has been demonstrated that contact of normal cells with hypoxanthine-guanine-phosphoribosyltransferase-deficient cells leads to correction of the abnormality in the mutant cells due to transfer of nucleotide into the cell (Subak-Sharpe *et al.*, 1969; Friedman *et al.*, 1968; Cox *et al.*, 1970). Hence, it seems possible that transplantation of cells with normal hypoxanthine-guanine phosphoribosyltransferase activity into patients lacking the enzyme might lead to a partial correction of the defect *in vivo*. Potential rejection of the cells with normal hypoxanthine-guanine phosphoribosyltransferase activity could be minimized if not totally avoided by using the patient's own HGPRT⁻ cells and selecting for back mutations by techniques described above. The patient could then be transplanted with his own cells, which would presumably (assuming mutagens are not used) be different only at the hypoxanthine-guanine phosphoribosyltransferase locus.

IX. Summary

In the present chapter I have reviewed what is known about the metabolic pathways and their regulation involved in the synthesis of purine and pyrimidine ribo- and deoxyribonucleotides. A series of human mutants in these pathways has been described. The nutritional requirements of most mammalian cells have been summarized briefly.

The importance of recently developed selective systems for and against various genetically determined alterations in these pathways has been emphasized with particular reference to cell hybridization.

REFERENCES

Abrams, R., Edmonds, M., and Libenson, L. (1965). Deoxyribosyl exchange activity associated with nucleoside phosphorylase. *Biochem. Biophys. Res. Commun.* **20**, 310–314.

Akedo, H., Nishihara, H., Shinkai, K., and Komatsu, K. (1970). Adenosine deaminases of two different molecular sizes in human tissues. *Biochim. Biophys. Acta* **212**, 189–191.

Albertini, R. J., and DeMars, R. (1970). Diploid azaguanine-resistant mutants of cultured human fibroblasts. *Science* **169**, 482–485.

Anderson, E. P., and Brockman, R. W. (1964). Feedback inhibition of uridine kinase by cytidine triphosphate and uridine triphosphate. *Biochim. Biophys. Acta* **91**, 380–386.

Anderson, J. H., and Sartorelli, A. C. (1968). Inosinic dehydrogenase of sarcoma 180 cells. *J. Biol. Chem.* **243**, 4762–4768.

Anderson, J. H., and Sartorelli, A. C. (1969). Inhibition of inosinic dehydrogenase of sarcoma 180 ascites cells by nucleotides and their analogs. *Biochem. Pharmacol.* **18**, 2747–2757.

Anderson, P. M., and Meister, A. (1966). Control of *Escherichia coli* carbamyl phosphate synthetase by purine and pyrimidine nucleotides. *Biochemistry* **5**, 3164–3169.

Appel, S. H. (1968). Purification and kinetic properties of brain orotidine 5'-phosphate decarboxylase. *J. Biol. Chem.* **243**, 3924–3929.

Askari, A., and Franklin, J. E., Jr. (1965). Effects of monovalent cations and ATP on erythrocyte AMP deaminase. *Biochim. Biophys. Acta* **110**, 162–173.

Atkinson, D. E., and Fall, L. (1967). Adenosine triphosphate conservation in biosynthetic regulation. *Escherichia coli* phosphoribosylpyrophosphate synthetase. *J. Biol. Chem.* **242**, 3241–3242.

Baer, H. P., and Drummond, G. I. (1968). Catabolism of adenine nucleotides by the isolated perfused rat heart (32614). *Proc. Soc. Exp. Biol. Med.* **127**, 33–36.

Baer, H. P., Drummond, G. I., and Duncan, E. L. (1966). Formation and deamination of adenosine by cardiac muscle enzymes. *Mol. Pharmacol.* **2**, 67–76.

Beltz, R. E. (1962). Comparison of the effects of X-radiation on the elevation of thymidine kinase and thymidylate synthetase during liver regeneration. *Biochem. Biophys. Res. Commun.* **9**, 78–83.

Bennett, L. L., Jr., Schnebli, H. P., Vails, M. H., Allan, P. W., and Montgomery, J. A. (1966). Purine ribonucleoside kinase activity and resistance to some analogs of adenosine. *Mol. Pharmacol.* **2**, 432–443.

Bertani, E., Haggmark, A., and Reichard, P. (1963). Enzymatic synthesis of deoxyribonucleotides. II. Formation and interconversion of deoxyuridine phosphates. *J. Biol. Chem.* **238**, 3407–3413.

Blakely, R. L., and Vitols, E. (1968). The control of nucleotide biosynthesis. *Annu. Rev. Biochem.* **37**, 201–224.

Brent, T. P., Butler, J. A. V., Crathorn, A. R. (1965). Variations in phosphokinase activities during the cell cycle in synchronous populations of HeLa cells. *Nature* (*London*) **207**, 176–177.

Bresnick, E. (1964). Regulatory control of pyrimidine biosynthesis in mammalian systems. *Advan. Enzyme Regul.* **2**, 213–236.

Bresnick, E. (1965). Early changes in pyrimidine biosynthesis after partial hepatectomy. *J. Biol. Chem.* **240**, 2550–2556.

Bresnick, E., and Blatchford, K. (1964). Inhibiton of dihydroorotase by purines and pyrimidines. *Biochim. Biophys. Acta* **81**, 150–157.

Bresnick, E., and Hitchings, G. H. (1961). Feedback control in Ehrlich ascites cells. *Cancer Res.* **21**, 105–109.

Bresnick, E., and Thompson, U. B. (1965). Properties of deoxythymidine kinase partially purified from animal tumors. *J. Biol. Chem.* **240**, 3967–3974.

Bresnick, E., Thompson, U. B., and Lyman, K. (1966). Aggregation of deoxythymidine kinase in dilute solutions: Properties of aggregated and disaggregated forms. *Arch. Biochem. Biophys.* **114**, 352–359.

Brockman, R. W. (1963). Mechanisms of resistance to anticancer agents. *Advan. Cancer Res.* **7**, 129–234.

Burger, R. M., and Lowenstein, J. M. (1970). Preparation and properties of 5'-nucleotidase from smooth muscle of small intestine. *J. Biol. Chem.* **245**, 6274–6280.

Canellakis, E. S. (1957a). Pyrimidine metabolism. II. Enzymatic pathways of uracil anabolism. *J. Biol. Chem.* **227**, 329–338.

Canellakis, E. S. (1957b). Pyrimidine metabolism. III. The interaction of the catabolic and anabolic pathways of uracil metabolism. *J. Biol. Chem.* **227**, 701–709.

Canellakis, E. S., Jaffe, J. J., Mantsavinos, R., and Krakow, J. S. (1959). Pyrimidine metabolism. IV. A comparison of normal and regenerating rat liver. *J. Biol. Chem.* **234**, 2096–2099.

Carter, C. E. (1951). Enzymatic evidence for the structure of desoxyribonucleotides. *J. Amer. Chem. Soc.* **73**, 1537–1539.

Caskey, C. T., Ashton, D. M., and Wyngaarden, J. B. (1964). The enzymology of feedback inhibition of glutamine phosphoribosylpyrophosphate amidotransferase by purine ribonucleotides. *J. Biol. Chem.* **239**, 2570–2579.

Center, M. S., and Behal, F. J. (1966). Calf intestinal 5′nucleotidase. *Arch. Biochem. Biophys.* **114**, 414–421.

Chu, E. H. Y. (1971). Mammalian cell genetics. III. Characterization of X-ray induced forward mutations in chinese hamster cell cultures. *Mutat. Res.* **11**, 23–34.

Cook, J. L., and Vibert, M. (1966). The utilization of purines and their ribosyl derivatives for the formation of adenosine triphosphate and guanosine triphosphate in the rabbit reticulocyte. *J. Biol. Chem.* **241**, 158–160.

Cox, R. P., Krauss, M. R., Balis, M. E., and Dancis, J. (1970). Evidence for transfer of enzyme product as the basis of metabolic cooperation between tissue culture fibroblasts of Lesch-Nyhan disease and normal cells. *Proc. Nat. Acad. Sci. U.S.* **67**, 1573–1579.

Craft, J. A., Dean, B. M., Watts, R. W. E., and Westwick, W. J. (1970). Studies on human erythrocyte IMP: Pyrophosphate phosphoribosyltransferase. *Eur. J. Biochem.* **15**, 367–373.

Creasey, W. A. (1963). Studies on the metabolism of 5-iodo-2′-deoxycytidine *in vitro*. Purification of nucleoside deaminase from mouse kidney. *J. Biol. Chem.* **238**, 1772–1776.

Creasey, W. A., and Handschumacher, R. E. (1961). Purification and properties of orotidylate decarboxylases from yeast and rat liver. *J. Biol. Chem.* **236**, 2058–2063.

Currie, R., Bergel, F., and Bray, R. C. (1967). Enzymes and cancer. Preparation and some properties of guanase from rabbit liver. *Biochem. J.* **104**, 634–638.

Davidson, J. D., Bradley, T. R., Roosa, R. A., and Law, L. W. (1962). Purine nucleotide pyrophosphorylases in 8-azaguanine sensitive and resistant leukemias. *J. Nat. Cancer Inst.* **29**, 789–803.

Davidson, R. L., and Ephrussi, B. (1965). A selective system for the isolation of hybrids between L cells and normal cells. *Nature (London)* **205**, 1170–1171.

Davidson, R. L., and Yamamoto, K. (1968). Regulation of melanin synthesis in mammalian cells as studied by somatic cell hybridization. II. The level of regulation of 3,4 dihydroxy phenylalanine oxidase. *Proc. Nat. Acad. Sci. U.S.* **60**, 894–901.

Davidson, R. L., Ephrussi, B., and Yamamoto, K. (1966). Regulation of pigment synthesis in mammalian cells as studied by somatic hybridization. *Proc. Nat. Acad. Sci. U.S.* **56**, 1437–1440.

Defendi, V., Ephrussi, B., Koprowski, H., and Yoshida, M. C. (1967). Properties of hybrids between polyoma-transformed and normal mouse cells. *Proc. Nat. Acad. Sci. U.S.* **57**, 299–305.

DeMars, R., and Hooper, J. L. (1960). A method of selecting for auxotrophic mutants of HeLa cells. *J. Exp. Med.* **111**, 559–572.

Dulbecco, R., Hartwell, L. H., and Vogt, M. (1965). Induction of cellular DNA synthesis by polyoma virus. *Proc. Nat. Acad. Sci. U.S.* **53**, 403–410.

Eagle, H. (1965). Metabolic controls in cultured mammalian cells. *Science* **148**, 42–51.

Edwards, M. J., and Maguire, M. H. (1970). Purification and properties of rat heart 5′-nucleotidase. *Mol. Pharmacol.* **6**, 641–648.

Eidinoff, M. L., Knoll, J. E., Marano, B., and Cheong, L. (1958). Purine studies. I. Effect of DON (6-diazo-5-oxo-L-norleucine) on incorporation of precursors into nucleic acid pyrimidines. *Cancer Res.* **18**, 105–109.

Eker, P. (1965). Activities of thymidine kinase and thymidine deoxyribonucleotide phosphatase during growth of cells in tissue culture. *J. Biol. Chem.* **240**, 2607–2611.

Eker, P. (1966). Studies on thymidine kinase of human liver cells in culture. *J. Biol. Chem.* **241**, 659–662.

Elford, H. L. (1968). Effect of hydroxyurea on ribonucleotide reductase. *Biochem. Biophys. Res. Commun.* **33**, 129–135.

Emmelot, P., and Bos, C. J. (1965). Differential effect of neuraminidase on the Mg^{2+}-ATPase, Na^+-K^+-Mg^{2+}-ATPase and 5′-nucleotidase of isolated plasma membranes. *Biochim. Biophys. Acta* **99**, 578–580.

Ensinck, J. W., Coombs, G. J., Williams, R. H., and Vallance-Owen, J. (1964). Studies *in vitro* of the transport of the A and B chains of insulin in serum. *J. Biol. Chem.* **239**, 3377–3384.

Epstein, C. J. (1970). Phosphoribosyltransferase activity during early mammalian development. *J. Biol. Chem.* **245**, 3289–3294.

Essner, E., Norikoff, A. B., and Quintana, N. (1965). Nucleoside phosphate activities in rat cardiac muscle. *J. Cell Biol.* **25**, 201–215.

Felix, J. S., and DeMars, R. (1969). Purine requirement of cells cultured from humans affected with Lesch-Nyhan syndrome (hypoxanthine-guanine phosphoribosyltransferase deficiency). *Proc. Nat. Acad. Sci. U.S.* **62**, 536–543.

Fiala, S., and Fiala, A. E. (1965). Deoxycytidylic acid deaminase in Ehrlich ascites tumor cells. *Cancer Res.* **25**, 922–932.

Flaks, J. G., Erwin, M. J., and Buchanan, J. M. (1957). Biosynthesis of the purines. XVI. The synthesis of adenosine 5′-phosphate and 5-amino-4-imidazolecarboxamide ribotide by a nucleotide pyrophosphorylase. *J. Biol. Chem.* **228**, 201–213.

Fox, I. H., and Kelley, W. N. (1971). Human Phosphoribosylpyrophosphate Synthetase. Distribution, Purification and Properties. *J. Biol. Chem.* **246**, 5739–5748.

Fox, I. H., and Kelley, W. N. (1972). Human Phosphoribosylpyrophosphate Synthetase. Kinetic mechanism and end-product inhibition. *J. Biol. Chem.* (in press).

Frearson, P. M., Kit, S., and Dubbs, D. R. (1965). Deoxythymidylate synthetase and deoxythymidine kinase activities of virus-infected animal cells. *Cancer Res.* **25**, 737–744.

Friedkin, M., and Roberts, D. (1954). The enzymatic synthesis of nucleosides. I. Thymidine phosphorylase in mammalian tissue. *J. Biol. Chem.* **207**, 245–256.

Friedman, T., Seegmiller, J. E., and Subak-Sharpe, J. H. (1968). Metabolic cooperation between genetically marked human fibroblasts in tissue culture. *Nature (London)* **220**, 272–274.

Friedman, T., Seegmiller, J. E., and Subak-Sharpe, J. H. (1969). Evidence against the existence of guanosine and inosine kinases in human fibroblasts in tissue culture. *Exp. Cell Res.* **56**, 425–429.

Fritzson, P. (1967). Dephosphorylation of pyrimidine nucleotides in the soluble fraction of homogenates from normal and regenerating rat liver. *Eur. J. Biochem.* **1**, 12–20.

Fujimoto, W. Y., and Seegmiller, J. E. (1970). Hypoxanthine-guanine phosphoribosyl-transferase deficiency activity in normal, mutant and heterozygote cultured human skin fibroblasts. Proc. Nat. Acad. Sci. U.S. 65, 577–584.

Gallo, R. C., and Breitman, T. R. (1968). The enzymatic mechanisms for deoxythymidine synthesis in human leukocytes. III. Inhibition of deoxythymidine phosphorylase by purines. J. Biol. Chem. 243, 4943–4951.

Gartler, S. M., and Pious, D. A. (1966). Genetics of mammalian cell cultures. Humangenetik 2, 83–114.

Gotto, A. M., Meikle, A. W., and Touster, O. (1964). Nucleoside metabolism in Ehrlich ascites tumor cell: Phosphorolysis of purine nucleosides. Biochim. Biophys. Acta 80, 552–561.

Gotto, A. M., Belkhode, M. L., and Touster, O. (1969). Stimulatory effects of inosine and deoxyinosine on the incorporation of uracil-2-¹⁴C, 5-fluorouracil-2-¹⁴C, and 5-bromouracil-2-¹⁴C into nucleic acids by Ehrlich ascites tumor cells in vitro. Cancer Res. 29, 807–811.

Green, H., Ephrussi, B., Yoshida, M., and Hamerman, D. (1966). Synthesis of collagen and hyaluronic acid by fibroblast hybrids. Proc. Nat. Acad. Sci. U.S. 55, 41–44.

Green, M., Pina, M., and Chagoya, V. (1964). Biochemical studies on adenovirus multiplications. V. Enzymes of deoxyribonucleic acid synthesis in cells infected by adenovirus and vaccinia virus. J. Biol. Chem. 239, 1188–1197.

Greenberg, G. R., and Somerville, R. L. (1962). Deoxyuridylate kinase activity and deoxyuridinetriphosphatase in Escherichia coli. Proc. Nat. Acad. Sci. U.S. 48, 247–257.

Greene, M. L., Boyles, J. R., and Seegmiller, J. E. (1970). Substrate stabilization: Genetically controlled reciprocal relationship of two human enzymes. Science 167, 887–889.

Hager, S. E., and Jones, M. E. (1965). Initial steps in pyrimidine synthesis in Ehrlich ascites carcinoma in vitro. I. Factors affecting the incorporation of ¹⁴C bicarbonate into carbon 2 of the uracil ring of the acid-soluble nucleotides of intact cells. J. Biol. Chem. 240, 4556–4563.

Hager, S. E., and Jones, M. E. (1967a). Initial steps in pyrimidine synthesis in Ehrlich ascites carcinoma in vitro. II. The synthesis of carbamyl phosphate by a soluble, glutamine dependent carbamyl phosphate synthetase. J. Biol. Chem. 242, 5667–5673.

Hager, S. E., and Jones, M. E. (1967b). A glutamine dependent enzyme for the synthesis of carbamyl phosphate for pyrimidine biosynthesis in fetal rat liver. J. Biol. Chem. 242, 5674–5680.

Harris, H. (1970). In "Cell Fusion," pp. 1–108. Harvard Univ. Press, Cambridge, Massachusetts.

Hartman, S. C. (1970). Purines and pyrimidines. Metab. Pathways 4, 1–68.

Hartman, S. C., Levenberg, B., and Buchanan, J. M. (1956). Biosynthesis of purines. XI. Structure, enzymatic synthesis, and metabolism of glycinamide ribotide and (α-N-formyl) glycinamide ribotide. J. Biol. Chem. 221, 1057–1070.

Hatfield, D., and Wyngaarden, J. B. (1964a). 3-Ribosylpurines. I. Synthesis of (3-ribosyluric acid) 5'-phosphate and (3-ribosylxanthine) 5'-phosphate by a pyrimidine ribonucleotide pyrophosphorylase of beef erythrocytes. J. Biol. Chem. 239, 2580–2586.

Hatfield, D., and Wyngaarden, J. B. (1964b). 3-Ribosylpurines. II. Studies on (3-ribosylxanthine) 5'-phosphate and on ribonucleotide derivatives of certain uracil analogs. J. Biol. Chem. 239, 2587–2592.

Henderson, J. F. (1962). Feedback inhibition of purine biosynthesis in ascites tumor cells. *J. Biol. Chem.* **237**, 2631–2635.

Henderson, J. F. (1963). Dual effects of ammonium chloride on purine biosynthesis *de novo* in Ehrlich ascites tumor cells *in vitro. Biochim. Biophys. Acta* **76**, 173–180.

Henderson, J. F., and Khoo, M. K. Y. (1965). On the mechanism of feedback inhibition of purine biosynthesis *de novo* in Ehrlich ascites tumor cells *in vitro. J. Biol. Chem.* **240**, 3104–3109.

Henderson, J. F., Caldwell, I. C., and Paterson, A. R. P. (1967). Decreased feedback inhibition in a 6-(methylmercapto) purine ribonucleoside-resistant tumor. *Cancer Res.* **27**, 1773–1778.

Henderson, J. F., Brox, L. W., Kelley, W. N., Rosenbloom, F. M., and Seegmiller, J. E. (1968a). Kinetics of hypoxanthine-guanine phosphoribosyltransferase. *J. Biol. Chem.* **243**, 2514–2522.

Henderson, J. F., Rosenbloom, F. M., Kelley, W. N., and Seegmiller, J. E. (1968b). Variations in purine metabolism of cultured skin fibroblasts from patients with gout. *J. Clin. Invest.* **47**, 1511–1516.

Herscovics, A., and Johnstone, R. M. (1964). Formate utilization in cell-free extracts of Ehrlich ascites cells. *Biochim. Biophys.* Acta **93**, 251–263.

Hershko, A., Wind, E., Razin, A., and Mager, J. (1963). Conversion of guanine to hypoxanthine in mammalian red blood cells. *Biochim. Biophys. Acta* **71**, 609–620.

Hershko, A., Razin, A., and Mager, J. (1969). Relation of the synthesis of 5-phosphoribosyl-1-pyrophosphate in intact red blood cells and in cell free preparations. *Biochim. Biophys. Acta* **184**, 64–76.

Hiatt, H. H., and Bojarski, T. B. (1960). Stimulation of thymidine kinase activity in rat tissues by thymidine administration. *Biochem. Biophys. Res. Commun.* **2**, 35–39.

Hill, D. L., (1970). Hypoxanthine phosphoribosyltransferase and guanine metabolism of adenocarcinoma 755 cells. *Biochem. Pharmacol.* **19**, 545–557.

Hill, D. L., and Bennett, L. L., Jr. (1969). Purification and properties of 5-phosphoribosyl pyrophosphate amidotransferase from adenocarcinoma, 755 cells. *Biochemistry* **8**, 122–130.

Ho, D. H. W., Luce, J. K.,and Frei, E., III. (1968). Distribution of purine ribonucleoside kinase and selective toxicity of 6-methylthiopurine ribonucleoside. *Biochem. Pharmacol.* **17**, 1025–1035.

Holmsen, H., and Rozenberg, M. C. (1968). Adenine nucleotide metabolism of blood platelets. I. Adenosine kinase and nucleotide formation from exogenous adenosine and AMP. *Biochim. Biophys. Acta* **155**, 326–341.

Hori, M., and Henderson, J. F. (1966a). Purification and properties of adenylate pyrophosphorylase from Ehrlich ascites tumor cells. *J. Biol. Chem.* **241**, 1406–1411.

Hori, M., and Henderson, J. F. (1966b). Kinetic studies of adenine phosphoribosyltransferase. *J. Biol. Chem.* **241**, 3404–3408.

Hori, M., Gadd, R. E. A., and Henderson, J. B. (1967). Inhibition and stimulation of adenine phosphoribosyltransferase by purine nucleotides. *Biochem. Biophys. Res. Commun.* **28**, 616–620.

Howard, W. J., and Appel, S. H. (1968). Control of purine biosynthesis. FGAR amidotransferase (abstract). *Clin. Res.* **16**, 344.

Hurlbert, R. B., and Kammen, H. O. (1960). Formation of cytidine nucleotides from uridine nucleotides by soluble mammalian enzymes: Requirements for glutamine and guanosine nucleotides. *J. Biol. Chem.* **235**, 443–449.

Inagaki, A., and Tatibana, M. (1970). Contol of pyrimidine biosynthesis in mammalian tissues. III. Multiple forms of aspartate transcarbamylase of mouse spleen. *Biochim. Biophys. Acta* **220**, 491–502.

Ipata, P. L. (1967). Studies on the inhibition by nucleoside triphosphates of sheep brain 5-nucleotidase. *Biochem. Biophys. Res. Commun.* **27**, 337–343.

Ipata, P. L. (1968). Sheep brain 5′-nucleotidase. Some enzymatic properties and allosteric inhibition by nucleoside triphosphates. *Biochemistry* **7**, 507–515.

Itoh, R., Mitsui, A., and Isushima, K. (1968). Properties of 5′-nucleotidase from hepatic tissue of higher animals. *J. Biochem. (Tokyo)* **63**, 165–169.

Ito, K., Nakanishi, S., Terada, M., and Tatibana, M. (1970). Control of pyrimidine biosynthesis in mammalian tissues. II. Glutamine-utilizing carbamyl phosphate synthetase of various experimental tumors: Distribution, purification, and characterization. *Biochim. Biophys. Acta* **220**, 477–490.

Ives, D. H. (1965). Evidence for thymidine diphosphate as the precursor of thymidine triphosphate in tumor. Transfer of the terminal phosphate of adenosine triphosphate to thymidylate. *J. Biol. Chem.* **240**, 819–824.

Jones, M. E., and Hager, S. E. (1966). Source of carbamyl phosphate for pyrimidine biosynthesis in mouse Ehrlich ascites cells and rat liver. *Science* **154**, 422.

Josan, V., and Krishman, P. S. (1968). Regulation of rat liver guanine amino hydrolase by GTP. *Biochem. Biophys. Res. Commun.* **31**, 299–302.

Kalckar, H. M. (1947). Differential spectrophotometry of purine compounds by means of specific enzymes. III. Studies of the enzymes of purine metabolism. *J. Biol. Chem.* **167**, 461–475.

Kao, F., Johnson, R. T., and Puck, T. D. (1969). Complementation analysis on virus-fused Chinese hamster cells with nutritional markers. *Science* **164**, 312–314.

Kara, J., and Weil, R. (1967). Specific activation of the DNA-synthesizing apparatus in contact-inhibited mouse kidney cells by polyoma virus. *Proc. Nat. Acad. Sci. U.S.* **57**, 63–70.

Kasbekar, D. K., Nagabhushanam, A., and Greenberg, D. M. (1964). Purification and properties of orotic acid-decarboxylating enzymes from calf thymus. *J. Biol. Chem.* **239**, 4245–4249.

Kelley, W. N. (1968). Hypoxanthine-guanine phosphoribosyltransferase deficiency in the Lesch-Nyhan syndrome and gout. *Fed. Proc., Fed. Amer. Soc. Exp. Biol.* **27**, 1047–1052.

Kelley, W. N. (1971a). Studies on the adenine phosphoribosyltransferase enzyme in human fibroblasts lacking hypoxanthine-guanine phosphoribosyltransferase. *J. Lab. Clin. Med.* **77**, 33–38.

Kelley, W. N. (1971b). Unpublished observations.

Kelley, W. N., and Meade, J. C. (1971). Studies on hypoxanthine-guanine phosphoribosyltransferase in fibroblasts from patients with the Lesch-Nyhan syndrome: Evidence for genetic heterogeneity. *J. Biol. Chem.* **246**, 2953–2958.

Kelley, W. N., and Wyngaarden, J. B. (1970). Effects of allopurinol and oxipurinol on purine synthesis in cultured human cells. *J. Clin. Invest.* **49**, 602–609.

Kelley, W. N., Rosenbloom, F. M., Henderson, J. F., and Seegmiller, J. E. (1967a). Xanthine phosphoribosyltransferase in man: Relationship to hypoxanthine-guanine phosphoribosyltransferase. *Biochem. Biophys. Res. Commun.* **28**, 340–345.

Kelley, W. N., Rosenbloom, F. M., Henderson, J. F., and Seegmiller, J. E. (1967b).

A specific enzyme defect in gout associated with overproduction of uric acid. *Proc. Nat. Acad. Sci. U.S.* **57**, 1735–1739.

Kelley, W. N., Levy, R. I., Rosenbloom, F. M., Henderson, J. F., and Seegmiller, J. E. (1968). Adenine phosphoribosyltransferase deficiency: A previously undescribed genetic defect in man. *J. Clin. Invest.* **47**, 2281–2289.

Kelley, W. N., Greene, M. L., Rosenbloom, F. M., Henderson, J. F., and Seegmiller, J. E. (1969). Hypoxantine-guanine phosphoribosyltransferase deficiency in gout. *Ann. Inter. Med.* **70**, 155–206.

Kelley, W. N., Fox, I. H., and Wyngaarden, J. B. (1970a). Regulation of purine biosynthesis in cultured human cells. I. Effects of orotic acid. *Biochim. Biophys. Acta* **215**, 512–516.

Kelley, W. N., Greene, M. L., Fox, I. H., Rosenbloom, F. M., Levy, R. I., and Seegmiller, J. E. (1970b). Effects of orotic acid on purine and lipoprotein metabolism in man. *Metab. Clin. Exp.* **19**, 1025–1035.

Kelley, W. N., Fox, I. H., and Wyngaarden, J. B. (1970c). Further evaluation of adenine phosphoribosyltransferase deficiency in man. Occurrence in a patient with gout. *Clin. Res.* **18**, 53.

Kim, B. K., Cha, S., and Parks, R. E., Jr. (1968). Purine nucleoside phosphorylase from human erythrocytes. I. Purification and properties. *J. Biol. Chem.* **243**, 1763–1770.

Kimball, A. P., Allinson, P. S., and Frymire, M. J. (1967). Purine ribonucleotide reductase: End product inhibition and stimulation of a mammalian enzyme *Proc. Soc. Exp. Biol. Med.* **125**, 1105–1108.

King, C. D., and Van Lancker, J. L. (1969). Molecular mechanisms of liver regeneration. VII. Conversion of cytidine to deoxycytidine in rat regenerating livers. *Arch. Biochem. Biophys.* **129**, 603–608.

Kit, S. (1970). Nucleotides and nucleic acids. *Metab. Pathways* **4**, 70–275.

Kit, S., Dubbs, D. R., Piekarski, L. J., and Hsu, T. C. (1963). Deletion of thymidine kinase activity from L cells resistant to bromodeoxyuridine. *Exp. Cell Res.* **31**, 297–312.

Kit, S., Dubbs, D. R., and Frearson, P. M. (1965). Decline of thymidine kinase activity in stationary phase mouse fibroblast cells. *J. Biol. Chem.* **240**, 2565–2573.

Kit, S., Dubbs, D. R., and Frearson, P. M. (1966a). Enzymes of nucleic acid metabolism in cells infected with polyoma virus. *Cancer Res.* **26**, 638–646.

Kit, S., Dubbs, D. R., Piekarski, L. J., DeTorres, R. A., and Melnick, J. L. (1966b). Acquisition of enzyme function by mouse kidney cells abortively infected with papovavirus SV40. *Proc. Nat. Acad. Sci. U.S.* **56**, 463–470.

Klenow, H. (1962). Further studies on the effect of deoxyadenosine on the accumulation of deoxyadenosine triphosphate and inhibition of deoxyribonucleic acid synthesis in Ehrlich ascites tumor cells *in vitro*. *Biochim. Biophys. Acta* **61**, 885–896.

Korbecki, M., and Plagemann, P. G. W. (1969). Competitive inhibition of uridine incorporation by 6-azauridine in uninfected and mengovirus-infected Novikoff hepatoma cells (34266). *Proc. Soc. Exp. Biol. Med.* **132**, 587–595.

Krakoff, I. H., Brown, N. C., and Reichard, P. (1968). Inhibition of ribonucleoside diphosphate reductase by hydroxyurea. *Cancer Res.* **28**, 1559–1565.

Krakow, G., and Vennesland, B. (1961). The equilibrium constant of the dihydroorotic dehydrogenase reaction. *J. Biol. Chem.* **236**, 142–144.

Krenitsky, T. A., and Papaioannow, R. J. (1969). Human hypoxanthine phosphoribosyltransferase. II. Kinetics and chemical modification. *J. Biol. Chem.* **244**, 1271–1277.

Krenitsky, T. A., Barclay, M., and Jacquez, J. A. (1964). Specificity of mouse uridine phosphorylase. Chromatography, purification, and properties. *J. Biol. Chem.* **239**, 805–812.

Krenitsky, T. A., Elion, G. B., Henderson, A. M., and Hitchings, G. H. (1968). Inhibition of human purine nucleoside phosphorylase. Studies with intact erythrocytes and the purified enzyme. *J. Biol. Chem.* **243**, 2876–2881.

Krenitsky, T. A., Neil, S. M., Elion, G. B., and Hitchings, G. H. (1969). Adenine phosphoribosyltransferase from monkey liver. *J. Biol. Chem.* **244**, 4779–4784.

Krooth, R. S. (1964). Properties of diploid cell strains developed from patients with an inherited abnormality of uridine biosynthesis. *Cold Spring Habor Symp. Quant. Biol.* **29**, 189–212.

Krooth, R. S. (1970). Studies on the regulation of uridine 5′-monophosphate synthesis in human diploid cells. In "Control Mechanisms in the Expression of Cellular Phenotypes" Vol. 9, pp. 43–68. Academic Press, New York.

Kumar, S. (1969). Guanine deaminase in developing rat brain. *Arch. Biochem. Biophys.* **130**, 693–694.

Kumar, S., Tewari, K. K., and Krishman, P. S. (1965). Guanine deaminase activity in rat brain and liver. *Biochem. J.* **95**, 797–802.

Kusano, T., Long, C., and Green, H. (1971). A new reduced human-mouse somatic cell hybrid containing the human gene for adenine phosphoribosyltransferase. *Proc. Nat. Acad. Sci. U.S.* **68**, 82–86.

Lacroute, F., Pierard, A., Grenson, M., and Wiame, J. M. (1965). The biosynthesis of carbamyl phosphate in *Saccharomyces cerevisiae*. *J. Gen. Microbiol.* **40**, 127–142.

Lagerkvist, U. (1958a). Biosynthesis of guanosine 5′-phosphate. I. Xanthosine 5′-phosphate as an intermediate. *J. Biol. Chem.* **233**, 138–142.

Lagerkvist, U. (1958b). Biosynthesis of guanosine 5′-phosphate. II. Amination of xanthosine 5′-phosphate by purified enzyme from pigeon liver. *J. Biol. Chem.* **233**, 143–140.

Larsson, A. (1963). Enzymatic synthesis of deoxyribonucleotides. III. Reduction of purine ribonucleotides with an enzyme system from *Escherichia coli* B. *J. Biol. Chem.* **238**, 3414–3419.

Larsson, A., and Reichard, P. (1966). Enzymatic synthesis of deoxyribonucleotides. IX. Allosteric effects in the reduction of pyrimidine ribonucleotides by the ribonucleoside diphosphate reductase system of *Escherichia coli*. *J. Biol. Chem.* **241**, 2533–2539.

Lee, Y.-P. (1957a). 5′-Adenylic acid deaminase. I. Isolation of the crystalline enzyme from rabbit skeletal muscle. *J. Biol. Chem.* **227**, 987–992.

Lee, Y.-P. (1957b). 5′-Adenylic acid deaminase, II. Homogeneity and physicochemical properties. *J. Biol. Chem.* **227**, 993–998.

Lee, Y.-P. (1957c). 5′-Adenylic acid deaminase. III. Properties and kinetic studies. *J. Biol. Chem.* **227**, 999–1007.

Lee, Y.-P. and Wang, M. H., (1968). Studies of the nature of the inhibitory action of inorganic phosphate, fluoride, and detergents on 5′-adenylic acid deaminase activity and on the activation by adenosine triphosphate. *J. Biol. Chem.* **243**, 2260–2265.

Levenberg, B. (1962). Role of L-glutamine as donor of carbamyl nitrogen for the enzymatic synthesis of citrulline in *Agricus bisporus*. *J. Biol. Chem.* **237**, 2590–2598.

Lindberg, B., Klenow, H., and Hansen, K. (1967). Some properties of partially purified mammalian adenosine kinase. *J. Biol. Chem.* **242**, 350–356.

Littlefield, J. W. (1963). The inosinic acid pyrophorylase activity of mouse fibroblasts partially resistant to 8-azaguanine. *Proc. Nat. Acad. Sci. U.S.* **50**, 568–576.

Littlefield, J. W. (1964a). Three degrees of guanylic acid-inosinic acid pyrophosphorylase deficiency in mouse fibroblasts. *Nature (London)* **203**, 1142–1144.

Littlefield, J. W. (1964b). Selection of hybrids from matings of fibroblasts *in vitro* and their presumed recombinants. *Science* **145**, 709–710.

Littlefield, J. W. (1969). Hybridization of hamster cells with high and low folate reductase activity. *Proc. Nat. Acad. Sci. U.S.* **62**, 88–95.

Lucas, Z. J. (1967). Pyrimidine nucleotide synthesis: Regulatory control during transformation of lymphocytes *in vitro*. *Science* **156**, 1237–1240.

Lukens, L. N., and Buchanan, J. M. (1957). Further intermediates in the biosynthesis of inosinic acid *de novo*. *J. Amer. Chem. Soc.* **79**, 1511–1513.

McDonald, J. A., and Kelley, W. N. (1971a). Unpublished observations.

McDonald, J. A., and Kelley, W. N. (1971b). Lesch-Nyhan syndrome: Altered kinetic properties of mutant enzyme. *Science* **171**, 689–691.

Magasanik, B., and Karihian, D. (1960). Purine nucleotide cycles and their metabolic role. *J. Biol. Chem.* **235**, 2672–2681.

Mainigi, K. D., and Bresnick, E. (1969). Inhibition of deoxythymidine kinase by beryllium. *Biochem. Pharmacol.* **18**, 2003–2007.

Maley, F., and Maley, G. F. (1960). Nucleotid interconversions. II. Elevation of deoxycytidylate deaminase and thymidine synthetase in regenerating rat liver. *J. Biol. Chem.* **235**, 2968–2970.

Maley, F., and Maley, G. F. (1961). Nucleotide interconversions. IV. Activities of deoxycytidylate deaminase and thymidylate synthetase in normal rat liver and hepatomas. *Cancer Res.* **21**, 1421–1426.

Maley, G. F., and Maley, F. (1964). The purification and properties of deoxycytidylate deaminase from chick embryo extracts. *J. Biol. Chem.* **239**, 1168–1176.

Maley, G. F., and Maley, F. (1968a). Regulatory properties and subunit structure of chick embryo deoxycytidylate deaminase. *J. Biol. Chem.* **243**, 4506–4512.

Maley, G. F., and Maley, F. (1968b). Active and inactive states of deoxycytidylate deaminase and their relation to subunit structure. *J. Biol. Chem.* **243**, 4513–4516.

Miech, R. P., and Parks, R. E., Jr. (1965). Adenosine triphosphate: Guanosine monophosphate phosphotransferase. *J. Biol. Chem.* **240**, 351–357.

Miech, R. P., York, R., and Parks, R. E., Jr. (1969). Adenosine triphosphate-guanine 5′-phosphate phosphotransferase II. Inhibition by 6-thioguanosine 5′-phosphate of the enzyme isolated from hog brain and sarcoma 180 ascites cells. *Mol. Pharmacol.* **5**, 30–37.

Migeon, B. R. (1968). Hybridization of somatic cells derived from mouse and Syrian hamster: Evolution of karyotype and enzyme studies. *Biochem. Genet.* **1**, 305–322.

Migeon, B. R. (1970). X-linked hypoxanthine-guanine phosphoribosyltransferase deficiency: Detection of heterozygotes by selective medium. *Biochem. Genet.* **4**, 377–383.

Migeon, B. R., and Childs, B. (1970). Hybridization of mammalian somatic cells. *Prog. Med. Genet.* **7**, 1–28.

Migeon, B. R., Der Kaloustian, K. M., Nyhan, W. L., Young, W. J., and Childs, B. (1968). X-linked hypoxanthine-guanine phosphoribosyltransferase deficiency: Heterozygote has two clonal populations. *Science* **160**, 425–427.

Migeon, B. R., Smith, S., and Teddy, C. (1969). The nature of thymidine kinase in mouse-human hybrid cells. *Biochem. Genet.* **3**, 583–590.

Miller, R. W., and Kerr, C. T. (1966). Dihydroorotate dehydrogenase. III. Interactions with substrates, inhibitors, artificial electron acceptors, and cytochrome. *J. Biol. Chem.* **241**, 5597–5604.

Miller, R. W., and Massey, V. (1965a). Dihydroorotate dehydrogenase. I. Some properties of the enzyme. *J. Biol. Chem.* **240**, 1453–1465.

Miller, R. W., and Massey, V. (1965b). Dihydroorotate dehydrogenase. II. Oxidation and reduction of cytochrome. *J. Biol. Chem.* **240**, 1466–1472.

Mizoburchi, K., and Buchanan, J. M. (1968). Biosynthesis of purines. XXX. Isolation and characterization of formyglycinamide ribonucleotide amidotransferase-glutamyl complex. *J. Biol. Chem.* **243**, 4853–4862.

Moore, E. C. (1967). A thioredoxin-thioredoxin reductase system from rat tumor. *Biochem. Biophys. Res. Commun.* **29**, 264–268.

Moore, E. C. (1969). The effects of ferrous ions and dithioerythritol on inhibition by hydroxyurea of ribonucleotide reductase. *Cancer Res.* **29**, 291–195.

Moore, E. C., and Hurlbert, R. B. (1966). Regulation of mammalian deoxyribonucleotide biosynthesis by nucleotides as activators and inhibitors. *J. Biol. Chem.* **241**, 4802–4809.

Moore, E. C., and Reichard, P. (1964). Enzymatic synthesis of deoxyribonucleotides. VI. The cytidine diphosphate reductase system from Novikoff hepatoma. *J. Biol. Chem.* **239**, 3453–3456.

Morrow, J. (1970). Genetic analysis of azaguanine resistance in an established mouse cell line. *Genetics* **65**, 279–287.

Mourad, N., and Parks, R. E., Jr. (1966). Erythrocytic nucleoside diphosphokinase. II. Isolation and kinetics. *J. Biol. Chem.* **241**, 271–278.

Munch-Peterson, A. (1960). Formation *in vitro* of deoxyadenosine triphosphate from deoxyadenosine in Ehrlich ascites tumor cells. *Biochem. Biophys. Res. Commun.* **3**, 392–396.

Muntz, J. A. (1953). The formation of ammonia in brain extracts. *J. Biol. Chem.* **201**, 221–233.

Murray, A. W. (1966a). Inhibition of purine phosphoribosyltransferases from Ehrlich ascites tumour cells by purine nucleotides. *Biochem. J.* **100**, 671–674.

Murray, A. W. (1966b). Purine phosphoribosyltransferase activities in rat and mouse tissues and in Ehrlich ascites tumour cells. *Biochem. J.* **100**, 664–670.

Murray, A. W. (1967). Studies on the nature of the regulation of purine nucleotides of adenine phosphoribosyltransferase and of hypoxanthine-guanine phosphoribosyltransferase from Ehrlich ascites tumor cells. *Biochem. J.* **103**, 271–279.

Murray, A. W. (1968). Some properties of adenosine kinase from Ehrlich ascites tumour cells. *Biochem. J.* **106**, 549–555.

Murray, A. W., and Friedrichs, B (1969). Inhibition of 5'-nucleotidase from Ehrlich ascites tumour cells by nucleoside triphosphates. *Biochem. J.* **111**, 83–89.

Murray, A. W., Elliott, D. C., and Atkinson, M. R. (1970). Nucleotide biosynthesis from preformed purines in mammalian cells: Regulatory mechanisms and biological significance. *Progr. Nucl. Acid Res. Mol. Biol.* **10**, 87–119.

Nadler, H. L., Chacko, C. M., and Rochneler, M. (1970). Interallelic complementation in hybrid cells derived from human diploid strains deficient in galactose-1-phosphate uridyl transferase activity. *Proc. Nat. Acad. Sci. U.S.* **67**, 976–982.

Nakamura, H., and Sugino, Y. (1966a). Metabolism of deoxyribonucleotides. III. Purification and some properties of nucleoside diphosphokinase of calf thymus. *J. Biol. Chem.* **241**, 4917–4922.

Nakamura, H., and Sugino, Y. (1966b). Metabolism of deoxyribonucleotides. II.

Enzymatic phosphorylation of deoxycytidylic acid in normal rat liver and rat ascites hepatoma cells. *Cancer Res.* **26**, 1425–1429.

Nakata, Y., and Bader, J. P. (1969). The uptake of nucleosides by cells in culture. II. Inhibition by 2-mercapto-1-(-4-pyridethyl) Benzimidazole. *Biochim. Biophys. Acta* **190**, 250–256.

Nierlich, D. P., and Magasanik, B. (1965a). Regulation of purine ribonucleotide synthesis by end product inhibition: The effect of adenine and guanine ribonucleotides on the 5′-phosphoribosylpyrophosphate amidotransferase in Aerobacter aerogenes. *J. Biol. Chem.* **240**, 358–365.

Nierlich, D. P., and Magasanik, B. (1965b). Phosphoribosylglycinamide synthetase of *Aerobacter aerogenes. J. Biol. Chem.* **240**, 366–374.

Nordenskjold, B. A., and Krakoff, I. H. (1968). Effects of hydroxyurea on polyoma virus replication. *Cancer Res.* **28**, 1686–1691.

Overgaard-Hansen, K. (1965). Metabolic regulation of the adenine nucleotide pool. I. Studies on the transient exhaustion of the adenine nucleotides by glucose in Ehrlich ascites tumor cells. *Biochim. Biophys. Acta* **104**, 330–347.

Paege, L. M., and Schlenk, F. (1952). Bacterial uracil riboside phosphorylase. *Arch. Biochem. Biophys.* **40**, 42–49.

Paterson, A. R. P. (1965a). The biosynthesis of extracellular ribonucleosides by ascites tumor cells *in vitro. Can. J. Biochem.* **43**, 257–269.

Paterson, A. R. P. (1965b). Enhancement of ribonucleoside synthesis in Ehrlich ascites tumor cells by arsenate and iodoacetate. *Can. J. Biochem.* **43**, 1693–1700.

Paterson, A. R. P. (1965c). Inhibition of ribonucleoside metabolism in Ehrlich ascites tumor cells by purine analogue ribonucleosides. *Can. J. Biochem.* **43**, 1701–1710.

Paterson, A. R. P., and Hori, A. (1963). Resistance to 6-mercaptopurine. III. Deletion of a 5′-nucleotidase in a 6-mercaptopurine-resistant subline of the Ehrlich ascites carcinoma. *Can. J. Biochem. Biophys.* **41**, 1339–1348.

Pfrogner, N. (1967). Adenosine deaminase from calf spleen. I. Purification. *Arch. Biochem. Biophys.* **119**, 141–146.

Pierard, A., Glansdorff, N., Mergeay, M., and Wiame, J. M. (1965). Control of the biosynthesis of carbamyl phosphate in *Escherichia coli. J. Mol. Biol.* **14**, 23–36.

Pierre, K. J., and LePage, G. A. (1968). Formation of inosine-5′-monophosphate by a kinase in cell-free extracts of Ehrlich ascites cells *in vitro. Proc. Soc. Exp. Biol. Med.* **127**, 432–440.

Pierre, K. J., Kimball, A. P., and LePage, G. A. (1967). The effect of structure on nucleoside kinase activity. *Can. J. Biochem.* **45**, 1619–1632.

Pinsky, L., and Krooth, R. S. (1967a). Studies on the genetic control of pyrimidine biosynthesis in human diploid cell strains. I. Effect of 6-azauridine on cellular phenotype. *Proc. Nat. Acad. Sci. U.S.* **57**, 925–932.

Pinsky, L., and Krooth, R. S. (1967b). Studies on the control of pyrimidine biosynthesis in human diploid cell strains. II. Effects of 5-azaorotic acid, barbituric acid and pyrimidine precursors on cellular phenotype. *Proc. Nat. Acad. Sci. U.S.* **57**, 1267–1274.

Pinto, B., and Touster, O. (1966). Separation and modification of the phosphorolytic and ribosyl transfer activities of the purine nucleoside phosphorylase of Ehrlich ascites tumor cells. *J. Biol. Chem.* **241**, 772–773.

Plagemann, P. G. W., and Roth, M. F. (1969). Permeation as the rate-limiting step in the phosphorylation of uridine and choline and their incorporation into macromolecules by Novikoff hepatoma cells. Competitive inhibition by phenethyl alcohol, persantin, and adenosine. *Biochemistry* **8**, 4782–4789.

Plagemann, P. G. W., and Shea, M. A. (1971). Transport as the rate limiting step in the incorporation of uridine into mangovirus ribonucleic acid in Novikoff rat hepatoma cells. *J. Virol.* **7**, 137–143.

Raska, K., Jr., and Cohen, E. P. (1967). Specific activity of several enzymes of nucleotide metabolism in mouse spleen after immunization. *Clin. Exp. Immunol.* **2**, 559–563.

Reem, G. H., and Friend, C. (1967). Phosphoribosylamidotransferase: Regulation of activity in virus-induced murine leukemia by purine nucleotides. *Science* **157**, 1203–1204.

Reichard, P., and Skold, O. (1957). Formation of uridine phosphates from uracil in extracts of Ehrlich ascites tumor. *Acta Chem. Scand.* **11**, 17–23.

Reichard, P., Skold, O., and Klein, G. (1959). Possible enzymatic mechanism for the development of resistance against fluorouracil in ascites tumours. *Nature (London)* **183**, 939–941.

Reyes, P. (1969). The synthesis of 5-fluorouridine 5′-phosphate by a pyrimidine phosphoribosyltransferase of mammalian origin. I. Some properties of the enzyme from P1534J mouse leukemic cells. *Biochemistry* **8**, 2057–2062.

Roberts, D., and Hall, T. C. (1969). Enzyme activities and deoxynucleoside utilization of leukemic leukocytes in relation to drug therapy and resistance. *Cancer Res.* **29**, 166–173.

Rosenbloom, F. M., Kelley, W. N., Henderson, J. F., and Seegmiller, J. E. (1967). Lyon hypothesis and X-linked disease. *Lancet* **2**, 305–306.

Rosenbloom, F. M., Henderson, J. F., Caldwell, I. C., Kelley, W. N., and Seegmiller, J. F. (1968a). Biochemical bases of accelerated purine biosynthesis *de novo* in human fibroblasts lacking hypoxanthine-guanine phosphoribosyltransferase. *J. Biol. Chem.* **243**, 1166–1173.

Rosenbloom, F. M., Henderson, J. F., Kelley, W. N., and Seegmiller, J. F. (1968b). Accelerated purine biosynthesis *de novo* in skin fibroblasts deficient in hypoxanthine-guanine phosphoribosyltransferase activity. *Biochim. Biophys. Acta* **166**, 258–260.

Rothman, I. K., Silber, R., Klein, K. M., and Becker, F. F. (1971). Nucleoside deaminase and adenosine deaminase activities in regenerating mouse liver. *Biochim. Biophys. Acta* **228**, 307–312.

Rozenberg, M. C., and Holmsen, H. (1968). Adenine nucleotide metabolism of blood platelets. II. Uptake of adenosine and inhibition of ADP-induced platelet aggregation. *Biochim. Biophys. Acta* **155**, 342–352.

Rubin, C. S., Balis, M. E., Piomelli, S., Berman, P. H., and Davies, J. (1969). Elevated AMP pyrophosphorylase activity in congenital IMP pyrophosphorylase deficiency (Lesch-Nyhan disease). *J. Lab. Clin. Med.* **74**, 732–741.

Rudolph, F. B., and Fromm, H. J. (1969). Initial rate studies of adenylosuccinate synthetase with product and competitive inhibitors. *J. Biol. Chem.* **244**, 3832–3839.

Salzman, J., DeMars, R., and Benke, P. (1968). Single allele expression at an X-linked hyperuricemia locus in heterozygous human cells. *Proc. Nat. Acad. Sci. U.S.* **60**, 545–552.

Salzman, N. P., Eagle, H., and Sebring, E. D. (1958). The utilization of glutamine, glutamic acid, and ammonia for the biosynthesis of nucleic acid bases in mammalian cell cultures. *J. Biol. Chem.* **230**, 1001–1012.

Sanford, K. K. (1963). Growth and characterization of mammalian tissue culture cells in chemically defined media. *In* "Proceedings of the Human Diploid Cell Strains," pp. 195–208. Opatya, Yugoslavia.

Scarano, E., Bonaduce, L., and DePetrocellis, B. (1960). The enzymatic deaminase of 6-aminopyrimidine deoxyribonucleotides. II. Purification and properties of a 6-aminopyrimidine deoxyribonucleoside 5'-phosphate deaminase from unfertilized eggs of sea urchins. *J. Biol. Chem.* **235**, 3556–3561.

Scarano, E., Bonaduce, L., and DePetrocellis, B. (1962). The enzymatic aminohydrolysis of 4-aminopyrimidine deoxyribonucleotides. *J. Biol. Chem.* **237**, 3742–3751.

Scarano, E., Geraci, G., and Rossi, M. (1967). Deoxycytidylate aminohydrolase. II. Kinetic properties. The activatory effect of deoxycytidine triphosphate and the inhibitory effect of deoxythymidine triphosphate. *Biochemistry* **6**, 192–201.

Schnebli, H. P., Hill, D. L., and Bennett, L. L., Jr. (1966). Purification and properties of adenosine kinase from human tumor cells of type H. Ep. No. 2. *J. Biol. Chem.* **242**, 1997–2004.

Seegmiller, J. E., Rosenbloom, F. M., and Kelley, W. N. (1967). An enzyme defect associated with a sex-linked human neurological disorder and excessive purine synthesis. *Science* **155**, 1682–1684.

Segal, H. L., and Brenner, B. M. (1960). 5'-Nucleotidase of rat liver microsomes. *J. Biol. Chem.* **235**, 471–474.

Setlow, B., and Lowenstein, J. M. (1967). Adenylate deaminase. II. Purification and some regulatory properties of the enzyme from calf brain. *J. Biol. Chem.* **242**, 607–615.

Setlow, B., and Lowenstein, J. M. (1968). Adenylate deaminase. IV. Nucleotide specificity of the enzyme from calf brain with special reference to guanosine triphosphate. *J. Biol. Chem.* **243**, 3409–3415.

Silber, R., Gabrio, B. W., and Huennekens, F. M. (1963). Studies on normal and leukemic leukocytes. VI. Thymidylate synthetase and deoxycytidylate deaminase. *J. Clin. Invest.* **42**, 1913–1921.

Sinha, D., and Ghosh, A. (1964). Cytochemical study of the suprarenal cortex of the pigeon under altered electrolytic balance. *Acta Histochem.* **18**, 222–229.

Skold, O. (1960). Uridine kinase from Ehrlich ascites tumor: Purification and properties. *J. Biol. Chem.* **235**, 3273–3279.

Skold, O. (1963). Studies on resistance against 5-fluorouracil IV. Evidence for an altered uridine kinase in resistant cells. *Biochim. Biophys. Acta* **76**, 160–162.

Smith, L. H., Huguley, C. M., Bain, J. A. (1966). Hereditary orotic aciduria. *In* "The Metabolic Basis of Inherited Disease" (J. B. Stanbury, Wyngaarden, J. B., and Frederickson, D. F. ed.), pp. 739–758. McGraw-Hill, New York.

Sneider, T. W., Potter, V. R., and Morris, H. P. (1969). Enzymes of thymidine triphosphate synthesis in selected Morris hepatomas. *Cancer Res.* **29**, 40–54.

Song, C. S., and Bodansky, O. (1967). Subcellular localization and properties of 5' nucleotidase in the rat liver. *J. Biol. Chem.* **242**, 694–699.

Song, C. S., Tandler, B., and Bodansky, O. (1967). Solubilization of 5'-nucleotidase from human liver and reassociation of its activity into membranous vesicles. *Biochem. Med.* **1**, 100–109.

Stock, T. L., Nakata, Y., and Bader, J. P. (1969). The uptake of nucleosides by cells in cultures. I. Inhibiton by heterologous nucleosides. *Biochim. Biophys. Acta* **190**, 237–249.

Subak-Sharpe, H., Burk, R. R., and Pitts, J. D. (1969). Metabolic cooperation between biochemically marked mammalian cells in tissue culture. *J. Cell Sci.* 4, 353–367.

Sugino, Y., Fanaka, F., and Miyoshi, Y. (1964). Latent deoxyuridine monophosphokinase in calf thymus. *Biochem. Biophys. Res. Commun.* 16, 362–367.

Sugino, Y., Teraoka, H., and Shimono, H. (1966). Metabolism of deoxyribonucleotides. I. Purification and properties of deoxycytidine monophosphokinase of calf thymus. *J. Biol. Chem.* 241, 961–969.

Sulkowski, E., Bjork, W., and Laskowski, M., Sr. (1963). A specific and nonspecific alkaline monophosphatase in the venom of *Bothrops atrox* and their occurrence in the purified venom phosphodiesterase. *J. Biol. Chem.* 238, 2477–2486.

Szybalski, W., and Szybalski, E. H. (1962). Drug sensitivity as a genetic marker for human cell lines. *Univ. Mich. Med. Bull.* 28, 277–293.

Tarr, H. L. A. (1964). Formation of purine and pyrimidine nucleosides, deoxynucleosides, and the corresponding mononucleotides by salmon milt extract nucleoside phosphorylase and nucleoside kinase enzymes. *Can. J. Biochem.* 42, 1535–1545.

Tatibana, M., and Ito, K. (1967). Carbamyl phosphate synthetase of the hematopoietic mouse spleen and the control of pyrimidine biosynthesis. *Biochem. Biophys. Res. Commun.* 26, 221–227.

Tatibana, M., and Kazuhiko, I. (1969). Control of pyrimidine biosynthesis in mammalian tissues. Partial purification and characterization of glutamine-utilizing carbamyl phosphate synthetase of mouse spleen and its tissue distribution. *J. Biol. Chem.* 244, 5403–5413.

Tay, B. S., Lilley, R. McC., Murray, A. W., and Atkinson, M. R. (1969). Inhibition of phosphoribosyl pyrophosphate amidotransferase from Ehrlich ascites tumor cells by thiopurine nucleotides. *Biochem. Pharmacol.* 18, 936–938.

Tomchick, R., Saslaw, L. D., and Waravdekar, V. S. (1968). Mouse kidney cytidine deaminase. Purification and properties. *J. Biol. Chem.* 243, 2534–2537.

Warren, L., Flaks, J. G., and Buchanan, J. M. (1957). Biosynthesis of purine. XX. Integration of enzymatic transformylation reactions. *J. Biol. Chem.* 229, 627–640.

Weissman, S. M., Smellie, R. M. S., and Paul, J. (1960). Studies on the biosynthesis of deoxyribonucleic acid by extracts of mammalian cells. IV. The phosphorylation of thymidine. *Biochim. Biophys. Acta* 45, 101–110.

Wong, P. C. L., and Murray, A. W. (1969). 5-Phosphoribosylpyrophosphate synthetase from Ehrlich ascites tumor cells. *Biochemistry* 8, 1608–1614.

Wyngaarden, J. B., and Ashton, D. M. (1959). The regulation of activity of phosphoribosylpyrophosphate amidotransferase by purine ribonucleotides: A potential feedback control of purine biosynthesis. *J. Biol. Chem.* 234, 1492–1496.

Wyngaarden, J. B., and Greenland, R. A. (1963). The inhibition of succinoadenylate kinosynthetase of *Escherichia coli* by adenosine and guanosine 5′ monophosphates. *J. Biol. Chem.* 238, 1054–1057.

Wyngaarden, J. B., and Kelley, W. N. (1972). Gout. *In* "The Metabolic Basis of Inherited Disease" (J. B. Stanbury, J. B. Wyngaarden, and D. S. Fredrickson, eds.), 3rd ed., Chapter 39. McGraw-Hill, New York 889–968.

Young, J. E., Prager, M. D., and Atkins, I. C. (1967). Comparative activities of aspartate transcarbamylase in various tissues of the rat (132224). *Proc. Soc. Exp. Biol. Med.* 125, 860–862.

Zicha, B., and Buric, L. (1969). Deoxycytidine and radiation response: Exceedingly
    high deoxycytidine aminohydrolase activity in human liver. *Science* **163**,
    191–192.
Zimmerman, M. (1964). Deoxyribosyl transfer. II. Nucleoside: Pyrimidine deoxyribo-
    syltransferase activity of three partially purified thymidine phosphorylases. *J.
    Biol. Chem.* **239**, 2622–2627.
Zimmerman, M., and Seidenberg, J. (1964). Deoxyribosyl transfer. I. Thymidine
    phosphorylase and nucleoside deoxyribosyltransferase in normal and malignant
    tissues. *J. Biol. Chem.* **239**, 2618–2621.
Zittle, C. A. (1946). Adenosine deaminase from calf intestinal mucosa. *J. Biol.
    Chem.* **166**, 499–503.

FATTY ACID, GLYCERIDE, AND

PHOSPHOLIPID METABOLISM

Arthur A. Spector

I. Introduction

Fatty acids, phospholipids, and glycerides are extremely important components of mammalian cells. A considerable percentage of the energy requirement of many cells is derived from fatty acid oxidation. Fatty acids also are incorporated into cellular phospholipids and glycerides.

Fatty acid required for these purposes may be synthesized within the cell or taken up from the extracellular fluid. Phospholipids are the major lipid constituents of membranes and therefore are concerned with cell structure and permeability. Triglycerides are a storage form of fatty acid. If the supply of fatty acid exceeds immediate needs, the excess is stored in the cell as triglyceride. In cells in culture the accumulated triglyceride often appears as cytoplasmic droplets or granules (Geyer, 1967; Mackenzie *et al.*, 1967a). Since fatty acids, phospholipids, and glycerides play vital roles in the nutrition and function of mammalian cells, it is most important for those working with cells in culture to have a thorough understanding of the metabolism of these lipids.

II. Origin of Lipids in Cells in Culture

Lipids account for an appreciable fraction of the dry weight of mammalian cells. For example, lipids comprise 16% of the dry weight of the Ehrlich ascites cell (Wallach *et al.*, 1960). In many culture systems there is a doubling of cell mass within 24–30 hours. Yet, mammalian cells are able to grow in a lipid-free medium, e.g., the L strain mouse fibroblast (Bailey, 1967), HeLa-S_3 (Harary et al., 1967), 3681 fibroblasts, P338-DR mouse lymphoma, and NCTC clone 3453 mouse salivary gland cells (V. J. Evans *et al.*, 1965). Cells grown in the absence of lipids have sufficient amounts of long-chain fatty acids to satisfy their growth and metabolic needs (Bailey, 1967; Harary *et al.*, 1967). Moreover, they appear identical to cells grown in media containing lipids except for the absence of granules in the cytoplasm (Lengle and Smith, 1970). These results clearly demonstrate that many mammalian cells in culture possess the ability to synthesize adequate amounts of fatty acids from nonlipid precursors. However, the fatty composition of cells grown in lipid-free media differs from that of cells grown in media containing serum (Geyer *et al.*, 1961). For example, linoleic and arachidonic acids, fatty acids that cannot be synthesized *de novo* by mammalian tissues, are markedly reduced or absent in cells grown in lipid-free media (Geyer *et al.*, 1961; Bailey, 1967; Harary *et al.*, 1967).

Although most mammalian cells are capable of synthesizing fatty acids from nonlipid substrates, they obtain most of their fat from the extracellular fluid if lipid is present in the culture medium. When cells are grown in a medium containing serum, the fatty acid composition of the cells reflects that of the serum (Geyer *et al.*, 1961; Boyle and Ludwig, 1962). Under these conditions, the bulk of the cell fatty acid probably is derived from triglycerides contained in the serum lipoproteins. Free fatty acids contained in the extracellular fluid are another important source of lipid for mammalian cells. Most of the fatty acid that is required by the Ehrlich ascites cell growing in the mouse peritoneal cavity is supplied by the ascites plasma free fatty acid (Spector, 1967a). Large amounts of free fatty acid also can be utilized by cells grown in culture (Geyer, 1967; Moskowitz, 1967; Howard and Kritchevsky, 1969). Indeed, the fatty acid composition of strain L fibroblasts can be altered markedly by addition of a single free fatty acid to the culture medium. For example, oleate comprised 60% of the cell fatty acids during growth in a lipid-free medium. When oleate was present in the medium, oleate accounted for 75% of the cell fatty acids. No linoleate was detected in the cell lipids under these conditions. However, when linoleate was added to the medium, the oleate content of the cell fatty acids decreased to 43% and linoleate accounted for 21% of the cell fatty acids (Geyer, 1967). Studies with radioactive substrates also demonstrate that the physiologically important free fatty acids contained in the extracellular fluid can be utilized by isolated mammalian cells, including Ehrlich cells (Fillerup et al., 1958; Medes et al., 1960; Spector and Steinberg, 1967a), polymorphonuclear leukocytes (Elsbach, 1964), alveolar macrophages (Elsbach, 1965), peritoneal macrophages (Day and Fidge, 1962), myocardial cells (Harary et al., 1967), WI-38 cells (Howard and Kritchevsky, 1969), HeLa cells (Geyer, 1967; Harary et al., 1967), erythrocytes (Goodman, 1958a; Donabedian and Karmen, 1967; Shohet et al., 1968), platelets (Spector et al., 1970), and sarcoma 180 cells (Lengle and Smith, 1970).

The accumulation of fat in cells in culture in the form of cytoplasmic droplets occurs when there is a high concentration of lipid in the medium. Intracellular fat accumulation in human aortic cells in culture was related directly to the concentration of serum triglycerides in the medium (Rutstein et al., 1964). Lipid deposition increased when serum was obtained from a donor who previously had eaten a fat meal, in which case the serum triglyceride concentration was elevated. On the other hand, lipid accumulation in human MAF cells in culture was dependent on the free fatty acid concentration of the serum added to the medium. Serum obtained after a prolonged fast contained a large amount of free fatty acid and produced excessive intracellular lipid depositions. In contrast, serum obtained after a carbohydrate meal contained much less free fatty acid and produced little lipid accumulation in the cells (Rutstein et al., 1967).

Table I illustrates the fatty acid composition of the lipid isolated from five mammalian cell lines. It also illustrates the changes that occur

TABLE I.

FATTY ACID COMPOSITION OF MAMMALIAN CELLS

Fatty Acid		Percentage of Total Cell Fatty Acid (%)						
Name	Structure[a]	Ehrlich Ascites[b]	L[c]	Rat Heart[a]	Rat Heart[f]	HeLa[g]	HeLa[i]	Aortic Intima[j]
Myristic	14:0	0.7	2.4	N.R.[e]	N.R.[e]	2.4	6.1	3.5
Palmitic	16:0	17.4	19.5	20.0	44.0	28.0	24.0	11.0
Palmitoleic	16:1	1.4	3.4	0.5	1.1	3.7	11.0	2.8
Stearic	18:0	18.1	24.7	18.0	32.0	18.0	8.8	29.3
Oleic	18:1	23.5	32.7	15.0	11.0	16.0	19.0	21.9
Linoleic	18:2	23.5	2.0	9.1	0.5	6.8[h]	0.6[h]	22.3
Linolenic	18:3	1.8	4.3	0.4	0.4	—	—	1.2
Arachidonic	20:4	9.9	5.7	18.0	2.2	14.0	1.1	7.7
Others	—	3.7	5.3	19.0	8.8	11.1	29.4	0.3

[a] Chain length: degree of unsaturation.
[b] Spector (1967a).
[c] Weinstein et al. (1969).
[d] After 4 days of incubation in serum (Harary et al., 1967).
[e] Not reported.
[f] After 4 days of incubation in lipid-free medium (Harary et al., 1967).
[g] After 7 days of incubation in fetal calf serum (Harary et al., 1967).
[h] Sum of 18:2 and 18:3. The two acids were not separated in these analyses.
[i] After 7 days of incubation in a lipid-free medium (Harary et al., 1967).
[j] Show Racer pigeon (Smith et al., 1965).

in cellular fatty acid composition when serum is omitted from the culture medium. These data emphasize the extent of utilization of exogenously supplied fatty acid. Therefore, to understand the mechanisms involved in the lipid nutrition of cells in culture, it is necessary to be familiar with the lipids that are present in serum and the mechanisms for transfer of these lipids to the cells.

III. Serum Lipids

Lipids are very poorly soluble in aqueous solutions. In spite of this, it is necessary to move large quantities of lipid from one organ to another, and this transport must occur through the plasma. The structure that accomplishes this feat is the lipid-protein complex. Protein serves as the carrier for the lipid, much as a boat might be employed to carry a person who cannot swim well. In many cases, phospholipids assist the protein moiety in solubilizing the lipid that must be transported. Two major classes of lipid transport proteins are present in serum. The first, albumin, carries free fatty acid. The second, lipoproteins, carries triglycerides and cholesterol. Table II contains information about the major lipoprotein classes present in serum (Fredrickson *et al.*, 1967; Schumaker and Adams, 1969). These data are derived from analyses of human serum. Differences between human and animal serum lipoproteins are discussed in Section III,B,4,b.

A. Free Fatty Acids

From 5 to 10% of the total fatty acid present in serum is in unesterified or "free" form. In human blood, the free fatty acid concentration may vary from 0.09 to 1.71 μEq/ml (Fredrickson and Gordon, 1958), but the usual concentration is between 0.4 and 0.7 μEq/ml (Eaton *et al.*, 1969). Most of the free fatty acid is composed of straight-chain, mono-carboxylic molecules containing from 12 to 20 carbons atoms. Palmitic, stearic, oleic, and linoleic acids comprise approximately 80% of the total free fatty acid fraction in human and most animal sera (Spector, 1971). Short-chain fatty acids, particularly acetate, also are present in serum (Mehadavan and Zieve, 1969). Although short-chain fatty acids may comprise as much as 10% of the total free fatty acid *molecules* in serum, they make up less than 2% of the free fatty acid *carbon atoms*. Hence, it is the long-chain free fatty acids that are important as a supply of

TABLE II.

MAJOR CLASSES OF SERUM LIPOPROTEINS[a]

Class	Electrophoretic Mobility[b]	Density Range (gm/ml)	S_f [c]	Major Lipid		Protein Content (%)	Percentage of Total Serum Lipoprotein (%)[d]	
				Type	Fraction of Total Lipid (%)		Male	Female
Chylomicrons	0	0.95	400	Triglyceride	85	2	<1	<1
Very low density	pre-β	0.95–1.006	20–400	Triglyceride	70	10	16	5
Low density	β	1.006–1.063	0–20	Cholesterol	57	20	50	40
High density	α_1	1.063–1.21	—	Phospholipids	60	50	34	55

[a] Data summarized from Fredrickson et al. (1967) and Schumaker and Adams (1969). These data refer to human serum.
[b] Barbital buffer, ionic strength 0.1, pH 8.6, descending paper electrophoresis.
[c] Svedberg flotation units at 1.063 gm/ml of NaCl.
[d] Serum obtained after an overnight fast.

lipid to tissues. Albumin is the transport protein for long-chain free fatty acid (Goodman, 1958b; Spector *et al.*, 1969), and more than 99% of the long-chain free fatty acid that is present in serum is bound to this protein. Under normal physiological conditions, the molar ratio of free fatty acid to albumin rarely exceeds 3, and the average binding constant is of the order of 10^6–10^7 M^{-1} (Goodman, 1958b; Spector *et al.*, 1969). In spite of being tightly bound to albumin, free fatty acid can be taken up rapidly by mammalian cells. The half-life of circulating free fatty acid is very short, varying from 1 to 3.9 minutes (Fredrickson and Gordon, 1958; Eaton *et al.*, 1969). It is likely that the fatty acid dissociates from albumin in the process of uptake (Spector, 1968).

Long-chain fatty acids that are stored as triglycerides in adipose cells are released into the blood as free fatty acids. Nutritional, nervous, and hormonal factors regulate the release of free fatty acid from the adipose tissue. Therefore the free fatty acid concentration of serum that is incorporated into a tissue culture medium may vary considerably, depending on when the donor ate last and the degree of stress to which the donor was subjected prior to bleeding. Another source of serum free fatty acid is the triglyceride contained in the lipoproteins. Free fatty acid is formed by hydrolysis when the triglycerides are cleared from the circulation. The hydrolysis probably is catalyzed by lipoprotein lipase (Engelberg, 1966), an enzyme or group of enzymes located in the capillary wall or on the parenchymal cell surface (Cunningham and Robinson, 1969). Lipoprotein lipase, also known as clearing-factor lipase, is released into the blood by heparin (Cherkes and Gordon, 1959). At least three enzymatic activities are contained in the material released following injection of heparin: triglyceride lipase, mono- and diglyceride lipase (Greten *et al.*, 1970), and lecithinase (Vogel and Bierman, 1968). Lipases present in the serum added to the culture media apparently catalyze the hydrolysis of the lipoprotein triglycerides, and the released free fatty acids are utilized by the cells in culture (Howard and Kritchevsky, 1969).

Long-chain free fatty acids can be added to either serum or albumin by incubation with warm solutions of fatty acid salts (Spector and Steinberg, 1966a) or with fatty acid absorbed on solid particles such as Celite (Spector and Hoak, 1969). To prepare albumin media that contain only a single fatty acid, the inherent fatty acid content of the albumin can be removed with charcoal (Chen, 1967). β-Lactoglobulin, a protein contained in bovine milk, can replace albumin as a fatty acid carrier for experimental studies (Spector and Fletcher, 1970). The long-chain free fatty acid concentration of serum or a protein solution can be measured by titration after the lipids are extracted into an organic solvent (Dole,

1956). Since the serum may contain relatively large quantities of lactate or other short-chain acids, the initial lipid extract should be washed prior to titration in order to obtain a more accurate value for long-chain free fatty acids (Trout *et al.*, 1960).

B. Lipoproteins

There are four major classes of serum lipoproteins (Table II). These are chylomicrons, very-low-density lipoproteins (VLDL or pre-beta), low-density lipoproteins (LDL or beta), and high-density lipoproteins (HDL or alpha). Recently, two additional subclasses of serum lipoproteins have been described: an S_f 20 lipoprotein (Fisher, 1970) and an S_f 0–3 lipoprotein class (Puppione *et al.*, 1970). Lipoproteins usually are isolated in bulk quantities by the ultracentrifugal method of Havel *et al.* (1955). An alternative procedure that can be used to isolate lipoproteins involves precipitation with high-molecular-weight dextran or heparin (Stokes *et al.*, 1967; Fredrickson *et al.*, 1967, Burstein *et al.*, 1970). The most commonly used method for separation of lipoproteins for qualitative analysis is the paper electrophoretic procedure of Lees and Hatch (1963).

1. CHYLOMICRONS

Chylomicrons are very big lipoprotein particles that contain large amounts of triglyceride. They are secreted into the lymph by the intestine following fat ingestion, and they enter the blood through the thoracic duct. Chylomicrons are the vehicles for transporting dietary fat from the intestine primarily to the liver and adipose tissue. Unless a metabolic abnormality exists, chylomicrons usually are removed from the blood within 8–12 hours after ingestion of fat. Therefore, serum that is obtained from a fasting donor should not contain an appreciable quantity of chylomicrons. When present, chylomicrons can be removed from serum by centrifugation at 10,000 g for 10–30 minutes.

2. VERY-LOW-DENSITY LIPOPROTEIN

VLDL is the vehicle for transport of triglycerides that are released by the liver. These lipoproteins are synthesized in the endoplasmic reticulum and secreted by the Golgi apparatus (A. L. Jones *et al.*, 1966; Mahley *et al.*, 1970). Recent evidence indicates that the intestine also is a site of synthesis and secretion of VLDL (Roheim *et al.*, 1966; Wind-

mueller and Levy, 1968; Ockner *et al.*, 1969). VLDL is a complex molecule containing triglyceride, LDL, and, possibly, HDL (Levy *et al.*,
1966; Nichols, 1969). Two additional aproproteins, one containing an
amino-terminal threonine and another containing an amino-terminal
serine, are present in VLDL. These D-peptides, which probably are components of apolipoprotein C (Gustafson *et al.*, 1966), account for approximately half of the total protein in VLDL (Brown *et al.*, 1969).
VLDL makes up 16% of the lipoproteins in fasting serum from normal
human males and 5% of the lipoproteins in normal female serum (Fredrickson *et al.*, 1968). VLDL can be removed from serum by centrifugation at 100,000 g for 16 hours.

3. Low-Density Lipoproteins

LDL is the major cholesterol-carrying lipoprotein of the serum. It
accounts for 50% of the lipoproteins present in fasting serum of normal
human males and 40% of the lipoproteins in the serum of normal females
(Fredrickson *et al.*, 1968). It is thought to contain only one apoprotein
species, the B protein. LDL may be a product of the catabolism of
VLDL (Nichols, 1969). Elevations in LDL have been associated with
the development of atherosclerosis (Fredrickson *et al.*, 1967; Lees, 1970),
a pathological process that is characterized by the accumulation of excessive amounts of lipid in the arterial intima.

Chylomicrons, VLDL, and LDL possess a micellar structure (Schumaker and Adams, 1969). They have a hydrophobic center containing
triglyceride and cholesterol esters surrounded by a hydrophilic coat of
phospholipids, unesterified cholesterol, and protein. This is in contrast
to the high-density lipoproteins which have a pseudomolecular structure
(Schumaker and Adams, 1969). They consist of lipid-protein subunits
that are associated in a definite quarternary structure.

4. High-Density Lipoproteins

There are two major subclasses of HDL: HDL_2, which is isolated
in the density range of 1.063–1.125 gm/ml, and HDL_3, which is isolated
between 1.125 and 1.21 gm/ml (Levy and Fredrickson, 1965; Scanu
et al., 1970). HDL comprises 34% of the lipoprotein present in fasting
serum from normal human males and 55% of those present in normal
female serum (Fredrickson *et al.*, 1968). The major lipids of HDL are
phospholipids, primarily lecithin and sphingomyelin, and cholesterol.
A third high-density lipoprotein, VHDL, also has been isolated. VHDL
has a density of greater than 1.21 gm/ml, and most of its lipid content

is lysolecithin (Schumaker and Adams, 1969). VHDL may be an artifact of the lipoprotein preparative procedures rather than a physiological entity. The metabolic role of high-density lipoproteins, including any that they may have in the lipid nutrition of mammalian cells, is unknown.

a. *Reconstitution and Labeling Procedures for Lipoproteins.* HDL are potentially useful tools for study of the role of serum lipoproteins in the nutrition and metabolism of cells in culture. When HDL is delipidated, the resulting apoproteins are water-soluble. Lipids, especially phospholipids and unesterfied cholesterol, will recombine with the delipidated residue (Scanu, 1967; Sodhi and Gould, 1967), and the product is structurally similar to native HDL (Scanu *et al.,* 1970). Moreover, purified lipids that are commercially available can be added to the delipidated product in place of those originally removed from the lipoprotein (Scanu *et al.,* 1970). These lipid-protein complexes which closely resemble native lipoproteins can be utilized as models for metabolic studies. Such work may help to determine whether the phospholipids or unesterified cholesterol contained in serum HDL are important substrates for cells in culture. Similar studies are more difficult to carry out with VLDL and LDL, for the delipidated apoproteins isolated from these lipoproteins are insoluble.

Trace quantities of radioactive lipids can be incorporated into intact lipoproteins by incubation with lipid-coated Celite (Avigan, 1959; Eaton *et al.,* 1969). In this way, the utilization of labeled lipids by cells in culture can be followed when these substrates are presented to the cells in the form of physiological lipoprotein complexes.

b. *Animal Sera.* The above information and the data listed in Table II were obtained from studies with human serum. Based on results obtained from rabbit (Lutton and Tsaltas, 1965) and rat serum (Koga *et al.,* 1969), it appears that similar lipids are present in animal sera. However, quantitative differences between human and animal sera definitely exist. Normal rabbit serum contains considerably less cholesterol than human serum, and a larger fraction of the cholesterol is esterified. On the other hand, the glyceride content is about the same as in the human. The lipoproteins in sera obtained from fasting animals, like those in fasting human serum, can be separated into three distinct density classes: VLDL, LDL, and HDL. However, in dog and most other animal serum, the HDL content is much greater than that of LDL (Sakagami and Zilversmit, 1961). This differs from the human where low-density lipoproteins account for 45% (female) to 65% (male) of the total serum lipoprotein content. In both dog and rat serum, ultracentrifugation at density 1.063 does not produce a complete separation of high- and low-density lipoproteins. Rat LDL is isolated between density 1.006 and 1.040, and HDL between density 1.063 and 1.21. The material isolated between density 1.040 and 1.063 contains a mixture of LDL and HDL.

Studies with bovine and rabbit serum albumin indicate that they bind long-chain free fatty acids in a manner that is qualitatively similar to human albumin (Spector *et al.*, 1969). The affinity of the albumins for palmitic acid was bovine > rabbit > human.

IV. Cellular Uptake of Serum Lipids

Mammalian cells can utilize the free fatty acids, phospholipids, and triglycerides that are present in serum. Although the mechanisms of uptake of these lipids are still under study, sufficient information already is available to describe the basic aspects of each process. Figure 1 illustrates the fundamental differences in the mechanisms by which free fatty acids, phospholipids, and triglycerides are taken up. These differences are as follows: Free fatty acid dissociates from albumin, and the unbound fatty acid anion binds to the cell surface. Phospholipids are transferred in the interaction between lipoproteins and the cell membrane. Triglycerides are hydrolyzed by lipases present either in the serum or the cell membrane, and the fatty acid that is released is taken up by the cell.

A. *Free Fatty Acids*

The initial step in the utilization of exogenous free fatty acid is binding to the cell in unesterified form (Fillerup *et al.*, 1958; Goodman, 1958a; Spector *et al.*, 1965; Shohet *et al.*, 1968). Most of the fatty acid binding sites appear to be located on or within the cell membrane (Spector, 1968). The human erythrocyte contains only one class of fatty acid binding sites, and they are located on the cell surface. There is one site per 420 $Å^2$ of erythrocyte surface area (Goodman, 1958a). The association constant for palmitate binding to these sites is $2.2 \times 10^6 \ M^{-1}$ (Goodman, 1958a), and this is approximately equal to the constant for palmitate binding to the secondary sites of human serum albumin (Goodman, 1958b). Fatty acid is rapidly transferred from albumin to the surface binding sites (Spector *et al.*, 1965). The actual time required for the transfer appears to be of the order of milliseconds or less. The binding

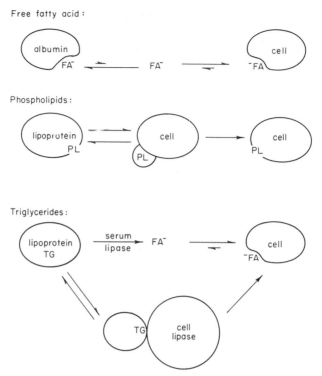

Fig. 1. Schematic representation of the mechanism of cellular uptake of serum lipids. The abbreviations are FA⁻, free fatty acid anions; PL, phospholipids; TG, triglyceride.

is reversible, and most of the fatty acid that is present at the surface sites can be removed quickly by exposure of the cell to a solution containing albumin (Spector, 1968). Studies on the binding of the methyl ester analogue suggest that the *major* interactions between palmitic acid and the cell binding sites are nonpolar (Kuhl and Spector, 1970). Double-label studies demonstrate that the fatty acid dissociates from the albumin carrier during uptake, and fatty acid can be taken up by the cell in the absence of a carrier protein (Spector *et al.*, 1965). Fatty acid that is bound to β-lactoglobulin also can be taken up by the cell (Spector and Fletcher, 1970). Hence, it appears that a specific interaction between serum albumin and the membrane binding site is not necessary for fatty acid transfer to the cell. Indeed, the fatty acid molecules actually incorporated probably are those that are free in solution and in equilibrium with the fatty acid bound to albumin. As fatty acid is taken up, the unbound pool is replenished continuously by fatty acid molecules that dissociate from albumin in an attempt to maintain the bound-free equilibrium (Spector, 1968).

1. FACTORS REGULATING FATTY ACID UPTAKE

The main factor that regulates the amount of free fatty acid that is taken up by the cell is the fatty acid/albumin molar ratio. As the molar ratio is raised, the quantity of fatty acid that is bound by the cell increases markedly (Spector, 1968; Spector et al., 1970). Even at the highest fatty acid/albumin molar ratios that occur physiologically, only a small fraction of the cell-binding capacity for fatty acid is saturated. At a given fatty acid/albumin molar ratio, uptake increases as the fatty chain length increases, i.e., laurate < palmitate < stearate (Spector and Steinberg, 1967a). Uptake decreases as the degree of unsaturation increases, i.e., stearate > oleate > linoleate (Spector and Steinberg, 1967a; Donabedian and Karmen, 1967; Spector et al., 1970). These findings are illustrated in Fig. 2. As contrasted with the magnitude of fatty acid binding, the subsequent rate of metabolism, e.g., oxidation and esterification, is somewhat greater for unsaturated than for saturated fatty acids (Spector and Steinberg, 1967a; Spector et al., 1970). Uptake by the binding sites is not affected appreciably by the presence of either metabolic inhibitors or glucose (Goodman, 1958a; Spector et al., 1965; Spector and Steinberg, 1966b). However, uptake is increased markedly when the pH of the incubation medium is lowered from 7.4 to 6.5 (Spector, 1969). The increased fatty acid uptake by the cells probably results from a pH-induced weakening of fatty acid binding to albumin (Spector et al., 1969). A pH-induced increase in fatty acid uptake may have produced the increase in cell lipid and the appearance of triglyceride-containing cytoplasmic granules that occurred in cells in culture when the pH of culture media was lowered from 7.4 to 6.9 (Mackenzie et al., 1967b). Recent studies with rat intestine suggest that the presence of calcium ions also may increase fatty acid uptake (Munday et al., 1969). In experiments lasting from 1 to 2 hours, the amount of fatty acid taken up and metabolized by Ehrlich ascites cells was not appreciably greater when a mixture of fatty acids was available as compared with only a single fatty acid (Spector, 1970).

2. PENETRATION INTO THE CELL

The mechanism by which fatty acid moves from the surface binding sites to the interior of the cell has not been definitely established. Certain

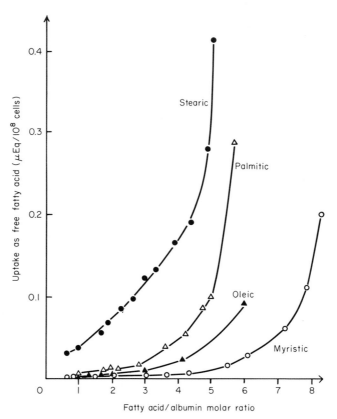

Fig. 2. Relationship between fatty acid/albumin molar ratio and uptake of fatty acid by the cell. Ehrlich cells were incubated for 2 minutes at 37°C with radioactive fatty acids bound to bovine albumin. The ordinate refers only to the radioactive fatty acid associated with the cells in *unesterified* form, *not* to the total amount that was incorporated into the cells.

of the available data are compatible with an energy-independent diffusion of fatty acid through the cell membrane (Stein and Stein, 1963; Vaughan *et al.*, 1964b; Strauss, 1966; Spector, 1968). Other studies indicate that the movement of fatty acid into the cell is energy-dependent (Shohet *et al.*, 1968). The proposed energy-dependent penetration mechanisms include the acylation of lysolecithin followed by enzymatic hydrolysis of the newly formed lecithin (Reshef and Shapiro, 1966), the formation of an acylcarnitine intermediate (Wittels and Hochstein, 1967), and the operation of a calcium ion pump (Munday *et al.*, 1969). These conflicting results may be resolved by the observation of Munday *et al.* (1969) that both energy-independent and energy-dependent transport processes occur but that the latter is operative only in the presence of calcium.

B. *Phospholipids*

Both major classes of phospholipids, phosphatides and sphingolipids, can be taken up *intact* by mammalian cells. Sphingomyelin and lecithin can be transferred from serum lipoproteins to erythrocytes (Reed, 1968). Phospholipid exchange occurs not only between serum lipoproteins and chick embryo fibroblasts (Peterson and Rubin, 1969), but also between individual fibroblasts growing in sheets on agar (Peterson and Rubin, 1970). Lysolecithin is taken up and subsequently converted to lecithin by Landschütz ascites cells (Stein and Stein, 1967). Glycerosphingolipids with Lewis blood group activity are taken up from serum lipoproteins by erythrocytes (Marcus and Cass, 1969). The only evidence against phospholipid uptake was obtained by Elsbach (1965) with rabbit leukocytes and alveolar macrophages. However, in these experiments, the radioactive phospholipid was presented to the cells as an emulsion, not as a physiological lipoprotein complex. The precise lipoprotein in serum that serves as the phospholipid donor to cells has not been identified, but it is reasonable to assume that this might be a function of the major phospholipid-carrying lipoproteins, HDL. It is likely that the phospholipids are taken up in the course of a lipoprotein-cell membrane interaction. Only 60% of the lecithin and 30% of the sphingomyelin in human erythrocytes exchanges with phospholipids present in extracellular lipoproteins (Reed, 1968).

C. *Triglycerides*

Triglycerides contained in serum lipoproteins also are taken up by mammalian cells (Bailey, 1967; Ontko, 1967). However, unlike phospholipids, they are hydrolyzed prior to uptake, and the released fatty acid actually is the species that is taken up. Two mechanisms for the hydrolysis have been proposed. The work of Howard and Kritchevsky (1969) indicates that hydrolysis occurs in the serum and is mediated by a serum lipase. On the other hand, studies with rat chylomicrons suggest that hydrolysis occurs at the cell surface while the lipoprotein is bound to the membrane (Higgins, 1967). Studies with a model complex, fatty acid methyl esters complexed to bovine albumin, demonstrate that this lipid ester also is hydrolyzed while attached to the cell (Kuhl and Spector, 1970).

The inability of intact triglycerides to cross cell membranes appears to be a general phenomenon. For example, dietary triglycerides are partially hydrolyzed prior to uptake by the intestine even though the hydrolysis products must be resynthesized into triglycerides in the mucosal cells. Triglycerides in adipocytes also must be hydrolyzed prior to release into the blood, i.e., as free fatty acid. On the other hand, triglycerides in the form of lipoprotein complexes can leave the hepatocyte and intestinal mucosal cell. However, these cells have a specialized secretory mechanism for the *lipoprotein*, not for the isolated triglyceride.

V. Lipid Biosynthesis

Most mammalian cells have the capacity to synthesize fatty acids from nonlipid precursors. They also can incorporate the fatty acids that are synthesized or taken up from the incubation medium into phospholipids and glycerides. The following subsections deal with the mechanisms involved in these processes. A summary of lipid biosynthetic pathways in mammalian cells is given in Table III.

TABLE III.

PATHWAYS FOR LIPID BIOSYNTHESIS

Lipid	Pathway	Substrates
Fatty acids	*De novo* (cytoplasm)	Acetyl ACP, malonyl ACP
	De novo [mitochondria (heart)]	Acetyl CoA
	De novo (mammary gland)	$D(-)$-β-hydroxybutyrate
	Elongation (mitochondria)	Acetyl CoA, acyl CoA
	Elongation (microsomes)	Malonyl CoA, acyl CoA
Sphingolipids	*De novo*	Sphingosine, acyl CoA
Phosphatides	*De novo*	L-Glycerol-3-phosphate (dihydroxyacetone phosphate), acyl CoA
	Acylation	Lysophosphatide, acyl CoA
	Condensation	Lysophosphatide
Alkyl ether	*De novo*	Dihydroxyacetone phosphate, fatty alcohol
Alk-l-enyl ether	Dehydrogenation	Alkyl acyl phosphoglyceride
Triglyceride	*De novo*	L-Glycerol-3-phosphate, acyl CoA
	Monoglyceride	2-Monoglyceride, acyl CoA

A. *Fatty Acids*

Two types of fatty acid synthesis occur in mammalian cells. One is *de novo* synthesis in which the entire fatty acid is produced from non-lipid precursors. The other is chain elongation in which 2-carbon atom units are added to a preexisting fatty acid.

1. Complete Synthesis: The *De Novo* Pathways

Three *de novo* pathways for fatty acid synthesis have been described. The malonate-acyl carrier protein pathway is present in the cell cytoplasm. It is active in liver, adipose tissue, and mammary gland, tissues that have a large capacity to synthesize fatty acid. A second pathway is present in heart mitochondria (Whereat *et al.*, 1967). This system utilizes acetate, not malonate, and some of the enzymes of the fatty acid oxidation pathway. The third pathway, in which β-hydroxybutyrate is the substrate, is present in mammary gland cells (Kinsella, 1970).

Malonate-Acyl Carrier Protein Pathway. This system exists in mammalian cells as a soluble multienzyme complex. The primary substrate is acetate, but only the two carbon atoms at the ω end of the fatty acid actually enter the chain as acetate. All of the other acetate units that are added are first converted to malonate by addition of bicarbonate (Wakil *et al.*, 1964). In the subsequent condensation, the newly added bicarbonate is released from malonate as CO_2. Hence, the fatty acid actually is built by successive additions of 2-carbon atom fragments. Rather than being wasteful, the loss of CO_2 from malonate provides the energy needed to drive the synthesis. Throughout the synthesis, the growing fatty acid chain is covalently bound to a sulfhydryl group of acyl carrier protein (ACP), a protein which functions as a coenzyme for the process (Majerus *et al.*, 1964; Wakil *et al.*, 1964). The initial substrate for this pathway, acetyl CoA, is generated intramitochondrially and probably is transferred into the cytoplasm as citrate (Spencer and Lowenstein, 1962). Acetyl CoA then is released from citrate through the action of ATP citrate lyase (Srere, 1959). The rate-limiting step in the malonate-ACP pathway is the initial one involving the conversion of acetate to malonate (Majerus and Vagelos, 1967). This reaction is catalyzed by acetyl CoA carboxylase, a biotin-containing enzyme complex that is regulated allosterically (Ryder *et al.*, 1967). Cells that cannot synthesize long-chain fatty acids, such as mature erythrocytes and leukocytes, lack acetyl CoA carboxylase (Pittman and Martin, 1966; Majerus and Lastra,

1967). NADPH is the cofactor for reduction reactions in the malonate-ACP pathway. Palmitic acid is the major end product of this pathway, apparently because the thiolase which cleaves the fatty acid from ACP exhibits maximal activity when the acyl chain contains 16 carbon atoms (Barnes and Wakil, 1968).

2. Chain Elongation Pathways

Two pathways for lengthening of a fatty acid chain have been reported. One is a mitochondrial system that utilizes acetyl CoA (Christ, 1967; Mooney and Barron, 1970). The major products are fatty acids containing 18 to 24 carbon atoms. A second elongation system is present in the endoplasmic reticulum. This system utilizes malonyl coenzyme A. However, CO_2 is released in the condensation so that the fatty acid still is lengthened in each step by only 2-carbon atoms. The microsomal system is stimulated by high concentrations of ATP (Landriscina *et al.*, 1970).

3. Desaturation

Mammalian cells are capable of introducing unsaturated bonds into a fatty acid. The fatty acid desaturase is present in the endoplasmic reticulum (P. D. Jones *et al.*, 1969). CoA derivatives of the fatty acids serve as the substrates, and oxygen is required for the reduction. NADH is the electron donor and NADH-cytochrome b_5 reductase is a component of this system (Holloway and Wakil, 1970). Important fatty acid conversions that occur in this desaturase system include the synthesis of oleate from stearate and the synthesis of palmitoleate from palmitate.

B. Essential Fatty Acids

Although mammalian cells can insert one or more unsaturated bonds into fatty acids, they cannot synthesize all the polyunsaturated fatty acids that normally are present. The acids that cannot be synthesized are obtained from dietary fat and are known as essential fatty acids. They are required for the well-being of the intact animal (Jorgensen, 1961). Cells in culture obtain essential fatty acids from serum when it is present in the incubation medium. The most important essential fatty acid is linoleate. Archidonate, another important essential fatty acid, is synthesized from linoleate by desaturation and chain elongation. However, HeLa cells cannot perform this conversion, and heart cells lose the ability to do this if they remain in culture for a prolonged period (Harary *et al.*, 1967). Linoleic acid cannot be synthesized by any mammalian tissue, for the mammalian desaturase cannot insert a second double bond into oleic acid in the proper position, i.e., between carbon atoms 12 and 13. The precise reason that essential fatty acids are required is not known. This may be related to the fact that arachidonic acid is the substrate for the synthesis of the hormones prostaglandin E_2 and $F_{2\alpha}$ (Bergström, 1967). In both HeLa and rat heart cells in culture, essential fatty acids are necessary for normal oxidative phosphorylation and respiratory control in mitochondria (Gerschenson *et al.*, 1967a,b). Linoleic acid also is a growth factor for CHD3 Chinese hamster cells (Ham, 1963) and macrophages in culture (Dubin *et al.*, 1965), but not for Chang human liver cells (Savchuck *et al.*, 1965). On the other hand, many cell lines grow in fat-free media (V. J. Evans *et al.*, 1965). L strain fibroblasts grow well in the absence of lipids even though they contain almost no essential fatty acids (Bailey, 1967). Yet, when they are grown in a medium supplemented with serum, they contain considerable quantities of linoleic acid (Bailey, 1967). These findings raise some questions about the need for essential fatty acids, at least in some mammalian cell lines. Cells grown in the absence of essential fatty acids accumulate 5,8,11-eicosatrienoic acid, a polyunsaturated 20-carbon atom acid that is synthesized by chain elongation and desaturation of oleate.

C. Phospholipids

The two major classes of phospholipids are the sphingolipids and phosphatides. The initial intermediate in sphingolipid synthesis is dihydrosphingosine which is formed by condensation of palmityl CoA with serine. This reaction requires pyridoxal phosphate (Braun *et al.*, 1970). The second intermediate, ceramide, is synthesized through the action of a specific *N*-acyltransferase (Morell and Radin, 1970). Ceramide is the precursor of gangliosides (Basu *et al.*, 1968) and sphingomyelin (Sribney and Kennedy, 1958). Ceramide containing hydrox fatty acid is the precursor of cerebrosides (Morell and Radin, 1969).

There are three pathways for phosphatide synthesis in mammalian cells. The first is the *de novo* pathway in which phosphatidic acid is an intermediate. Both of the fatty acid residues are supplied as acyl CoA. The triose phosphate backbone to which the fatty acids are esterified is either L-glycerol-3-phosphate (Fallon and Lamb, 1968) or dihy-droxyacetone phosphate (Hajra, 1968). These intermediates usually are derived from glucose through the glycolytic pathway. This fact intro-duces a very important point concerning the evaluation of fatty acid synthesis from radioactive glucose in mammalian cells. Unless the lipid that is produced is checked carefully by saponification, one cannot con-clude that incorporation of glucose radioactivity into cell lipids indicates fatty acid synthesis. For example, considerable quantities of glucose radioactivity are incorporated into lipids, particularly into phosphatides, by Ehrlich ascites cells. However, analysis of the synthesized lipids dem-onstrates that almost all the radioactivity is present in the glycerol moiety and not in fatty acids (Spector and Steinberg, 1966b; Spector, 1967a). Ehrlich cells have a very limited capacity for *de novo* fatty acid synthesis but a large capacity for *de novo* phosphatide synthesis (Spector and Steinberg, 1967a). The phosphatidic acid intermediate that is formed in the *de novo* pathway can be converted to serine, ethanolamine, inositol, glycerol, and choline phosphatides. The latter, lecithin, is the most abundant phosphatide in many mammalian cells (Wallach *et al.*, 1960; Weinstein *et al.*, 1969). It is synthesized by conversion of phospha-tidic acid to 1,2-diglyceride, which then condenses with cytidine diphos-phate choline (Kennedy *et al.*, 1959).

A second pathway for phosphatide biosynthesis involves the incorpora-tion of a single fatty acid residue into a preformed lysophosphatide (Lands, 1960). Acyl CoA serves as the substrate, and the fatty acid usually is esterified at the 2-carbon atom of a 1-lysophosphatide. This is the major pathway for incorporation of fatty acid into phospholipids in erythrocyte membranes (Oliveira and Vaughan, 1964). In mammalian lecithin, the 2 position contains a much higher percentage of polyunsatu-rated fatty acids than the 1 position. This probably occurs because the acyltransferase which incorporates fatty acids into the 2 position has a greater affinity for polyunsaturated acyl CoA derivatives (Waku and Lands, 1968).

A third mechanism for phosphatide biosynthesis involves the conden-sation of two lysophosphatide molecules (Erbland and Marinetti, 1965). Studies with erythrocytes suggest that this is not an important pathway for phosphatide synthesis in the intact cell (Shohet and Nathan, 1970).

Phospholipids are synthesized throughout the cell cycle in growing neoplastic mast cells (Bergeron *et al.*, 1970). Synthesis occurs through the *de novo* pathway (Pasternak and Bergeron, 1970). The rate of syn-thesis begins to increase during the G1 phase (Warmsley and Pasternak, 1970), and the phospholipid content of the cells doubles during the S phase (Bergernon *et al.*, 1970).

Alkyl and Alk-1-enyl Ethers

Phosphatides and glycerides containing ether-linked fatty acids are present in mammalian cells (Snyder and Wood, 1968). Dihydroxyacetone phosphate and fatty alcohols are substrates for the biosynthesis of these alkyl ethers (Hajra, 1969; Wykle and Snyder, 1970). Fatty acids are precursors of the alcohols (Wood and Healy, 1970). Dihydroxyacetone phosphate is esterified initially with fatty acid in the 1 position and the acyl residue then exchanges with the alcohol (Hajra, 1970). After reduction of the keto group, the resulting 1-alkylglycerol-3-phosphate is acylated in the 2 position (Hajra, 1970). Alk-1-enyl ethers are formed from intact alkyl acyl phosphoglycerides by dehydrogenation (Wood and Healy, 1970).

D. Triglycerides

Two pathways exist for triglyceride synthesis. The first is identical to the *de novo* phosphatide synthetic pathway up to 1,2-diglyceride formation. At this point, the diglyceride combines with a third acyl CoA to form triglyceride. The second pathway involves a 2-monoglyceride intermediate. This is acylated in both the 1 and 3 positions by acyl CoA (Senior and Isselbacher, 1962). The monoglyceride pathway is the major mechanism for the resynthesis of triglycerides from dietary fat in the intestinal mucosa. Whether this pathway also is operative in cells in culture remains to be elucidated.

Triglyceride synthesis in Ehrlich cells is stimulated when the incubation medium contains high concentrations of glucose and free fatty acid (Spector and Steinberg, 1966b). Free fatty acids contained in the culture medium also are incorporated into triglycerides by HeLa cells (Geyer, 1967) and Earle's L strain mouse cells (Moskowitz, 1967). The triglycerides present in the cytoplasmic droplets that form in these and other cells in culture are synthesized within the cell (Mackenzie *et al.*, 1967a). The main source of fatty acid for this synthesis is the culture medium (Mackenzie *et al.*, 1967a; Moskowitz, 1967).

VI. Lipid Utilization

Except in special cases, lipids are used by cells for two purposes. They are either incorporated into membranes or oxidized as a source of energy. In the latter case, the fatty acid may be oxidized immediately or stored intracellularly, usually as triglyceride, until the need for additional energy arises. Very little is known about the mechanisms of membrane biosynthesis. Therefore, Section VI is devoted almost entirely to fatty acid oxidation and the release of fatty acid from intracellular storage forms.

A. Fatty Acid Oxidation

Fatty acid can be oxidized in four ways, which are listed in Table IV. The major pathway is β oxidation in which acetyl CoA is formed. The acetyl CoA units are degraded subsequently in the tricarboxylic acid cycle to CO_2 and H_2O, and approximately 45% of the energy is trapped in the form of ATP. In each sequence in the β-oxidation pathway, the acyl CoA substrate is shortened by 2-carbon atoms. In α oxidation, a process that occurs in the endoplasmic reticulum, the fatty acid is shortened by only one carbon atom. Fatty acid, not the CoA derivative, is the substrate for α oxidation. A hydroxyl group is inserted on the second carbon atom, and the carboxyl group then is released as carbon dioxide (Mead and Levis, 1963; Levis and Mead, 1964). This is the pathway for formation of the α-hydroxy and odd-number carbon atom fatty acids that are present in nervous tissue. α oxidation also is

TABLE IV.

PATHWAYS FOR FATTY ACID OXIDATION

Oxidative Pathway	Substrate	Product	Cofactor Requirement	Subcellular Localization
β	Acyl CoA[a]	Acetyl CoA	ATP, CoA, Mg^{2+}, carnitine, FAD, NAD	Mitochondria
α	Fatty acid	CO_2	ATP, NAD, ascorbate, Fe^{2+}	Microsomes
ω	Fatty acid	ω-COOH-fatty acid	NADPH, cytochrome P 450	Microsomes
δ	Acyl CoA	Butyryl CoA	[b]	[c]

[a] Abbreviations: CoA, coenzyme A; ATP, adenosine triphosphate; FAD, flavin adenine dinucleotide; NAD, diphosphopyridine nucleotide; NADPH, reduced triphosphopyridine nucleotide.

[b] Unknown. It appears that they should be similar to the cofactors needed for β oxidation.

[c] Unknown. It is likely that this, like β oxidation, occurs in the mitochondrion. This oxidative pathway has been described only in mammary gland tissue.

necessary for the degradation of branched-chain fatty acids such as phytanic, which cannot be oxidized by the β pathway unless they initially are decarboxylated (Mize et al., 1969). ω oxidation occurs in the endoplasmic reticulum, and the fatty acid rather than a CoA derivative is the substrate. First, a hydroxyl group is inserted on the carbon atom farthest removed from the carboxyl group, the ω-carbon. This is oxidized subsequently to a carboxyl group so that a dicarboxylic acid is formed (Preiss and Bloch, 1964). Both α and ω oxidation are minor pathways for the physiologically important straight-chain fatty acids (Antony and Landau, 1968). The recently described δ-oxidation pathway in which butyryl CoA is produced may be peculiar to mammary gland cells (Dimick et al., 1969). It should be noted that mammary gland cells also have a unique pathway for de novo fatty acid synthesis which utilizes D($-$)-β-hydroxybutyrate as the substrate (Kinsella, 1970). Only the main pathway, β oxidation, will be considered in further detail.

1. FATTY ACID ACTIVATION

The first step in the β oxidation of fatty acids is activation, i.e., conversion of the fatty acid to the acyl CoA derivative. This reaction is catalyzed by an acyl CoA synthetase. Two types of enzymes have been described: ATP- and GTP-dependent synthetases (Kornberg and Pricer, 1953; Galzigna et al., 1967). It is thought currently that the ATP-dependent synthetase is much more important for the β-oxidation system (Van Tol, 1969). ATP-dependent long-chain acyl CoA synthetase is firmly bound to membranes, and it is located in the mitochondria, microsomes, and surface membrane of the cell. In liver, the surface membrane contains the highest activities of this activating enzyme (Pande and Mead, 1968). However, it is likely that at least some of the fatty acid that is oxidized is activated at the outer mitochondrial membrane (Skrede and Bremer, 1970), the site of the ATP-dependent synthetase in mitochondria (Haddock et al., 1970).

2. CARNITINE ACYLTRANSFERASE

Acyl CoA derivatives cannot reach the site of the β-oxidase system within the mitochondrial matrix. To cross the mitochondrial "barrier," probably the inner membrane, the acyl group is transferred to carnitine by carnitine palmityltransferase (Fritz, 1963; Bremer, 1963). This is a mitochondrial enzyme (Norum and Bremer, 1967), and it is associated primarily with the inner membrane (Yates and Garland, 1970). This enzyme apparently has three functions: to transfer the acyl group from

CoA to carnitine, to translocate the acylcarnitine ester across the inner membrane, and to return the acyl group to CoA at the site of the β-oxidation system in the matrix (Fritz and Marquis, 1965). Carnitine is a cofactor for the β-oxidation system and stimulates fatty acid oxidation in homogenates and isolated mitochondria (Fritz, 1963). However, carnitine does not increase fatty acid oxidation in the intact mammalian cell even though it enters the cell (Spector, 1967b). Apparently, there is sufficient carnitine in the cell to support the maximum rate of fatty acid oxidation that can be attained.

3. Mechanism and Regulation of β-Oxidation

When the acyl CoA derivative reaches the site of the fatty acid oxidase, the process of releasing acetyl CoA units begins. The reactions involved are listed in Table V. Energy in the form of ATP is derived from the oxidation of the $FADH_2$ and NADH that are formed in the dehydrogenation reactions. However, most of the energy is obtained from the combustion of the acetyl CoA units. In the thiolytic cleavage, an acyl CoA containing two carbon atoms less than the original substrate is formed. This and the subsequent acyl CoA units that are generated reenter the β-oxidation sequence until the entire chain is degraded to acetyl CoA.

The rate of oxidation of fatty acid added to a suspension of mammalian cells is regulated in part by the molar ratio of fatty acid to albumin in the incubation medium (Spector and Steinberg, 1965). When the molar ratio is increased over the usual physiological range, i.e., 0.5–2.0, CO_2 formation from the added fatty acid increases linearly. If more

TABLE V.

THE β-Oxidation Mechanism

Reaction	Substrate	Product	Enzyme	Cofactor
First dehydrogenation	Acyl CoA	α,β-unsaturated	Acyl dehydrogenase	FAD
Hydration	α,β-unsaturated acyl CoA	L-β-hydroxy	Enoyl hydrase	
Second dehydrogenation	L-β-hydroxy-acyl CoA	β-keto	β-Hydroxyacyl-dehydrogenase	NAD
Thiolytic clevage	β-ketoacyl CoA	Acetyl CoA + acyl CoA[a]	Thiolase	CoA

[a] The fatty acid chain is 2-carbon atoms shorter than the initial substrate.

fatty acid is added to further raise the molar ratio, CO_2 formation increases much more slowly and approaches a maximum. This is quite different from fatty acid uptake which continues to increase as the molar ratio is raised (see Fig. 2). Hence, the rate-limiting factor in the oxidation of exogenous fatty acid is not the ability of the cell to bind fatty acid; it occurs at some point subsequent to binding. An important question raised by these observations is whether the total oxygen consumption of the cell also is increased when large amounts of fatty acid are available. If so, the exogenous fatty acid concentration may be an important regulator of cellular respiration. Studies with guinea pig leukocytes indicated that oxygen utilization increased markedly as the fatty acid concentration was raised (W. H. Evans and Mueller, 1963). A similar result occurred with rat myocardium (Challoner and Steinberg, 1966). In contrast, little increase in oxygen consumption occurred in Ehrlich cells as the fatty acid/albumin molar ratio was raised (Spector and Steinberg, 1967b). The respiratory quotient of Ehrlich cells incubated with fatty acid was approximately 0.8, indicating that a sizable fraction of the oxidative substrate was lipid. However, it also was 0.8 in a lipid-free medium (Spector and Steinberg, 1965). This suggests that fatty acid present in the cells was utilized as the oxidative substrate when exogenous fatty acid was unavailable. Isotopic experiments have supported this viewpoint. As the availability of exogenous fatty acid increased, there was a corresponding decrease in the quantity of endogenous lipid that was oxidized (Fig. 3). Therefore, at least in certain cells, a regulatory mechanism exists to ensure a relatively stable respiratory rate independently of the availability of exogenous fatty acid.

Addition of glucose to the incubation medium reduces the oxidation of exogenous fatty acid in both Ehrlich and HeLa cells (Spector and Steinberg, 1966b; Geyer, 1967). A large percentage of the fatty acid uptake is channeled into phospholipids and glycerides when glucose is present. This control mechanism also ensures that cellular energy production will remain relatively constant independently of exogenous substrate availability. In addition, it provides for storage of substrate that is available but not needed to satisfy immediate energy needs. Results from studies with intact tissues suggest that all cells may not possess the glucose-fatty acid regulatory mechanism. For example, addition of glucose does not reduce fatty acid oxidation in skeletal muscle (Eaton and Steinberg, 1961).

All the commonly occurring long-chain fatty acids are readily oxidized by leukocytes and Ehrlich cells (Elsbach, 1964; Spector and Steinberg, 1967a). In both HeLa and Ehrlich cells, the 16- and 18-carbon atom acids are oxidized more rapidly than myristate or laurate (Geyer, 1967;

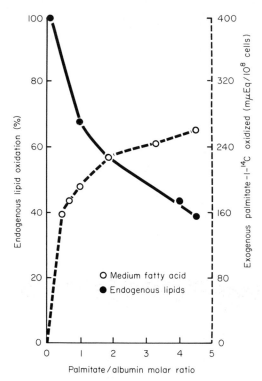

Fig. 3. Effect of exogenous fatty acid concentration on endogenous fatty acid oxidation. Ehrlich cells were incubated for 1 hour at 37°C. In the first case (open circles), cells were incubated with palmitate-1-^{14}C, and $^{14}CO_2$ production is plotted as a function of the palmitate/albumin molar ratio. In the second case (closed circles), cells labeled with palmitate-1-^{14}C were employed. They were incubated with palmitate that was *not* radioactive. The $^{14}CO_2$ that was produced was plotted as a function of the palmitate/albumin molar ratio of the solution with which the labeled cells were incubated. In this case, $^{14}CO_2$ production is calculated as a percentage of that occurring when no exogenous palmitate was available.

Spector and Steinberg, 1967a). Neither octanoate nor acetate is oxidized to any appreciable extent by Ehrlich cells. In the case of acetate, this may be due to a deficiency of acetyl CoA synthetase (Hepp *et al.*, 1966).

B. *Metabolism of Intracellular Lipids*

A mammalian cell can utilize a portion of its endogenous fat if the need for lipid substrate arises. These intracellular lipids constitute a metabolically active lipid pool (Geyer, 1967). In L strain fibroblasts, lipids taken up from the culture medium initially enter the transient pool. From this pool, they exchange with medium lipids, are oxidized, or are incorporated into relatively stable intracellular lipids which comprise the permanent lipid pool. Lipids are lost from the permanent pool primarily through cell death and disintegration (Geyer, 1967).

Results with Ehrlich ascites cells are in agreement with these interpretations. When Ehrlich cells were harvested from mice that previously were fed palmitic acid-1-^{14}C, radioactivity was recovered in cell lipids (Medes and Weinhouse, 1958). When these labeled cells were incubated *in vitro*, radioactive CO_2 was produced. The specific radioactivity of the CO_2 was much higher than that of the cell fatty acids, suggesting that the labeled palmitate did not equilibrate with the total fatty acid content of the cell prior to being oxidized. Similar results were observed with tissues slices obtained from rats fed palmitic acid-1-^{14}C (Volk *et al.*, 1952). Hence, there appears to be a small, metabolically active lipid pool that does not equilibrate with the bulk of the cell lipids. This interpretation remained open to some question because palmitate was the only labeled fatty acid that was studied. However, this concept was supported by subsequent studies with labeled palmitate, stearate,

TABLE VI.

COMPARISON OF UTILIZATION OF TOTAL INTRACELLULAR ESTERIFIED FATTY ACID AND NEWLY INCORPORATED RADIOACTIVE ESTERIFIED FATTY ACIDS[a]

Experiment Number	Cell Lipid	Percentage Decrease[b] (%)	Ratio of Decrease (Radioactivity/Total Fatty Acid)
1–4	Total esterified fatty acid	1.3[c]	—
1	Esterified palmitate-l-^{14}C	8.1	6.2
2	Esterified stearate-l-^{14}C	11.6	8.9
3	Esterified oleate-l-^{14}C	11.4	8.8
4	Esterified linoleate-l-^{14}C	6.2	4.8

[a] Ehrlich ascites cells were labeled by intraperitoneal injection of a radioactive fatty acid into tumor-bearing mice. The cells were harvested 1 hour after the isotope was injected, and after washing they were incubated for 1 hour at 37°C in a lipid-free medium (Spector and Steinberg, 1967b). A separate cell preparation was tested with each radioactive fatty acid.

[b] After 1 hour of incubation *in vitro*.

[c] Average value from all four experiments.

oleate, and linoleate, the four fatty acids that constitute 83% of the total fatty acid of the Ehrlich cell. As shown in Table VI, only 1.3% of the *total* esterified fatty acid contained in the Ehrlich cell was depleted during a 1-hour incubation in the absence of added substrate. Under these conditions, from 6.2 to 11.6% of the newly esterified radioactive fatty acid was depleted. These data confirm the existence of a rapidly turning-over pool of intracellular lipid esters.

Much of the fatty acid that comprises the metabolically active pool of the Ehrlich cell is contained in phospholipids, particularly lecithin (Spector and Steinberg, 1967b). Yet, most of the phospholipids of this cell are part of the relatively stable, permanent lipid pool. In this and other cells, triglycerides also are a component of the active pool. Triglycerides of rabbit granulocytes supply fatty acid for lecithin synthesis during phagocytosis (Elsbach and Farrow, 1969). In L cells, there is a progressive shift in newly synthesized fatty acid from triglycerides to phospholipids (Geyer, 1967). Likewise, in sarcoma 180 cells, newly incorporated labeled palmitate is depleted rapidly from triglycerides during the postlabeling period (Lengle and Smith, 1970). Fatty acids contained in intracellular lipid esters are hydrolyzed and pass into the intracellular free fatty acid pool prior to reutilization (Spector and Steinberg, 1966a). A schematic representation which illustrates the metabolism of intracellular lipids is presented in Fig. 4.

The present experimental evidence indicates that only a single *intracellular* free fatty acid pool exists. Therefore, in Fig. 4, both exogenous fatty acid and fatty acid derived from endogenous lipid esters are shown to pass into the same pool. The cell free fatty acid content varies, ranging

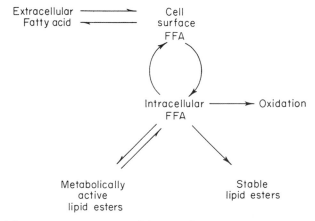

Fig. 4. Schematic representation of fatty acid utilization in mammalian cells.

from 1% of the total lipid in L cells to 4.7% in HeLa cells (Howard and Kritchevsky, 1970). In absolute amounts, the free fatty acid content varies from 0.14 μmole/10^8 L5178Y cells to 3.4 μmoles/10^8 skin fibroblasts (Howard and Kritchevsky, 1970). In Fig. 4 the pathways for fatty acid entrance and efflux are shown to be different. The pathway for efflux is energy-independent (Spector and Steinberg, 1966a), whereas at least some evidence indicates that movement of fatty acid from the cell surface to the intracellular pool requires energy (Shohet et al., 1968). Cell free fatty acid comprises a second metabolically active lipid pool. In the Ehrlich ascites cell, most of the free fatty acid content is derived from the incubation medium. This may not be true in all cell lines, for about 75% of the free fatty acid contained in diaphragm is derived from endogenous lipid esters (Schonfeld, 1968).

1. Cellular Lipases

The hydrolysis of endogenous esterified fatty acid is catalyzed by intracellular lipases. These enzymes are listed in Table VII. The role of the heparin-released enzymes in exogenous lipid utilization has been discussed in Section III. These enzymes probably are located on the surface of many cells. A triglyceride lipase also is located in the lysosomes (Gudor et al., 1969; Hayase and Tappel, 1970). In cells such as adipocytes, this enzyme is activated hormonally. This occurs indirectly, for the hormones actually activate adenyl cyclase, not the triglyceride lipase. The cyclic AMP that is produced by adenyl cyclase in turn activates the lipase (Butcher et al., 1968). The hydrolysis of the first fatty acid residue is the rate-limiting step in triglyceride degradation. Both diglycerides and monoglycerides appear to be degraded by a single enzyme, monoglyceride lipase (Vaughan et al., 1964a).

Intracellular phosphatides are degraded by two types of phospholipases. Since lecithin usually is employed experimentally as the substrate, these enzymes are often called lecithinases. However, it is likely that phosphatides other than lecithin also are degraded by these or very similar enzymes. Phospholipase A (or A_2) which hydrolyzes the β- or 2-fatty acid, is located in the outer mitochondrial membrane (Nachbaur and Vignais, 1968) and in lysosomal membranes (Rahman, 1970). Phospholipase B (or A_1) hydrolyzes fatty acids present in the α or 1 position and is present in the endoplasmic reticulum (Waite and Van Deenen, 1967). Lysophospholipases are soluble enzymes and therefore probably are located in the cell cytoplasm (Waite and Van Deenen, 1967). Only those sphingolipases that are fairly well characterized are listed in Table VII.

TABLE VII.

CELLULAR ENZYMES THAT HYDROLYZE LIPID ESTERS

Enzymes	Substrate	Product	Special Properties
Cell Surface			
Lipoprotein lipase	Exogenous triglycerides	FA, diglyceride	Released by heparin
Monoglyceride lipase	Exogenous monoglycerides, diglycerides	FA, glycerol	Released by heparin
Phospholipase B[a]	Exogenous phosphatides	FA, 2-lysophosphatide	Released by heparin
Intracellular			
Triglyceride lipase	Triglycerides	FA, diglycerides	Activated hormonally[a]
Monoglyceride lipase	Monoglycerides, diglycerides	FA, glycerol	
Phospholipase A[b]	Phosphatides[e]	FA, 1-lysophosphatide	
Phospholipase B[c]	Phosphatides	FA, 2-lysophosphatide	
Lysophospholipase	Lysophosphatide	FA, glycerylphosphoryl (base)[f]	Ca^{2+}-stimulated
Sphingomyelinase	Sphingomyelin	Ceramide, phosphorylcholine	Deficient in Niemann-Pick disease
Ceramide trihexosidase	Ceramide trihexoside	Galactose, ceramide hexoside	Deficient in Fabry's disease
Ceramide monohexosidase	Ceramide monohexoside	Hexose, ceramide	
β-Glucosidase	Glucocerebroside	Ceramide, glucose	Deficient in Gaucher's disease
β-Galactosidase	Ganglioside GM$_1$	Galactose	Deficient in generalized gangliosidosis
Hexoaminidase	Ganglioside GM$_2$	N-acetyl galactosamine	Deficient in Tay-Sachs disease
Sulfatidases[d]	Sulfatide	Galactosulfatide	Deficient in metachromatic leuko-dystrophy

[a] Active against lecithin and phosphatidylethanolamine.
[b] Also called phospholipase A$_2$ or lecithinase A$_2$.
[c] Also called phospholipase A$_1$ or lecithinase A$_1$.
[d] Other sphingolipases exist, but they are not well characterized at this time.
[e] Most studies have employed lecithin as the substrate. However, phosphatidylthanolamine also is degraded, and it is likely that other phosphatides can serve as substrates for this and the other phosphatidases.
[f] The nature of the base, e.g., choline, ethanolamine, etc., depends on the lysophosphatide that is degraded.
[g] In many cells, the enzyme is activated by cyclic AMP which is formed following hormonal activation of adenyl cyclase (Butcher et al., 1968).

2. Regulation of Endogenous Lipid Utilization

The utilization of endogenous fatty acid decreases when fatty acid is available in the extracellular fluid (see Fig. 3). Two mechanisms which explain this effect have been reported. In adipocytes, lipolysis is inhibited by the presence of free fatty acid in high concentrations (Rodbell, 1965). A different mechanism for the fatty-acid-induced lipid sparing effect occurs in Ehrlich cells (Spector and Steinberg, 1966a). The decrease in endogenous fatty acid oxidation probably results from mixing of endogenous and exogenous fatty acid in the intracellular free fatty acid pool. The total amount of fatty acid that is oxidized remains relatively constant even when the cell free fatty acid pool expands. In contrast, the increase in pool size stimulates fatty acid incorporation into lipid esters. The net effect is a sparing of intracellular lipids due to increased reesterification (Spector, 1968). Glucose availability also produces sparing of endogenous lipids (Spector and Steinberg, 1966b). This also results primarily from increased reesterification, not reduced lipolysis. Indeed, a continuing breakdown and oxidation of some endogenous lipid occurs even under conditions that produce a large net increase in cell lipid (Spector and Steinberg, 1967b).

VII. Summary and Conclusions

Many mammalian cells in culture can synthesize sufficient quantities of fatty acid from nonlipid substrates to satisfy their growth and metabolic needs. However, if adequate amounts of fat are present in the culture medium, the lipid requirements of the cell usually are satisfied by uptake of the exogenous fats. Serum is the most common source of lipids in culture media. Lipids are present in serum as albumin-bound free fatty acids and as components of lipoproteins. Free fatty acids and at least some phospholipids are taken up intact by the cells, whereas triglycerides are hydrolyzed prior to uptake. Certain polyunsaturated fatty acids, known as essential fatty acids, cannot be synthesized by mammalian cells. These fatty acids must be supplied in the culture medium. However, the role of essential fatty acid in the isolated mammalian cell is not completely understood, and certain cells can function and multiply even if they contain very little or no essential fatty acid.

Lipids are utilized by the cell in three ways. They are oxidized as an energy source, serve as structural components of cell membranes, and serve as intracellular reservoirs of substrate. If the need for energy arises, a small portion of the cell lipid ester content can be degraded and metabolized. These esters, together with the small amount of free fatty acid present in the cell, constitute a metabolically active lipid pool. The remainder of the intracellular lipid is relatively stable and is vital for the maintenance of membrane structure and function.

ACKNOWLEDGMENTS

This investigation was supported by research grants from the National Institutes of Health (HE-11,485 and HE-14,781) and the Iowa Heart Association. The author is a Research Career Development Awardee of the National Heart and Lung Institute (5-K04-HE 20,338).

REFERENCES

Antony, G. J., and Landau, B. R. (1968). Relative contribution of α-, β- and ω-oxidative pathways to *in vitro* fatty acid oxidation in rat liver. *J. Lipid Res.* **9**, 267–269.

Avigan, J. (1959). A method for incorporating cholesterol and other lipids into serum lipoproteins *in vitro*. *J. Biol. Chem.* **234**, 787–790.

Bailey, J. M. (1967). Cellular lipid nutrition and lipid transport. *Lipid Metab. Tissue Cult. Cells, Symp., 1966* pp. 85–113.

Barnes, E. M., Jr., and Wakil, S. J. (1968). Studies on the mechanism of fatty acid synthesis. XIX. Preparation of general properties of palmityl thioesterase. *J. Biol. Chem.* **243**, 2955–2962.

Basu, S., Kaufman, B., and Roseman, S. (1968). Enzymatic synthesis of ceramide-lactose by glycosyltransferases from embryonic chicken brain *J. Biol. Chem.* **243**, 5802–5804.

Bergeron, J. J. M., Warmsley, A. M. H., and Pasternak, C. A. (1970). Phospholipid synthesis and degradation during the life-cycle of P815Y mast cells synchronized with excess of thymidine. *Biochem. J.* **119**, 489–492.

Bergström, S. (1967). Prostaglandins: Members of a new hormonal system. *Science* **157**, 382–391.

Boyle, J. J., and Ludwig, E. H. (1962). Analysis of fatty acids of continuously cultured mammalian cells by gas-liquid chromatography. *Nature (London)* **196**, 893–894.

Braun, P. E., Morell, P., and Radin, N. S. (1970). Synthesis of C_{18}- and C_{20}-dihydrosphingosines, ketodihydrosphingosines and ceramides by microsomal preparations from mouse brain. *J. Biol. Chem.* **245**, 335–341.

Bremer, J. (1963). Carnitine in intermediary metabolism. The biosynthesis of palmitylcarnitine by cell subfractions. *J. Biol. Chem.* **238**, 2774–2779.

Brown, W. V., Levy, R. I., and Fredrickson, D. S. (1969). Studies of the proteins in human plasma very low density lipoproteins. *J. Biol. Chem.* **244**, 5687–5694.

Burstein, M., Scholnick, H. R., and Morfin, R. (1970). Rapid method for the isolation of lipoproteins from human serum by precipitation with polyanions. *J. Lipid Res.* **11**, 583–595.

Butcher, R. W., Baird, C. E., and Sutherland, E. W. (1968). Effects of lipolytic and anti-lipolytic substances on adenosine-3′,5′-monophosphate levels in isolated fat cells. *J. Biol. Chem.* **243**, 1705–1712.

Challoner, D. R., and Steinberg, D. (1966). Effect of free fatty acid on the oxygen consumption of perfused rat heart. *Amer. J. Physiol.* **210**, 280–286.

Chen, R. F. (1967). Removal of fatty acids from serum albumin by charcoal treatment. *J. Biol. Chem.* **242**, 173–181.

Cherkes, A., and Gordon, R. S., Jr. (1959). The liberation of lipoprotein lipase by heparin from adipose tissue incubated *in vitro*. *J. Lipid Res.* **1**, 97–101.

Christ, E. J. V. J. (1967). Fatty acid synthesis in mitochondria. Elongation of short-chain fatty acids and formation of unsaturated long-chain fatty acids. *Biochim. Biophys. Acta* **152**, 50–62.

Cunningham, V. J., and Robinson, D. S. (1969). Clearing factor lipase in adipose tissue. Distinction of different states of the enzyme and the possible role of the fat cell in the maintenance of tissue activity. *Biochem. J.* **112**, 203–209.

Day, A. J., and Fidge, N. H. (1962). The uptake and metabolism of C^{14}-labeled fatty acids by macrophages *in vitro*. *J. Lipid Res.* **3**, 333-338.

Dimick, P. S., Walker, N. J., and Patton, S. (1969). Lipid metabolism. Evidence of a δ-oxidation pathway for saturated fatty acids. *Biochem. J.* **111**, 395–399.

Dole, V. P. (1956). A relation between non-esterified fatty acids in plasma and the metabolism of glucose. *J. Clin. Invest.* **35**, 150–154.

Donabedian, R. K., and Karmen, A. (1967). Fatty acid transport and incorporation into human erythrocytes *in vitro*. *J. Clin. Invest.* **46**, 1017–1027.

Dubin, I. N., Czernobilsky, B., and Herbst, B. (1965). Effects of albumin and linoleic acid on growth of macrophages in tissue culture. *J. Nat. Cancer Inst.* **34**, 43–51.

Eaton, R. P., and Steinberg, D. (1961). Effects of medium fatty acid concentration, epinephrine and glucose on palmitate-1-^{14}C oxidation and incorporation into neutral lipids by skeletal muscle *in vitro* *J. Lipid Res.* **2**, 376 -382.

Eaton, R. P., Berman, M., and Steinberg, D. (1969). Kinetic studies of plasma free fatty acid and triglyceride metabolism in man. *J. Clin. Invest.* **48**, 1560–1579.

Elsbach, P. (1964). Comparison of uptake of palmitic, stearic, oleic and linoleic acid by polymorphonuclear leukocytes. *Biochim. Biophys. Acta* **84**, 8–17.

Elsbach, P. (1965). Uptake of fat by phagocytic cells. An examination of the role of phagocytosis. II. Rabbit alveolar macrophages. *Biochim. Biophys. Acta* **98**, 420–431.

Elsbach, P., and Farrow, S. (1969). Cellular triglyceride as a source of fatty acid for lecithin synthesis during phagocytosis. *Biochim. Biophys. Acta* **176**, 438–441.

Engelberg, H. (1966). The effect of a heparin antagonist on fasting serum triglycerides in man. *J. Atheroscler. Res.* **6**, 240–246.

Erbland, J. F., and Marinetti, G. V. (1965). The metabolism of lysolecithin in rat-liver particulate systems. *Biochim. Biophys. Acta* **106**, 139–144.

Evans, V. J., Bryant, J. C., Kerr, H. A., and Schilling, E. L. (1965). Chemically defined media for cultivation of long-term cell strains from four mammalian species. *Exp. Cell Res.* **36**, 439–474.

Evans, W. H., and Mueller, P. S. (1963). Effects of palmitate on the metabolism of leukocytes from guinea pig exudate. *J. Lipid Res.* **4**, 39–45.

Fallon, H. J., and Lamb, R. G. (1968). Acylation of sn-glycerol-3-phosphate by cell fractions of rat liver. *J. Lipid Res.* **9**, 652–660.

Fillerup, D. L., Migliori, J. C., and Mead, J. F. (1958). The uptake of lipoproteins by ascites tumor cells. The fatty acid-albumin complex. *J. Biol. Chem.* **233**, 98–101.

Fisher, W. R. (1970). The characterization and occurrence of an S_f 20 serum lipoprotein. *J. Biol. Chem.* **245**, 877–884.

Fredrickson, D. S., and Gordon, R. S., Jr. (1958). The metabolism of albumin-bound C^{14}-labeled unesterified fatty acids in normal human subjects. *J. Clin. Invest.* **37**, 1504–1515.

Fredrickson, D. S., Levy, R. I., and Lees, R. S. (1967). Fat transport in lipoproteins—an integrated approach to mechanisms and disorders. *N. Engl. J. Med.* **276**, 32–44.

Fredrickson, D. S., Levy, R. I., and Lindgren, F. T. (1968). A comparison of heritable abnormal lipoprotein patterns as defined by two different techniques. *J. Clin. Invest.* **47**, 2446–2457.

Fritz, I. B. (1963). Long-chain carnitine acyltransferase and the role of acylcarnitine derivatives in the catalytic increase of fatty acid oxidation induced by carnitine. *J. Lipid Res.* **4**, 279–288.

Fritz, I. B., and Marquis, N. R. (1965). The role of acylcarnitine esters and carnitine palmityltransferase in the transport of fatty acyl groups across mitochondrial membranes. *Proc. Nat. Acad. Sci. U.S.* **54**, 1226–1233.

Galzigna, L., Rossi, C. R., Sartorelli, L., and Gibson, D. M. (1967). A guanosine triphosphate dependent acyl coenzyme A synthetase from rat liver mitochondria. *J. Biol. Chem.* **242**, 2111–2115.

Gerschenson, L. E., Mead, J. F., Harary, I., and Haggerty, D. F., Jr. (1967a). Studies on the effects of essential fatty acids on growth rate, fatty acid composition, oxidative phosphorylation and respiratory control of HeLa cells in culture. *Biochim. Biophys. Acta* **131**, 42–49.

Gerschenson, L. E., Harary, I., and Mead, J. F. (1967b). Studies *in vitro* of single beating rat-heart cells. X. The effect of linoleic and palmitic acids on beating and mitochondrial phosphorylation. *Biochim. Biophys. Acta* **131**, 50–58.

Geyer, R. P. (1967). Uptake and retention of fatty acids by tissue culture cells. *Lipid Metab. Tissue Cult. Cells, Symp., 1966* pp. 33–47.

Geyer, R. P., Bennett, A., and Rohr, A. (1961). Fatty acids of the triglycerides and phospholipids of HeLa cells and strain L fibroblasts. *J. Lipid Res.* **3**, 80–83.

Goodman, D. S. (1958a). The interaction of human erythrocytes with sodium palmitate. *J. Clin. Invest.* **37**, 1729–1735.

Goodman, D. S. (1958b). The interaction of human serum albumin with long-chain fatty acid anions. *J. Amer. Chem. Soc.* **80**, 3892–3898.

Greten, H., Levy, R. I., Fales, H., and Fredrickson, D. S. (1970). Hydrolysis of diglyceride and glyceryl monoester diethers with "lipoprotein lipase." *Biochim. Biophys. Acta* **210**, 39–45.

Guder, W., Weiss, L., and Wieland, O. (1969). Triglyceride breakdown in rat liver. The demonstration of three different lipases. *Biochim. Biophys. Acta* **187**, 173–185.

Gustafson, A., Alaupovic, P., and Furman, R. H. (1966). Studies of the composition and structure of serum lipoproteins. Separation and characterization of phospholipid-protein residues obtained by partial delipidization of very low density lipoproteins of human serum. *Biochemistry* **5**, 632–640.

Haddock, B. A., Yates, D. W., and Garland, P. B. (1970). The localization of some coenzyme A dependent enzymes in rat liver mitochondria. *Biochem. J.* **119**, 565–573.

Hajra, A. K. (1968). Biosynthesis of phosphatidic acid from dihydroxyacetone phosphate. *Biochem. Biophys. Res. Commun.* **33**, 929–935.

Hajra, A. K. (1969). Biosynthesis of alkyl-ether containing lipid from dihydroxyacetone phosphate. *Biochem. Biophys. Res. Commun.* **37**, 486–492.

Hajra, A. K. (1970). Acyl dihydroxyacetone phosphate: Precursor of alkyl ethers. *Biochem. Biophys. Res. Commun.* **39**, 1037–1044.

Ham, R. G. (1963). Albumin replacement by fatty acids in clonal growth of mammalian cells. *Science* **140**, 802–803.

Harary, I., Gerschenson, L. E., Haggerty, D. F., Jr., Desmond, W., and Mead, J. F. (1967). Fatty acid metabolism and function in cultured heart and HeLa cells. *Lipid Metabl. Tissue Cult. Cells, Symp., 1966* pp. 17–31.

Havel, R. J., Eder, H. A., and Bragdon, J. H. (1955). The distribution and chemical composition of ultracentrifugally separated lipoproteins in human serum. *J. Clin. Invest.* **34**, 1345–1353.

Hayase, K., and Tappel, A. L. (1970). Specificity and other properties of lysosomal lipase of rat liver. *J. Biol. Chem.* **245**, 169–175.

Hepp, D., Prusse, E., Weiss, H., and Wieland, O. (1966). Essigaure als end Produkt des Aeroben Krebstoffwechsels. *Biochem. Z.* **344**, 87–102.

Higgins, J. A. (1967). Forces involved in chylomicron binding by isolated cells of rat liver. *J. Lipid Res.* **8**, 636–641.

Holloway, P. W., and Wakil, S. J. (1970). Requirement for reduced diphosphopyridine nucleotide-cytochrome b_5 reductase in stearyl coenzyme A desaturation. *J. Biol. Chem.* **245**, 1862–1865.

Howard, B. V., and Kritchevsky, D. (1969). The source of cellular lipid in the human diploid cell strain WI-38. *Biochim. Biophys. Acta* **187**, 293–401.

Howard, B. V., and Kritchevsky, D. (1970). Free fatty acids in cultured cells. *Lipids* **5**, 49–55.

Jones, A. L., Ruderman, N. B., and Herrera, M. G. (1966). An electron microscopic study of lipoprotein production and release by the isolated perfused rat liver. *Proc. Soc. Exp. Biol. Med.* **123**, 4–9.

Jones, P. D., Holloway, P. W., Peluffo, R. O., and Wakil, S. J. (1969). A requirement for lipids by the microsomal stearyl coenzyme A desaturase. *J. Biol. Chem.* **244**, 744–754.

Jorgensen, E. A. (1961). Essential fatty acids. *Physiol. Rev.* **41**, 1–51.

Kennedy, E. P., Borkenhagen, L. F., and Smith, S. W. (1959). Possible metabolic functions of deoxycytidine diphosphate choline and deoxycytidine diphosphate ethanolamine. *J. Biol. Chem.* **234**, 1998–2000.

Kinsella, J. E. (1970). Biosynthesis of lipids from [2-^{14}C]acetate and D(-)-β-hydroxy[1,3-^{14}C] butyrate by mammary cells from bovine and rat. *Biochim. Biophys. Acta* **210**, 28–38.

Koga, S., Horowitz, D. L., and Scanu, A. M. (1969). Isolation and properties of lipoproteins from normal rat serum. *J. Lipid Res.* **10**, 577–588.

Kornberg, A., and Pricer, W. E., Jr. (1953). Enzymatic synthesis of the coenzyme A derivatives of long-chain fatty acids. *J. Biol. Chem.* **204**, 329–343.

Kuhl, W. E., and Spector, A. A. (1970). Uptake of long-chain fatty acid methyl esters by mammalian cells. *J. Lipid Res.* **11**, 458–465.

Landriscina, C., Gnoni, G. V., and Quagliariello, E. (1970). Mechanisms of fatty acid synthesis in rat-liver microsomes. *Biochim. Biophys. Acta* **202**, 405–415.

Lands, W. E. M. (1960). Metabolism of glycerolipids. II. The enzymatic acylation of lysolecithin. *J. Biol. Chem.* **235**, 2233–2237.

Lees, R. S. (1970). Immunoassay of plasma low-density lipoproteins. *Science* **169**, 493–495.

Lees, R. S., and Hatch, F. T. (1963). Sharper separation of lipoprotein species by paper electrophoresis in albumin-containing buffer. *J. Lab. Clin. Med.* **61**, 518–528.

Lengle, E., and Smith, J. L. (1970). Lipid metabolism of sarcoma 180 cells cultured on lipid deficient medium. *Fed. Proc., Fed. Amer. Soc. Exp. Biol.* **29**, 1018.

Levis, G. M., and Mead, J. F. (1964). An α-hydroxy acid decarboxylase in brain microsomes. *J. Biol. Chem.* **239**, 77–80.

Levy, R. I., and Fredrickson, D. S. (1965). Heterogeneity of plasma high density lipoproteins. *J. Clin. Invest.* **44**, 426–441.

Levy, R. I., Lees, R. S., and Fredrickson, D. S. (1966). The nature of pre-beta (very low density) lipoproteins. *J. Clin. Invest.* **45**, 63–77.

Lutton, C. E., and Tsaltas, T. T. (1965). Plasma lipid variations and fatty acid composition in normal male and female adult rabbits. *Proc. Soc. Exp. Biol. Med.* **118**, 1048–1051.

Mackenzie, C. G., Mackenzie, J. B., and Reiss, O. K. (1967a). Regulation of cell lipid metabolism and accumulation. V. Quantitative and structural aspects of triglyceride accumulation caused by lipogenic substances. *Lipid Metab. Tissue Cult. Cells, Symp. 1966* pp. 63–83.

Mackenzie, C. G., Mackenzie, J. B., and Reiss, O. K. (1967b). Increase in cell lipid and cytoplasmic particles in mammalian cells cultured at reduced pH. *J. Lipid Res.* **8**, 642–645.

Mahley, R. W., Bersot, T. P., LeQuire, V. S., Levy, R. I., Windmueller, H. G., and Brown, W. V. (1970). Identity of very low density lipoprotein apoproteins of plasma and liver Golgi apparatus. *Science* **168**, 380–382.

Majerus, P. W., and Lastra, R. R. (1967). Fatty acid biosynthesis in human leukocytes. *J. Clin. Invest.* **46**, 1596–1602.

Majerus, P. W., and Vagelos, P. R. (1967). Fatty acid biosynthesis and the role of acyl carrier protein. *Advan. Lipid Res.* **5**, 1–35.

Majerus, P. W., Alberts, A. W., and Vagelos, P. R. (1964). The acyl carrier protein of fatty acid synthesis: Purification physical properties, and substrate binding site. *Proc. Nat. Acad. Sci. U.S.* **51**, 1231–1238.

Marcus, D. M., and Cass, L. E. (1969). Glycosphingolipids with Lewis blood group activity: Uptake by human erythrocytes. *Science* **164**, 533–535.

Mead, J. F., and Levis, G. M. (1963). A one carbon degradation of the long-chain fatty acids of brain sphingolipids. *J. Biol. Chem.* **238**, 1634–1636.

Medes, G., and Weinhouse, S. (1958). Metabolism of neoplastic tissue. XIII. Substrate competition in fatty acid oxidation in ascites tumor cells. *Cancer Res.* **18**, 352–359.

Medes, G., Thomas, A. J., and Weinhouse, S. (1960). Metabolism of neoplastic tissue. XV. Oxidation of exogenous fatty acids in Lettre-Ehrlich ascites tumor cells. *J. Nat. Cancer Inst.* **24**, 1–12.

Mehadevan, V., and Zieve, L. (1969). Determination of volatile free fatty acids of human blood. *J. Lipid Res.* **10**, 338–341.

Mize, C. E., Avigan, J., Steinberg, D., Pittman, R. C., Fales, H. M., and Milne, G. W. A. (1969). A major pathway for the mammalian oxidative degradation of phytanic acid. *Biochim. Biophys. Acta* **176**, 720–739.

Mooney, L. A., and Barron, E. J. (1970). Cofactor requirements and general characteristics of soluble fatty acid elongating system from mitochondria. *Biochemistry*
9, 2138–2143.

Morell, P., and Radin, N. S. (1969). Synthesis of cerebroside by brain from cytidine
disphosphate galactose and ceramide containing hydrox fatty acid. *Biochemistry*.
8, 506–512.

Morell, P., and Radin, N. S. (1970). Specificity in ceramide biosynthesis from
long-chain bases and various fatty acyl coenzyme A's by brain microsomes.
J. Biol. Chem. 245, 342–350.

Moskowitz, M. S. (1967). Fatty acid-induced steatosis in monolayer cell cultures.
Lipid Metab. Tissue Cult. Cells, Symp., 1966 pp. 49–62.

Munday, K. A., Parsons, B. J., and York, D. A. (1969). The transport of palmitic
acid across intestinal brush border membranes. *Biochem. J.* 114, 66P.

Nachbaur, J., and Vignais, P. M. (1968). Localization of phospholipase A₂ in outer
membrane of mitochondria. *Biochem. Biophys. Res. Commun.* 33, 315–320.

Nichols, A. V. (1969). Functions and inter-relationships of different classes of plasma
lipoproteins. *Proc. Nat. Acad. Sci. U.S.* 64, 1128–1137.

Norum, K. R., and Bremer, J. (1967). The localization of acyl coenzyme A-carnitine
acyltransferase in rat liver cells. *J. Biol. Chem.* 242, 407–411.

Ockner, R. K., Hughes, F. B., and Isselbacher, K. J. (1969). Very low density
lipoproteins in intestinal lymph: Origin, composition and role in lipid transport
in the fasting state. *J. Clin. Invest.* 48, 2079–2088.

Oliveira, M. M., and Vaughan, M. (1964). Incorporation of fatty acids into phospholipids of erythrocyte membranes. *J. Lipid Res.* 5, 156–162.

Ontko, J. A. (1967). Chylomicron, free fatty acid and ketone body metabolism
of isolated liver cells and liver homogenates. *Biochim. Biophys. Acta* 137, 13–22.

Pande, S. V., and Mead, J. F. (1968). Long-chain fatty acid activation in subcellular
preparations of rat liver. *J. Biol. Chem.* 243, 352–361.

Pasternak, C. A., and Bergeron, J. J. M. (1970). Turnover of mammalian phospholipids. Stable and unstable components in neoplastic mast cells. *Biochem. J.*
119, 473–480.

Peterson, J. A., and Rubin, H. (1969). The exchange of phospholipids between
cultured chick embryo fibroblasts and their growth medium. *Exp. Cell Res.*
58, 365–378.

Peterson, J. A., and Rubin, H. (1970). The exchange of phospholipids between
cultured chick embryo fibroblasts as observed by autoradiography. *Exp. Cell
Res.* 60, 383–392.

Pittman, J. G., and Martin, D. B. (1966). Fatty acid biosynthesis in human erythrocytes. Evidence in mature erythrocytes for an incomplete long chain fatty
acid synthesizing system. *J. Clin. Invest.* 45, 165–172.

Preiss, B., and Bloch, K. (1964). ω-Oxidation of long-chain fatty acids in rat liver.
J. Biol. Chem. 239, 85–88.

Puppione, D. L., Forte, G. M., Nichols, A. V., and Strisower, E. H. (1970). Partial
characterization of serum lipoproteins in the density interval 1.04–1.06 g/ml.
Biochim. Biophys. Acta 202, 392–395.

Rahman, Y. E. (1970). Evidence of a membrane-bound phospholipase A in rat
liver lysosomes. *Biochem. Biophys. Res. Commun.* 38, 670–677.

Reed, C. F. (1968). Phospholipid exchange between plasma and erythrocytes in
man and the dog. *J. Clin. Invest.* 47, 749–760.

Reshef, L., and Shapiro, B. (1966). Depletion and regeneration of fatty acid-absorbing capacity of adipose tissue and liver particles. *Biochim. Biophys. Acta* **125**, 456–464.

Rodbell, M. (1965). Modulation of lipolysis in adipose tissue by fatty acid concentration in fat cells. *Ann. N.Y. Acad. Sci.* **131**, 302–313.

Roheim, P. S., Gidez, L. I., and Eder, H. A. (1966). Extrahepatic synthesis of lipoproteins of plasma and chyle: Role of the intestine. *J. Clin. Invest.* **45**, 297–300.

Rutstein, D. D., Castelli, W. P., Sullivan, J. C., Newell, J. M., and Nickerson, R. J. (1964). Effects of fat and carbohydrate ingestion in human beings on serum lipids and intracellular lipid deposition in tissue culture. *N. Engl. J. Med.* **271**, 1–11.

Rutstein, D. D., Castelli, W. P., and Nickerson, R. J. (1967). Effect of carbohydrate ingestion in human on intracellular lipid deposition in tissue culture. *Amer. J. Clin. Nutr.* **20**, 98–107.

Ryder, E., Gregolin, C., Chang, H. C., and Lane, M. D. (1967). Liver acetyl CoA carboxylase. Insight into the mechanism of activation by tricarboxylic acids and acetyl CoA. *Proc. Nat. Acad. Sci. U.S.* **57**, 1455–1462.

Sakagami, T., and Zilversmit, D. B. (1961). Separation of dog serum lipoproteins by ultracentrifugation, dextran sulfate precipitation and paper electrophoresis. *J. Lipid Res.* **2**, 271–277.

Savchuck, W. B., Lockhard, W. L., and Long, H. W. (1965). Proliferation of cultured liver cells in the presence of lysine and arginine salts of fatty acids. *Exp. Cell Res.* **37**, 169–174.

Scanu, A. (1967). Binding of human serum high density lipoprotein apoprotein with aqueous dispersions of phospholipids. *J. Biol. Chem.* **242**, 711–719.

Scanu, A., Cump, E., Toth, J., Koga, S., Stiller, E., and Albers, L. (1970). Degradation and reassembly of a human serum high-density lipoprotein. Evidence for differences in lipid affinity among three classes of polypeptide chains. *Biochemistry* **9**, 1327–1335.

Schonfeld, G. (1968). Uptake and esterification of palmitate by rat diaphragm in vitro. *J. Lipid Res.* **9**, 453–459.

Schumaker, V. N., and Adams, G. H. (1969). Circulating lipoproteins. *Annu. Rev. Biochem.* **38**, 113–136.

Senior, J. R., and Isselbacher, K. J. (1962). Direct esterification of monoglycerides with palmityl coenzyme A by intestinal epithelial subcellular fractions. *J. Biol. Chem.* **237**, 1454–1459.

Shohet, S. B., and Nathan, D. G. (1970). Incorporation of phosphatide precursors from serum into erythrocytes. *Biochim. Biophys. Acta* **202**, 202–205.

Shohet, S. B., Nathan, D. G., and Karnovsky, M. L. (1968). Stages in the incorporation of fatty acids into red blood cells. *J. Clin. Invest.* **47**, 1096–1108.

Skrede, S., and Bremer, J. (1970). The compartmentation of CoA and fatty acid activating enzymes in rat liver mitochondria. *Eur. J. Biochem.* **14**, 465–472.

Smith, S. C., Strout, R. G., Dunlop, W. R., and Smith, E. C. (1965). Fatty acid composition of cultured aortic cells from white carneau and show racer pigeons. *J. Atheroscler. Res.* **5**, 379–387.

Snyder, F., and Wood, R. (1968). The occurrence and metabolism of alkyl an alk-1-enyl ethers of glycerol in transplantable rat and mouse tumors. *Cancer Res.* **28**, 972–978.

Sodhi, H. S., and Gould, R. G. (1967). Combination of delipidized high density lipoprotein with lipids. *J. Biol. Chem.* **242**, 1205–1210.

Spector, A. A. (1967a). The importance of free fatty acid in tumor nutrition. *Cancer Res.* **27**, 1580–1586.

Spector, A. A. (1967b). Effect of carnitine on free fatty acid utilization in Ehrlich ascites tumor cells. *Arch. Biochem. Biophys.* **122**, 55–61.

Spector, A. A. (1968). The transport and utilization of free fatty acid. *Ann. N.Y. Acad. Sci.* **149**, 768–783.

Spector, A. A. (1969). Influence of pH of the medium on free fatty acid utilization by isolated mammalian cells. *J. Lipid Res.* **10**, 207–215.

Spector, A. A. (1970). Free fatty acid utilization by mammalian cell suspensions. Comparison between individual fatty acids and fatty acid mixtures. *Biochim. Biophys. Acta* **218**, 36–43.

Spector, A. A. (1971). Metabolism of free fatty acids. *Progr. Biochem. Pharmacol.* **6**, 130–176.

Spector, A. A., and Fletcher, J. E. (1970). Binding of long-chain fatty acids to β-lactoglobulin. *Lipids* **5**, 403–411.

Spector, A. A., and Hoak, J. C. (1969). An improved method for the addition of long-chain free fatty acid to protein solutions. *Anal. Biochem.* **32**, 297–302.

Spector, A. A., and Steinberg, D. (1965). The utilization of unesterified palmitate by Ehrlich ascites tumor cells. *J. Biol. Chem.* **240**, 3747–3753.

Spector, A. A., and Steinberg, D. (1966a). Release of free fatty acid from Ehrlich ascites tumor cells. *J. Lipid Res.* **7**, 649–656.

Spector, A. A., and Steinberg, D. (1966b). Relationship between fatty acid and glucose utilization in Ehrlich ascites tumor cells. *J. Lipid Res.* **7**, 657–663.

Spector, A. A., and Steinberg, D. (1967a). The effect of fatty acid structure on utilization by Ehrlich ascites tumor cells. *Cancer Res.* **27**, 1587–1594.

Spector, A. A., and Steinberg, D. (1967b). Turnover and utilization of esterified fatty acids in Ehrlich ascites tumor cells. *J. Biol. Chem.* **242**, 3057–3062.

Spector, A. A., John, K., and Fletcher, J. E. (1969). Binding of long-chain fatty acids to bovine serum albumin. *J. Lipid Res.* **10**, 56–67.

Spector, A. A., Steinberg, D., and Tanaka, A. (1965). Uptake of free fatty acid by Ehrlich ascites tumor cells. *J. Biol. Chem.* **240**, 1032–1041.

Spector, A. A., Hoak, J. C., Warner, E. D., and Fry, G. L. (1970). Utilization of long-chain free fatty acids by human platelets. *J. Clin. Invest.* **49**, 1489–1496.

Spencer, A. F., and Lowenstein, J. M. (1962). The supply of precursors for the synthesis of fatty acids. *J. Biol. Chem.* **237**, 3640–3648.

Srere, P. A. (1959). The citrate cleavage enzyme. I. Distribution and purification. *J. Biol. Chem.* **234**, 2544–2547.

Sribney, M., and Kennedy, E. P. (1958). The enzymatic synthesis of sphingomyelin *J. Biol. Chem.* **233**, 1315–1322.

Stein, O., and Stein, Y. (1963). Metabolism of fatty acids in the isolated perfused rat heart. *Biochim. Biophys. Acta* **70**, 517–530.

Stein, O., and Stein, Y. (1967). Utilization of lysolecithin by Landshutz ascites tumor *in vivo* and *in vitro*. *Biochim. Biophys. Acta* **137**, 232–239.

Stokes, R. P., Jacobsson, A., and Walton, K. W. (1967). The isolation of low-density (β) lipoprotein from small volumes of human serum. *J. Atheroscler. Res.* **7**, 187–196.

Strauss, E. W. (1966). Electron microscopic study of intestinal fat absorption in vitro from mixed micelles containing linolenic acid, monoolein and bile salt. *J. Lipid Res.* **7**, 307–323.

Trout, D. L., Estes, E. H., Jr., and Friedberg, S. J. (1960). Titration of free fatty acids of plasma. A study of current methods and a new modification. *J. Lipid Res.* 1, 199–202.

Van Tol, A. (1969). On fatty acid activation in rat liver mitochondria. *Biochim. Biophys. Acta* 176, 414–416.

Vaughan, M., Berger, J. E., and Steinberg, D. (1964a). Hormone sensitive lipase and monoglyceride lipase activities in adipose tissue. *J. Biol. Chem.* 239, 401–409.

Vaughan, M., Steinberg, D., and Pittman, R. (1964b). On the interpretation of studies measuring uptake and esterification of [1-^{14}C]palmitic acid by rat adipose tissue *in vitro*. *Biochim. Biophys. Acta* 84, 154–166.

Vogel, W. C., and Bierman, E. L. (1968). Evidence for *in vivo* activity of post-heparin plasma lecithinase in man. *Proc. Soc. Exp. Biol. Med.* 127, 77–80.

Volk, M. E., Millington, R. H., and Weinhouse, S. (1952). Oxidation of endogenous fatty acids of rat tissues *in vitro*. *J. Biol. Chem.* 195, 493–501.

Waite, M., and Van Deenen, L. L. M. (1967). Hydrolysis of phospholipids and glycerides by rat-liver preparations. *Biochim. Biophys. Acta* 137, 498–517.

Wakil, S. J., Pugh, E. L., and Sauer, F. (1964). The mechanism of fatty acid synthesis. *Proc. Nat. Acad. Sci. U.S.* 52, 106–114.

Waku, K., and Lands, W. E. M. (1968). Control of lecithin biosynthesis in erythrocyte membranes. *J. Lipid Res.* 9, 12–18.

Wallach, D. F. H., Soderberg, J., and Bricker, L. (1960). The phospholipids of Ehrlich ascites carcinoma cells. Composition and intracellular distribution. *Cancer Res.* 20, 397–402.

Warmsley, A. M. H., and Pasternak, C. A. (1970). The use of conventional and zonal centrifugation to study the life cycle of mammalian cells. Phospholipid and macro-molecular synthesis in neoplastic mast cells. *Biochem. J.* 119, 498–499.

Weinstein, D. B., Marsh, J. B., Glick, M. C., and Warren, L. (1969). Membranes of animal cells. IV. Lipids of the L cell and its surface membrane. *J. Biol. Chem.* 244, 4103–4111.

Whereat, A. F., Hull, F. E., Orishimo, M. W., and Rabinowitz, J. L. (1967). The role of succinate in the regulation of fatty acid synthesis by heart mitochondria. *J. Biol. Chem.* 242, 4013–4022.

Windmueller, H. G., and Levy, R. I. (1968). Production of β-lipoprotein by intestine in the rat. *J. Biol. Chem.* 243, 4878–4884.

Wittels, B., and Hochstein, P. (1967). The identification of carnitine palmityltransferase in erythrocyte membranes. *J. Biol. Chem.* 242, 126–130.

Wood, R., and Healy, K. (1970). Tumor Lipids. Biosynthesis of plasmalogens. *J. Biol. Chem.* 245, 2640–2648.

Wykle, R. L., and Snyder, F. (1970). Biosynthesis of an 0-Alkyl analogue of phosphatidic acid and 0-alkylglycerols via 0-alkylketone intermediates by microsomal enzymes of Ehrlich ascites tumor. *J. Biol. Chem.* 245, 3047–3058.

Yates, D. W., and Garland, P. B. (1970). Carnitine palmitoyltransferase activities (EC 2.3.1.-) of rat liver mitochondria. *Biochem. J.* 119, 547–552.

9

CELLULAR STEROL METABOLISM

George H. Rothblat

I. Introduction

Tissue culture cells provide an excellent experimental model for the study of sterol metabolism. A variety of cells is available that can be studied as homogeneous populations. Experiments with cells in culture can be conducted under conditions which would be impossible to obtain in more complex systems. The requirement by cells for sterol, coupled with the ability to incorporate, release, and synthesize sterol, affords an opportunity to study these phenomena as integrated reactions designed to ensure a critical intracellular level of a necessary metabolite.

Although the use of cell cultures for studies of sterol metabolism offers many advantages over studies *in vivo*, there are a number of factors which must be considered in the interpretation of data obtained from tissue culture cells. Most investigations have been conducted on "dedifferentiated" cells which do not carry out many of the specialized functions found in differentiated cells. Moreover, cells in culture are rapidly proliferating, and until recently, data were obtained mainly from heteroploid cells of malignant origin. Another prime consideration is that all studies on cell sterol metabolism have been conducted in closed systems (for further discussion, see Chapter 2, Volume 2) in which cell mass is increasing at the expense of exogenous sterol and sterol precursors. Thus, in such closed systems, the rates of sterol influx, efflux, and synthesis will presumably change during the growth cycle of the cells.

A major problem encountered by the investigator attempting to perform lipid studies in cell systems has been the difficulty of presenting exogenous lipids to the cells. All the techniques utilized have had limitations. Methods employing an inert carrier such as Celite (Avigan, 1959) or the addition of labeled sterol of high specific activity to serum in solvents (Rose, 1968) assume that the added sterol will be equally distributed among all the serum lipoproteins. This assumption may not always be valid, and disproportionate labeling of a rapidly exchangeable pool could greatly influence the results obtained from various exchange experiments. Other experiments in which sterols are added to various protein carriers, such as albumin or delipidized serum proteins (Rothblat et al., 1968a; Rothblat and Buchko, 1971) must be interpreted with some caution, since the techniques available probably do not permit the reassembly of serum lipoproteins in a completely natural form.

Perhaps the most physiological condition in which sterol can be presented to cells is through the use of serum labeled in vivo. After labeled sterol has been incorporated in vivo into serum lipoproteins, the preferential utilization of added labeled compound by the cells is of no concern. Serum labeled in vivo, however, has a number of disadvantages. The specific activity of serum labeled by cholesterol feeding is low, particularly when compared to the specific activities which can be obtained by in vitro labeling techniques. A second and more serious objection is that serum labeled in vivo contains both radioactive free and esterified sterol. The use of this serum which contains two compounds labeled with the same isotope and capable of being interconverted by either serum or cell enzymes complicates any experiments designed to study sterol metabolism. The difficulties are even more pronounced in the case of esterified sterol. At present there is no satisfactory method for solubilizing cholesteryl ester in a biologically acceptable form other than through in vivo feeding techniques. This restriction is reflected in the lack of information concerning cellular cholesteryl ester metabolism.

II. Cellular Sterol Content and Intracellular Location

A. Free and Ester Sterol Content

Table I lists the sterol content of some cell lines and strains grown under a variety of conditions. Data originally presented in terms of cellular dry weight were converted to protein, assuming 50% of the cellular dry weight to be protein. The values presented in Table I compare favorably with similar data reported for other cell lines (Bole and Castor, 1964; J. M. Bailey et al., 1959). The total amount of cellular sterol ranges from 15 to 40 μg/mg of protein and comprises about 10% of the total cell lipids. With few exceptions the cells studied contain primarily unesterified sterol with cholesteryl ester/free cholesterol ratios less than 1.0. Three cell lines have been shown to have a cholesteryl ester/cholesterol ratio greater than 1.0: the L5178Y line of mouse lymphoblasts (Rothblat et al., 1967), a line derived from human mesothelioma carcinoma (Carruthers, 1966), and the MAF cell line when grown on lipemic rabbit serum (P. J. Bailey and Keller, 1971). In the case of the L5178Y and MAF cells, this ratio is dependent on the type of serum in which the cells are grown. With L5178Y cells, changing the type of serum markedly influenced the esterified sterol content without significantly altering the content of free cholesterol (Rothblat et al., 1967). The addition of lipemic rabbit serum to the culture medium stimulated the synthesis and accumulation of cholesteryl esters in MAF cells (P. J. Bailey and Keller, 1971). The metabolism of cholesteryl esters is discussed in Section VI.

The function and metabolic fate of sterol in tissue culture cell systems have not been studied extensively. In the few cell lines which have been studied the cholesterol incorporated by cells from the culture medium remains unchanged. No extensive catabolism of cholesterol occurs in MBIII (J. M. Bailey, 1967) or L5178Y mouse lymphoblasts (Rothblat, et al., 1966) or in L cell mouse fibroblasts (Rothblat 1968). In addition, when labeled sterols such as cholestanol, β-sitosterol, and desmosterol were incorporated by L cells, no metabolic products arising from the metabolism of these sterols could be detected (Rothblat and Burns, 1971). However, L cells and L5178Y cells are capable of esterifying a small percentage of the cholesterol or other sterol taken up from the growth medium (Rothblat and Kritchevsky, 1967; Rothblat and Burns, 1971). In contrast to sterols, steroids such as cholestanone and Δ⁴-cholestenone are, according to preliminary evidence, actively metabolized by L cells (Rothblat and Buchko, 1971). The end products

TABLE I.

CHOLESTEROL CONTENT OF TISSUE CULTURE CELLS

Cell	Origin	Level of Serum in Growth Med. (%)	Cholesterol Content (µg/mg Cell Protein)				References
			Free (C)	Ester (CE)	CE/C	Total Chol. as % of Total Lipid	
HeLa	Human epitheloid carcinoma	20 horse	15.0	5.0	0.33	8.7	Mackenzie et al. (1964a)
L	Mouse connective tissue	20 rabbit	14.0	6.0	0.43	9.3	Mackenzie et al. (1964a)
		20 horse	21.0	6.0	0.29	13.9	Mackenzie et al. (1964a)
		20 rabbit	26.0	8.0	0.31	10.0	Mackenzie et al. (1964a)
MBIII	Mouse lymphosarcoma	5 fetal calf	22.0	3.2	0.15		Rothblat et al. (1967)
		20 human placental cord serum	46.0	0.6	0.01		J. M. Bailey (1968)
L5178Y	Mouse lymphoblast	7.5 horse	12.0	20.0	1.7	13.5	Rothblat et al. (1967)
		7.5 fetal calf	13.0	10.0	0.8		Rothblat et al. (1967)
EB2	Human lymphoblast	7.5 calf	12.0	31.0	2.7		Rothblat et al. (1967)
Chang liver	Human liver	10 fetal calf	30.0	7.0	0.22	12.0	Rothblat et al. (1967)
		20 horse	19.0	4.0	0.21	11.3	Mackenzie et al. (1964a)
Rat liver		20 rabbit	17.0	6.0	0.35	11.8	Mackenzie et al. (1964a)
		20 horse	18.0	4.0	0.22	11.4	Mackenzie et al. (1964a)
Rabbit liver		20 rabbit	17.0	5.0	0.29	11.3	Mackenzie et al. (1964a)
		20 horse	20.0	7.0	0.35	7.0	Mackenzie et al. (1964a)
		20 rabbit	22.0	9.0	0.41		Mackenzie et al. (1964a)
WI38	Human embryonic lung	10 calf	42.0	6.0	0.14	11.6	Howard and Kritchevsky (1969)
WI38VA13A	Human embryonic lung (SV_{40} transformed)	10 calf	51.8	5.9	0.11	16.0	Howard and Kritchevsky (1969)
MDBK	Bovine kidney	10 calf	29.4	3.4	0.11	13.7	Klenk and Choppin (1970)
MK	Primary monkey kidney	10 calf	15.7	0.5	0.03	6.4	Klenk and Choppin (1969)
BHK21-F	Baby hamster kidney	10 calf	20.3	2.7	0.14	5.9	Klenk and Choppin (1969)
HaK	Adult hamster kidney	10 calf	16.2	1.7	0.11	8.3	Klenk and Choppin (1969)
MAF	Human embryonic skin and muscle	10 rabbit	56	17	0.30		P. J. Bailey and Keller (1971)
MAF		10 rabbit (lipemic)	210	330	1.57		P. J. Bailey and Keller (1971)

of such reactions have not been identified. Many more cells, however, both primary and established, must be carefully studied before we can generalize that all tissue culture cells do not actively metabolize sterols.

B. Intracellular Location of Sterol

The intracellular location of sterol is of considerable interest but has not received thorough investigation. One of the major questions remaining to be resolved is the extent to which both free and esterified sterol are maintained within the cell, either associated with cellular membranes or stored in lipid vacuoles. Mackenzie et al. (1964b), in a study of five cell lines, concluded that cholesterol is confined almost entirely to unit membrane fractions. These investigators (Mackenzie et al., 1966, 1967a,b) observed that the lipid-rich particles found in many cells contained mainly triglycerides and had only low levels of sterol (see Chapter 8, Volume 1). In L cells, approximately 95% of the cellular sterol can be sedimented in conjunction with membrane structures (Rothblat, 1968). Although all cellular membranes probably contain sterol, the surface membrane contains more than other membrane fractions. In whole L cells, sterol comprises 10% of the total lipids, whereas in purified surface membranes this value rises to 20% (Weinstein and Marsh, 1969). In HeLa cells, the plasma membrane has 154 μg of cholesterol per 1 mg of protein, while the value for the smooth internal membranes is 43 μg/mg of protein (Bosmann et al., 1968). The molar ratios of cholesterol/phospholipid for some representative cell lines are shown in Table II.

Although cells cultured under normal conditions may not accumulate sterol in vacuoles or lipid particles, cells can be forced to accumulate excess unesterified sterol which probably is not membrane-associated. Microscopic examination of L cells grown in medium supplemented with free cholesterol (J. M. Bailey, 1961, 1967) showed that these cells contained cholesterol microcrystals. The addition of excess free cholesterol to the culture was also shown to result in the formation of Liebermann-Burchard positive granules in human aorta cells in culture (Rutstein et al., 1958) and in lipid deposition and necrosis in both human and canine aorta cells (Curtis and Galvin, 1963). By the combined use of autoradiography and electron microscopy, Robertson (1967) demonstrated the incorporation of exogenous labeled free cholesterol into lipid vacuoles in human arterial cells in culture. Robertson (1962, 1967) also showed the progressive movement of labeled free cholesterol from cell membrane "ghosts" into the "microsomal-cell sap."

The intracellular distribution of esterified sterol among subcellular

TABLE II.

CHOLESTEROL/PHOSPHOLIPID RATIOS FOR CELLS AND PLASMA MEMBRANES

| Cell | Molar Ratio (Cholesterol/Phospholipid) | | References |
	Whole Cells	Plasma Membrane	
L cells (mouse connective tissue)	0.26	0.69	Weinstein and Marsh (1969)
HeLa (human epitheloid carcinoma)	0.19	1.05	Bosmann *et al.* (1968)
HaK (adult hamster kidney)	0.15	0.51	Klenk and Choppin (1969)
MDBK (bovine kidney)	0.25	0.75	Klenk and Choppin (1970)
MK (primary monkey kidney)	0.15	0.81	Klenk and Choppin (1969)
BHK21-F (baby hamster kidney)	0.15	0.68	Klenk and Choppin (1969)

organelles has received even less study than has that of free cholesterol, probably because of the low cellular level of esterified sterol in most cells maintained in culture. Robertson (1968) studied human arterial cells in culture and found 65–86% of the total cholesterol isolated from cell vacuoles to be esterified, while studies on L5178Y mouse lymphoblasts have revealed that after cell disruption 82% of the total cell cholesterol ester could be recovered in a "debris" fraction consisting of membrane and nuclear fragments and other material which pelleted under low centrifugal force (Rothblat *et al.*, 1967). A difference in the intracellular distribution of free and esterified sterol in these cells was also demonstrated.

The observation that the few cell lines which have been studied do not extensively modify sterol incorporated from the medium, together with the clearly established association of unesterified sterol with cellular membranes, indicates that the free sterol in tissue culture cells functions primarily as an integral unit in the formation of functional cell membranes. If sterol is required for other than structural purposes, there is, as yet, no indication of these functions. An understanding of the function of steryl esters in tissue culture cells must await further studies on their intracellular location and synthesis.

III. Sterol Flux

Sufficient data have been obtained from studies on sterol metabolism to allow the formulation of a model cell system which explains much of the information available. This system is based on the following premises: (1) all mammalian cells in culture require unesterified sterol for the maintenance of cell integrity and for the continuation of cell growth, and (2) the sterol in these cells is not extensively changed to other products by oxidation or esterification. In such a simplified system three factors would contribute to the maintenance of the required internal cellular sterol concentration: (1) the incorporation of exogenous sterol from the medium by the cells (influx), (2) the release of cellular sterol to the culture medium (efflux), and (3) the ability of the cell to contribute to its own internal sterol pool through *de novo* synthesis. These three processes are integrated in a metabolic system designed to ensure the maintenance of an essential cellular concentration.

Two general approaches have been used to study sterol metabolism in tissue culture cells: (1) those which measure cellular sterol content in cells maintained under various growth conditions and (2) those which study the influx, efflux, and synthesis as individual steps in the cellular utilization of sterol. Although results obtained by the two techniques often complement each other, many studies demonstrate that changes in cellular influx, efflux, and synthesis do not always lead to changes in sterol content.

Under normal growth conditions exogenous sterol is obtained by cells from the serum in the culture medium. The sterol in the serum of all species is essentially cholesterol, present as unesterified (free) cholesterol and cholesteryl esters. The ratio of free to ester cholesterol in most sera is approximately 1:3 and cholesterol is solubilized in serum by association with serum lipoproteins which also carry other classes of lipid (see Chapter 8, Volume 1). Sera from different species vary in their total sterol content, as well as in the composition of the individual lipoprotein classes. Considerable individual variation can also be encountered within a single species.

The free sterol content of cells responds only to a limited degree to changes in the concentration of sterol in the growth medium. Thus, varying either the type of serum or its concentration elicits only a limited change in cellular free sterol levels (Table I).

This lack of cellular response to changes in serum is evident from experiments in which MBIII mouse lymphoblasts demonstrated no change in sterol content when grown in medium in which serum cholesterol was added in concentrations ranging from 16 to 80 mg/100 ml (J. M. Bailey, 1961, 1967). This same cell line did contain higher sterol levels when grown in medium containing rabbit serum than did similar cells maintained in medium containing human serum, even though the rabbit serum itself had a lower cholesterol content than the human serum (J. M. Bailey, 1961). Cultivation of cells in sterol-free medium may reduce the cellular level of sterol. Although L cells grown in defined medium had a sterol content similar to that of cells grown for one passage in medium containing serum (J. M. Bailey, 1964a; Anderson et al., 1969), continued cultivation of L cells in sterol-free medium supplemented with delipidized calf serum protein yielded cells with a sterol content of 12 μg/mg of cell protein, whereas similar cells grown on 5% calf serum contained 20 μg of sterol per milligram of cell protein (Rothblat, 1968). Also, both WI38 and WI38VA13A cells grown for extended periods in medium supplemented with delipidized protein (Rothblat et al., 1971) had less sterol (WI38 cells, 22 μg/mg of protein; WI38VA13A cells, 19 μg/mg of protein) than that previously reported for these cells when grown in the presence of 10% calf serum (Howard and Kritchevsky, 1969). These observations would suggest that continued cultivation of cells in serum-containing medium allows an accumulation of sterol above that which is synthesized by the cells and which is presumably the minimum amount needed for continued growth.

Although increasing the serum concentration in the medium does not generally produce an accumulation of large excesses of cellular sterol, intracellular sterol deposition can result when cells are grown under conditions in which the normal association of sterol to serum lipoproteins is modified. Thus, J. M. Bailey (1961) found that when free cholesterol was added to the growth medium as a Tween emulsion, a preferential uptake of the added sterol resulted in a marked increase in cellular sterol levels. Microscopic examination of these cells demonstrated the presence of small cholesterol microcrystals within the cells. The addition of unbound free cholesterol to the culture medium was shown to result in the formation of Liebermann-Burchard positive granules in human aorta cells (Rutstein et al., 1958) and in lipid deposition and necrosis in both human and canine aorta cells (Curtis and Galvin, 1963). Intracellular lipidosis also resulted when aorta cells (Rutstein et al., 1958) or mouse fibroblasts (Maca and Rose, 1971) were grown in medium containing only serum β-lipoproteins.

An assessment of the data available indicates that a number of processes such as influx, efflux, and synthesis must be operating in a tissue culture cell system to ensure proper maintenance of internal sterol concentrations. These mechanisms require the interaction of cells with serum lipoproteins. At least part of the balance maintained by cells with serum is due to the composition of the serum lipoproteins and the relative concentrations of the various classes of lipoproteins. A more complete understanding of the relationships between cells and serum can be derived from an examination of the principles involved in the influx, efflux, and synthesis of cellular sterol.

A. *Sterol Influx*

The observation that unbound free cholesterol was preferentially incorporated by MBIII mouse lymphoblasts led J. M. Bailey (1961) to conclude that the rate of cholesterol uptake was closely linked to the association of the sterol with other serum components. This conclusion has gained further support from sterol uptake studies conducted with L5178Y mouse lymphoblasts using delipidized serum protein as a carrier of exogenous cholesterol (Rothblat *et al.*, 1968a). These experiments demonstrated that the addition of phospholipids to an incubation medium containing delipidized protein and unesterified cholesterol could greatly reduce the incorporation of sterol over a 5-hour incubation period. Various phospholipids differed in their ability to reduce influx, with the order of the efficiency of inhibition being sphingomyelin > lecithin > phosphatidyl ethanolamine > phosphatidyl serine. The amount of cholesterol incorporated by the cells was a function of the cholesterol/phospholipid/protein ratio in the incubation medium. Of the three components, the phospholipid/sterol ratio was of prime importance. These results suggest that the various classes of lipoprotein present in serum differ in their ability to supply free sterol to cells, and that such differences are linked to the relative composition of each class of lipoprotein.

In addition to the observation that varying the ratio of individual lipoprotein components influenced free cholesterol adsorption, it was found that proportionately increasing the concentrations of all the components resulted in increased influx (Rothblat *et al.*, 1968a). In terms of native serum, this indicates that increasing serum concentration in the culture medium should result in an increased influx of sterol. Studies with L5178Y cells have shown that the influx of unesterified cholesterol increases as the concentration of serum in the medium is increased (Rothblat *et al.*, 1966).

Sterols other than cholesterol can be incorporated by cells, although the rate of incorporation is not similar among all sterols. Over a 48-hour incubation period in a delipidized protein system, L cells incorporated

β-sitosterol at one third and cholestanol at two thirds the rate of equivalent exogenous concentrations of cholesterol (Rothblat and Burns, 1971). The incorporation of both of these sterols followed the same general pattern as that of cholesterol; i.e., the rate of influx depended on the phospholipid and protein concentration in the medium.

The addition of glycerides or various free fatty acids to the cholesterol-protein complex had no marked effect on free cholesterol incorporation (Rothblat et al., 1968a). Bailey also observed that the addition of either saturated or unsaturated fatty acid had no effect on the cellular cholesterol content of MBIII cells growing in medium containing added free cholesterol (J. M. Bailey, 1961, 1967). His observations do not agree, however, with those obtained by Rutstein et al. (1958), who used human aortic cells. They demonstrated histologically that the intracellular depositon of lipid caused by adding cholesterol to the culture medium was completely inhibited by the simultaneous addition of linolenic acid. Deposition of lipid within the cell was increased by the simultaneous addition of stearic acid. In a different series of experiments with aorta cells in culture under similar experimental conditions, Curtis and Galvin (1963) demonstrated that a wide variety of compounds, including vitamins, steroids, phospholipids, and fatty acids, could reduce or eliminate lipid deposition in the cells, as shown by histological examination. However, conclusions based only on histological examination may be misleading, since this experimental method cannot distinguish between true uptake and differential mobilization of lipids within the cell. It is also possible that the differences seen in similar studies on mouse lymphoblasts and human aorta cells may reflect fundamental differences in the metabolism of various cells in culture.

In studies on arterial cells maintained in culture, Robertson (1961, 1962, 1967) demonstrated that the surface charge of the cell also plays a role in sterol incorporation. He found that the presence of cationic compounds such as protamine sulfate or methylated bovine serum albumin increased the cellular incorporation of isotopically labeled free cholesterol, whereas anionic compounds such as sodium stearate, n-acetylated bovine serum albumin, and heparin reduced free cholesterol uptake. In addition, Robertson (1967) found that the presence of ATP in the culture medium stimulated the cellular uptake of labeled free cholesterol, as did a reduction in oxygen concentration in the medium. In contrast to the data on arterial cells in culture, which suggest enzyme-mediated uptake of free cholesterol, it was observed (Rothblat et al., 1966, 1967) that heating L5178Y cells (60°C for 30 minutes) reduced incorporation of free cholesterol by only $30 \pm 5\%$ over a 1-hour incubation period. It was also found that exposure of the cells to both

KCN and NaF did not reduce the uptake of labeled sterol in short-term experiments (Rothblat *et al.*, 1967).

B. Sterol Efflux

The ability of cells to release sterol to the culture medium was demonstrated by J. M. Bailey (1964b, 1965, 1967) in a series of experiments conducted on L cells and MBIII cells. Using cultures prelabeled with radioactive cholesterol, he demonstrated that there was a continuous excretion of cholesterol into the culture medium. At the end of the growth cycle approximately 60% of the labeled free sterol in the cells at the start of the experiment could be recovered from the culture medium. The cellular excretion of sterol was dependent on some component in serum since no sterol was released into serum-free medium. Release of sterol into the medium was related to the concentration and the species of serum used, with rabbit serum being the least efficient in promoting excretion. In other studies, cholesterol synthesized by L cells from mevalonate-^{14}C and acetate-^{14}C was also released into the culture medium. Fractionation of serum on ion exchange cellulose columns showed that most of the excretion-catalyzing ability of serum was recovered in the α-globulin fraction. This association with the α-globulin fraction of serum may be explained by studies in which delipidized rabbit serum was used as the acceptor of excreted free cholesterol from prelabeled L5178Y cells (Burns and Rothblat, 1969). These studies have shown that the amount of free sterol released into the culture medium from prelabeled cells is significantly increased when phospholipids are added to the culture medium together with either delipidized serum protein or bovine serum albumin. Similar results were observed for the efflux of endogenously synthesized sterol. The presence of exogenous sterol was not required for the release of cellular sterol, nor did the presence of exogenous free or esterified cholesterol (15 μg/ml) influence efflux values in L or L5178Y cells. Thus the release of cellular sterol into medium containing phospholipid and protein is not, in a strict sense, a simple exchange reaction, since no exogenous sterol is necessary for the replacement of sterol lost from the cell.

These delipidized serum experiments (Burns and Rothblat, 1969) showed that the rate of sterol efflux was directly related to the phospholipid/protein ratios and to the total concentration of phospholipid-protein complex in the culture medium. Thus, as in the case of influx, the data indicate that cellular sterol efflux is intimately related to the composition of the individual classes of serum lipoproteins and also to the concentration of these molecules in the medium.

IV. Sterol Synthesis

Numerous studies have demonstrated that many tissue culture lines and strains are able to incoporate labeled precursors into sterol or are capable of continued growth in the absence of an exogenous source of cellular sterol. Acetate incorporation into the sterols of cells grown in serum has been demonstrated with U12-72 human fibroblasts (Berliner et al., 1958), L5178Y mouse fibroblasts (Rothblat et al., 1967), aorta cells (Robertson, 1967), WI38 human diploid cells (Howard and Kritchevsky, 1969), and WI38VA13A, SV40-transformed WI38 cells (Howard and Kritchevsky, 1969). The inhibitory effect of serum on cellular sterol synthesis was demonstrated by J. M. Bailey (1964a, 1966, 1967) using MBIII and L cells. By growing these cells on chemically defined medium, he demonstrated that the cells were capable of synthesizing all the sterol needed for continued growth and that the addition of serum or cholesterol to the medium greatly reduced the incorporation of acetate into cellular sterol. Data on sterol synthesis in WI38 and WI38VA13A cells indicate that the amount of sterol synthesized in the presence of serum was less than 10% of that necessary for cell growth (Howard and Kritchevsky, 1969). It has also been shown that sterol synthesis in mouse liver parenchymal cells (Waymouth et al., 1971) and in human skin fibroblasts (Avigan et al., 1970) was reduced by the presence of serum or cholesterol in the growth medium.

In cell cultures, acetate and glucose are efficient precursors of synthesized sterol, whereas mevalonic acid is rather poorly incorporated into synthesized sterols (J. M. Bailey, 1967; Rothblat and Burns, 1971); this poor incorporation of mevalonate is probably due to permeability factors rather than an inability to be metabolized by cellular enzymes.When homogenates prepared from L cells, WI38 cells, or WI38VA13A cells are fortified with ATP and an NADPH generating system, they readily incorporate labeled mevalonate into sterol (Rothblat, 1968).

Studies have been conducted on the effect of exogenous cholesterol on the incorporation of various precursors into the synthesized sterol of cells grown in medium supplemented with delipidized serum proteins. When cholesterol (30 μg/ml) was present in the medium together with lecithin and delipidized protein, acetate and glucose incorporation into the sterol of L cells was reduced by 70–80%, whereas mevalonate incorporation was inhibited by approximately 20% (Rothblat and Burns, 1971). Studies on human skin fibroblasts have demonstrated that with the substitution of delipidized serum protein for whole serum in the medium, the incorporation of label from acetate or ^3H_2O into cellular nonsaponifiable lipids was stimulated to a greater extent than was that of mevalonate (Avigan et al., 1970). This increase in the incorporation of acetate into nonsaponifiable lipids was greater than the increase of acetate incorporation into fatty acids or $^{14}CO_2$. No change in the synthesis of amino acids was noted (Avigan et al., 1970). The results from these experiments are consistent with data obtained in liver, which indicate a cholesterol-induced block after β-hydroxy-2-methylglutarate (HMG) and prior to mevalonate (Siperstein and Fagan, 1966; White and Rudney, 1970). These observations, however, differ somewhat from those obtained from studies on L cells grown in a defined medium (J. M. Bailey, 1966, 1967). In these investigations the addition of cholesterol to the culture medium reduced acetate incorporation into cellular sterol to a greater extent than glucose incorporation. Acetate incorporation into other cellular lipids was also inhibited by the presence of cholesterol in the growth medium. It was suggested that the major effect of exogenous cholesterol in this cell system was at the level of the acetate-activating enzyme (J. M. Bailey, 1966, 1967). Thus all the studies indicate that the negative feedback elicited in tissue culture cells by the presence of exogenous cholesterol occurs at a metabolic step prior to mevalonic acid. Other minor blocks seem to be present at steps following mevalonic acid. It is also possible that the differences observed within a single cell line may reflect changes induced in the sublines selected to grow in defined media.

A number of steroids other than cholesterol, when added to the growth medium, will reduce the incorporation of acetate into L cell sterols. Table III shows comparative measurements of both cellular sterol synthesis and cell growth when various steroids were added at levels up to 40 μg/ml (Rothblat and Buchko, 1971). The results are expressed as a percent of control cultures grown in the absence of exogenous sterol. Under the conditions used in this experiment, an increase in concentration of exogenous cholesterol resulted in a concomitant decrease in the incorporation of acetate into cellular sterol. At the highest level of cholesterol tested (40 μg/ml) sterol synthesis was 17% that of controls. Lathosterol (cholest-7-ene-3β-ol) was somewhat less effective than cholesterol in reducing synthesis from acetate, while desmosterol (cholest-5,24-diene-3β-ol) was more efficient. All the compounds that reduced de novo synthesis also reduced cellular growth (as measured by protein). For some of the steroids this reduction in cellular growth was evident only at a concentration of 40 μg/ml, while compounds such as coprostanol (5β-cholestan-3β-ol)—and to an even greater extent, Δ^4-cholestene-3-one—were clearly toxic. Preliminary investigations indicate that both of the ketosteroids shown in Table III were metabolized to as yet unidentified products (Rothblat, 1968).

A number of compounds had little or no effect on L cell sterol biosynthesis when added at concentrations up to 40 μg/ml (Table III). The C_{28} and C_{29} phytosterols, the saturated sterol cholestanol (5α-cholestan-3β-ol), cholesteryl ester, and the hydrocarbon cholestane had no pronounced effect on sterol synthesis or cell growth. A comparative study of cholesterol, cholestanol, and β-sitosterol metabolism in L cells (Rothblat and Burns, 1971) has indicated that the failure of sitosterol to elicit a reduction in sterol synthesis may be related to its relatively slow rate of incorporation, compared to that of cholesterol. Although cholestanol is also incorporated at somewhat lower rates than cholesterol, it has been suggested (Rothblat and Burns, 1971) that the different effects of these two compounds on sterol synthesis may not be entirely related to differences in their rates of incorporation, but also to the inability of cholestanol to interact as efficiently at the enzymatic level. This lack of response to some sterols can be overcome; it was observed that labeled acetate incorporation into the nonsaponifiable lipids of human skin fibroblasts was reduced when the cells were incubated in medium containing 0.2 mg/ml and 1 mg/ml of cholestanol, β-sitosterol, and stigmasterol (Avigan et al., 1970).

Although cholesterol has been assumed to be the major product of sterol synthesis in cell cultures, a detailed examination of the sterols present in a number of cell lines grown in medium supplemented with delipidized serum protein demonstrated that approximately 90% of the sterol recovered from L cell mouse fibroblasts was desmosterol (cholest-5,24-diene-3β-ol) (Rothblat et al., 1970). A minor sterol ($<$10%) was also present in L cells and has been tentatively identified as cholest-5,7,24-triene-3β-ol (Rothblat et al., 1970). No significant amount of cholesterol was present and apparently the sterol-Δ^{24} reductase, which normally converts desmosterol to cholesterol, is missing or inoperative in L cells. When L cells were grown in medium containing serum or added unesterified cholesterol, the cholesterol was incorporated and de novo synthesis of desmosterol was reduced, thus yielding cells containing both cholesterol and desmosterol. The proportion of either sterol present was directly related to the amount of cholesterol incorporated from the medium. The presence of desmosterol in L cells produced no overt deleterious effects on these cells. Thus desmosterol seems to be an effective substitute for cholesterol in cells. Another possible example of the ability

TABLE III.

INFLUENCE OF STEROIDS ON CELLULAR GROWTH AND DESMOSTEROL SYNTHESIS IN L CELLS (% OF CONTROL)[a]

Compound	5 µg		20 µg		40 µg	
	Synthesis	Growth	Synthesis	Growth	Synthesis	Growth
Cholesterol, Δ^5	76 ± 3	95 ± 4	32 ± 3	92 ± 3	17 ± 2	60 ± 5
Desmosterol, $\Delta^{5,24}$	46 ± 6	98 ± 5	7 ± 1	69 ± 7	5 ± 1	63 ± 3
Lathosterol, Δ^7	104 ± 11	97 ± 9	66 ± 4	97 ± 6	30 ± 9	67 ± 4
7-Dehydrocholesterol, $\Delta^{5,7}$	87 ± 2	95 ± 3	30 ± 3	86 ± 5	9 ± 1	71 ± 8
Cholestanone, 3 keto, 5αH	47 ± 8	105 ± 9	29 ± 7	82 ± 9	22 ± 6	69 ± 13
Coprostanol, 5βH	90 ± 3	105 ± 2	75 ± 10	40 ± 4	41 ± 6	29 ± 2
Δ^4-Cholestene-3-one	18 ± 7	26 ± 2	—	—	—	—
Cholestanol, 5αH	103 ± 1	103 ± 2	96 ± 7	102 ± 6	87 ± 4	86 ± 8
β-Sitosterol, Δ^5, 24 Et	96 ± 3	106 ± 3	95 ± 3	107 ± 6	77 ± 7	105 ± 9
Campesterol, Δ^5, 24 Me	109 ± 5	94 ± 4	89 ± 8	92 ± 8	72 ± 7	90 ± 10
Stigmasterol, $\Delta^{5,22}$, 24 Et	114 ± 3	100 ± 1	102 ± 2	92 ± 7	113 ± 9	84 ± 8
Ergosterol, $\Delta^{5,7,22}$, 24 Me	94 ± 2	106 ± 3	100 ± 4	97 ± 5	94 ± 6	76 ± 6
Cholesteryl oleate, Δ^5, 3 OR	107 ± 5	95 ± 10	104 ± 4	92 ± 4	98 ± 5	92 ± 4
Cholestane	116 ± 7	108 ± 6	104 ± 5	108 ± 9	96 ± 6	99 ± 3

[a] Controls contained 5 mg/ml of delipidized calf serum protein plus 20 µg/ml of lecithin. Values are the average of at least four experiments ± SE.

of desmosterol to replace cholesterol can be seen in the studies on primary diploid human fibroblasts (Holmes *et al.*, 1969). These cells were unable to grow in the absence of exogenous sterol but would grow when cholesterol or desmosterol was added to the growth medium. In this study, however, as with some of the compounds listed in Table III, it was not known if the added sterols were metabolized by the cells.

Other cell lines have been examined and found to synthesize cholesterol (>90%), such as WI38 (human diploid embryonic lung cells), WI38VA13A [simian virus 40 (SV40)-transformed WI38], WIRL cells (primary rat liver cells), WIRL-SV40 (SV40-transformed WIRL), HEP-2 (human carcinoma), KB (human carcinoma), Nil-2 (spontaneously transformed hamster), and ST-3 (murine adenocarcinoma). Although the L cell line is the only one which has been shown to synthesize a sterol other than cholesterol, only a limited number of lines have been carefully examined and it should not be surprising to find that other lines, when grown in a sterol-free medium, contain sterol other than cholesterol. Since such a sterol would probably result from a loss in one or more of the enzymatic steps leading to cholesterol, the compounds which accumulated would probably be cholesterol precursors. Which of these precursors have the ability to support cellular growth is not presently known. The observation that 7-dehydrocholesterol (cholest-5,7-diene-3β-ol) would not support the growth of primary human skin fibroblasts which required sterol (Holmes *et al.*, 1969) suggests that the number of cholesterol precursors which could be utilized by cells may be limited.

V. Regulation of Sterol Synthesis by Sterol Flux

The data derived from studies on sterol content and metabolism in tissue culture cell systems demonstrate that cells possess a homeostatic mechanism for maintaining proper levels of sterol to ensure continued cell growth. This cellular sterol level would be a function of (1) the influx of exogenous sterol, which is regulated in large part by the composition (cholesterol-phospholipid-protein) and concentration of serum lipoprotein in the growth medium; (2) the efflux of cellular sterol, which is also a function of the composition and concentration of exogenous lipoproteins; and (3) *de novo* sterol biosynthesis, which is regulated by an effective feedback mechanism.

These observations led to the formulation of a theoretical scheme

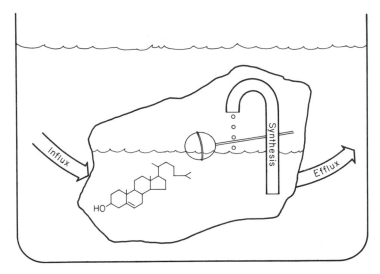

Fig. 1. Maintenance of cellular sterol concentration through the interaction of influx, efflux, and *de novo* synthesis.

for the cellular metabolism of unesterified sterol in tissue culture cells (Fig. 1). In this hypothetical system the flux of free sterol in the cell would be regulated by the amount and composition of the various serum lipoproteins in the culture medium. In addition, the amount of sterol synthesized by the cells would be regulated by the internal levels of unesterified sterol, which, in turn, would be a function of both the cellular influx and efflux. This hypothesis predicts that the amount of sterol synthesized and the distribution of this synthesized sterol between cells and medium would be influenced by the level of cholesterol, phospholipid, and protein present in the culture fluid. Thus, by growing cells under various conditions affecting influx and efflux of unesterified sterol, it should be possible to influence the biosynthesis and distribution of cellular free sterol.

Experiments designed to test this hypothesis have been conducted using acetate-^{14}C incorporation into L cell sterols as a measure of sterol biosynthesis (Rothblat, 1969). When the L cells were grown in increasing concentrations of calf serum, the amount of sterol synthesized by the cells decreased markedly as the level of whole serum in the medium increased (Fig. 2). This reduction occurred in both the synthesized sterol recovered from the cells and from the growth medium. In addition, the percent of the total radioactive sterol recovered in the medium increased as the serum increased. At the higher levels of serum, 80 to 90% of the labeled sterol was recovered in the medium. When whole

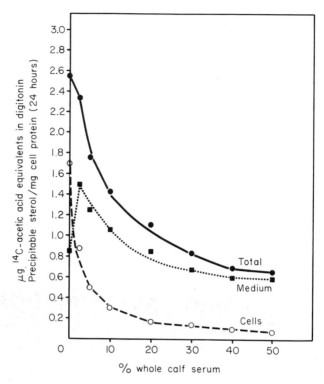

Fig. 2. Effect of calf serum on L cell sterol biosynthesis. Zero percent sample contained 5 mg/ml of delipidized calf serum protein. Average of four determinations. (From Rothblat, 1969.)

serum was added in increasing concentrations, exogenous sterol was taken up by the cells and an inhibition of synthesis occurred. Efflux continued, however, and increased with elevated serum concentrations. Thus, even as biosynthesis of sterol was reduced because of increased uptake of exogenous sterol, release was increased and a larger percent of the synthesized sterol was recovered in the culture medium. Under these conditions, increased serum concentration in the medium resulted in decreased synthesis and also promoted an increased rate in the flux of cellular sterol.

In a similar experiment L cells were grown in medium with increasing concentrations of lecithin and a constant level of delipidized serum protein. Since no exogenous cholesterol was added to the medium, it was necessary for the cells to synthesize all the sterol needed for growth. The values for total sterol synthesis (cell plus medium) show that the concentration of phospholipid in the culture medium can markedly influence the amount of sterol synthesized, with almost twice as much sterol synthesized in cells grown in medium containing 200 μg/ml of lecithin as in controls containing no lecithin (Fig. 3). When the distribution of the labeled sterol was determined, a small, but significant, increase was found in the amount of labeled sterol recovered from the cells; however, the effect of lecithin was most pronounced on the amount of newly synthesized sterol found in the medium. In the absence of exogenous sterol, increasing phospholipid concentrations in the medium led to increased sterol synthesis, with most of the increase occurring in sterol recovered from the medium. Thus as the protein-phospholipid complex in the medium was increased, more cellular sterol was lost

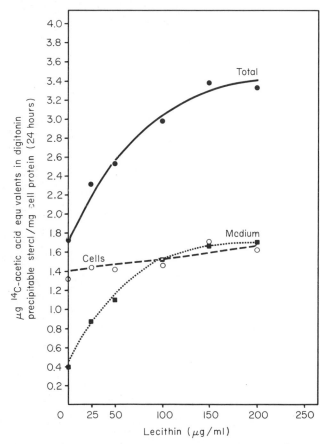

Fig. 3. Effect of lecithin on L cell sterol biosynthesis. Incubation medium contained 5 mg/ml of delipidized calf serum protein. Average of four determinations. (From Rothblat, 1969.)

to the medium. In the absence of an exogenous sterol, this loss caused a decrease in internal cellular sterol concentration to which the cell responded by an elevated rate of sterol synthesis in an effort to replace lost sterols.

The addition of increasing concentrations of free cholesterol to a culture medium containing constant levels of protein and phospholipid results in a concomitant decrease in the synthesis of cellular sterol. Figure 4 shows comparative results obtained from studies of L cells, WI38 cells, and WI38VA13A cells (Rothblat, 1969; Rothblat and Buchko, 1971; Rothblat *et al.*, 1971). Total sterol synthesis (cell plus medium) is presented as a percent of that obtained in control cultures grown in medium containing no added free sterol. The response of these three cell lines to exogenous cholesterol differed, with the WI38 cells exhibiting a greater

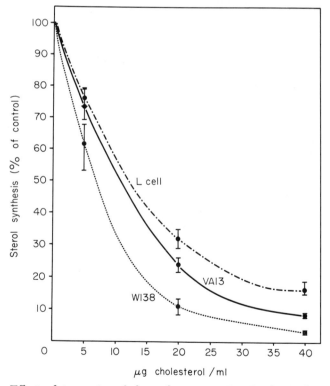

Fig. 4. Effect of increasing cholesterol concentration in the medium on cellular sterol biosynthesis in L cells, WI38VA13A cells, and WI38 cells. Growth medium contained 5 mg/ml of delipidized calf serum protein and 20 μg/ml of lecithin. Average of four determinations. (Data from Rothblat and Buchko, 1971; Rothblat *et al.*, 1971.)

reduction in sterol biosynthesis than WI38VA13A cells (Rothblat *et al.*, 1971). This observation is consistent with that of Howard and Kritchevsky (1969), who observed that WI38VA13A cells synthesized greater amounts of sterol than did WI38 cells in experiments with "resting" serum-starved cells. Sterol biosynthesis in L cells was less responsive to exogenous cholesterol than either WI38 cells or WI38VA13A cells (Fig. 3). These data suggest that there are differences in the efficiency of the sterol feedback mechanism among different cells. Further studies are needed to determine whether these differences are the result of varying enzymatic responses within individual cells or differences in rates of sterol exchange (influx and efflux) among cells. Another example of the possible effect of flux on sterol synthesis within a single cell line can be seen in the results obtained from prednisolone-treated HeLa cells (Dell'Orco and Melnykovych, 1970). Treatment of the cells with this hormone produced an increase in sterol synthesis which was attributed to the ability of prednisolone to reduce sterol influx without affecting a similar change in sterol efflux.

VI. Cholesteryl Ester Metabolism

Because of a lack of satisfactory experimental conditions for the *in vitro* addition of labeled esterified sterol into serum in any truly physiological manner, there is less information available concerning the factors which influence cholesteryl ester incorporation and utilization by tissue culture cells as compared to that available concerning the incorporation of free sterol. The information available suggests that esterified cholesterol is, for the most part, not utilized by tissue culture cells as a completely separate lipid class, but rather that the ester is hydrolyzed to its sterol and fatty acid components, which are then utilized by the cells.

Although most tissue culture cells which have been examined contain much smaller amounts of esterified than free sterol, they have been shown to incorporate exogenous cholesteryl ester from the growth medium. Studies on L5178Y cells exposed to rabbit serum labeled *in vivo* by cholesterol-^{14}C feeding (Rothblat *et al.*, 1966, 1967) show that serum cholesteryl ester is taken up by cells in culture. It was observed that esterified sterol was incorporated by cells at a rate considerably lower than that found with free cholesterol; however, no specific data were obtained on the nature of the uptake process.

Additional data demonstrating the cellular incorporation of esterified sterol using MBIII cells have been presented by J. M. Bailey (1961). Although almost all the sterol in these cells was present as free cholesterol, during the growth of the cells most of the decrease in sterol in the medium occurred in the ester fraction. Depletion of the cholesteryl esters from the medium occurred during cellular growth, but isotopic studies indicated that free cholesterol was being incorporated by the cells, even though this uptake was not paralleled by a depletion of free sterol from the culture medium. These data suggested that the cholesteryl esters were hydrolyzed by cellular enzymes to free cholesterol and fatty acids, after which the free cholesterol was excreted into the culture medium. Data confirming hydrolysis of cholesteryl esters by tissue culture cells were obtained from comparative studies on L cells and L5178Y cells (Rothblat et al., 1967, 1968b). Growing cultures of cells were exposed to cholesteryl-4-^{14}C esters which had been added to the culture medium in trace amounts. The data obtained showed that after a 24-hour incubation period, from 9 to 54% of the total label in the cells could be recovered as free cholesterol. It was evident that tissue culture cells could hydrolyze esterified sterol and that the amount of cholesteryl ester hydrolyzed differed in various cells.

Even though L cells possess the ability to hydrolyze cholesteryl esters taken up from the medium, the addition of cholesteryl oleate to cultures grown in medium containing delipidized serum protein had no effect on de novo synthesis of sterol (Table III). The slow rate of incorporation of exogenous ester by these cells probably limits the amount of cholesterol which could contribute to the cellular sterol pool and which would feed back on de novo sterol synthesis.

Although considerable excretion of free sterol from cells can be demonstrated, studies with L5178Y and L cells have shown that very little esterified sterol is returned to the growth medium (Rothblat and Kritchevsky, 1967), even though the L5178Y cell line contains considerable quantities of esterified sterol. Whereas very little esterified sterol was excreted from these cells, it was demonstrated that the free cholesterol liberated from cholesteryl esters by cell hydrolysis was recovered in the medium. This observation confirms the speculation of J. M. Bailey (1961) that free cholesterol levels of this medium could be maintained during the growth of cells in culture through the cellular hydrolysis of esterified sterol followed by the excretion of the free cholesterol obtained upon hydrolysis.

Additional experiments using cholesteryl palmitate-1-^{14}C have been conducted to determine the metabolic fate of the fatty acid liberated upon cellular cholesteryl ester hydrolysis (Rothblat et al., 1967, 1968b). It was found that in both L cells and L5178Y cells a major portion of the fatty acid which was released from the cholesteryl ester could be recovered in the cellular phospholipid fraction. All other lipids were also labeled, and small, but detectable, amounts of ^{14}C were recovered as $^{14}CO_2$ and in the defatted cellular residue.

Cell cultures may also synthesize some esterified sterol even when the cells are grown in the presence of exogenous cholesteryl ester. Thus, acetate incorporation into cellular cholesteryl esters has been demonstrated in primary chicken cells and in HeLa and Chang liver cells (Halevy and Geyer, 1961). This study did not, however, determine if the labeled acetate was incorporated into both the cholesterol and fatty acid moieties of the ester. The cholesteryl ester synthesis from acetate in cells grown in serum-containing medium probably represents only a small percentage of the actual cellular capability for ester synthesis. As in the case of other lipid classes, acetate incorporation into cholesteryl esters of L cells is reduced by 60% when serum is added to the growth medium (J. M. Bailey, 1964a). Other studies on L5178Y cells grown in serum-containing medium showed that only 7.5% of the total cellular sterol synthesized from labeled acetate was recovered as ester (Rothblat et al., 1967). L cells have also been shown to esterify only a small percentage (1%) of the desmosterol obtained through de novo synthesis (Rothblat and Burns, 1971).

Exogenous unesterified sterol is also incorporated into cellular steryl esters. In both L cells and L5178Y cells 1–15% of the exogenous free cholesterol which had been incorporated by the cells was recovered as ester (Rothblat and Kritchevsky, 1967). Cholestanol and β-sitosterol also undergo only very limited esterification in L cells (Rothblat and Burns, 1971). Although these investigations have suggested that cellular esterification of exogenous sterol occurs only to a limited extent under normal growth conditions, studies on the MAF line of diploid fibroblasts have indicated that sterol esterification is greatly stimulated by the addition of lipemic rabbit serum to the growth medium (P. J. Bailey and Keller, 1971). The presence of lipemic serum resulted in a 4-fold increase in the cellular concentration of unesterified cholesterol and a 19-fold increase in cholesteryl esters. Isotopic studies using acetate-^{14}C indicated that the major portion of the radioactivity could be recovered from the fatty acid moiety of the cholesteryl esters. It was postulated that the presence of lipemic rabbit serum stimulated increased fatty acid synthesis, leading to an increased rate of esterification of the free cholesterol which had been derived from the medium (P. J. Bailey and Keller, 1971). The mechanism by which lipemic serum stimulates steryl ester synthesis in some cell lines has not been elucidated and remains an important area for future investigations.

VII. Summary

A hypothetical scheme for the study of sterol metabolism in tissue culture cells (Fig. 5) has three major elements: (1) influx of exogenous sterol, (2) efflux of sterol, and (3) cellular synthesis of sterol. All these processes can contribute to the sterol pool which is required for the maintenance of membrane integrity. Two of these three elements are largely controlled by extracellular factors; the rate of influx and efflux of unesterified sterol seems to be a function of the serum lipoproteins present in the growth medium. The cells' primary response to changes in sterol flux is at the level of sterol synthesis. When the amount of sterol obtained from the medium is not sufficient to meet cellular requirements or when efflux of cellular sterol is great, the sterol deficit is replaced by synthesized sterol.

The rates of unesterified sterol influx and efflux seem to be directly related to the concentration and composition of the serum lipoproteins (Rothblat *et al.*, 1968a; Burns and Rothblat, 1969). A concept similar to the one proposed by Gurd (1960) for the exchange of lipids between lipoproteins could be applied to both the uptake and release of free cholesterol in tissue culture cells. According to this hypothesis, diffusion of free cholesterol would take place when serum lipoproteins come in contact with cell membrane lipoproteins and form a collision complex.

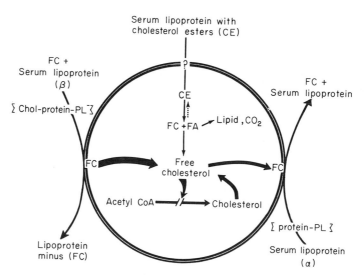

Fig. 5. Hypothetical scheme for the metabolism of free and esterified sterol in tissue culture cells.

The composition of the particular lipoproteins involved in this reaction would determine the partition coefficient of free cholesterol within the collision complex, which, in turn, would govern the rate and direction of diffusion of cholesterol.

In this hypothetical system influx and efflux of free sterol would be independent processes, each regulated to a great extent by different classes of serum lipoproteins. Because of their composition (protein-phospholipid-cholesterol), the β-lipoproteins would mediate the uptake of free sterol, while the α-lipoproteins would play the principal role in sterol efflux. Such a system could explain the accumulation of cholesterol in cells grown in medium supplemented with β-lipoproteins (Rutstein *et al.*, 1958; Maca and Rose, 1971) and the stimulation of sterol efflux by the α-globulin fraction of serum (J. M. Bailey, 1965, 1967). In addition, the protein moiety of the serum lipoproteins probably is more than just a simple carrier of lipid and may specifically interact with cell surface lipoproteins.

The mechanism by which steryl esters gain entrance to the cell is unknown. Esters are taken up, but the limited data available indicate that the rate of incorporation is considerably slower than with free cholesterol. Differences have been observed between the incorporation of cholesteryl esters and free cholesterol (Roheim *et al.*, 1965; Sodhi and Gould, 1967). The incorporation of cholesteryl esters at a less exchangeable site on the lipoprotein could result in differences in the rates of exchange of free and esterified sterol (Margolis, 1969), which, in turn, might result in a slow rate of transfer of esters from serum lipoproteins to plasma membrane lipoprotein or might make it necessary for the cell to incorporate the entire lipoprotein to obtain exogenous ester.

The cellular metabolism of cholesteryl ester in this system is primarily that of ester hydrolysis, with the subsequent liberation of free cholesterol and fatty acids. The fatty acids liberated upon ester hydrolysis serve as a carbon and energy source. It is also possible that the liberated cholesteryl ester fatty acid is selectively utilized by the cells. Such a selective utilization may conserve the cholesteryl ester fatty acids which generally contain high percentages of polyunsaturated fatty acid. In this proposed system, the synthesis of cholesteryl esters under normal growth conditions would occur to only a limited extent. However, some growth conditions, such as the presence of lipemic serum in the culture medium, may lead to the cellular accumulation of fatty acids (see Chapter 8, Volume I) or free cholesterol. When this occurs some cells may have the capacity to esterify this excess sterol.

The free cholesterol produced by the cellular hydrolysis of incorporated exogenous cholesteryl ester pools with other cellular free sterol is then used for membrane synthesis or returned to the culture medium. If the serum in the medium contains an active lecithin-cholesterol acyltransferase (Glomset, 1968), the excreted cholesterol can serve as a substrate for the formation of newly esterified cholesterol, thus replacing that which has been depleted through cellular cholesteryl ester hydrolysis. However, in a closed tissue culture system which generally does not utilize fresh serum, enzymatic activity would be low or non-existent, and little or no esterified sterol would be regenerated.

This hypothetical system explains much of the information which has been obtained from studies on the metabolism of sterols by mammalian cells in culture. It may also serve as a framework for designing future experiments to solve many of the questions which remain unanswered.

ACKNOWLEDGMENT

Supported, in part, by U.S. Public Health Service Research Grant R01-HE-09103 from the National Heart and Lung Institute. This manuscript was prepared during the tenure of an Established Investigatorship of the American Heart Association.

REFERENCES

Anderson, R. E., Cumming, R. B., Walton, M., and Snyder, F. (1969). Lipid metabolism in cells grown in tissue culture: O-alkyl, O-alk-I-enyl, and acyl moieties of L-M cells. *Biochim. Biophys. Acta* **176**, 491–501.

Avigan, J. (1959). A method for incorporating cholesterol and other lipides into serum lipoproteins *in vitro. J. Biol. Chem.* **234**, 787–790.

Avigan, J., Williams, C. D., and Blass, J. P. (1970). Regulation of sterol synthesis in human skin fibroblast cultures. *Biochim. Biophys. Acta* **218**, 381–384.

Bailey, J. M. (1961). Lipid metabolism in cultured cells. I. Factors affecting cholesterol uptake. *Proc. Soc. Exp. Biol. Med.* **107**, 30–35.

Bailey, J. M. (1964a). Lipid metabolism in cultured cells. V. Comparative lipid nutrition in serum and in lipid-free chemically defined medium. *Proc. Soc. Exp. Biol. Med.* **115**, 747–750.

Bailey, J. M. (1964b). Lipid metabolism in cultured cells. III. Cholesterol excretion process. *Amer. J. Physiol.* **207**, 1221–1225.

Bailey, J. M. (1965). Lipid metabolism in cultured cells. IV. Serum alpha globulins and cellular cholesterol exchange. *Exp. Cell Res.* **37**, 175–182.

Bailey, J. M. (1966). Lipid metabolism in cultured cells. VI. Lipid biosynthesis in serum and synthetic growth media. *Biochim. Biophys. Acta* **125**, 226–236.

Bailey, J. M. (1967). Cellular lipid nutrition and lipid transport. In "Lipid Metabolism in Tissue Cultured Cells," (G. H. Rothblat and D. Kritchevsky, eds.), Wistar Symp. Monograph No. 6, pp. 85–113, Wistar Inst. Press, Philadelphia, Pa.

Bailey, J. M. (1968). Personal communication.

Bailey, J. M., Gey, G. O., and Gey, M. K. (1959). Utilization of serum lipids by cultured mammalian cells. *Proc. Soc. Exp. Biol. Med.* **100**, 686–692.

Bailey, P. J., and Keller, D. (1971). The deposition of lipids from serum into cells cultured *in vitro. J. Atheroscler. Res.* 13, 333–343.

Berliner, D. L., Swim, H. E., and Dougherty, T. F. (1958). Synthesis of cholesterol by a strain of human uterine fibroblasts propagated *in vitro. Proc. Soc. Exp. Biol. Med.* 99, 51–53.

Bole, G. G., and Castor, C. W. (1964). Characterization of lipid constituents of human "fibroblasts" cultivated *in vitro. Proc. Soc. Exp. Biol. Med.* 115, 174–179.

Bosmann, H. B., Hagopian, A., and Eylar, E. H. (1968). Cellular membranes: The isolation and characterization of the plasma and smooth membranes of HeLa cells. *Arch. Biochem. Biophys.* 128, 51–69.

Burns, C. H., and Rothblat, G. H. (1969). Cholesterol excretion by tissue culture cells: Effect of serum lipids. *Biochim. Biophys. Acta* 176, 616–625.

Carruthers, C. (1966). Phosphatide composition of normal, hyperplastic and malignant squamous epithelium. *Oncologia* 20, 167–177.

Curtis, R. G., and Galvin, M. P. (1963). Application of a tissue culture technique to screen compounds for the prevention of lipoid deposition in aorta cells. *Aust. J. Exp. Biol.* 41, 687–694.

Dell'Orco, R. T., and Melnykovych, G., (1970). Lipid composition of heteroploid cells in culture: Effects of prednisolone. *J. Cell. Physiol.* 76, 101–106.

Glomset, J. A. (1968). The plasma lecithin: Cholesterol acyltransferase reaction. *J. Lipid Res.* 9, 155–167.

Gurd, F. R. N. (1960). Association of lipids with proteins. *In* "Lipid Chemistry" (D. J. Hanahan, ed.), pp. 208–259. John Wiley and Sons, Inc., New York.

Halevy, S., and Geyer, R. P. (1961). Comparison of lipid metabolism of chicken embryo organs and cells in culture. *Proc. Soc. Exp. Biol. Med.* 108, 6–9.

Holmes, R., Helms, J., and Mercer, G. (1969). Cholesterol requirement of primary diploid human fibroblasts. *J. Cell Biol.* 42, 262–271.

Howard, B. V., and Kritchevsky, D. (1969). The source of cellular lipid in the human diploid cell strain WI-38. *Biochim. Biophys. Acta* 187, 293–301.

Klenk, H., and Choppin, P. W. (1969). Lipids of plasma membranes of monkey and hamster kidney cells and of parainfluenza virions grown in these cells. *Virology* 38, 255–268.

Klenk, H., and Choppin, P. W. (1970). Plasma membrane lipids and parainfluenza virus assembly. *Virology* 40, 939–947.

Maca, R. D., and Rose, K. D. (1971). The role of beta-lipoprotein, cholesterol, and various sera in tissue culture intracellular lipidosis. *Proc. Soc. Exp. Biol. Med.* 136, 457–460.

Mackenzie, C. G., Mackenzie, J. B., and Reiss, O. K. (1964a). Regulation of cell lipid metabolism and accumulation. III. The lipid content of mammalian cells and the response to the lipogenic activity of rabbit serum. *Exp. Cell Res.* 36, 533–547.

Mackenzie, C. G., Mackenzie, J. B., Reiss, O. K., and Philpott, D. E. (1964b). The lipid content and structure of cells. *Fed. Proc., Fed. Amer. Soc. Exp. Biol.* 23, 375 (abstr.).

Mackenzie, C. G., Mackenzie, J. B., Reiss, O. K., and Philpott, D. E. (1966). Regulation of cell lipid metabolism and accumulation. IV. The isolation and composition of cytoplasmic lipid-rich particles. *Biochemistry* 5, 1454–1461.

Mackenzie, C. G., Mackenzie, J. B., and Reiss, O. K. (1967a). Regulation of cell lipid metabolism and accumulation. V. Quantitative and structural aspects of triglyceride accumulation caused by lipogenic substances. In "Lipid Metabolism in Tissue Culture Cells," (G. H. Rothblat and D. Kritchevsky, eds.), Wistar Symp. Monograph No. 6, pp. 63–83. Wistar Inst. Press, Philadelphia, Pa.

Mackenzie, C. G., Mackenzie, J. B., and Reiss, O. K. (1967b). Increase in cell lipid and cytoplasmic particles in mammalian cells cultured at reduced pH. *J. Lipid Res.* **8**, 642–645.

Margolis, S. (1969). Structure of very low and low density lipoproteins. *In* "Structural and Functional Aspects of Lipoproteins in Living Systems" (E. Tria and A. M. Scanu, eds.), pp. 369–424. Academic Press, New York.

Robertson, A. L. (1961). Effects of heparin on the uptake of lipids by isolated human and animal arterial endothelial type cells. *Angiology* **12**, 525–534.

Robertson, A. L. (1962). The uptake of labelled lipoproteins by isolated human and animal endothelial type cells. *Proc. Int. Pharmacol. Meet., 1st, 1961* Vol. 2, p. 193.

Robertson, A. L. (1967). Transport of plasma lipoproteins and ultrastructure of human arterial intimacytes in culture. In "Lipid Metabolism in Tissue Culture Cells," (G. H. Rothblat and D. Kritchevsky, eds.), Wistar Symp. Monograph No. 6, pp. 115–128, Wistar Inst. Press, Philadelphia, Pa.

Robertson, A. L. (1968). Personal communication.

Roheim, P. S., Miller, L., and Eder, H. A. (1965). The formation of plasma lipoproteins from apoprotein in plasma. *J. Biol. Chem.* **240**, 2994–3001.

Rose, H. (1968). Studies on the equilibration of radioisotopic cholesterol with human serum lipoprotein cholesterol *in vitro. Biochim. Biophys. Acta* **152**, 728–741.

Rothblat, G. H. (1968). Unpublished observation.

Rothblat, G. H. (1969). The effect of serum components on sterol biosynthesis in L-cells. *J. Cell. Physiol.* **74**, 163–170.

Rothblat, G. H., Hartzell, R. W., Mialhe, H., and Kritchevsky, D. (1966). The uptake of cholesterol by L5178Y tissue culture cells. Studies with free cholesterol. *Biochim. Biophys. Acta* **116**, 133–145.

Rothblat, G. H., and Buchko, M. K. (1971). The effect of exogenous steroid on sterol synthesis in L-cell mouse fibroblasts. *J. Lipid Res.* **12**, 647–652.

Rothblat, G. H., and Burns, C. H. (1971). A comparison of the metabolism of cholesterol, cholestanol and β-sitosterol in L-cell mouse fibroblasts. *J. Lipid Res.* **12**, 653–661.

Rothblat, G. H., and Kritchevsky, D. (1967). The excretion of free and ester cholesterol by tissue culture cells. Studies with L5178Y and L-cells. *Biochim. Biophys. Acta* **144**, 423–429.

Rothblat, G. H., Hartzell, R., Mialhe, H., and Kritchevsky, D. (1967). Cholesterol metabolism in tissue culture cells. In "Lipid Metabolism in Tissue Culture Cells." (G. H. Rothblat and D. Kritchevsky, eds.) Symp. Monograph No. 6, pp. 129–149, Wistar Inst. Press, Philadelphia, Pa.

Rothblat, G. H., Buchko, M., and Kritchevsky, D. (1968a). Cholesterol uptake by L5178Y tissue culture cells: Studies with delipidized serum. *Biochim. Biophys. Acta* **164**, 327–338.

Rothblat, G. H., Hartzell, R., Mialhe, H., and Kritchevsky, D. (1968b). Cholesterol ester metabolism in tissue culture cells. *Progr. Biochem. Pharmacol.* **4**, 317–324.

Rothblat, G. H., Burns, C. H., Conner, R. L., and Landrey, J. R. (1970). Desmosterol as the major sterol in L-cell mouse fibroblasts grown in sterol-free culture medium. *Science* **169**, 880–882.

Rothblat, G. H., Boyd, R., and Deal, C. (1971). Cholesterol biosynthesis in WI-38 and WI-38VA13A tissue culture cells. *Exp. Cell Res.* **67**, 436–440.

Rutstein, D. D., Ingenito, E. F., Craig, J. M., and Martinelli, M. (1958). Effects of linolenic and stearic acids on cholesterol-induced lipoid deposition in human aortic cells in tissue culture. *Lancet* **1**, 545–552.

Siperstein, M. D., and Fagan, V. M. (1966). Feedback control of mevalonate synthesis by dietary cholesterol. *J. Biol. Chem.* **241**, 602–609.

Sodhi, H. S., and Gould, R. G. (1967). Combination of delipidized high density lipoprotein with lipids. *J. Biol. Chem.* **242**, 1205–1210.

Waymouth, C. H., Chen, W., and Wood, B. G. (1971). Characteristics of mouse liver parenchymal cells in chemically defined media. *In Vitro* **6**, 371 (abstr.).

Weinstein, D. B., and Marsh, J. B. (1969). Membranes of animal cells. IV. Lipids of the L-cell and its surface membrane. *J. Biol. Chem.* **244**, 4103–4111.

White, L. W., and Rudney, H. (1970). Regulation of 3-hydroxy-3-methylglutarate and mevalonate biosynthesis by rat liver homogenates. Effects of fasting, cholesterol feeding, and triton administration. *Biochemistry* **9**, 2725–2731.

HUMAN DIPLOID CELL CULTURES: THEIR

USEFULNESS IN THE STUDY OF GENETIC

VARIATIONS IN METABOLISM

William J. Mellman
Vincent J. Cristofalo

I. Introduction

There is now abundant evidence that fibroblastic cell cultures can
be derived at will from human tissues; these cultures maintain both

chromosomal stability and the expression of many genetically determined biochemical properties throughout a finite period of time *in vitro* that consists of a number of subcultivations carried out over several months. Within this period of stable cell growth, cells proliferate at reproducible rates, and, under suitably controlled conditions, reliable analyses of various biochemical functions can be made. Thus, in these cell cultures it is possible to study questions about human cell physiology impossible to test in the intact human. Even if man could be studied by methods used with laboratory animals, such as those requiring large amounts of radioactive tracers and serial or complete sampling of organs or tissues, the cell culture system offers the unique advantage of rigid environmental control, as is now available with techniques for continuous medium perfusion.

It is difficult to establish that any *in vitro* observation of cell function is an accurate reflection of the *in vivo* state, and some of the best evidences are comparative studies of cell cultures derived from "normal" and "abnormal" individuals, particularly those with genetic aberrations in biochemical functions. Biochemical genetic variations are usually first identified by direct study of the individual's tissues. Biopsied tissues from affected subjects are then established in cell culture and tested for expression of the particular biochemical trait. If the genetic variation is expressed, these cell cultures can be used to explore the biological significance of such mutations to the cell.

The mechanism by which the synthesis and degradation of specific proteins are regulated is among the important areas in mammalian biology about which little is known. Cell cultures are nearly ideal tools for examining the general question of regulation, since environmental manipulation of the medium is readily accomplished. Studies of regulation should be particularly fruitful when comparative studies are made of normal and of mutant cell cultures with genetic alteration of the specific protein under examination.

The role of hormonal regulators of protein activity has not been able to be tested in human fibroblast cultures to the same extent as in cell lines of tumor origin. There is some recent evidence that there is a "repressor" in fibroblast cell cultures that affects inducibility of protein activity by hormones. In experiments that address this point, certain nonfibroblastic cell lines in which protein can be induced by hormones lose this property when fused with fibroblastic cell lines (Schneider and Weiss, 1971).

As an example of the usefulness of human cell cultures in the study of human disease, experiments with fibroblast cultures from humans afflicted with the X-linked form of hyperuricemia led to the precise identification of this disorder of purine biosynthetic control as a deficiency of the salvage pathway enzyme, hypoxanthine-guanine phosphoribosyltransferase (HGPRT) (Seegmiller *et al.*, 1967). The cells in culture were found to be resistant to inhibition of purine synthesis by 6-mercaptopurine (6MP), an effect known to occur in nonmutant cell cultures. This inhibitor requires the acquisition of a phosphoribosyl moiety to be effective, and the enzyme involved is HGPRT. Such an enzyme deficiency was identified in the cells in question, and the primary genetic effect was confirmed by direct analysis of tissues from affected individuals. This experience with human diploid fibroblast cultures showed that cells in culture can be exploited in the identification of the primary mechanism of genetic disorders of the intact organism and served to confirm by use of this mutant cell type that the regulation of macromolecular synthesis as analyzed in cell cultures is in fact a meaningful correlate of the *in vivo* situation.

The lymphoblastoid cell culture is a second type of human cell culture currently of considerable interest to investigators, although experience with these cultures is still limited. Interest in their genetics was aroused, just as in the case of the fibroblast cultures, by the finding that such cell lines were diploid in their chromosome composition, and this diploidy remains quite stable even with long-term culture (Kohn *et al.*, 1968). Lymphoblastoid cultures have two potential advantages over diploid fibroblast cultures: They have unlimited life *in vitro* in that they proliferate indefinitely without change in their growth properties, and they grow as mass suspension cultures. These lines were first derived from lymphoid tumors, and more recently they have been derived from the peripheral blood leukocytes of normal individuals. Although the issue is not completely resolved, there is a widely held view that all lymphoblastoid cell lines have an episome essential to the growth of these cells, and this growth factor is probably of viral origin, albeit perhaps in an incomplete form. The number of genetic markers shown to be expressed in lymphoblastoid lines from individuals with genetic variations are still small compared to those described for fibroblast lines, but those genetic variations described have been found to persist with long-term cultivation.

Amniotic cell cultures are now under intensive scrutiny as a tool in the prenatal diagnosis of genetic diseases, both those displaying gross chromosomal aberrations in the cells of affected individuals and those produced by biochemical genetic mutations.

Since most of the basic knowledge of biochemical genetic expression in cell culture comes from studies of fibroblast cultures derived from tissue samples of individuals after birth, fetal and nonfetal cell cultures are now being compared in terms of biochemical expression as are cultures of fetal organs and those of the desquamated fetal and amniotic cells aspirated from the pregnant uterus.

In human somatic cell genetics, cell cultures have been used not only for diagnostic purposes, but also in attempts to analyze mutational or recombinational events that might occur in somatic cells. These investigations require either the ability to derive clones of single cells in adequate numbers for appropriate assays or the use of selective growth media that permit either survival or identification of cell types within a population of cells in culture.

In this chapter we shall deal with our present knowledge of the genetic expressions of human diploid fibroblast cell cultures and shall provide background information on normal metabolism, nutritional requirements, and growth properties of these cell lines. Those genetic variations expressed in fibroblast cells will be described, as well as the preliminary data now available on lymphoblastoid cell lines and amniotic cell cultures.

II. Historical Development

Major developments in human somatic cell genetics occurred in the 1960s. Unquestionably a key stimulant to the intense activity now evident in human cell genetics was the observation first of Tjio and Puck (1958) and later of Hayflick and Moorhead (1961) that cells can be grown with relative ease from human tissues and that the resulting cell cultures maintain the chromosomal characteristics of their tissues of origin [i.e., they are diploid or homoploid in their chromosomal complement (Krooth, 1965)]. Both Swim and Parker (1957) and Hayflick and Moorhead (1961) noted that diploid cells could not be cultivated for indefinite periods; Hayflick and Moorhead, however, emphasized that cells of fibroblastic morphology are capable of prolonged culture under normal conditions but have a definite and limited life-span.

Puck et al. (1958) demonstrated that single cells from human diploid cell cultures could be cloned. They applied techniques that they, and other investigators, had developed in the 1950s in unsuccessful attempts to obtain homogeneous clonal derivatives from mammalian cell lines with chromosomally heterogeneous generations (i.e., heteroploid cell lines). The potential value of cloning techniques to human genetics became recognized in 1963 when Davidson et al. demonstrated in culture the clonal nature of an X-linked biochemical variant of glucose-6-phosphate dehydrogenase (G6PDH). They separated two biochemically distinct types of clones from the skin cells in culture of a female heterozygote for a G6PDH electrophoretic variant. This *in vitro* support for the Lyon hypothesis of X-gene activation was followed by similar information with three other X-linked loci responsible for disorders of metabolism, HGPRT deficiency, mucopolysaccharidosis (Hunter type), and Fabry's disease. Puck *et al.* (1958) correctly predicted the importance of the availability of techniques for long-term cultures of human cells with chromosome stability when they stated that "biochemical and immunological differences in single cell lines taken from different individuals may now be investigated with greater confidence about the meaningfulness of the results."

There has been a modest delay in utilizing available human cell culture technology, in part explained by the belief of many biochemists that the limited metabolism of fibroblast cells *in vivo* seriously restricts the available genetic variations expressed biochemically in cell culture. Investigators now realize that there is ample opportunity to explore biochemical genetics in the fibroblast cell culture system, because lines of fibroblastic type are stable in their chromosome complement (i.e., are "diploid" cells) and express their genetically determined phenotype throughout the life of the culture.

III. Human Diploid Cultures

A. Tissue of Origin

The skin has been the principal source of tissue from which human cell lines have been derived for biochemical-genetic analysis. Human cell cultures derived from other tissues, however, such as fetal lung (Hayflick and Moorhead, 1961), have been developed and are capable of serial passage.

Fibroblast cells readily grow in abundance from the edges of small bits of human skin that are placed into a culture vessel with either a glass or plastic surface. The nutritional requirements for initial or primary culturing of human skin have not been rigorously established. Thus far, chemically defined media cannot be used exclusively in establishing cultures from skin. Some nondialyzable serum components are essential, and although fetal calf serum is now probably the most popular source of serum factors, selected human and nonfetal bovine sera can be used with success in establishing these cell lines.

B. Nutritional Requirements

The experiments for determining the nutritional requirements of cells in culture have in general used as their criterion the proliferation of cells either in terms of increases in cell numbers, DNA, or total cell protein. More recently questions have been directed to the effect of medium components on other parameters, including cell survival, protein turnover, and the synthesis of specific macromolecules, such as glycogen, mucopolysaccharides, and specific enzymes. The following information is based principally on experiments that have mainly been concerned with cell growth.

1. SERUM

The high fetuin and low γ-globulin and fat content of fetal calf serum have been the basis for advocating this type of serum as a medium supplement (Puck *et al.*, 1958). Many procedures call for the use of embryo extracts of various sources in primary cell explantation, but these do not appear to be essential for primary culture of biopsied human skin.

Kruse *et al.* (1969) obtained multiple-layered cultures of human diploid fibroblasts derived from tonsillar tissue that became nonmitotic but remained viable when the serum concentration was reduced below 0.1%. They suggested that the serum may be critical only to cell proliferation. Although this has not been tested in human cells, Jainchill and Todaro (1970) have shown that medium containing serum, freed of γ-globulin by alcohol precipitation, supports survival of live 3T3 mouse cells for at least 4 weeks but does not support cell division. The addition of the precipitated protein or whole scrum to the medium results in a rapid resumption of cell division. SV40-transformed cells derived from 3T3 cultures grow almost as well in the γ-globulin-free medium as they do in regular medium.

2. CHEMICALLY DEFINED PORTION

The basal medium of Eagle (BME) (1955) was formulated to contain the essential growth requirements for established mammalian heteroploid cell lines. There is considerable empirical evidence that the requirements described by Eagle for two heteroploid cell lines (HeLa, a human cervical carcinoma line, and L, a mouse fibroblast line) include all the essential nutrients for diploid fibroblast cell lines derived from human skin.

A variety of other formulations are currently used for the serial cultivation of human diploid fibroblasts, yet there are still no experimental
data that establish the requirements for the additional components of
these other media. Jacobs (1966) found that human fetal diploid cell
lines require only cystine and glutamine as amino acid supplements
to media that contain 10% unheated and undialyzed calf serum. He
also recorded the amino acid content of the sera used in these
experiments.

Eagle and Piez (1962) noted that there are requirements for additional
nutrients when the cell density of heteroploid culture is below a critical
level. For example, at sufficiently high population densities, a variety
of human heteroploid cell lines are able to grow in cystine-free medium.
Eagle et al. (1966) found that, regardless of cell density, 10 human
diploid cell strains could not grow in the absence of cystine. Although
there is cystathionase activity in some human heteroploid cell lines,
this activity was not detectable in the human diploid fibroblast lines.

Most "superior" media suggested for human diploid cells have been
recognized as such because of their effect on the ability to clone the
cells. Thus it is reasonable to suspect that there are population-dependent requirements for human diploid cells still undefined (Eagle and
Piez, 1962; Ham, 1963; Waymouth, 1967; and Chapter 2, Volume 1).

The essential need in cell culture for the amino acids arginine, cystine,
glutamine, histidine, and tyrosine, considered nonessential to the human
organism, suggests the inactivity of certain specific enzymatic pathways
in cell cultures. Yet as Salzman (1961) has suggested, these amino acids
are "essential" only for cell growth and survival and not necessarily
for other functions. The requirement for arginine (or citrulline) has
now been observed in human diploid cells by Tedesco and Mellman
(1967). The ability to substitute citrulline for arginine has been supported by evidence obtained in a mutant cell line with a defect in
argininosuccinate synthetase. In addition, two extraneous factors have
been identified that affect the arginine concentration in culture medium.
Kihara and De La Flor (1968) have noted that fetal calf serum contains
significant heat-stable arginase activity which may seriously deplete the
arginine of the medium during the course of incubation with cells. This
activity was either absent or insignificant in other types of sera used
in support of cell growth. Schimke and Barile (1964) have identified
significant degradation rates of the arginine in media contaminated with
mycoplasma. These two observations provide possible explanations for
reports of unusual arginine requirements by cell cultures of various
types.

More recently, Griffiths (1970a) has studied the amino acid metabolism of the human diploid cell line, WI38. In general, he reported that the media formulation designated Eagle's minimal essential medium (MEM) supported growth better than BME. Eagle's MEM contains higher concentrations of amino acids and a vitamin supplement similar to that of BME except for the absence of biotin and the presence of inositol. Griffiths' data also show that within certain limits the concentration of amino acids can be limiting for cell proliferation. The most rapidly utilized amino acid was glutamine. Other than glutamine, arginine, leucine, isoleucine, and cystine were the most rapidly utilized. These were followed by lysine, histidine, and valine. Very large amounts of glutamine were used, and although glutamine is known to be very labile in cell culture medium, it did not become growth-limiting over the period of the experiment.

The requirement for folic acid by human diploid cells seems certain in view of the experience with cells deficient in HGPRT in which there is evidence *in vitro* of an increase of purine biosynthetic activity in the presence of folic acid. In the report of Felix and DeMars (1969), control cells grew maximally in the folic acid contributed by serum and a medium containing 4.5×10^{-8} M folic acid, while HGPRT-deficient cells required 1.5×10^{-6} M folic acid in the medium for maximal growth.

Other vitamins not specifically added to culture media may be essential or important for cell growth and metabolism. Ascorbic acid, for example, has been identified as a requirement for collagen synthesis in human fibroblast cultures. Certain media formulations, such as that described by McCoy et al. (1959), include ascorbic acid and thus obviate the dependence of cell cultures on the ascorbate content of the serum in the medium. Vitamin B_{12} may also be essential for cells in culture (Sanford et al., 1963; Mangum et al., 1969) and is probably present in adequate amounts in the serum supplement of the medium. Human diploid fibroblast cultures have the capacity to synthesize and accumulate vitamin B_{12} coenzyme. Cell lines from two boys with vitamin-B_{12}-dependent methylmalonic aciduria were found to have low vitamin B_{12} coenzyme levels. The levels became normal when the mutant cells were grown in media that contained 25,000 pg/ml of vitamin B_{12} (Rosenberg et al., 1969).

C. Growth Properties of Human Diploid Fibroblast Cultures

The fibroblast-like cells that grow from explanted skin fragments can be serially propagated by detaching them from their growing surface and converting them into a suspension of rounded cells. They are then diluted by dispersing them into several flasks of the same size or to a single, larger-surfaced vessel. Trypsin is the agent most commonly used for detaching and dispersing, although pronase has been used (Weinstein, 1966; Sullivan and Schafer, 1966). When the dispersed cells are returned to medium, containing protein, they attach again to the growth surface, resume their fibroblastic shape, and after a brief, still poorly defined, lag period usually enter a period of rapid cell division.

The length of this period is dependent on the density of the cell inoculum. At the end of the period there is a slower phase of cell proliferation that may persist as long as there is no critical nutritional deficiency in the medium due to the metabolite depletion or accumulation. The serial passage of cells by detachment, dispersion, and dilution, followed by a rapid and then a slow growth period, occurs predictably in most skin fibroblast cultures whether derived from fetal skin or the skin of infants, children, or adult human subjects. Hayflick and Moorhead (1961) described a characteristic pattern of serial passage of human diploid cell strains on the basis of their experience with fibroblast-like cell lines derived from fetal lung tissue. One of these, the WI38 cell line, is a strain of human diploid fibroblasts developed by Hayflick (1965) from fetal lung and has been the source of cells for extensive investigation of diploid human fibroblasts.

Considerable data, both recorded and unrecorded, have now accumulated on the growth of nonfetal human skin lines because these are the lines which are the source of cells with genetic variations. A number of observers have noted that fetal skin cultures proliferate more rapidly and are less tolerant of prolonged periods of maintenance in monolayers without trypsinization and passage once they have reached a stage of confluent growth. Todaro et al. (1963) compared adult skin and fetal lung fibroblast cell cultures and found that the fetal cells had a more rapid growth rate than the adult skin cells. Hayflick (1965), using comparable lung cultures, found a dramatic difference between cell lines derived from human embryos and those from human adults. More recently, Goldstein et al. (1969) found that an inverse correlation does exist between the age of the donor and the number of population doublings that a series of skin cultures is able to undergo. Martin et al. (1970), in an extensive study in which over 100 mass cultures of fibroblast-like human diploid cells were derived from a variety of donors with ages between 1 and 10 and 80 and 90 years, found a highly significant negative regression of growth potential as a function of the age of the donor. Goldstein et al. (1969) and Martin et al. (1970) have shown that cells from patients with progeria or with Werner's syndrome, both diseases associated with premature aging, have a reduced proliferative capacity as compared with control cells from normal donors of the same ages. In addition, cells derived from diabetic individuals are less able to grow and survive in culture, as manifested by a reduced plating efficiency.

Condon *et al.* (1971) have recently noted that cell cultures derived from fetal skin differ from those of nonfetal origin in regard to glycolysis. Their data revealed that during rapid cell proliferation there is increased activity of the pentose phosphate pathway, with less metabolized glucose appearing as lactate; nonfetal cells did not show this increase in shunt activity during the period of rapid cell growth.

Maciera-Coelho and Pontén (1969) have shown that after only a few passages adult skin cells show characteristics of their cell cycle similar to those of degenerating fetal cell cultures. In addition, Cristofalo *et al.* (1967) have shown that adult lung cells have lysosomal enzyme characteristics after only 12–14 passages similar to those of fetal cells after 35–50 passages.

Whatever the significance of these differences between cultures derived from fetal and adult tissues, multiple layers of cells will form from cell inocula of either fetal lung or nonfetal skin cultures when proper nutritional conditions are maintained.

Kruse and Miedema (1965a) have attempted to separate the phenomenon of slow growth due to contact inhibition from that due to nutritional deficiency. Using a continuous perfusion system that maintains constant conditions in the culture throughout the time of the experiment, they have estimated the number of cells that constitute a single confluent layer (monolayer equivalent). Under these conditions cell proliferation is at first rapid. Then, after one to two monolayer equivalents have been established, cell proliferation continues, but at a reduced rate (Chapter 2, Volume 2).

Elsdale and Foley (1969) have illustrated the morphogenetic aspects of multilayering in petri dish cultures of human fetal lung fibroblasts. They have shown that attachment of cells to the surface of the dish is not dependent on collagen formation but that multilayering requires collagen synthesis. Collagenase in the medium prevented multilayering of human fetal lung fibroblasts. They suggest that fetal lung cultures are different from cultures of fetal gut fibroblasts. The latter show little or no tendency toward multilayering. There are no reported comparisons of collagen synthesis and collagenase activity in human fibroblast cultures of various tissue origins that might biochemically account for differences in multilayer growth.

D. The Nature of the Degenerative Phase

It is now relatively well established that populations of normal human diploid cells can proliferate in culture for only finite periods of time (Hayflick and Moorhead, 1961; Hayflick, 1965; Hayakawa, 1969; Martin *et al.*, 1970). Typically, after explantation there is a period of rapid cell proliferation during which the cell cultures can be subcultivated relatively often. This is followed by a period of declining proliferative capacity when the cells become granular, debris accumulates, and ultimately the culture is lost. Hay (1967) and Cristofalo (1972a) have reviewed various aspects of this phenomenon. The reader is referred to these reviews and to the work of Swim and Parker (1957), Hayflick and Moorhead (1961), and others in establishing the generality of this observation. In summary, the results showed that a variety of human tissues from fetal, neonatal, and adult individuals was incapable of unlimited proliferation. Hayflick and Moorhead (1961) have shown that this degeneration is not related, at least in any simple, direct way, to inadequate nutrition, pH variation, toxic metabolic products, or microcontaminants. They concluded that the limited life-span of these cells is probably programmed and might reflect a cellular expression of aging.

A number of parameters of cell physiology have been examined in serially passaged human fibroblasts in attempts to account for the occurrence of the degenerative phase of human diploid fibroblast cultures. Table I summarizes these studies. The parameters investigated include carbohydrate metabolism, respiration, some aspects of lipid metabolism, amino acid and protein metabolism, nucleic acid content and turnover, and lysosomes and lysosomal enzymes. Of all these studies, only a few of these parameters showed significant changes with age. In carbohydrate metabolism a decline in activity of the pentose phosphate pathway, both in terms of the incorporation of labeled glucose into nucleic acids (Cristofalo *et al.*, 1972a) and in terms of several pentose phosphate enzyme activities (Cristofalo, 1970) was noted. Other metabolic differences noted between cells in the active proliferative and senescent phase of growth include increased glycogen (Cristofalo *et al.*, 1970) and lipid content (Kritchevsky and Howard, 1966) and the loss of collagen synthetic function in the latter phase (Macek *et al.*, 1967; Houck *et al.*, 1971). The fraction of cells involved in DNA synthesis (after a short pulse of radio-labeled thymidine) was decreased and the generation time was increased in later passage subcultures. The time required for DNA synthesis and mitosis seemed to be unchanged; however, only

TABLE I

SOME METABOLIC PROPERTIES OF DIPLOID CELLS STUDIED DURING AGING[a]

Parameter	Variation with Cell Age[b]	References
Glycolysis	0	Cristofalo and Kritchevsky (1966); Cristofalo, et al. (1972b).
Glycolytic enzymes	0, −	Cristofalo (1970); Cristofalo et al. (1972b); Cristofalo, et al., (1972); Wang et al. (1970).
Pentose phosphate shunt	−	Cristofalo (1970); Cristofalo et al. (1972a).
Permeability to glucose	0	Hay and Strehler (1967)
Glycogen content	+	Cristofalo et al. (1970)
Mucopolysaccharide synthesis	−	Kurtz and Stidworthy (1969)
Respiration	0	Cristofalo and Kritchevsky (1966)
Respiratory enzymes	0	Hakami and Pious (1968); Wang et al. (1970)
Lipid content	+	Kritchevsky and Howard (1966); Hay and Strehler (1967)
Lipid synthesis	+	Chang (1962); Yuan and Chang (1969)
Protein content	+, 0	Cristofalo et al. (1970); Wang et al. (1970)
Permeability to amino acids	0	Hay and Strehler (1967)
Transaminases	0, −	Wang et al. (1970)
Glutamic dehydrogenase	0	Wang et al. (1970)
Nucleohistone content	0	Ryan and Quinn (1971)
Collagen synthesis	−	Macek et al. (1967); Houck et al. (1971)
DNA content	0, −	Hay and Strehler (1967); Cristofalo and Kritchevsky (1969)
RNA content	+	Cristofalo and Kritchevsky (1969)
Nucleic acid synthesis	−	Macieira-Coelho et al. (1966a,b)
RNA turnover	+	Michl and Svobodová (1970)
Lysosomes and lysosomal enzymes	+	Cristofalo et al. (1967); Robbins et al. (1970); Wang et al. (1970); Brandes et al. (1972)
Alkaline phosphatase	0	Cristofalo et al. (1967); Wang et al. (1970)

[a] Reprinted with permission from Cristofalo (1972a).

[b] +, increase with age; −, decrease with age; 0, no change.

a small proportion of the older cells synthesized DNA, and some older-passage cells apparently synthesized DNA but did not go on to divide (Maciera-Coelho et al., 1966a,b).

It appears that viable cells in late passage have a low rate of cell proliferation, yet they are actively synthesizing and, presumably, degrading RNA, protein, and other macromolecules.

The DNA content of late-passage human fibroblasts did not differ from early passage cells (approximately 8–15 μg/10^6 cells) (Tedesco and Mellman, 1966; Cristofalo and Kritchevsky, 1969). The RNA content, however, has been shown to increase during aging in culture (Cristofalo and Kritchevsky, 1969).

Macek *et al.* (1967) and Houck *et al.* (1971) have reported that the rate of collagen synthesis declines during clonal senescence of diploid fibroblast-like cells. Kritchevsky and Howard (1966) have shown a disproportionate accumulation of lipid in senescent WI38 cells.

IV. Genetic Variations in the Metabolism of Human Diploid Fibroblast Cells

It is now clear that a number of biosynthetic and degradative processes, nonessential to their growth and survival, can be detected in fibroblast cell cultures. Those mutant cell lines that have been described will be reported here along with the available data on the concerned areas of metabolism in human diploid fibroblast culture.

Human diploid fibroblast cells grow in a reproducible and probably optimal manner with the medium described by Eagle (1955); this has suggested that the metabolism of human diploid cells is superficially, at least, similar to that of HeLa cells for which the medium was designed. The biochemistry of mammalian cells was thoroughly reviewed by Levintow and Eagle (1961). In their survey they developed a comparison between the metabolism of serially propagable cells, with the capacity to multiply indefinitely in culture, and cultured tissue explants or "primary" cultures of cells freshly isolated from tissues. The latter two classes of cells were considered as a single category. In view of the development, since the time of that review, of an awareness of human diploid cells as a class of cell cultures distinct from the human mixoploid cell lines, we shall deal with the data that attempt to differentiate between the biochemistry of human heteroploid cell lines capable of serial and indefinite culture and the human diploid cell lines, which must be construed as "primary" cultures of human tissues with the capacity of serial propagation but with only a finite life-span in continuous culture.

Some confusion exists in the literature of human cell culture biochemistry because many studies have attempted to characterize the metabolic differences between continuously propagable, presumably heteroploid, cell lines derived from "normal" and "malignant" human tissue. These two categories of cells in culture will be considered as a single entity in this discussion and will be contrasted with cell cultures derived from nonmalignant tissues that are diploid. For a discussion of these differences, the reader is referred to Hayflick (1965).

A. Carbohydrate Metabolism

Definitive reviews of the carbohydrate metabolism of mammalian cells in culture have been published by Levintow and Eagle (1961), by Paul (1965), and in Chapter 4, Volume 1. Although many minor differences in the carbohydrate metabolism appear among various permanently propagable cell lines, several properties seem to be common to all the cells studied. Among these are the following: (1) Serially cultivated cells degrade glucose to either carbon dioxide or lactic acid, the latter pathway predominating in most cases. (2) Other sugars, for example, mannose, fructose, and to a certain degree galactose, can be utilized in place of glucose for cell growth. (3) The rates of respiration and glycolysis are influenced by variables, such as pH (Danes et al., 1963; Zwartouw and Westwood, 1958), length of time in culture, and species of serum (Phillips and McCarthy, 1956; Phillips and Andrews, 1960). Despite the wide variety in origin of cell cultures, a striking similarity exists in the rates of respiration and glycolysis of the cell lines studied and this has been considered to be a characteristic of cells in vitro (M. Green et al., 1958).

Respiration and glycolysis are fundamental and essential to the growth and survival of human diploid cells, as well as the mixoploid cell lines. In both, glucose represents the principal source of carbon in the medium and thus occupies a central position in their cellular metabolism. The interconversions of glucose provide energy through glycolysis and respiration, building blocks for the various macromolecules synthesized by the cell, including the ribose and deoxyribose elements of the nucleic acids, and the reducing power, principally as NADPH, necessary for the synthesis of a wide variety of biologically important compounds. Glucose is rapidly metabolized by the cells, with most of the glucose carbon appearing in the medium as lactic acid. Glucose metabolism has been shown to be essential for the growth and survival of human diploid cells. Several investigators have noted that complete removal of this simple sugar from the supportive medium leads to an arrest of growth, followed by cell death (Kruse, 1965; Cristofalo, 1965).

Among the first published reports on the carbohydrate metabolism of human diploid cells were those showing that the rate of proliferation of the cells directly paralleled their rate of glucose utilization (Cristofalo and Kritchevsky, 1965). This same observation was made simultaneously by Kruse and Miedema (1965a,b), who used a perfusion system for the growth of their cells. This observation that glucose utilization is more rapid during a rapid proliferative phase of the culture has been subsequently confirmed in perfusion systems by Cristofalo (1972b).

In general the carbohydrate and energy metabolism of diploid cells is qualitatively similar to that of mixoploid cell lines. For example, although diploid cell line WI38 uses glucose preferentially, these cells can utilize mannose, fructose, or galactose in place of glucose (Cristofalo and Kritchevsky, 1965). In experiments by Eagle *et al.* (1958), these hexoses permitted the growth of nine serially propagable cell lines of mammalian origin at a rate comparable to glucose. In diploid cells, mannose and glucose behave similarly, with most of the hexose utilized appearing as lactic acid. Fructose and galactose are converted to lactic acid more slowly but are able to support the growth of WI38 cells. Cristofalo (1970; Cristofalo and Kritchevsky, 1965, 1966) noted that the direction of flux of glucose carbons is primarily toward glycolysis in the human diploid cell line WI38. Both WI38 cells and the SV40-transformed sister cell line derived from them have essentially similar rates of glucose utilization and lactic acid production. This holds true not only at the usual concentration of glucose present in culture medium (5–10 mM), but also over a range in concentration of 2–30 mM. At the higher concentration levels, however, there is some indication that the transformed cells can metabolize more glucose and probably have a reduced sensitivity to very large amounts of glucose (for a discussion of the relationship between glycolysis and malignancy, see Chapter 4, Volume 1).

As with heteroploid cells, the concentration of the glucose in the medium has an effect on the rate of glucose utilization. In experiments in which the glucose concentration varied from 2 to 30 mM, Cristofalo *et al.* (1972b) have shown an increasing rate of glucose utilization by WI38 cells up to a concentration of about 20 mM and then a leveling off of the rate. Lactate production, on the other hand, shows maximum values at a concentration of approximately 5 mM glucose, and higher levels of glucose have no effect on the rate of lactic acid production. Cristofalo (1970) and Dunaway and Smith (1970) have assayed the maximal activities of a number of key enzymes of the glycolytic pathway in WI38 cells. In general, their enzyme activities are characteristic of cells with relatively high glycolytic rates: Fructose diphosphatase activity is very low; lactate dehydrogenase and phosphoglucose isomerase activities are relatively high, and hexokinase and phosphofructokinase show activities typical of their "pacemaker" function in glycolysis. Similar data to those reported for WI38 cells have been described for a series of glycolytic enzymes both in other human diploid fibroblasts and in lymphoblast cell lines (Mellman and Kohn, 1970).

UDPG-4-epimerase activity has been detected in broken cell preparations of both human diploid and heteroploid cell lines. NAD is an essential cofactor for this enzyme and it is inhibited by NADH. Both pH and the ratio of NAD to NADH critically influence its activity. Baugh and Tytell (1967), however, have suggested that supplemental pyruvate allows human diploid cells to grow with galactose as a carbon source by reducing the NADH level via lactic dehydrogenase.

Tedesco and Mellman (1969a) have described several systems in human diploid fibroblasts that can generate NADH in a test tube and consequently cause inhibition of UDPG-4-epimerase activity. UDPG-4-epimerase activity, along with UDPG pyrophosphorylase, generates UDP galactose from endogenous carbohydrates. UDP galactose is the donor of galactosyl residues for a variety of complex oligosaccharides formed in mammalian cells.

a. The Hexose Monophosphate Shunt. One of the functions of glucose metabolism is to supply reducing power for various synthetic processes, as well as the building blocks from which these syntheses proceed. A major pathway in these functions is the pentose phosphate or Warburg-Dickens pathway. Most of the work relevant to this pathway in human diploid cells has been concerned with the activities of the various enzymes involved. Cristofalo (1970) has analyzed the rate of incorporation of glucose carbon into the ribose and deoxyribose of the nucleic acids of WI38 cells. These same authors have compared the maximal activities of several key enzymes of both the oxidative and nonoxidative limbs of the hexose monophosphate shunt. As with most tissues, glucose-6-dehydrogenase activity was higher than 6-phosphogluconate dehydrogenase activity, and transaldolase activity was higher than transketolase. Thus, 6-phosphogluconate dehydrogenase appears to be rate-limiting for the oxidative pathway, while transketolase appears to be rate-limiting for the entire sequence. Glucose-6-phosphate dehydrogenase has been studied extensively because of the considerable genetic polymorphism of this enzyme in man. Racker (1956) noted that high concentrations of glucose inhibit the tricarboxylic acid cycle in ascites tumor cells and stimulate the hexose monophosphate shunt. No comparable studies have been made with human diploid cells. Condon *et al.* (1971) have compared $^{14}CO_2$ production from glucose-1-^{14}C and from glucose-6-^{14}C. Their results have shown that, in both fetal and nonfetal cell types, proportionately more $^{14}CO_2$ was produced from ^{14}C-labeled glucose than from glucose-6-^{14}C but that the ratio C-1/C-6 was greater in the fetal lines analyzed soon after subcultivation. The amount of $^{14}CO_2$ produced by cells incubated with glucose labeled in the 2 position was similar to that of the 6 position and suggests that there was no significant recycling of glucose by either cell type.

b. Glycogen Metabolism. Human diploid cells store their carbohydrates in the form of glycogen, and depending on the culture conditions, different amounts of glycogen can be detected. Assays of the glycogen content of WI38 cells gave a mean value of 15.8 $\mu g/mg$ dry weight of cells (9.65 $\mu g/10^6$ cells) (Cristofalo *et al.*, 1970). Glycogen storage was thought to reflect the normality or malignancy of heteroploid cell lines, although this now appears to be an oversimplified view, since the rate of cell multiplication and the availability of cell nutrients probably account for the absence of glycogen in certain cell lines. Alpers *et al.* (1963) have studied glycogen metabolism in HeLa cells and found that both phosphorylase and glycogen synthetase were significantly more active in cells maintained in medium with ample concentrations of glucose. Wu (1959) found that glycogen constituted about 5% of the dry weight of HeLa cells under some conditions. The data of Cristofalo *et al.* (1970) for diploid cells showed a value of 1.5–2% of the dry weight of these cells.

DiMauro *et al.* (1969) have studied a number of control diploid fibroblast cultures from biopsied adult skin and found that glycogen content is minimal during the active proliferating phase of growth. Glycogen content reaches a maximum several days after confluence is reached, provided there is ample glucose in the medium. When glucose is removed from the medium while the rest of the medium content is kept intact, glycogen is depleted over the next 72 hours. After glycogen is depleted, there is a rapid rise in cell death; glycogen therefore appears to be the major source of cell nutrition in the absence of glucose in the medium. Cells depleted of glycogen repleted this glycogen within 24 hours after glucose was added to the medium, although factors in serum were required for maximal glycogen accumulation (DiMauro and Mellman, 1972).

c. Mucopolysaccharide Metabolism. The acid mucopolysaccharide content of human diploid fibroblast cells derived from skin has been studied extensively since 1966, when abnormalities of mucopolysaccharide metabolism were noted in cell lines from patients with Hurler's syndrome. The indicators of this abnormality were the finding of metachromatic granules in cultured fibroblasts by Danes and Bearn (1966a) and of excessive incorporation of radioactive precursors by Matalon and Dorfman (1966).

Metachromasia of uncertain significance has been reported in cells from individuals with a variety of other diseases and from healthy controls (Taysi *et al.*, 1969; Matalon and Dorfman, 1969). Although atypical metachromasia occurs in certain cell lines, it is still unclear what factors of cell growth and metabolism affect the level of metachromasia of various preparations of the same cell line. Vitamin A (alcohol) in cultured skin fibroblasts reduces the degree of metachromasia (Danes and Bearn, 1966b).

Hamerman *et al.* (1965) noted that an adult strain of human diploid fibroblasts produced hyaluronate throughout its life-time in culture in amounts comparable to those produced by established mouse lines and to freshly cultivated human fibroblasts. Infection of these fibroblasts with SV40 led to transformation of the cells and to a marked diminution of hyaluronate synthesis.

1. Genetic Errors of Carbohydrate Metabolism

a. Galactose Metabolism

(1) *Galactose-l-Phosphate Uridyltransferase.* Cell cultures have been produced from individuals with mutations affecting the activity of both galactokinase and galactose-l-phosphate uridyltransferase (transferase). The latter enzyme is markedly deficient or absent in cell extracts from patients with galactosemia (Russell and DeMars, 1967; Tedesco and Mellman, 1969b), a rare autosomal recessive disease that is responsible for the accumulation of galactose-l-phosphate in cells when the affected person is on a diet containing lactose. Krooth and Weinberg (1960) observed that fibroblast cell cultures from normal individuals could convert galactose-1-^{14}C to $^{14}CO_2$ and that galactosemic cell cultures could not. They also found that galactose did not support the growth of galactosemic cell lines, while it did support the growth of other human diploid cells.

More recently Krooth and Sell (1970) have pointed out that control cells will grow in medium where galactose is substituted for glucose only when the cells are able to grow in medium containing less than 5 mg/100 ml of glucose, a level that is growth-limiting for human diploid fibroblast cells plated at low density and during periods of rapid proliferation. Therefore, the use of the galactosemia mutation in selective experiments *in vitro* has generally been unsatisfactory. Normal cells will survive when galactose is the only hexose in the medium, while galactosemic cells will not. This property has been exploited by Nadler *et al.* (1970) in experiments where they isolated hybrid cells produced by Sendai virus fusion of two different galactosemic cell lines that demonstrated transferase activity. They selected the heterokaryons with transferase activity by alternately growing their cultures in glucose-free medium and medium containing both glucose and galactose. After several weeks of alternating their cultures in these two media, they could recover hybrid cells with functional transferase activity, presumably by some type of molecular complementation.

Russell and DeMars (1967) have indicated that there is a functional transferase in the rapid proliferative phase of subcultured galactosemic fibroblasts. This finding is consistent with the recent evidence of Tedesco and Mellman (1971) that there is an inactive enzyme in cells from galactosemic individuals; they demonstrated this finding by immunological methods.

Galactose does not seem to alter the growth or survival of galactosemic cell cultures provided either glucose or accumulated glycogen is present in the cells. One report, however, has described the loss of electron-dense material from the rough endoplasmic reticulum of galactosemic cells; this ultrastructural change was different from those that occur with the same cells in a hexose-free environment (Miller *et al.*, 1968).

(2) *Galactokinase.* Galactokinase activity has been measured in fibroblast cultures and has been shown to be independent of the transferase genotype (Tedesco and Mellman, 1969b). The enzyme in human diploid fibroblasts is inhibited by galactose-1-phosphate and at least one cell line has been described that expresses the genotype of the donor who was a galactokinase-deficient heterozygote (Beutler *et al.*, 1971). Cell lines from galactokinase-deficient homozygotes will probably prove to be better galactose mutant cells for somatic cell genetic experiments, although some of the same problems cited above for transferase-deficient mutants will apply in terms of selective media where galactose is substituted for glucose. Comparisons of galactokinase and transferase activities in cell lysates suggest that galactokinase is the rate-limiting step in galactose utilization by cells in culture. Since galactokinase is considerably less active than hexokinase (Mellman and Kohn, 1970), the limited capacity of cell cultures to metabolize galactose is probably at the kinase level.

b. Glucose Metabolism

(1) *Glucose-6-Phosphate Dehydrogenase (G6PD).* Cultures from hemizygous males and homozygous females carrying the X-linked gene for the Mediterranean form of G6PD deficiency have reduced levels of this enzyme (Gartler *et al.*, 1962; Nitowsky *et al.*, 1965), while the

Negro form of G6PD deficiency is not expressed in fibroblast cell cultures by decreased activity. Electrophoretic variants, most notably the faster migrating variant ("A band") commonly found among Negro individuals, are expressed in cell culture and have proved to be a useful qualitative cell marker (Davidson *et al.*, 1963; DeMars and Nance, 1964; Peterson *et al.*, 1968). Davidson *et al.* (1963) demonstrated that clones derived from a Negro female heterozygous for the A and B G6PD polymorphism expressed only one electrophoretic mobility, either A or B, and their experiment has been used as a model demonstration of the Lyon hypothesis of X-gene inactivation as expressed in man.

Wajntal and DeMars (1967) have described a direct staining method, using the tetrazolium dye as a visible marker of G6PD activity with intact cells in culture, and DeMars (1968) has applied this technique to the study of a temperature-sensitive G6PD mutant. Siniscalco *et al.* (1969) have also used this method to identify "positive" and "negative" cells in hybridization experiments between a G6PD-deficient and HGPRT-deficient cell cultures.

(2) *Pyruvate Decarboxylase* (Blass *et al.*, 1970). Cells from a 9-year-old boy with elevated levels of pyruvic acid in the blood revealed a deficiency of pyruvate decarboxylase, as assayed in broken cell preparations as well as by the conversion of pyruvate-l-^{14}C to ^{14}CO$_2$ by trypsinized cell suspensions from fibroblast cultures. Cells from the parents of this child oxidized pyruvate to a degree intermediate between those of the patient and control cells. Thiamine pyrophosphate enhanced the activity of pyruvate decarboxylase activity of sonicated cell preparations by 20–30%. The activity of mutant cells was proportionately affected by coenzyme additions, suggesting that this mutation is not one of coenzyme synthesis or availability to the apoenzyme.

(3) *Phosphohexose Isomerase* (Krone *et al.*, 1970). Skin cells in culture from two patients with an anemia associated with deficiency of phosphohexose isomerase were studied. The cells from homozygotes had 53% of the activity of cells from controls, and cell cultures from heterozygotes could not be distinguished from controls on the basis of enzyme quantitation. The mutant enzyme was found to be less stable than wild-type enzyme to heating at 45°C, and it migrates more rapidly in starch gel electrophoresis. Cells from both homozygotes and heterozygotes demonstrated these mutant properties of the enzyme.

c. Glycogen Metabolism. Glycogen storage diseases of man are markedly heterogeneous in their specific biochemical cause, a reflection of the remarkably complex regulation of glycogen synthesis and degradation in mammalian tissues. Some of the enzymes whose defects lead to abnormal glycogen storage are expressed in cell cultures. The observation that glycogen synthetase activity in human fibroblast cultures is influenced by glycogen content suggests that cell cultures may be applied to the study of human mutations where there is disturbed glycogen regulation (DiMauro et al., 1972).

Type II glycogenosis is associated with a deficiency of lysosomal α-1,4-glucosidase; this activity as well as an α-1,6-glucosidase with pH optimum of 4.0 are deficient in cell cultures from affected individuals (Nitowsky and Grunfeld, 1967; Dancis et al., 1969a; Brown et al., 1970). There are excessive accumulations of metachromatic material in cell cultures from affected individuals with Type II glycogenosis (Taysi et al., 1969; Tenconi et al., 1970) and enlarged lysosomal-like bodies filled with glycogen (Hug et al., 1971). Increased total cell glycogen contents have been found (Zacchello et al., 1969), although these cells were found to mobilize their glycogen when glucose was removed from the culture medium (DiMauro et al., 1969).

Type III glycogenosis is now recognized to be a highly variable disorder, both from a clinical and biochemical standpoint. In one report the fibroblast cultures from a patient with this disease were low in debrancher (α-1,6-glucosidase) activity, the enzyme generally believed to be involved (Justice et al., 1970).

Type IV glycogenosis is associated with branching enzyme deficiency (α-1,6-glucan, α-1,4-glucan, 6-glycosyltransferase). In one report there is deficiency of this enzyme in a cell line from the skin of an affected infant (Howell et al., 1971).

Type V glycogenosis is a disease caused by muscle phosphorylase deficiency. Skin fibroblasts cultured from a patient with this disease did not evidence phosphorylase deficiency (DiMauro et al., 1972).

Fibroblast cultures from patients with cystic fibrosis, an autosomal recessive disorder, display abnormalities which indicate that they are expressing their genetic defect. Although the primary defect is still elusive, among the findings reported are excessive metachromatic staining (Danes and Bearn, 1968a), increased total acid mucopolysaccharides Matalon and Dorfman, 1968), and increased glycogen content when subcultures are permitted to grow undisturbed after they reach confluence (Pallavacini et al., 1970).

d. Mucopolysaccharides. Fibroblast cultures have been invaluable in the development of our understanding of the genetic disorders associated with excessive accumulations of mucopolysaccharides. This subject has been recently reviewed by Neufeld and Fratantoni (1970), who describe the development of the hypothesis that these accumulations of mucopolysaccharides reflect errors in their degradation. Various types of mucopolysaccharidoses have been differentiated *in vitro* by "correcting" one form of the defect with cells from another form: Two cell lines of the same form of the disease are not mutually correctable. These findings have led to the postulation that there are specific protein factors deficient or defective in abnormal cells in culture, different from one form of the disease to another, and that these factors readily diffuse between cells in culture. A preliminary report suggests that fibroblast cultures from patients with Hurler's disease have diminished l-iduronidase activity (Matalon *et al.*, 1971).

e. Other Complex Carbohydrates
Gangliosides

(a) *Gm_1 Gangliosidosis.* There are now two recognizable clinical forms of GM_1 gangliosidosis. Cell cultures from the two forms show different degrees of deficiency of lysosomal β-galactosidase activity (Sloan *et al.*, 1969; Pinsky *et al.*, 1970). The Type I form expresses a more profound deficiency of this enzyme in cells in culture.

(b) *GM_2 Gangliosidosis.* Cells in culture from the three recognized GM_2 storage diseases show enzymatic defects (Okada *et al.*, 1971). Tay-Sachs and juvenile GM_2 gangliosidosis have decreased levels of hexosaminidase A, and cells from a patient with Sandhoff disease was found to have markedly low levels of both hexosaminidase A and B.

(c) *Krabbe's Disease* (Galactocerebrosidosis). This rare degenerative neurological disease has recently been found to be deficient in a specific β-galactosidase (galactocerebroside β-galactosidase). This enzymatic deficiency was expressed in fibroblasts cultured from an affected infant (Suzuki and Suzuki, 1971).

(d) *Gaucher's Disease* (Glucocerebrosidosis). This is also a disorder of cerebroside metabolism which results in glucocerebroside accumulations in various tissues. Fibroblasts cultured from affected individuals have a marked deficiency of β-glucosidase activity. Of interest is a report of the influence of culture medium conditions on the quantitation of this enzyme in cells cultured from both normal and affected persons (Beutler *et al.*, 1971).

(e) *Fabry's Disease* (Ceramide Trihexosidosis). This rare disease develops slowly and tends to be diagnosed in adult life; its genetic determinant is located on the X chromosome. The defective enzyme is ceramide trihexosidase, an α-galactosidase (Brady *et al.*, 1971). Skin fibroblasts cultured from affected persons accumulate the trihexoside, galactosyl-galactosyl glucosyl ceramide. They also contain an increased content of acid mucopolysaccharides of normal qualitative distribution

(Matalon *et al.*, 1969). Clonal derivatives of cell lines from females heterozygous for this deficiency are of two types, negative and positive for α-galactosidase deficiency, a result consistent with the Lyon hypothesis (Romeo and Migeon, 1970).

B. Amino Acid Metabolism

As mentioned earlier, there has been no rigorous study of the amino acid requirements of human diploid fibroblast cells, although there is empirical evidence that these cells grow optimally in media that contain only the amino acids that Eagle (1955) found were essential for the growth of HeLa cells. There may, however, be amino acids in the serum supplement of media which are also essential or at least useful to human fibroblast cells.

Elsewhere in Volume 1 (Chapter 6) is a comprehensive review of the work on amino acid metabolism. There are a few brief reports which describe amino acid transport in human diploid cells. Platter and Martin (1966) showed that human fibroblast-like cells accumulated tryptophan. Mahoney and Rosenberg (1970) studied the uptake of α-aminoisobutyric acid by human fibroblasts and found that the processes by which it was accumulated depended on cellular energy. Inhibitors such as ouabain, cyanide, and *p*-chloromercuribenzoate significantly impaired uptake.

Griffiths (1970b) showed that the presence of insulin lowered the rate of amino acid uptake but caused higher protein yields, suggesting a more efficient use by the cell of the amino acid pool, which is probably related to the increased glucose uptake effected by insulin.

A number of so-called essential amino acids may be replaced by their precursors or metabolites. Citrulline can substitute for arginine and there are data both from growth characteristics and direct enzymatic evidence that the citrulline to arginine pathway is intact in diploid cell culture (Tedesco and Mellman, 1967). Ornithine is probably inactive as a presursor of arginine in diploid cells. This was noted by Eagle for HeLa cells (Eagle, 1959). Ornithine transcarbamylase has not been detected in human diploid fibroblasts. Homocysteine can replace methionine as an essential amino acid in cells. Methionine is then generated by the vitamin-B_{12}-dependent methyltransferase reaction (Mudd *et al.*, 1970). The keto acids corresponding to the essential amino acids are effective substitutes for their respective amino acids in HeLa cells but have not been specifically tested in human diploid cells.

Cystine has been established as a requirement of human diploid fibroblast cells. Eagle *et al.* (1966) have shown in studies of 10 different cell lines that homocysteine could be formed from methionine, that most cell lines could form cystathionine from homocysteine and serine, and that none could synthesize cystine. This is in contrast to the experience with a number of heteroploid human cell lines where cystine was produced in adequate amounts to permit the cells to grow with homocysteine or cysthathionine substituted for cystine.

Collagen synthesis has been detected in both diploid and heteroploid fibroblast cell lines with the measure of collagen synthesis being the amount of labeled hydroxyproline incorporated into collagen (Goldberg and Green, 1968; H. Green *et al.*, 1966). Heteroploid fibroblast cell lines have considerably more active collagen synthesis than nonfibroblastic types, such as HeLa, while lymphoblastic-type cell lines that grow in suspension are completely inactive. The variability of collagen-forming capacity in cultures of different cell types is not thought to be due to differences in the activity of proline hydroxylase, the enzyme responsible for the conversion of prolyl residues in nascent collagen to hydroxyproline. Lymphoma cell lines, for example, contained significant proline hydroxylase activity (Goldberg and Green, 1969).

Schafer *et al.* (1967) found that ascorbic acid supplementation enhanced the amount of collagen produced by human diploid fibroblasts, while the addition of dimethylsulfoxide to the medium of growing human fibroblasts resulted in a slowing of both cell growth and hydroxyproline accumulation (Stenchever *et al.*, 1967).

GENETIC ERRORS OF AMINO ACID METABOLISM

a. Branched-Chain Amino Acid Decarboxylase. Cells cultured from patients with maple syrup urine disease and a milder disease, intermittent branched-chain ketonuria, demonstrate, respectively, absent and decreased decarboxylase activity of the three amino acids leucine, isoleucine, and valine (Dancis *et al.*, 1969b; Sigman and Gartler, 1966).

b. Lysine-Ketoglutarate Reductase. The activity of this enzyme is reduced in skin fibroblasts grown from three siblings with hyperlysinemia. The metabolic defect in this disorder expressed in cell cultures involves the saccharopine pathway concerned with the degradation of lysine (Dancis *et al.*, 1969c).

c. Cystathionine Synthetase. Cultures of skin from patients with homocystinuria, a disorder of the methionine to cysteine pathway, have very low or undetectable levels of cystathionine synthetase activity (Uhlendorf and Mudd, 1968).

d. B_{12} Coenzyme Metabolism. Studies of fibroblasts cultured from the skin of a patient with abnormalities of sulfur amino acid and methylmalonic acid metabolism indicated that there was a defect in the metabolism of vitamin B_{12}. Unlike control cells, mutant cells were unable to grow normally when the methionine in the medium was replaced by homocystine. The addition of large amounts of hydroxyl-B_{12} to the medium improved mutant cell growth in the homocystine-containing medium. Evidence was presented that the activities of N^5-methyltetrahydrofolate methyltransferase and methylmalonyl coenzyme A carbonyl mutase were abnormal in mutant cells; both enzymes are known to require B_{12} derivatives as cofactors (Mudd *et al.*, 1970).

e. Cystinosis. The specific biochemical lesion in cystinosis, a disease accompanied by cystine accumulation in tissues, is unknown; cultured fibroblasts from patients with this disorder have been found to accumulate free cystine (Schneider *et al.*, 1967). This cystine is apparently unavailable to cells, since they require the same cystine in the medium for optimal growth as normal cells, even after they have accumulated cystine. The excess of cellular cystine is thought to be compartmentalized in lysosomes (Schulman and Bradley, 1971). Dithiothreitol was found to remove cystine from cystinotic fibroblasts (Goldman *et al.*, 1970).

f. Argininosuccinase. Fibroblast cells from patients with argininosuccinic aciduria are deficient in argininosuccinase (Shih *et al.*, 1969).

g. Argininosuccinate Synthetase. A cell line from a patient with citrullinemia has been described with the normal maximal enzyme velocity (V_{max}) for argininosuccinate synthetase, but a markedly decreased affinity (increased K_m) for citrulline. These cells grew poorly when citrulline was substituted for arginine, in contrast to control cell lines that have similar growth in citrulline-substituted medium (Tedesco and Mellman, 1967).

Preliminary reports have now appeared that describe other citrullinemic cell lines with decreased argininosuccinate synthetase activity (Scott-Emuakpor *et al.*, 1971).

h. Propionyl CoA Carboxylase. Cultured fibroblasts from a young girl with ketotic hyperglycinemia were unable to oxidize propionate-^{14}C to $^{14}CO_2$, a property of control cells. Both control and mutant cells could metabolize methylmalonate-^{14}C and succinate-^{14}C to CO_2. Propionyl CoA carboxylase activity was absent in mutant cells, and there was no effect of biotin, a known cofactor for this reaction, on the activity of deficient cells (Hsia *et al.*, 1971).

i. Methylmalonyl Coenzyme A Carbonylmutase (MMA-CoA Mutase). MMA-CoA mutase catalyzes the isomerization of methylmalonyl CoA to succinyl CoA. This is a B_{12} enzyme that requires coenzyme (deoxyadenosylcobalamin). Diseased infants with both vitamin-B_{12}-responsive and -unresponsive methylmalonic aciduria have been recognized. Corresponding expressions are noted in cell cultures; the cells from patients with the B_{12}-responsive form of the disease have low contents of B_{12} enzyme, while cells from the unresponsive variety have normal coenzyme content. Both types of mutant cells have deficient enzyme activity; the activity from the B_{12}-responsive form can be corrected *in vitro* by supplementation of the growth medium with vitamin B_{12} (Morrow *et al.*, 1969; Rosenberg *et al.*, 1969).

C. Lipid Metabolism

Very little information concerning lipid synthesis and intermediary metabolism in diploid cells in culture is available. The lipids of WI38 human cells were characterized by Kritchevsky and Howard (1966). They measured the lipid fractions of these cells during rapidly growing and "stationary" phases and found in both situations that phospholipids made up about 70% of the total lipids. Cholesterol and free fatty acids accounted for over 60% of the neutral lipids. The majority of the phospholipid compartment was lecithin, although phosphatides of ethanolamine, inositol and serine, and sphingomyelin and lysolecithin were also present. Some differences were noted in the distribution of components in the phospholipid fractions between rapid and stationary phases of growth.

Howard and Kritchevsky (1969) did find some cholesterol formation in WI38 cells, but estimated that it probably was not adequate to sustain growth. More recently, Rothblat *et al.* (1972) have studied sterol synthesis in WI38 cells which had been grown in a sterol-free medium containing 5 mg/ml of delipidized calf serum protein. The culture had the ability to synthesize all the sterol necessary for continued growth. Gasliquid chromatography revealed that the synthesized sterol was cholesterol and that all the sterol could be recovered as free cholesterol; no cholesterol esters were detected.

When free cholesterol was added back to the growth medium in increasing concentrations, the cells exhibited a negative feedback response for sterol biosynthesis. In contrast, Holmes *et al.* (1969) identified a cholesterol requirement for human diploid fibroblasts derived from both cartilage and skin. The cells did not survive when lipoprotein was removed from the serum of the medium. This subject is discussed in Chapter 9, Volume 1.

Disorders of mucopolysaccharide metabolism have been associated with abnormal lipid storage *in vivo;* lipid abnormalities also occur in cultured fibroblasts from certain individuals with mucopolysaccharidosis (Matalon *et al.*, 1968).

1. GENETIC DISORDERS OF LIPID METABOLISM

A number of genetic disorders of metabolism, expressed in cell culture, that involve lipid-containing macromolecules have already been described under carbohydrate defects. These were storage diseases of glycolipids where the degradative disorder affects the sugar moieties; the remaining lipidoses are listed here even though their biochemical defect does not always directly relate to the lipid portion of the affected molecule.

a. Refsum's Disease. Refsum's disease is a lipid storage disease identified as a defect in the α oxidation of phytanic acid. Skin fibroblasts from affected persons are unable to oxidize phytanic acid to pristanic acid, while the latter compound is readily oxidized by mutant cells. Phytanic acid is believed to be derived in man from dietary sources (Herndon *et al.*, 1969a,b).

b. Metachromatic Leukodystrophy (MLD). This is a disease caused by a deficiency in arylsulfatase A which leads to abnormal storage of sulfatides in both the central and peripheral nervous systems. Cells in culture accumulate cerebroside sulfates and are deficient in arylsulfatase A (Porter *et al.*, 1969; Kaback and Howell, 1970).

When arylsulfatase A is added to the growth medium of MLD cells, the deficient cells take up the enzyme and retain arylsulfatase A activity for at least 11 days after the enzyme is removed from the medium. These "corrected" cells no longer store cerebroside sulfate (Porter *et al.*, 1971).

c. Neiman-Pick's Disease. Neiman-Pick's disease is a syndrome in which sphingomyelin and cholesterol accumulate in tissues. Sphingomyelinase activity of cultured bone marrow and skin fibroblasts from affected persons with the two recognized forms of the disease is markedly deficient (Sloan *et al.*, 1969). These cells have a higher sphingomyelin content than cells from normal individuals (Holtz *et al.*, 1964).

D. Nucleic Acid Metabolism

The synthesis and degradation of purine and pyrimidine nucleotides in cell cultures has been reviewed in Chapter 7, Volume 1.

The influence of proliferative activity on the rates of macromolecule formation has been examined by Levine et al. (1965). They studied cultures derived from fetal bone and the skin of infants; they found that DNA synthesis was reduced to 5–15% of maximum after the cells reached confluence and RNA synthesis was correspondingly reduced. Free polyribosomes were abundant in the cytoplasm of diploid cells during the rapid proliferative phase of the culture, and in the postconfluent phase of growth they were markedly reduced. Comings and Okada (1970) have compared the ultrastructure of human fibroblasts during active proliferation and at confluency. Rhode and Ellem (1968) studied three fetal lung lines and found that DNA synthesis almost completely ceased several days after the cells reached confluence, at which time there was still significant RNA synthesis.

There are no reported data on the relative specific activities of enzymes concerned with DNA synthesis in human diploid fibroblasts cells during various proliferative phases of subcultures. Other cell cultures, such as primary cultures of trypsinized rabbit kidney cortex cells, have a specific activation of DNA polymerase and thymidine kinase at the time of DNA formation in vitro (Lieberman et al., 1963). In a heteroploid, human, liver, cell line, thymidine kinase activity was highest during the period of rapid cell proliferation (Eker, 1966).

Glutamine has been found to be a precursor of nucleic acid bases in heteroploid mammalian cell cultures (Salzman et al., 1958). The critical requirement for glutamine in human diploid fibroblast cells suggests that this amino acid contributes to nucleic acid synthesis in these cells as well.

The cellular uptake of polymerized mammalian DNA has been described for heteroploid mammalian cell cultures (Gartler, 1960) and has been reported to produce a heritable transformation of a biochemical trait in a human heteroploid cell line (Szybalska and Szybalski, 1962). It is still to be established whether human diploid cells can incorporate mammalian DNA. DNA extracted from SV40 has been used to infect and transform human diploid cells (Aaronson and Todaro, 1969).

DISORDERS OF PURINE AND PYRIMIDINE METABOLISM

Considerable insight into nucleic acid metabolism of human diploid fibroblast cells has come from studies of cell cultures from biopsied skin of patients with two human mutations: orotic aciduria and Lesch-Nyhan syndrome (or X-linked hyperuricemia). These are discussed in detail in Chapter 7, Volume 1.

In brief, human diploid fibroblasts cultured from normal individuals do not require preformed nucleic acid bases for cell proliferation, which is consistent with the evidence that pathways for the *de novo* synthesis of nucleotides exist in these cells. With cells from individuals homozygous for the autosomal recessive trait that produces the disease orotic aciduria, Krooth (1964) was able to show that the addition of the pyrimidine nucleotides (uridine and cytidine) to the medium significantly enhanced their proliferative capacity, while the growth of control cells was unaffected by supplementation with these nucleotides.

Low levels of orotidylic pyrophosphorylase and orotidylic decarboxylase activity were detected in diploid fibroblast cells from two patients with orotic aciduria. Adenosine was found to inhibit the activity of orotidylic decarboxylase in these mutant cells. Cells from a heterozygote for this disorder had intermediate levels of both enzymes. The activity of two other enzymes in the biosynthetic pathway of uridine monophosphate synthesis (dihydroorotase and dihydroorotic acid dehydrogenase) were normal in cells with the mutant genotype (Howell *et al.*, 1967; Wuu and Krooth, 1968).

Studies performed in skin fibroblast cells in culture from a child with the X-linked recessive neurological disorder associated with an overproduction of uric acid identified the defective enzyme of purine metabolism responsible for this disease. Seegmiller *et al.* (1967) observed that purine synthesis in normal human skin fibroblast cells was blocked by the addition of 6-mercaptopurine to the culture medium, while cells from a patient with the Lesch-Nyhan syndrome were resistant to the purine analogue. Lesch-Nyhan cells are resistant to 6-mercaptopurine and were found to lack the enzyme HGPRT that converts the free bases guanine and hypoxanthine to their respective nucleotides. HGPRT-deficient cultured fibroblasts synthesize four times as much purines as control cells (Rosenbloom *et al.*, 1968).

E. Miscellaneous Metabolic Disorders

1. Uroporphoryrinogen III Cosynthetase

In congenital erythropoietic porphyria the activity of uroporphoryrinogen III cosynthetase is low in the red cell. Assays of cell cultures from the skin of three individuals affected with this disorder revealed 19, 28, and 44% of control activity for this enzyme (Romeo *et al.*, 1970).

2. CATALASE

Cell lines from patients with the so-called Japanese and Swiss forms of acatalasia are catalase-deficient. The Japanese type has virtually no detectable activity in cells in culture, while the Swiss form has some residual activity (Krooth *et al.*, 1962; Sadamato, 1966).

3. DNA REPAIR ACTIVITY

Cells from patients with xeroderma pigmentosum are sensitive to UV irradiation and show 0–25% of the normal amount of DNA repair synthesis after UV irradiation (Cleaver, 1970).

4. ACID PHOSPHATASE

Absence of acid phosphatase activity was noted in the lysosomal fraction of cells in culture from an infant who died of a bizarre metabolic disease. This same deficiency was found in the postmortem tissues of the infant and in the amniotic cells in culture from a subsequent pregnancy of the mother of the index case (Nadler and Egan, 1970).

A summary of the above-described mutant cell lines, along with some of their key characteristics, appears in Table II.

F. Lymphoblastoid Cell Lines

Cell lines capable of indefinite *in vitro* life have been developed from human peripheral blood leukocytes by a variety of techniques (Broder *et al.*, 1970; Henle *et al.*, 1967). Most lines derived from normal subjects are chromosomally diploid and there is generally persistent diploidy over a long period of *in vitro* life (Kohn *et al.*, 1968). These cell lines all appear to harbor a herpes-like virus, although the expression of infection with this virus is variable from cell line to cell line and within the life of the same cell line. Because they grow in suspension, they are relatively easy to handle in the laboratory, and these lines are now under intensive study. Information on their ability to express genetic variation is still limited, but at least one valuable marker has been reported in a line derived from a patient with hypoxanthine-guanine phosphoribosyltransferase deficiency (Lesch-Nyhan disease) (Choi and Bloom, 1970; see also Chapter 5, Volume 2).

A phosphoglucomutase polymorphism has been observed in long-term lymphoid cell cultures (Conover *et al.*, 1970). A current major investi-

TABLE II.

HUMAN BIOCHEMICAL DEFECTS EXPRESSED IN SKIN AND AMNIOTIC CELL CULTURE

Enzyme or Metabolic Defect	Genetic Disease	Prenatal Diagnosis Reported with Amniotic Cell Cultures
A. Carbohydrate metabolism		
Galactokinase	Galactosemia (galactokinase deficiency) (1)[a]	
Galactose-1-PO$_4$ uridyltransferase	Galactosemia (classic transferase deficiency) (2,3)	+ (4)
Glucose-6-PO$_4$ dehydrogenase	Nonspherocytic hemolytic anemia (Mediterranean form of G6PD deficiency) (5,6)	
Phosphohexose isomerase	Nonspherocytic hemolytic (7)	
Pyruvate decarboxylase	Intermittent movement disorder (8)	
α-1,4-Glucosidase (lysosomal)	Glycogenosis, Type II (Pompe's disease) (9,10)	+ (11,12)
α-1,6-Glucosidase	Glycogenosis, Type III (Debrancher deficiency) (13)	
α-1,6-Glucan, α-1,4-glucan, 6-glycosyltransferase	Glycogenosis, Type IV (Brancher deficiency) (14)	
Defective MPS degradation (?)	Mucopolysaccharidoses (15,16) I. Hurler's syndrome II. Hunter's syndrome III. San Filippo syndrome IV. Morquio's syndrome V. Scheie syndrome VI. Maroteaux-Lamy syndrome	+ (17)
β-Galactosidase	GM$_1$ gangliosidosis (18,19)	
Hexosaminidase A	Tay-Sachs disease; Juvenile GM$_2$ gangliosidosis (20)	+ (21)
Hexosaminidase A and B	Sandhoff disease (20)	
Galactocerebroside β-galactosidase	Krabbe's disease (galactocerebrosidosis) (22)	+ (23)
β-Glucosidase	Gaucher's disease (glucocerebrosidosis) (24)	
Ceramidetrihexosidase	Fabry's disease (ceramidetrihexosidosis) (25)	+ (25)
B. Amino acid metabolism		
Branched-chain amino acid decarboxylase	Maple syrup urine disease (26)	
Lysine-ketoglutarate reductase	Hyperlysinemia (28)	
Cystathionine synthetase	Homocystinuria (29)	
B$_{12}$ coenzyme	B$_{12}$-Dependent methylmalonic aciduria (30) B$_{12}$-Responsive homocystinuria and methylmalonic aciduria (31)	

TABLE II. (*Continued*)

HUMAN BIOCHEMICAL DEFECTS EXPRESSED IN SKIN AND AMNIOTIC CELL CULTURE

Enzyme or Metabolic Defect	Genetic Disease	Prenatal Diagnosis Reported with Amniotic Cell Cultures
Cystine transport (lysosomal) (?)	Cystinosis (32)	
Argininosuccinase	Argininosuccinic aciduria (33)	
Argininosuccinate synthetase	Citrullinemia (34)	
Propionyl CoA carboxylase	Ketotic hyperglycinemia (35)	
Methylmalonyl coenzyme A mutase	Methylmalonic acidemia (36)	+ (37)
C. Lipid metabolism		
α oxidation of phytanic acid	Refsum's disease (38)	
Arylsulfatase A	Metachromatic leukodystrophy (39,40)	+ (41)
Sphingomyelinase	Neiman-Pick disease (42)	
D. Nucleic acid metabolism		
Hypoxanthine-guanine phosphoribosyltransferase	Lesch-Nyhan syndrome (hyperuricemia) (43)	+ (44)
Orotidylic pyrophophorylase and decarboxylase	Oroticaciduria (45)	
DNA repair	Xeroderma pigmentosum (46)	
E. Other		
Uroporphoryrinogen III co-synthetase	Congenital erythropoietic porphyria (47)	
Catalase	Acatalasia (Japanese and Swiss types) (48)	
Acid phosphatase	Acid phosphatase deficiency (49)	+ (49)

a Numbers indicate references as follows: (1) Monteleone *et al.*, 1971; (2) Russell and DeMars, 1967; (3) Tedesco and Mellman, 1969b; (4) Nadler, 1968; (5) Gartler *et al.*, 1962; (6) Nitowsky *et al.*, 1965; (7) Krone *et al.*, 1970; (8) Blass *et al.*, 1970; (9) Dancis *et al.*, 1969a; (10) Nitowsky and Grunfeld, 1967; (11) Cox *et al.*, 1970; (12) Nadler and Messina, 1969; (13) Justice *et al.*, 1970; (14) Howell *et al.*, 1971; (15) Matalon and Dorfman, 1969; (16) Neufeld and Fratantoni, 1970; (17) Fratantoni *et al.*, 1969; (18) Pinsky *et al.*, 1970; (19) Sloan *et al.*, 1969; (20) Okada *et al.*, 1971; (21) O'Brien *et al.*, 1971; (22) Suzuki and Suzuki, 1971; (23) Suzuki *et al.*, 1971; (24) Beutler *et al.*, 1971; (25) Brady *et al.*, 1971; (26) Dancis *et al.*, 1969a; (27) Milunsky *et al.*, 1970; (28) Dancis *et al.*, 1969b; (29) Uhlendorf and Mudd, 1968; (30) Rosenberg *et al.*, 1969; (31) Mudd *et al.*, 1970; (32) Schneider *et al.*, 1967; (33) Shih *et al.*, 1969; (34) Tedesco and Mellman, 1967; (35) Hsia *et al.*, 1971; (36) Morrow *et al.*, 1969; (37) Morrow *et al.*, 1970; (38) Herndon *et al.*, 1969a; (39) Porter *et al.*, 1969; (40) Kaback and Howell, 1970; (41) Nadler and Gerbie, 1970; (42) Sloan *et al.*, 1969; (43) Rosenbloom *et al.*, 1968; (44) DeMars *et al.*, 1969; (45) Howell *et al.*, 1967; (46) Cleaver, 1970; (47) Romeo *et al.*, 1970; (48) Krooth *et al.*, 1962; (49) Nadler and Egan, 1970.

gative interest with these cell cultures relates to their immunogloblin-producing capacity (Takahaski *et al.*, 1969).

G. Amniotic Cell Cultures

An important clinical dividend from studies of genetic variation in cells in culture has been the development of prenatal diagnosis based on analysis of amniotic cells in culture. These cells can be obtained by amniocentesis as early as the fourteenth to eighteenth week of pregnancy. The current investigations and clinical applications of amniotic cell cultures have been recently reviewed (Milunsky *et al.*, 1970).

V. Summary

This discussion has described the current state of our knowledge concerning the biochemical function of human diploid cells in culture. This is still a subject where specific information is fragmentary. Yet the evidence is now abundant, and we hope that this chapter has so presented it, that these cells have had, and should have an even more important role in the investigation of human genetic and, more broadly, human biological problems.

ACKNOWLEDGMENT

Supported, in part, by U.S. Public Health Service Research Grants HD-00588, HD-15545, and HD-02721 from the National Institute of Child Health and Human Development.

REFERENCES

Aaronson, S. A., and Todaro, G. J. (1969). Human diploid cell transformation by DNA extracted from the tumor virus SV40. *Science* **166**, 390–391.

Alpers, J. B., Wu, R., and Racker, E. (1963). Regulatory mechanisms in carbohydrate metabolism. VI. Glycogen metabolism in HeLa cells. *J. Biol. Chem.* **238**, 2274–2280.

Baugh, C. L., and Tytell, A. A. (1967). Propagation of the human diploid cell WI-38 in galactose medium. *Life Sci.* **6**, 371–380.

Beutler, E., Kuhl, W., Trinidad, F., Teplitz, R., and Nadler, H. (1971). β-glucosidase activity in fibroblasts from homozygotes and heterozygotes for Gaucher's disease. *Amer. J. Hum. Genet.* **23**, 62–66.

Blass, J. P., Avigan, J., and Uhlendorf, B. W. (1970). A defect in pyruvate decarboxylase in a child with an intermittent movement disorder. *J. Clin. Invest.* **49**, 423–432.

Brady, R. O., Uhlendorf, B. W., and Jacobson, C. B. (1971). Fabry's disease: Antenatal detection. *Science* **172**, 174–175.

Brandes, D., Murphy, D. G., and Anton, E. (1972). *J. Ultrastruc. Res.* In press.

Broder, S. W., Glade, P. R., and Hirschhorn, K. (1970). Establishment of long-term lines from small aliquots of normal lymphocytes. *Blood* **35**, 539–542.

Brown, B. I., Brown, D. H., and Jeffrey, P. L. (1970). Simultaneous absence of alpha-1,4-glucosidase and alpha-1,6-glucosidase activities (pH 4) in tissues of children with type II glycogen storage disease. *Biochemistry* **9**, 1423–1428.

Chang, R. S. (1962). Metabolic alterations with senescence of human cells. *Arch. Intern. Med.* **110**, 563–568.

Choi, K. W., and Bloom, A. D. (1970). Biochemically marked lymphocytoid lines—establishment of Lesch-Nyhan cells. *Science* **170**, 89–90.

Cleaver, J. E. (1970). DNA repair and radiation sensitivity in human (xeroderma pigmentosum) cells. *Int. J. Radiat. Biol.* **18**, 557.

Comings, D. E., and Okada, T. A. (1970). Electron microscopy of human fibroblasts in tissue culture during logarithmic and confluent stages of growth. *Exp. Cell Res.* **61**, 295–301.

Condon, M. A. A., Oski, F. A., and DiMauro, S. A., and Mellman, W. J. (1971). Glycolytic differences between fetal and non-fetal fibroblast lines. *Nature (London)* **229**, 214–215.

Conover, J. H., Hathaway, P., Glade, P. R., and Hirschhorn, K. (1970). Persistence of phosphoglucomutase (PGM) polymorphism in long-term lymphoid lines. *Proc. Soc. Exp. Biol. Med.* **133**, 750.

Cox, R. P., Douglas, G., Hutzler, J., Lynfield, J., and Dancis, J. (1970). In-utero detection of Pompe's Disease. *Lancet* **1**, 893.

Crisofalo, V. J. (1965). Unpublished data.

Cristofalo, V. J. (1970). Metabolic aspects of aging in diploid cells. *In* "Aging in Cell and Tissue Culture" (E. Holečková and V. J. Cristofalo, eds.), pp. 83–119. Plenum Press, New York.

Cristofalo, V. J. (1972a). Animal cell cultures as a model system for the study of aging. *Advan. Gerontol. Res.* **4**, 45–79.

Cristofalo, V. J. (1972b). Manuscript in preparation.

Cristofalo, V. J., and Kritchevsky, D. (1965). Growth and glycolysis in the human diploid cell strain WI-38. *Proc. Soc. Exp. Biol. Med.* **118**, 1109–1113.

Cristofalo, V. J., and Kritchevsky, D. (1966). Respiration and glycolysis in the human diploid cell strain WI-38. *J. Cell. Comp. Physiol.* **67**, 125–132.

Cristofalo, V. J., and Kritchevsky, D. (1969). Cell size and nucleic acid content in the diploid human cell line WI-38 during aging. *Med. Exp.* **19**, 313–320.

Cristofalo, V. J., Kabakjian, J. R., and Kritchevsky, D. (1967). Heterogeneity of acid phosphatase in the human diploid cell strain WI-38. *Proc. Soc. Exp. Biol. Med.* **126**, 649–653.

Cristofalo, V. J., Howard, B. V., and Kritchevsky, D. (1970). The biochemistry of human cells in culture. *Res. Progr. Org.-Biol. Med. Chem.*, 95–146.

Cristofalo, V. J., Opalek, A., and Baker, B. B. (1972a). In preparation.

Cristofalo, V. J., Baker, B. B., and Godfrey, S. S. (1972b). In preparation.

Dancis, J., Hutzler, J., and Lynfield, J. (1969a). Absence of acid maltase in glycogenesis type 2 (Pompe's disease) in tissue culture. *Amer. J. Dis. Child.* **117**, 108–111.

Dancis, J., Hutzler, J., and Cox, R. P. (1969b). Enzyme defect in skin fibroblasts in intermittent branched-chain ketonuria and in maple syrup urine disease. *Biochem. Med.* **2**, 407–411.

Dancis, J., Hutzler, J., and Cox, R. P. (1969c). Familial hyperlysinemia with lysine-ketoglutarate reductase insufficiency. *J. Clin. Invest.* **48**, 1447–1452.

Danes, B. S., and Bearn, A. G. (1966a). Hurler's syndrome. A genetic study in cell culture. *J. Exp. Med.* **123**, 1–16.

Danes, B. S., and Bearn, A. G. (1966b). Hurler's syndrome. Effect of retinol (vitamin A alcohol) on cellular mucopolysaccharides in cultured human skin fibroblasts. *J. Exp. Med.* **124**, 1181–1198.

Danes, B. S., Broadfoot, M. M., and Paul, J. (1963). A comparative study of respiratory metabolism in cultured mammalian cell strains. *Exp. Cell Res.* **30**, 369–378.

Davidson, R. G., Nitowsky, H. M., and Childs, B. (1963). Demonstration of two populations of cells in the human female heterozygous for glucose-6-phosphate dehydrogenase variants. *Proc. Nat. Acad. Sci. U.S.* **50**, 481–485.

DeMars, R. (1968). A temperature-sensitive glucose-6-phosphate dehydrogenase in mutant cultured human cells. *Proc. Nat. Acad. Sci. U.S.* **61**, 562–569.

DeMars, R., and Nance, W. E. (1964). Electrophoretic variants of glucose-6-phosphate dehydrogenase and the single-active-X in cultivated human cells. *Wistar Inst. Symp. Monogr.* **1**, 35–48.

DeMars, R., Sarto, G., Felix, J. S., and Benke, P. (1969). Lesch-Nyhan mutation: Prenatal detection with amniotic fluid cells. *Science.* **164**, 1303–1305.

DiMauro and Mellman (1972). In preparation.

DiMauro, S., Mellman, W. J., Oski, F., and Baker L. (1969). Glycogen and hexose metabolism in fibroblast cultures from galactosemic and glycogenesis Type II patients. *Pediat. Res.* **3**, 368.

De Mauro *et al.* (1972). Unpublished data.

Dunaway, G. A., and Smith, E. C. (1970). Comparative enzymology of human diploid and virus-transformed cells. *Ann. Okla. Acad. Sci.* **1**, 84–93.

Eagle, H. (1955). Nutrition needs of mammalian cells in tissue culture. *Science.* **122**, 501–504.

Eagle, H. 1959. Amino acid metabolism in mammalian cell cultures. *Science.* **130**, 432–437.

Eagle, H., and Piez, K. A. 1962. The population-dependent requirement by cultured mammalian cells for metabolites which they can synthesize. *J. Exp. Med.* **116**, 29–43.

Eagle, H., Barban, S., Levy, M., and Schulze, H. O. (1958). The utilization of carbohydrates by human cell cultures. *J. Biol. Chem.* **235**, 551–558.

Eagle, H., Washington, C., and Friedman, S. M. (1966). The synthesis of homocystine, cystathionine, and cystine by cultured diploid and heteroploid human cells. *Proc. Nat. Acad. Sci. U.S.* **56**, 156–163.

Eker, P. (1966). Studies on thymidine kinase of human liver cells in culture. *J. Biol. Chem.* **241**, 659–662.

Elsdale, T., and Foley, R. (1969). Morphogenetic aspects of multi-layering in petri dish cultures of human fetal lung fibroblasts. *J. Cell Biol.* **41**, 298–311.

Felix, J. S., and DeMars, R. (1969). Purine requirement of cells cultured from humans affected with Lesch-Nyhan syndrome (hypoxanthine-guanine phosphoribosyltransferase deficiency). *Proc. Nat. Acad. Sci. U.S.* **62**, 536–543.

Fratantoni, J. C., Neufeld, E. F., and Uhlendorf, B. W. (1969). Intrauterine diagnosis of the Hurler and Hunter syndromes. *N. Engl. J. Med.* **280**, 686–688.

Gartler, S. M. (1960). Demonstration of cellular uptake of polymerized RNA in mammalian cell cultures. *Biochem. Biophys. Res. Commun.* **3**, 127–131.

Gartler, S. M., Gandini, E., and Cepellini R. (1962). Glucose-6-phosphate dehydrogenase deficient mutant in human cell cultures. *Nature* (*London*) **193**, 602–603.

Goldberg, G., and Green, H. (1968). The synthesis of collagen and protocollagen hydroxylase by fibroblastic and non-fibroblastic cell lines. *Proc. Nat. Acad. Sci. U.S.* **59**, 1110–1115.

Goldberg, B., and Green, H. (1969). Relation between collagen synthesis and collagen proline hydroxylase activity in mammalian cells. *Nature* (*London*) **221**, 267–268.

Goldman, H., Scriver, C. R., Aaron, K., and Pinsky, L. (1970). Use of dithiothreitol to correct cystine storage in cultured cystinotic fibroblasts. *Lancet* **1**, 811–812.

Goldstein, S., Littlefield, J. W., and Soeldner, J. S. (1969). Diabetes mellitus and aging: Diminished plating efficiency of cultured human fibroblasts. *Proc. Nat. Acad. Sci. U.S.* **64**, 155–160.

Green, H., Goldberg, B., and Todaro, G. J. (1966). Differentiated cell types and the regulation of collagen synthesis. *Nature* (*London*) **212**, 631–633.

Green, M., Henle, G., and Deinhardt, F. (1958). Respiration and glycolysis of human cells grown in tissue culture. *Virology* **5**, 206–219.

Griffiths, J. B. (1970a). The quantitative utilization of amino acids and glucose and contact inhibition of growth in cultures of the human diploid cell WI-38. *J. Cell Sci.* **6**, 739–749.

Griffiths, J. B. (1970b). The effect of insulin on the growth and metabolism of the human diploid cell, WI-38. *J. Cell Sci.* **7**, 575–585.

Hakami, N., and Pious, D. A. (1968). Mitochondrial enzyme activity in "senescent" and virus-transformed human fibroblasts. *Exp. Cell Res.* **53**, 135–138.

Ham, R. G. (1963). An improved nutrient solution for diploid Chinese hamster and human cell lines. *Exp. Cell Res.* **29**, 515–526.

Hamerman, D., Todaro, G. J., and Green, H. (1965). The production of hyaluroniate by spontaneously established cell lines and viral transformed lines of fibroblastic origin. *Biochim. Biophys. Acta* **101**, 343–351.

Hay, R. J. (1967). Cell and tissue culture in aging research. *Advan. Gerontol. Res.* **2**, 121–158.

Hay, R. J., and Strehler, B. L. (1967). The limited growth span of cell strains isolated from the chick embryo. *Exp. Gerontol.* **2**, 123–235.

Hayakawa, M. (1969). Progressive changes of the growth characteristics of the human diploid cells in serial cultivation *in vitro*. *Tohoku J. Exp. Med.* **98**, 171–179.

Hayflick, L. (1965). The limited *in vitro* lifetime of human diploid cell strains. *Exp. Cell Res.* **37**, 614–636.

Hayflick, L., and Moorhead, P. S. (1961). The serial cultivation of human diploid cell strains. *Exp. Cell Res.* **25**, 585–621.

Henle, W., Diehl, V., Kohn, G., zurHausen, H., and Henle, G. (1967). Herpes-type virus and chromosome marker in normal leukocytes after growth with X-irradiated Burkitt cells. *Science* **157**, 1064–1065.

Herndon, J. H., Steinberg, D., and Uhlendorf, B. W. (1969a). Refsum's disease: Defective oxidation of phytanic acid in tissue cultures derived from homozygotes and heterozygotes. *N. Engl. J. Med.* **281,** 1034–1038.

Herndon, J. H., Steinberg, D., and Uhlendorf, B. W. (1969b). Refsum's disease: Characterization of the enzyme defect in cell culture. *J. Clin. Invest.* **48,** 1017–1032.

Holmes, R., Helms J., and Mercer, G. (1969). Cholesterol requirement of primary diploid human fibroblasts. *J. Cell Biol.* **42,** 262–271.

Holtz, A. I., Uhlendorf, B. W., and Fredrickson, D. S. (1964). Persistence of a lipid defect in tissue cultures derived from patients with Niemann-Pick disease. *Fed. Proc., Fed. Amer. Soc. Exp. Biol.* **23,** 128.

Houck, J. C., Sharma, V. K., and Hayflick, L. (1971). Functional failures of cultured human diploid fibroblasts after continued population doublings. *Proc. Soc. Exp. Biol. Med.* **137,** 331–333.

Howard, B. V., and Kritchevsky, D. (1969). The source of cellular lipid in the human diploid cell strain WI-38. *Biochim. Biophys. Acta* **187,** 293–301.

Howell, R. R., Klinenberg, J. R., and Krooth, R. S. (1967). Enzyme studies on diploid cell strains developed from patients with hereditary orotic aciduria. *Johns Hopkins Med. J.* **120,** 81–88.

Howell, R. R., Kaback, M. M., and Brown, B. I. (1971). Type IV glycogen storage disease—branching enzyme deficiency in skin fibroblasts and possible heterozygote detection. *J. Pediat.* **78,** 638–642.

Hsia, Y. E., Scully, K. J., and Rosenberg, L. E. (1971). Inherited propionyl-CoA carboxylase deficiency in "ketotic hyperglycinemia." *J. Clin. Invest.* **50,** 127–130.

Hug, G., Schubert, W. K., and Soukup, S. (1971). Ultrastructure and enzymatic deficiency of fibroblast cultures in Type II glycogenosis. *Pediat. Res.* **5,** 107–112.

Jacobs, J. P. 1966. A simple method for the propagation and maintenance of human diploid cell strains. *Nature (London)* **210,** 100–101.

Jainchill, J. L., and Todaro, G. J. (1970). Stimulation of cell growth in vitro by serum with and without growth factor. *Exp. Cell Res.* **59,** 137–146.

Justice, P., Ryan, C., Hsia, D. Y., and Krmpotik, E. (1970). Amylo-1,6-glucosidase in human fibroblasts. Studies in Type III glycogen storage disease. *Biochem. Biophys. Res. Commun.* **39,** 301–306.

Kaback, M. M., and Howell, R. R. (1970). Infantile metachromatic leukodystrophy: Heterozygote detection in skin fibroblasts and possible applications to intrauterine diagnosis. *N. Engl. J. Med.* **282,** 1336–1340.

Kihara, H., and De La Flor, S. D. (1968). Arginase in fetal calf serum. *Proc. Soc. Exp. Biol. Med.* **129,** 303–304.

Kohn, G., Diehl, V., Mellman, W. J., Henle, W., and Henle, G. (1968). C-group chromosome marker in long-term leucocyte cultures. *J. Nat. Cancer Inst.* **41,** 795–804.

Kritchevsky, D., and Howard, B. V. (1966). The lipids of the human diploid cell strain WI-38. *Ann. Med. Exp. Biol. Fenn.* **44,** 343–347.

Krone, W., Schneider, G., Schulz, D., Arnold, H., and Blume, K. G. (1970). Detection of phosphohexose isomerase deficiency in human fibroblast cultures. *Humangenetik* **10,** 224–230.

Krooth, R. S. (1964). Properties of diploid cell strains developed from patients with an inherited abnormality of uridine biosynthesis. *Cold Spring Harbor Symp. Quant. Biol.* **29,** 189–212.

Krooth, R. S. (1965). The future of mammalian cell genetics. *Birth Defects, Orig. Art. Ser.* 1, 21–56.

Krooth, R. S., and Sell, E. K. (1970). The action of Mendelian genes in human diploid cell strains. *J. Cell Physiol.* 76, 311–330.

Krooth, R. S., and Weinberg, A. N. (1960). Properties of galactosemic cells in culture. *Biochem. Biophys. Res. Commun.* 3, 518–524.

Krooth, R. S., Howell, R. R., and Hamilton, H. H. (1962). Properties of acatalasic cells growing *in vitro*. *J. Exp. Med.* 115, 313–328.

Kruse, P. F. (1965). Personal Communication.

Kruse, P. F., and Miedema, E. (1965a). Production and characterization of multiple-layered populations of animal cells. *J. Cell Biol.* 27, 273–279.

Kruse, P. F., and Miedema, E. (1965b). Glucose uptake related to proliferation of animal cells *in vitro*. *Proc. Soc. Exp. Biol. Med.* 119, 1110–1112.

Kruse, P. F., Whittle, W., and Miedema, E. (1969). Mitotic and non-mitotic multiple-layered perfusion cultures. *J. Cell Biol.* 42, 113–121.

Kurtz, M. J., and Stidworthy, G. H. (1969). Enzymatic sulfation of mucopolysaccharides as a function of age in cultured rat gut fibroblasts. *Proc. Int. Congr. Gerontol., 8th*, Vol. II, p. 49.

Levine, E. M., Becker, Y., Boone, C. W., and Eagle, H. (1965). Contact inhibition, macromolecular synthesis, and polyribosomes in cultured human diploid fibroblasts. *Proc. Nat. Acad. Sci. U.S.* 53, 350–356.

Levintow, L., and Eagle, H. (1961). Biochemistry of cultured mammalian cells. *Annu. Rev. Biochem.* 30, 605–640.

Lieberman, I., Abrams, R., Hunt, N., and Ove, P. (1963). Levels of enzyme activity and deoxyribonucleic acid synthesis in mammalian cells cultured from the animal. *J. Biol. Chem.* 238, 3955–3962.

McCoy, T. A., Maxwell, M., and Kruse, P. F., Jr. (1959). Amino acid requirements of the Novikoff hepatoma *in vitro*. *Proc. Soc. Exp. Biol. Med.* 100, 115–120.

Macek, M., Hurych, J., and Chvapil, M. (1967). The collagen protein formation in tissue cultures of human diploid strains. *Cytologia* 32, 426–443.

Macieira-Coelho, A., and Pontén, J. (1969). Analogy in growth between late passage human embryonic and early passage human adult fibroblasts. *J. Cell Biol.* 43, 374–377.

Macieira-Coelho, A., Pontén, J., and Philipson, L. (1966a). The division cycle and RNA synthesis in diploid human cells at different passage levels *in vitro*. *Exp. Cell Res.* 42, 673–684.

Macieira-Coelho, A., Pontén, J., and Philipson, L. (1966b). Inhibition of the division cycle in confluent cultures of human fibroblasts *in vitro*. *Exp. Cell Res.* 43, 20–29.

Mangum, J. H., Murray, B. K., and North, J. A. (1969). Vitamin B_{12} dependent methionine biosynthesis in cultured mammalian cells. *Biochemistry* 8, 3496–3499.

Mahoney, M. J., and Rosenberg, L. E. (1970). Uptake of α-aminoisobutyric acid by cultured human fibroblasts. *Biochim. Biophys. Acta* 219, 500.

Martin, G. M., Sprague, C. A., and Epstein, C. J. (1970). Replicative life span of cultivated human cells. Effect of donor's age, tissue and genotype. *Lab. Invest.* 23, 86–92.

Matalon, R., and Dorfman, A. (1966). Hurler's syndrome: Biosynthesis of acid mucopolysaccharides in tissue culture. *Proc. Nat. Acad. Sci. U.S.* 56, 1310–1316.

Matalon, R., and Dorfman, A. (1968). Acid mucopolysaccharides in cultured fibroblasts of cystic fibrosis of the pancreas. *Biochem. Biophys. Res. Commun.* 33, 954–958.

Matalon, R., and Dorfman, A. (1969). Acid mucopolysaccharides in cultured human fibroblasts. *Lancet* 2, 838–841.

Matalon, R., Cifonelli, J. A., and Zellweger, H. (1968). Lipid abnormalities in a variant of Hurler syndrome. *Proc. Nat. Acad. Sci. U.S.* 59, 1097–1102.

Matalon, R., Dorfman, A., Dawson, G., and Sweeley, C. C. (1969). Glycolipid and mucopolysaccharide abnormality in fibroblasts of Fabry's disease. *Science* 164, 1522–1523.

Matalon, R., Cifonelli, J. A., and Dorfman, A. (1971). L-iduronidase in cultured human fibroblasts and liver. *Biochem. Biophys. Res. Commun.* 42, 340–345.

Mellman, W. J., and Kohn, G. (1970). Human cell cultures: Their use in the investigation and diagnosis of disease. *Med. Clin. N. Amer.* 54, 701–712.

Michl, J., and Svobodová, J. (1970). RNA and DNA metabolism in aging cultured cells. *In* Aging in Cell and Tissue Culture (E. Holečková and V. J. Cristofalo, eds.), pp. 133–146. Plenum Press, New York.

Miller, L. R., Gordon, G. B., and Bensch, K. G. (1968). Cytologic alterations in hereditary metabolic disorders. I. The effects of galactose on galactosemic fibroblasts *in vitro. Lab. Invest.* 19, 428–436.

Milunsky, A., Littlefield, J. W., Kanfer, J. N., Kolodny, E. H., Shih, V. E., and Atkins, L. (1970). Prenatal genetic diagnosis (first of three parts). *N. Engl. J. Med.* 283, 1370–1381.

Monteleone, J. A., Beutler, E., Monteleone, P. L., Utz, C. L., and Casey, E. C. (1971). Galactokinase deficiency in a child: Studies of a kindred. *J. Pediat.* 78, 1067–1068 (abstr.).

Morrow, G., 3rd, Mellman, W. J., and Barness, L. A. (1969). Propionate metabolism in cells cultured from a patient with methylmalonic acidemia. *Pediat. Res.* 3, 217–254.

Morrow, G., 3rd, Schwarz, R. H., Hallock, J. A., and Barness, L. A. (1970). Prental detection of methylmalonic acidemia. *J. Pediat.* 77, 120–123.

Mudd, S. H., Uhlendorf, B. W., Hinds, K. R., and Levy, H. L. (1970). Deranged B₁₂ metabolism: Studies of fibroblasts grown in tissue cultures. *Biochem. Med.* 4, 215–239.

Nadler, H. L. (1968). Antenatal diagnosis of hereditary disorders. *Pediatrics* 42, 912–917.

Nadler, H. L., and Egan, T. J. (1970). Deficiency of lysosomal acid phosphatase. A new familial metabolic disorder. *N. Engl. J. Med.* 282, 302–307.

Nalder, H. L., and Gerbie, A. B. (1970). Role of amniocentesis in the intrauterine detection of genetic disorders. *N. Engl. J. Med.* 282, 596–599.

Nadler, H. L., and Messina, A. M. (1969). In-utero detection of type-II glycogenosis (Pompe's disease). *Lancet* 2, 1277–1278.

Nadler, H. L., Chacko, C. M., and Rachmeler, M. (1970). Interallelic complementation in hybrid cells derived from human diploid strains deficient in galactose-1-phosphate uridyl transferase activity. *Proc. Nat. Acad. Sci. U.S.* 67, 976–982.

Neufeld, E. F., and Frantantoni, J. C. (1970). Inborn errors of mucopolysaccharide metabolism. *Science* 169, 141–146.

Nitowsky, H. M., and Grunfeld, A. (1967). Lysosomal X-glucosidase in Type II glycogenesis activity in leukocytes and cell cultures in relation to genotype. *J. Lab. Clin. Med.* 69, 472–484.

Nitwosky, H. M., Davidson, R. G., Soderman, D. D., and Childs, B. (1965). Glucose-6-phosphate dehydrogenase activity of skin fibroblast cultures fron enzyme-deficient subjects. *Johns Hopkins Med. J.* **117**, 363–373.

O'Brien, J. S., Okada, S., Fillerup, D. L., Veath, M. L., Adornato, B., Brenner, P. H., and LeRoy, J. B. (1971). Tay-Sachs disease: Prenatal diagnosis. *Science* **172**, 61–64.

Okada, S., Veath, M. L., LeRoy, J., and O'Brien, J. S. (1971). Ganglioside GM_2 storage diseases: Hexaminidase deficiencies in cultured fibroblasts. *Amer. J. Hum. Genet.* **23**, 55.

Pallavacini, J. C., Weismann, U., Uhlendorf, W. B., and di Sant Agnese, P. A. (1970). Glycogen content of tissue culture fibroblasts from patients with cystic fibrosis and other heritable disorders. *J. Pediat.* **77**, 280–284.

Paul, J. (1965). Carbohydrate and energy metabolism. *In* "Cells and Tissues in Culture" (E. N. Willmer, ed.), Vol. 1, pp. 239–276. Academic Press, New York.

Peterson, W. D., Stulberg, C. S., Swanborg, N. K., and Robinson, A. R. (1968). Glucose-6-phosphate dehydrogenase isoenzymes in human cell cultures determined by sucrose-agar gel and cellulose acetate zymograms. *Proc. Soc. Exp. Biol. Med.* **128**, 772–776.

Phillips, H. J., and Andrews, R. V. (1960). Instability of metabolic quotients obtained from tissue cultures. *Proc. Soc. Exp. Biol. Med.* **103**, 160–163.

Phillips, H. J., and McCarthy, H. L. (1956). Oxygen uptake and lactate formation of HeLa cells. *Proc. Soc. Exp. Biol. Med.* **93**, 573–576.

Pinsky, L., Powell, E., and Callahan, J. (1970). GM_1-gangliosidosis types 1 & 2—enzymatic differences in cultured fibroblasts. *Nature (London)* **228**, 1093–1095.

Platter, H., and Martin, G. M. (1966). Tryptophane transport in cultures of human fibroblasts. *Proc. Soc. Exp. Biol. Med.* **123**, 140–143.

Porter, M. T., Fluharty, A. L., and Kihara, H. (1969). Metachromatic leukodystrophy: Arylsulfatase-A deficiency in skin fibroblast cultures. *Proc. Nat. Acad. Sci. U.S.* **62**, 887–891.

Porter, M. T., Fluharty, A. L., and Kihara, H. (1971). Correction of abnormal cerebroside sulfate metabolism in cultured metachromatic leukodystrophy fibroblasts. *Science* **172**, 1263–1265.

Puck, T. T., Cieciura, S. J., and Robinson, A. (1958). Genetics of somatic mammalian cells. III. Long-term cultivation of euploid cells from human and animal subjects. *J. Exp. Med.* **108**, 945–955.

Racker, E. (1956). Carbohydrate metabolism in ascites tumor cells. *Ann. N.Y. Acad. Sci.* **63**, 1017–1021.

Rhode, S. L., 3rd, and Ellem, K. A. (1968). Control of nucleic acid synthesis in human diploid cells undergoing contact inhibition. *Exp. Cell Res.* **53**, 184–204.

Robbins, E., Levine, E. M., and Eagle, H. (1970). Morphologic changes accompanying senescence of cultured human diploid cells. *J. Exp. Med.* **131**, 1211–1222.

Romeo, G., and Migeon, B. R. (1970). Genetic inactivation of the α-galactosidase locus in carriers of Fabry's disease. *Science* **170**, 180–181.

Romeo, G., Kaback, M. M., and Levin, E. Y. (1970). Uroporphyrinogen III cosynthetase activity in fibroblasts from patients with congenital erythropoietic porphyria. *Biochem. Genet.* **4**, 659.

Rosenberg, L. E., Lilljeqvist, A., Hsia, Y. E., and Rosenbloom, F. M. (1969). Vitamin B_{12} dependent methylmalonicaciduria: Defective B_{12} metabolism in cultured fibroblasts. *Biochem. Biophys. Res. Commun.* **37**, 607–614.

Rosenbloom, F. M., Henderson, J. F., and Caldwell, I. C. (1968). Biochemical bases of accelerated purine biosynthesis de novo in human fibroblasts lacking hypoxanthine-guanine phosphoribosyl transferase. *J. Biol. Chem.* 243, 1166–1173.

Rothblat, G. H., Boyd, R., and Deal, C. (1971). Cholesterol biosynthesis in WI-38 and WI-38VA13A tissue culture cells. *Exp. Cell Res.* 67, 436–440

Russell, J. D., and DeMars, R. (1967). UDP-glucose: α-D-galactose-1-phosphate uridyltransferase activity in cultured human fibroblasts. *Biochem. Genet.* 1, 11–24.

Ryan, J. M., and Quinn, L. Y. (1971). Nucleohistone content during aging in tissue culture. *In Vitro* 6, 269–273.

Sadamoto, M. (1966). Nature of cultured cells of the skin from acatalasemic individuals with Takahara's disease. *Acta Med. Okayama* 20, 193–202.

Salzman, N. P. (1961). Animal cell cultures. *Science* 133, 1559–1565.

Salzman, N. P., Eagle, H., and Sebring, E. D. (1958). The utilization of glutamine, glutamic acid, and ammonia for the biosynthesis of nucleic acid bases in mammalian cell cultures. *J. Biol. Chem.* 230, 1001–1012.

Sanford, K. K., Dupree, L. T., and Covalesky, A. B. (1963). Biotin, B$_{12}$ and other vitamin requirements of a strain of mammalian cells grown in chemically defined medium. *Exp. Cell Res.* 31, 345–375.

Schafer, I. A., Silverman, L., and Sullivan, J. C. (1967). Ascorbic acid deficiency in cultured human fibroblasts. *J. Cell Biol.* 34, 83–95.

Schimke, R. T., and Barile, M. F. (1964). Arginine breakdown in mammalian cell culture contaminated with pleuro-pneumonia-like organisms (PPLO). *Exp. Cell Res.* 30, 593–596.

Schneider, J. A., and Weiss, M. C. (1971). Expression of differentiated functions in hepatoma cell hybrids. I. Tyrosine aminotransferase in hepatoma-fibroblast hybrids. *Proc. Nat. Acad. Sci. U.S.* 68, 127–131.

Schneider, J. A., Rosenbloom, F. M., and Bradley, K. H. (1967). Increased free-cystine content of fibroblasts cultured from patients with cystinosis. *Biochem. Biophys. Res. Commun.* 29, 527–531.

Schulman, J. D., and Bradley, K. H. (1971). Cystinosis: Therapeutic implications of *in vitro* studies of cultured fibroblasts *J. Pediat.* 78, 833–836.

Scott-Emuakpor, A., Higgins, J. V., and Kohrman, A. F. (1971). Citrullinemia: A new case, with implications concerning adaptation to defective urea synthesis. *Abstr., 81st Ann. Meet. Soc. Pediatr. Res.* p. 201.

Seegmiller, J. E., Rosenbloom, F. M., and Kelley, W. N. (1967). Enzyme defect associated with a sex-linked human neurological disorder and excessive purine synthesis. *Science* 155, 1682–1684.

Shih, V., Littlefield, J. W., and Moser, H. W. (1969). Argininosuccinase deficiency in fibroblasts cultured from patients with argininosuccinic aciduria. *Biochem. Genet.* 3, 81–83.

Sigman, B., and Gartler, S. M. (1966). The absence of inactivation at two autosomal loci. *Humangenetik* 2, 372–377.

Siniscalco, M., Klinger, H. P., and Eagle, H. (1969). Evidence for intergenic complementation in hybrid cells derived from two human diploid strains each carrying an X-linked mutation. *Proc. Nat. Acad. Sci. U.S.* 62, 793–799.

Sloan, H. R., Uhlendorf, B. W., Jacobson, C. B., and Fredrickson, D. S. (1969). Beta-galactosidase in tissue culture derived from human skin and bone marrow. Enzyme defect in M$_1$ gangliosidosis. *Pediat. Res.* 3, 532–537.

Stenchever, M. A., Hopkins, A. L., and Sipes, J. (1967). Dimethyl sulfoxide and related compounds. Some effects on human fibroblasts *in vitro. Proc. Soc. Exp. Biol. Med.* **126**, 270–273.

Sullivan, J. C., and Schafer, I. A. (1966). Survival of pronase-treated cells in tissue culture. *Exp. Cell Res.* **43**, 676.

Suzuki, Y., and Suzuki, K. (1971). Krabbe's globoid cell leukodystrophy: Deficiency of galactocerebrosidase in serum, leukocytes, and fibroblasts. *Science* **171**, 73–75.

Suzuki, K., Schneider, E. L., and Epstein, C. J. (1971). *In utero* diagnosis of globoid cell leukodystrophy (Krabbe's disease). *Biochem. Biophys. Res. Commun.* **45**, 1363–1366.

Swim, H. E., and Parker, R. F. (1957). Culture characteristics of human fibroblasts propagated serially. *Amer. J. Hyg.* **66**, 235–243.

Szybalska, E. H., and Szybalski, W. (1962). Genetics of human cell lines. IV. DNA-mediated heritable transformation of a biochemical trait. *Proc. Nat. Acad. Sci. U.S.* **48**, 2026–2034.

Takahashi, M., Takagi, N., Yagi, Y., Moore, G. E., and Pressman, D. (1969). Immunoglobin production in cloned sublines of a human lymphocytoid cell line. *J. Immunol.* **102**, 1388–1393.

Taysi, K., Kistenmacher, M. L., Punnett, H. H., and Mellman, W. J. (1969). The limitations of metachromasia as a diagnostic aid in pediatrics. *N. Engl. J. Med.* **281**, 1108–1111.

Tedesco, T. A., and Mellman, W. J. (1966). DNA assay as a measure of cell number in preparations from monolayer cell cultures and peripheral blood leucocytes. *Exp. Cell Res.* **45**, 230–232.

Tedesco, T. A., and Mellman, W. J. (1967). Agininosuccinate synthetase activity and citrulline metabolism in cells cultured from a citrullinemic subject. *Proc. Nat. Acad. Sci. U.S.* **57**, 829–834.

Tedesco, T. A., and Mellman, W. J. (1969a). Inhibition of mammalian uridine-diphosphoglucose 4-epimerase by the dithiothreitol-stimulated formation of NADH. *Biochim. Biophys. Acta* **191**, 144–154.

Tedesco, T. A., and Mellman, W. J. (1969b). Galactose-1-phosphate uridyl transferase and galactokinase activity in cultured human diploid fibroblasts and peripheral blood leukocytes. I. Analysis of transferase genotypes by the ratio of the activity of the two enzymes. *J. Clin. Invest.* **48**, 2390–2397.

Tedesco, T. A., and Mellman, W. J. (1971). Galactosemia: Evidence for a structural gene mutation. *Science* **172**, 727–728.

Tenconi, R., Baccichetti, C., Zacchello, F., and Sartori, E. (1970). Metachromasia in cultured fibroblasts of subjects with glycogenosis type II. *Experientia* **26**, 1238–1239.

Tjio, J. H., and Puck, T. T. (1958). Genetics of somatic mammalian cells. II. Chromosomal constitution of cells in tissue culture. *J. Exp. Med.* **108**, 259–268.

Todaro, G. J., Wolman, S. R., and Green, H. (1963). Rapid transformation of human fibroblasts with low growth potential into established cell lines by SV40. *J. Cell. Comp. Physiol.* **62**, 257–265.

Uhlendorf, B. W., and Mudd, S. H. (1968). Cystathionine synthase in tissue culture drived from human skin: Enzyme defect in homocystinuria. *Science* **160**, 1007–1009.

Wajntal, A., and DeMars, R. (1967). A tetrazolium method for distinguishing between cultured human fibroblasts having either normal or deficient levels of glucose-6-phosphate dehydrogenase. *Biochem. Genet.* **1**, 61–64.

Wang, K. M., Rose, N. R., Bartholomew, E. A., Balzer, M., Berde, K., and Foldvary,

M. (1970). Changes of enzymic activities in human diploid cell line WI-38 at various passages. *Exp. Cell Res.* **61**, 357–364.

Waymouth, C. (1967). Somatic cells *in vitro:* Their relationship to progenitive cells and to artificial milieux. *Nat. Cancer Inst., Monogr.* **26**, 1–21.

Weinstein, D. (1966). Comparison of pronase and trypsin for detachment of human cells during serial cultivation. *Exp. Cell Res.* **43**, 234–236.

Wu, R. (1959). Leakage of enzymes from ascites tumor cells. *Cancer Res.* **19**, 1217–1222.

Wuu, K., and Krooth, R. S. (1968). Dihydroorotic acid dehydrogenase activity of human diploid cell strains. *Science* **160**, 539–541.

Yuan, G. C., and Chang, R. S. (1969). Effect of hydrocortisone on age-dependent changes in lipid metabolism of primary human amnion cells *in vitro*. *Proc. Soc. Exp. Biol. Med.* **130**, 934–936.

Zacchello, F., Tenconi, R., and Baccichetti, C. (1969). Glycogenosis type II—glycogen storage in cell cultures from muscle. *Experientia* **25**, 1316–1317.

Zwartuow, H. T., and Westwood, J. C. N. (1958). Factors affecting growth and glycolysis in tissue cultures. *Brit. J. Exp. Pathol.* **39**, 529–539.

11

COMPLEX CARBOHYDRATES OF MAMMALIAN

CELLS IN CULTURE

Paul M. Kraemer

I. Introduction *

Only a few years ago, complex carbohydrates (i.e., glycoproteins, glycolipids, and glycosaminoglycans) were usually thought of, if at all,

* Abbreviations: Glc, glucose; Gal, galactose; Fuc, fucose; Man, mannose; Xyl, xylose; GlcNAc, N-acetyl glucosamine; GalNAc, N-acetyl galactosamine; NANA, N-acetyl neuraminic acid; GlcUA, glucuronic acid; IdUA, iduronic acid; Ser, serine; Thr, threonine; NH-Asp, asparagine; Hyl, hydroxylysine.

as the extracellular products of particular differentiated cells. Investigators with a generally "cellular" orientation were often aware that many serum proteins are glycoproteins, that mucins are rich in carbohydrate, that brain tissue contains gangliosides, that the "ground substances" of tissues are partly susceptible to hyaluronidase, and that cartilage formation involves secretion of chondroitin sulfates. Furthermore, almost all the structural information concerning the complex carbohydrates was then derived from studies of these extracellular or differentiated materials, since only in that way was it possible to obtain a sufficient sample of a single molecular species.

A contemporary view is quite different from the above. Complex carbohydrates are now recognized to be vital constituents of the cellular level of life. All animal cells, including cells in culture, make them, and each cell makes them in considerable variety. For instance, the mouse L cell in culture, after more than 20 years in culture, continues to synthesize a wide spectrum of glycolipids, including gangliosides (Weinstein et al., 1970; Yogeeswaran et al., 1970), numerous glycoproteins (Warren and Glick, 1968) including collagen (Green and Goldberg, 1965), and glycosaminoglycans (Aleo et al., 1967) including heparan sulfate (Kraemer, 1971b). A similar impression of complex carbohydrate variety is obtained when any other homogeneous population of cells in culture, such as HeLa cells or leukocytes or various types of early-passage material, is reviewed. This is not to imply that all cells can synthesize all the major types; for instance, it has been reported that lymphocytes and lymphoma cells do not synthesize collagen (Green and Goldberg, 1965). On the other hand, it now appears probable that all animal cells in culture synthesize heparan sulfate, suggesting a general life function for this species (Dietrich and De Oca, 1970; Kraemer, 1971b). In some cases, cells in culture synthesize enormous amounts of a particular complex carbohydrate [such as the hyaluronic acid production of synovial cells in culture (Clarris and Fraser, 1968) or the immunoglobulin production of myeloma cell lines (e.g., Schenkein and Uhr, 1970a)]; in such cases, it may appear as if only the one variety was made, since the separations and analyses are overwhelmed by the preponderant differentiated species. However, careful examination invariably shows that the usual range of complex carbohydrates is, in fact, present.

The ability of cells to synthesize complex carbohydrates has several implications. First, glycosylation of various materials may be a prerequisite to the process of secretion (Eylar, 1965). This idea implies that the general mechanism of secretion of macromolecules (excluding certain specialized forms of secretion involving zymogen granule formation) involves recognition of carbohydrate moieties for passage through membranes. Second, it now appears clear that complex carbohydrates are integral constituents of cellular membranes in general and plasma membranes in particular (see reviews of Winzler, 1970; Kraemer, 1971a). Why this is so, in the general case, is not clear; it may have something to do with the general mechanism of binding of enzymes to membranes (R. G. Spiro, 1970) or with the ever-present dynamic turnover of living cellular membranes (Warren, 1969). Third, the plasma membrane not only contains complex carbohydrate as do other cellular membranes but is, in fact, enormously enriched in these species. Much of the excess, which may constitute a major portion of the total cellular complex carbohydrate, is concentrated on the external face of the plasma membrane. The generality of this carbohydrate-rich cell surface *in vivo* has been emphasized by Rambourg and Leblond's studies (1967) of rat tissues and appears to be true for all cells in culture as well, whether grown in suspension or as monolayers.

Studies of cells in culture have been vital to the recent general advances made in this field. The reasons are similar to the reasons for usefulness of these systems for the study of other macromolecules such as nucleic acids. They include the advantages of working with defined, homogeneous populations of cells, growing under defined conditions, with many of the complexities of the animal removed. Provision of radioactive precursors results in very high specific-activity products, compared to those achieved by *in vivo* labeling. In some cases, metabolism can be studied in synchronized populations of cells or cells in the process of reaggregating or expressing differentiated functions. There is no doubt that these *in vitro* systems may yield some results which reflect the peculiarities of the mode of cultivation; nevertheless, many basic biological mechanisms appear to be similar to those operative *in vivo*.

II. General Structural Features of Mammalian Complex Carbohydrates

There are a large number of recent reviews on the structure of complex carbohydrates, and a detailed outline will not be attempted here. The reader is referred to glycoprotein reviews by Montgomery (1970), R. G. Spiro (1970), and Winzler (1972); glycolipid reviews by McCluer (1968), McKibbin (1970), and Svennerholm (1970); glycosaminoglycan reviews by Meyer (1966), Cifonelli (1968), and Jeanloz (1970); and general complex carbohydrate reviews by Ginsburg and Neufeld (1969) and Kraemer (1971a). A rudimentary synopsis of general structural features follows.

Native mammalian complex carbohydrates consist of carbohydrate chains covalently linked to protein (glycoproteins and glycosaminoglycans) or to lipid (glycolipids). The monosaccharide residues of glycoproteins and glycolipids include glucose, galactose, mannose, fucose, N-acetyl galactosamine, N-acetyl glucosamine, and sialic acid (generic term for acylated neuraminic acids); glycosaminoglycans (commonly called mucopolysaccharides) contain N-acetyl glucosamine, N-acetyl galactosamine, glucuronic acid, iduronic acid, galactose, and xylose. The last two sugars form the linkage region to peptide with the structure

$$\beta\text{-Glc-UA } 1 \rightarrow 3\beta\text{-Gal } 1 \rightarrow 3\beta\text{-Gal } 1 \rightarrow 4\beta\text{-Xyl-}O\text{-serine}$$

Most of the well-studied glycolipids are sphingoglycolipids with linkage between glucose and ceramide. Three types of linkage are known for glycoproteins: β-GlcNAc-NH-aspartate ("plasma-type"), α-GalNAc-O-Ser (Thr) ("mucin-type"), and β-Gal-O-hydroxylysine ("collagen-type"). The sugar chains of all mammalian complex carbohydrates are oriented with the reducing terminal of each sugar toward the aglycone moiety. The chain length, in the case of the glycoproteins and glycolipids, varies from 1 to about 20 sugars; chain branching is the rule for all but the smallest chains, and the chain sequence and linkages are invariably complex. Glycosaminoglycans, by contrast, often have very long unbranched sugar chains that consist primarily of alternating copolymers of uronic acid and hexosamine and that may be sulfated (chondroitin sulfates, heparan sulfates) or unsulfated (hyaluronic acid). Structures intermediate between classic glycoproteins and glycosaminoglycans are known (keratosulfates and sulfated glycoproteins), in which case the linkage is of glycoprotein type rather than the glycosaminoglycan type. At the macromolecular level, several types of sugar-chain multiplicity patterns are recognized, and, in some cases, various hybrid structures can occur. Some sugars, such as sialic acid and fucose, are generally found only as chain terminal groups, but all other sugars can be terminal as well. For a number of reasons, complex carbohydrate molecules generally show more heterogeneity than other cellular macromolecules (Gottschalk, 1969); hence, there is a certain vagueness to the question of how many entities occur. There is some reason to believe, however, that structural "fine tuning" is highly relevant to function at the cellular level.

III. Biosynthesis

A. *General Features*

There is now general agreement concerning the basic features of complex carbohydrate biosynthesis in higher animals (recent reviews include Ginsburg, 1964; Leloir, 1964; Warren, 1966; Roseman, 1968; Ginsburg and Neufeld, 1969; Stoolmiller and Dorfman, 1969; R. G. Spiro, 1970; Kraemer, 1971a). A brief synopsis of these processes follows.

1. Activation of Monosaccharides

The immediate precursors of the heteropolymeric sugars are the respective nucleotide sugars, of which the normal forms are UDP derivatives of glucose, galactose, xylose, glucuronic acid, iduronic acid, N-acetyl glucosamine, and N-acetyl galactosamine; GDP derivatives of mannose and fucose; and CMP derivatives of the sialic acids. All of these, under "physiological" conditions, are formed from fructose-6-P by a large series of enzymatically catalyzed reactions. The activation reactions, catalyzed by pyrophosphorylases, are of general form

nucleoside triphosphate + sugar-1-P → nucleotide sugar + inorganic P

2. Incorporation into Heteropolymers

Biosynthesis of each sugar chain is catalyzed by batteries of specific glycosyl transferases which mediate the stepwise, one-sugar-at-a-time, linkages between the reducing end of the activated monosaccharides and a nonreducing group of the terminal sugar of the growing chain. It appears likely that each individual glycosyl transferase possesses three-way specificity: (1) a high degree of specificity for a particular nucleotide sugar; (2) a high degree of specificity in terms of the particular sugar-sugar- or sugar-peptide, -lipid linkage formed; and (3) specificity which varies in broadness, among the individual glycosyl transferases, for numerous features of the incomplete acceptor structure.

3. Relationship of Glycosylation to Membranes

The intracellular site of complex carbohydrate formation has invariably been found to be membrane-associated. This has been true in a very large number of studies whether the localization of nascent glycosylated product was determined or the localization of the glycosyl transferases was the parameter studied. Beyond this, however, there is little uniformity of opinion as to whether any particular cellular membrane system is the exclusive or even major site for all complex carbohydrates.

A good deal of the information on which the above principles are based was obtained from a variety of materials: from whole animals, organs, tissues, and cell-free systems. Not all the individual metabolic pathways have been shown to be important in cells in culture. In particular, little is known about the relative importance of alternate, reversible, and coordinated pathways. That such information may be of great interest and may be readily attainable is suggested by Kalckar's studies of the enzyme responsible for conversion of UDP-Glc to UDP-Gal (Kalckar, 1965; Robinson et al., 1966). These workers have suggested that in cells in culture coordinated metabolic controls may exist between energy-producing machinery (aerobic glycolysis) and the cellular sociology as influenced by cell-surface complex carbohydrates. Such a notion would neatly bring together classic views of neoplasia (Warburg, 1956) with more recent views concerning the cell-surface architecture. In any event, recent studies have indicated that the cell in culture provides an extremely useful system for examination of these biosynthetic processes. Therefore, further details of these processes, biased in favor of the conditions and purposes of cell culture studies, are presented below.

B. Intermediary Metabolism

Figure 1 illustrates the general pathways, evidently operative in all animal cells in culture, by which the 10 activated monosaccharides are formed from fructose-6-P. Hence, under the usual culture conditions, glucose of the medium provides the carbon skeleton for all monosaccharide residues of all the glycoproteins, glycosaminoglycans, and glycolipids of animal cells in culture. It seems likely that, under these conditions, all 10 nucleotide sugars are always maintained at low but sufficient concentrations—low concentrations in the sense that only minor withdrawal pressure is put upon the general nucleotide pools of the cell but sufficient concentration to permit a wide variety of complex carbohydrate biosyntheses. The primary regulatory device appears to be feedback inhibition by the nucleotide sugars of early, unique steps in their own respective biosynthetic pathways. Three of these feedback inhibitions are indicated in Fig. 1 and are widely accepted to exist

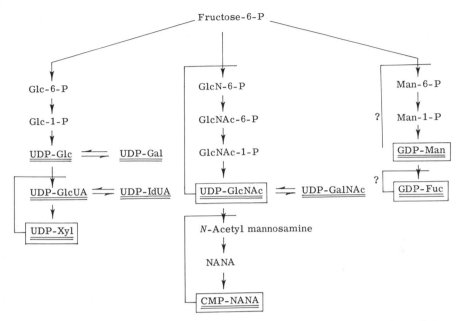

Fig. 1. Formation pathways for animal nucleotide sugars. Steps inhibited by feedback inhibition by nucleotide sugars are indicated.

in animal material in general: (1) CMP-NANA inhibits UDP-N-acetyl glucosamine-2-epimerase (Kornfeld *et al.*, 1964); (2) UDP-N-acetyl glucosamine inhibits glutamine-D-fructose-6-phosphate transaminase (Kornfeld *et al.*, 1964); and (3) UDP-xylose inhibits UDP-glucose dehydrogenase (Neufeld and Hall, 1965). In addition, GDP-fucose and GDP-mannose probably regulate their own formation independently of each other (R. H. Kornfeld and Ginsburg, 1966; Kaufman and Ginsburg, 1968).

In addition to the pathways illustrated in Fig. 1, a variety of other enzymes (or enzymes that are less specific) are present in cells in culture that permit incorporation of exogenous sugars other than glucose into complex carbohydrates. In general, such exogenous sugars are not greatly metabolized to CO_2 or converted into other species such as amino acids or fatty acids by cells in culture *provided that* (1) they are present in low concentration (e.g., "labeling" concentration), (2) adequate glucose is also provided, and (3) the time period studied is of relatively short duration. Hence, complex carbohydrates can often be labeled without serious dispersion of label that occurs upon provision of radioactive glucose, by providing one of these other radioactive sugars.

It would be convenient, therefore, if each of the 10 normally occurring nucleotide sugars could be specifically and exclusively labeled in whole cells by exogenous provision of the respective radioactive monosaccharides. This is not possible for a variety of reasons. For one, transport into the cell of the uronic acids, sialic acids, and acetylated amino sugars is apparently sluggish or absent; this also appears to be true of most of the phosphorylated intermediates and the nucleotide sugars themselves. In another situation, exogenous xylose is also not incorporated as such, probably because of absence of the appropriate pyrophosphorylase. Finally, since several of the nucleotide sugars, in addition to serving as direct precursors for incorporation into heteropolymers, are also precursors to other nucleotide sugars, expansion and labeling of pools of the entire set occur. HeLa cells, for instance, readily convert GDP-mannose-^{14}C to GDP-fucose-^{14}C (Bosmann et al., 1968a); hence, exogenous mannose could be incorporated as both mannose and fucose.

Radioactive glucosamine has generally been the precursor of choice for cell culture studies. As indicated in Fig. 1, the radioactivity from glucosamine is incorporated as N-acetyl glucosamine, N-acetyl galactosamine, and sialic acid (S. Kornfeld and Ginsburg, 1966). Furthermore, provision of glucosamine, which is readily transported, acetylated, and phosphorylated, results in bypass of the normal control point (feedback inhibition of glutamine-fructose-6-P transaminase), resulting in rapid inflation of the intracellular UDP-N-acetyl hexosamine pools and consequent rapid dilution of cold endogenous pools. In fact, high concentrations of exogenous glucosamine (2 mM) are toxic to cells in culture, presumably due to the excessive demands created for uridine-containing nucleotides in general (S. Kornfeld and Ginsburg, 1966). At lower concentrations, provision of exogenous radioactive glucosamine has proved to be very useful as a "block" label for the complex carbohydrates of cells in culture. Dispersion and breakdown of label are often small or nonexistent. For instance, following a 4-hour pulse of HeLa cells with glucosamine-1-^{14}C, no significant loss of $^{14}CO_2$ occurred until the culture was grown in "chase" medium for several days (S. Kornfeld and Ginsburg, 1966). With CHO Chinese hamster cells labeled for one generation with 0.1 μC/ml of glucosamine-1-^{14}C, isolated and purified histone Fl (a nonglycosylated protein) contained no radioactivity at all (Kraemer and Gurley, 1969). On the other hand, white blood cell cultures have been reported to convert glucosamine to glucose in Eagle's minimal medium (S. Kornfeld and Gregory, 1968). However, according to Hayden et al. (1970), this was probably due to the inadvertent presence of many platelets. Since the incorporated radioactivity appears as N-acetyl hexosamines and sialic acids, almost all animal complex carbohydrates become labeled, the known exceptions being collagen-type glycoproteins (glycosylated with hexoses only) and some of the simpler glycolipids.

In some studies, it would be desirable to label only sialic acid residues. Since uptake of sialic acids themselves is denied, one might suppose that provision of radioactive mannosamine would achieve this end. However, it appears that, while exogenous mannosamine is transported and phosphorylated, acetylation is not accomplished and, hence, no incorporation into macromolecules occurs (Raisys and Winzler, 1970). Some workers, therefore, have utilized exogenous radioactive N-acetyl mannosamine. As expected, uptake is sluggish (Brady et al., 1969); however, specific sialic acid labeling has been demonstrated in Tay-Sachs brain ganglioside isolated from rats given intracerebral injections of N-acetyl-^{3}H mannosamine (Kolodny et al., 1970). In some instances, exogenous mannosamine has been demonstrated to have physiological effects, unexplained by the above, such as in the cell aggregation system of Oppenheimer et al. (1969). However, in the latter example, the mannosamine may operate merely as an amine donor. In any case, there has been no demonstration that mannosamine or N-acetyl mannosamine occur as residues of animal complex carbohydrates.

While radioactive glucosamine has proved to be an excellent "block" label for animal complex carbohydrates, exogenous radioactive fucose has frequently been reported to be incorporated only as fucose (Kaufman and Ginsburg, 1968). For instance, in a recent study of BHK hamster cells, Buck et al. (1970) reported that less than 5% of the incorporated radioactivity from supplied fucose-^{3}H and fucose-^{14}C was dispersed to other materials. Thus, despite the fact that mammalian tissues and organs possess the enzymes for conversion of fucose to CO_2 and dispersion of carbons to other materials (Yuen et al., 1970), as a practical matter, use of exogenous radioactive fucose for specific labeling of fucose-containing complex carbohydrates of cells in culture has proved feasible.

In general, the specificity of labeling from exogenous precursors of constituent monosaccharides of complex carbohydrates synthesized by cells in culture will depend on the particular purposes and conditions of the study. For instance, in a study of the viral membrane glycoprotein of Sindbis virus, grown in either chick embryo fibroblasts or BHK Syrian hamster cells, Strauss et al. (1970) write, "The specificity of labeling with the various carbohydrates was greater than anticipated. Fucose, glucosamine and galactose appear to be highly specific (during the 10-hour growth cycle of the virus). Mannose is less specific, but the metabolic fate of this compound appears to involve conversion into amino acids rather than into other carbohydrates. Glucose is the least specific, but again it appears to be converted only to mannose and amino acids; almost no labeled galactose or glucosamine is observed in the glucose-labeled virus."

Bosmann and Winston's study (1970) of total incorporation of glucosamine-^{14}C and fucose-^{3}H supplied as a 30-minute pulse to L5178Y murine lymphoma cells showed that radioactivity from glucosamine appeared as glucosamine (70%), galactosamine (20%), and sialic acid (10%); radioactivity from fucose appeared as fucose (85%), mannose (10%), and galactose (5%). However, various cellular fractions of these totals were quite distinctive; for instance, secreted glycoproteins had 50% of their label from glucosamine as sialic acid and 50% as glucosamine, while all the fucose-derived label appeared as fucose in this material. Most of the galactosamine labeled from glucosamine-^{14}C was in the cellular glycolipid fraction.

C. Glycosyl Transferases

The 10 nucleotide sugars, which as the immediate precursors of hetero-polymeric carbohydrates, can be considered as the substrates for 10 families of glycosyl transferases. The members of each transferase family have a common high specificity for both base and sugar moieties of their nucleotide sugar, while the individual members are distinguished one from another by the following characteristics: (1) *acceptor specificity,* including both the nature of the sugar, amino acid, or lipid terminal to which the substrate sugar will be attached, as well as other micro- or macromolecular characteristics of the entire acceptor molecule; (2) *particular glycoside bond formed* under the influence of the transferase; (3) *intracellular localization* of the transferase. Five of the 10 families include members whose function it is to *initiate* glycosylation of an aglycone peptide or lipid structure. These five are the transferases that attach (1) xylose to serine residues of peptide in the formation of glycosaminoglycans, (2) galactose to hydroxylysine residues of collagen and galactose to ceramide in the formation of some sphingoglyco-lipids, (3) glucose to ceramide in the formation of most sphingoglyco-lipids, (4) GlcNAc to asparagine residues of protein in the formation of plasma-type glycoproteins, and (5) GalNAc to Ser (Thr) residues in the formation of mucin-type glycoproteins. Nine of the 10 families (the exception being the xylosyl transferase) also include members that participate in the formation of sugar-sugar linkages, of which almost

50 have been reported to occur in animal material. On this basis alone, the glycosyl transferase families would at present consist of about (in addition to initiating enzymes) 5 sialyl transferases, 6 fucosyl, 12 galactosyl, 5 mannosyl, 7 N-acetyl glucosaminosyl, 5 N-acetyl galactosaminosyl, 4 glucuronosyl, 1 glucosyl, and 1 iduronosyl transferase, for a total of 46 transferases. These numbers are derived as in the following example for the galactosyl transferase family: Galactose is known to be attached, in various animal materials, β-Gal 1 → 3, 4, or 6 Gal, α-Gal 1 → 3 or 4 Gal, β-Gal 1 → 3, 4, or 6 GlcNAc, β-Gal 1 → 3 or 4 GalNAc, β-Gal 1 → 4 Xyl, and β-Gal 1 → 4 Glc, for a total of 12 linkages.

Many workers believe that each sugar-sugar or sugar-aglycone linkage requires a separate transferase (e.g., Roseman, 1970). Opinion varies on the question of whether different transferases are required for catalyzing the same linkage when it occurs in different complex carbohydrates or different parts of the same sugar chain. For instance, of the 20 known sugar-sugar linkages that occur in animal sphingoglycolipids, at least 12 of them are also known to occur in animal glycoproteins. Blood-group substances, which occur in the same individual in both glycolipid and glycoprotein form, provide the inference that the same genetically determined transferases catalyze the formation of both types of complex carbohydrate (Ginsburg and Neufeld, 1969). Modification of the *apparent* acceptor specificity of a particular glycosyl transferase could be a function of intracellular localization of both the transferase and endogenous acceptors. In addition, various microenvironmental factors such as different ion or cofactor levels at the intracellular sites of glycosylation could have the same effect. For instance, it is known that α-lactalbumin regulates the acceptor specificity of a galactosyl transferase of mammary gland (Brew et al., 1968; Ebner and Schanbacher, 1970).

In practice, the presence of glycosyl transferase activity is experimentally determined by transfer of the radioactive sugar moiety from a particular added labeled nucleotide sugar to either endogenous or added acceptor in a cell-free system. Nevertheless, if the product is not completely characterized, the presence of one or more members of a family is inferred. Complete characterization of the product permits identification of the subfamilial classes, and if, in addition, the preparation is tested with a suitable spectrum of exogenous acceptors and varying conditions, individual members can be characterized. At the family level, workers in recent years have detected activity for all 10 families using a variety of animal material. In addition, many studies have contributed information at the subfamily and individual level. For instance, the sialyl transferases have been extensively studied in many types of material by Roseman and his co-workers (reviewed by Roseman, 1968). The cumulative total impression gained from this work is that, while the total number of glycosyl transferases operative at the cellular level is still not known, the number is probably in excess of 50 and may well be considerably greater. Since the number of different complex carbohydrate structures produced by animals is also very large, these findings tend to confirm the currently popular ideas on the mechanisms that determine complex carbohydrate structure (i.e., mechanisms based on batteries of genetically determined glycosyl transferases rather than on template-derived information).

Studies of cells in culture have added greatly to our knowledge of glycosyl transferases, their functions *in situ*, and their role in cytodifferentiation, morphogenesis, and malignant transformation (some of these studies will be considered further in other sections). The transferases studied have included members of many families; for instance, in HeLa cells, Bosmann, Eylar, and co-workers have distinguished two glycoprotein fucosyl transferases (Fuc $1 \rightarrow 4$ GlcNAc, Fuc $1 \rightarrow 2$ Gal) (Bosmann *et al.*, 1968a), collagen glucosyl transferase (Glc $1 \rightarrow 2$ Gal), glycoprotein galactosyl transferase (Gal $1 \rightarrow ?$ GlcNAc), and initiation GalNAc transferase (GalNAc $1 \rightarrow$ Ser-peptide) (Hagopian *et al.*, 1968) as well as mannosyl and fucosyl transferases that catalyze the formation of glyceroglycolipids (Bosmann, 1969c). The last are of some interest in several respects: First, mannose-containing glycolipids have not often been reported to occur in animal material—in any case, neither fucose nor mannose glyceroglycolipids are known. Bosmann (1969c) postulates that they may be membrane components, in distinction to the mannose-containing, vitamin A-like glycolipid that has been postulated to be an intermediate in the transfer of mannose from GDP-Man to secreted mannose-containing glycoproteins (Caccam *et al.*, 1969; Zatz and Barondes, 1969; De Luca *et al.*, 1970). Caccam *et al.* (1969) failed to find the latter type of mannolipid in HeLa cells, which they interpreted as related to the low secretory activity of their HeLa cells.

Another interesting facet of Bosmann and Eylar's studies of glycosyl transferases of cells in culture concerns the enzymes that glycosylate collagen (Gal $\rightarrow$ Hyl, Glc $1 \rightarrow 2$ Gal). Green and co-workers have shown that cells in culture vary in their ability to synthesize collagen, varying from high activity (fibroblastic cells) to low activity (HeLa cells) to indetectable biosynthetic capability (lymphoma cells) (Green and Goldberg, 1965; Green *et al.*, 1966). Bosmann (1970b) and Bosmann and Eylar (1968a,b) found little relationship between Green's data and the activity of these transferases with either endogenous or exogenous acceptors. This suggests that these transferases may function to glycosylate collagen and other complex carbohydrates as well.

D. The Cellular Anatomy of Glycosylation

Well over 50 studies have been published in the last decade concerning the intracellular sites of complex carbohydrate biosynthesis [a large number of these studies have been recently reviewed (Kraemer, 1971a)]. These studies nearly all concluded that glycosylation of complex carbohydrates is a membrane-associated process. It is membrane-associated in at least two respects: when whole cells are provided with exogenous radioactive monosaccharides, labeled glycosylated *products* are, at first at least, found to be associated with subcellular membrane fractions rather than free in the cell sap; alternatively, when subcellular fractions are analyzed for *glycosyl transferase activities,* by provision of radioactive nucleotide sugars (in a suitable buffer), again membrane fractions possess the bulk of the activity. Thus the process of glycosylation, in a wide variety of systems, is sequestered from the general cytoplasm (cell sap) at least for a time. Furthermore, the process is largely postribosomal [mitochondria apparently have autonomous complex carbohydrate biosynthetic processes (Bosmann and Martin, 1969)], since many kinetic studies have shown that, following the addition of puromycin, cessation of incorporation of amino acids into polypeptides precedes cessation of sugar incorporation into glycoproteins or mucopolysaccharides. Finally, there is agreement that either labeled glycosylated *products* or *glycosyl transferase activities* can be liberated from these subcellular membrane fractions (which, in most cases, are vesicular and include membranes plus cisternal contents) by nonpolar detergents such as Triton X-100. In general, these findings have been interpreted in the context of the ideas developed by the Rockefeller group concerning the origins and functions of the endoplasmic reticular membranes and other organelles derived from these membranes (Dallner *et al.,* 1966; Redman *et al.,* 1966; Jamieson and Palade, 1967). Briefly, these studies have suggested that the nascent polypeptides from membrane-bound polysomes are transported across the rough endoplasmic reticular membrane, released into the cisternal lumen of that organelle, and then move through the cisternal spaces of first rough, then smooth, endoplasmic reticulum and then into Golgi structures. Various portions of the latter structures can apparently distribute their stored and/or concentrated contents to cell sap, plasma membrane, or extracellular space, although the details of these distribution phenomena are not yet clear. Numerous studies of the intracellular sites of glycosylation largely agree that it is within these sequestered passageways that the bulk of the carbohydrate additions occurs.

Differences between the results of the various studies are largely concerned with the relative importance of the various parts of this structurally and biogenically continuous membranous apparatus. These studies have utilized a variety of tissues such as rat liver, sheep thyroid, or Erhlich ascites cells, as well as cells in culture such as HeLa and various types of fibroblasts. Among the studies with cells in culture, there has also been considerable difference in reported results and conclusions. Uhr and his colleagues have studied the synthesis and glycosylation of immunoglobulin G by P_3K cells, an established mouse myeloma line (Uhr and Schenkein, 1970; Sherr and Uhr, 1970), and have concluded that glycosylation begins while the nascent polypeptides are still attached to polysomes and continues, after release into the cisternal lumen, during passage through both rough and smooth endoplasmic reticulum (their smooth membrane fraction included elements of the Golgi complex). Only glucosamine was significantly incorporated into polysome-bound and rough endoplasmic reticular immunoglobulin, while glucosamine and galactose were both incorporated in the smooth membrane fraction. This was consistent with the view that formation of the initial sugar-peptide linkage (GlcNAc-NH-Asp) should precede further sugar additions, which would include both GlcNAc and Gal. Their results with the P_3K cell line were entirely consistent with their more detailed studies of the serially transplantable LPC_1 plasmacytoma (Sherr and Uhr, 1969; Schenkein and Uhr, 1970a; Zagury et al., 1970) in which electron microscopic autoradiography and the effect of puromycin were also studied. Studies with another myeloma indicated that mannose is also incorporated in rough membranes, along with GlcNAc, as would be expected considering the "core-sugar" composition of this type of glycoprotein, while fucose residues, being terminal, were added just before exit from the cell (Melchers and Knopf, 1967). The general finding that peripheral sugars are added at a more distal site (in respect to the ribosomal origin of the polypeptide chain) has been frequently noted in various studies [e.g., chondroitin sulfate (Horwitz and Dorfman, 1968), thyroglobulin (R. G. Spiro and Spiro, 1966)]. Schubert's studies (1970) with a cloned tissue culture line of an IgA-producing mouse myeloma (S194) were at variance with the above in that no carbohydrate was detected on polysome-bound heavy chains and a temporal lag between release from polysomes and the first carbohydrate additions was evident.

Eylar and co-workers have studied the localization of glycosylation in HeLa cells using both of the two general approaches. In their first paper (Hagopian et al., 1968), microsomal membrane subfractions were assayed for three different glycosyl transferases using both endogenous and added acceptors: (1) polypeptide-GalNAc transferase, an "initiating" transferase for certain mucin-type glycoproteins; (2) a galactosyl transferase that forms Gal $1 \rightarrow 4$ GlcNAc . . . linkages in glycoproteins; and (3) collagen-glucosyl transferase that catalyzes the α-Glc $1 \rightarrow 2$ Gal . . . linkage of animal collagen disaccharides. The first two activities were found predominantly in smooth membrane (including Golgi), while the last was found in their plasma membrane fraction. [The collagen-galactosyl transferase has also been found in the plasma membrane fraction of cultured fibroblasts (Bosmann, 1969a).] They proposed that this difference related to the ultimate fate of the glycoprotein: ribosomal products destined for incorporation into cellular membranes became glycosylated in the smooth endoplasmic reticulum (or Golgi apparatus), while collagen became glycosylated as an immediate prelude to secretion through the plasma membrane. In a second paper (Bosmann et al., 1968a), two fucosyl transferases of HeLa cells (Fuc $1 \rightarrow 4$ GlcNAc, Fuc $1 \rightarrow 2$ Gal) were also found to be localized in the smooth membrane (plus Golgi) fraction; they suggested that this fraction might contain a multienzyme complex for the glycosylation of membrane glycoproteins.

Their studies of the kinetics of localization of exogenous radioactive glucosamine and fucose, using intact HeLa cells (Bosmann et al., 1969), concern both glycoprotein and glycolipid materials. After 30 minutes of labeling, no great difference was found between glucosamine and fucose (both were found predominantly in smooth membrane fractions as either glycolipid or glycoprotein) while, at that time, leucine-^{3}H was predominantly associated with rough membranes. No evidence for ribosome-bound sugar could be detected. Upon "chase" with unlabeled glucosamine, the radioactive glucosamine of both glycolipid and glycoprotein moved progressively to the plasma membrane fraction. These workers propose that this migration of label may represent the formation and integration of large membrane subunits into the plasma membrane. That such a process is consistent with the capabilities of the smooth membrane-Golgi system is indicated by Claude's studies (1970) of lipoprotein granule synthesis in rat liver cells.

By contrast to the conclusions of Eylar and his co-workers with HeLa cells, studies with KB cells and chick embryo fibroblasts have suggested that rough membranes and ribosomes themselves participate in glucosamine incorporation into glycoproteins (Got et al., 1968; Louisot et al., 1970). That different cells might have somewhat different "glycosylation-flow" characteristics is also indicated by Hagopian and Eylar's report (1969) that the polypeptidyl-GalNAc and glycoprotein-

Gal transferases, which in HeLa cells are predominantly localized in smooth membranes, appear to be associated with the plasma membranes of bovine submaxillary gland cells.

The relative importance of the Golgi apparatus, as a subdivision of what many studies isolate as "smooth membranes," continues to be controversial. There is no doubt that incorporated sugars appear in the Golgi apparatus rather quickly (Neutra and Leblond, 1966) and that a large portion of the total cellular activity of at least some glycosyl transferases is present in isolated Golgi-rich fractions of liver (Fleischer et al., 1969; Morre et al., 1969; Fleischer and Fleischer, 1970; Schachter et al., 1970). In thyroid, mannose-^{3}H incorporation into the sugar "core" of thyroglobulin appears to occur in the rough endoplasmic reticulum, while galactose-^{3}H and fucose-^{3}H incorporation into the peripheral part of the sugar chain appears to be localized in the Golgi apparatus (Whur et al., 1969; Herscovics, 1970). On the other hand, the kinetics of movement of materials through these membranous channels, as well as the fluctuating availability of endogenous acceptors from point to point, make this type of study difficult to interpret.

Many workers in this field believe that glycosyl transferases are tightly bound to membrane (Hagopian and Eylar, 1969; Schachter et al., 1970; R. G. Spiro, 1970). If this is true, then measurement of the glycosyl transferase activity of purified membrane fractions, using defined exogenous acceptors, should give an unambiguous answer concerning the in vivo site of glycosylation of materials moving near these enzymes attached to the luminal surfaces of the membranous channels. Such an answer would also be based on the assumption that the membrane components themselves move very slowly relative to soluble cisternal contents (Omura et al., 1967; Widnell and Siekevitz, 1967). While there is no doubt that active glycosyl transferases in soluble form can be prepared (e.g., Bartholomew and Jourdian, 1966; Horwitz et al., 1970; Schenkein and Uhr, 1970b), in practice, membrane fractions are generally assayed in the presence of detergents such as Triton X-100 (e.g., Hagopian et al., 1968). The enhancement of activity with detergents may serve to permit exogenous acceptors and added radioactive nucleotide sugar substrates to approach the transferases by breaking up the vesicles of the membrane preparation. However, such enhancement of activity does not necessarily distinguish between transferase activity sequestered on the luminal surface of the vesicles as tightly membrane-bound enzyme and activity sequestered inside the vesicles but otherwise free. According to Schachter et al. (1970), the distinction can be made by ultrasonic treatment of the membrane preparation, in which case entrapped radioactive product should be released (Helgeland, 1965), while membrane-bound enzymes should remain attached. However, collagen and thyroglobulin glycosyl transferases have been reported to be releasable with mild ultrasonic oscillation (R. G. Spiro and Spiro, 1968, M. J. Spiro and Spiro, 1968a,b), and Hagopian et al. (1968) have noted that collagen-glycosyl transferases have different binding relationship to plasma membrane than their other transferases have to smooth membranes. The implications of this uncertainty concerning the relationship of glycosyl transferases to membranes is as follows: the ribosomal products are thought to be progressively glycosylated in transit through the membranous channel; if the glycosyl transferases are *also* moving through the channel as they operate (although not necessarily at the same rate), then their apparent localization will be subject to the vagaries of movement, storage, pathway length, tortuosity, etc. Hence, definitive localization of the actual site of glycosylation (i.e., the site of conjunction of transferase and an acceptor which is appropriate at that instant) will be very difficult. Some idea of the timing of one aspect of the problem can be gained from Glaumann and Ericsson's study (1970) of the migration of newly synthesized albumin (which is apparently not glycosylated) through the membranes of liver cells. These workers found, after leucine-^{14}C administration, that peak radioactivity of albumin was at the bound ribosomes in 2 minutes, rough membrane channels at 6 minutes, smooth channels at 15 minutes, and Golgi apparatus at 20 minutes. However, the Golgi zone began to contain radioactive albumin at 9 minutes and by 20 minutes appeared to have, already, the greatest mass of newly synthesized albumin. Studies that involve localization of radioactivity after many hours (e.g., Frot-Coutaz et al., 1968) would seem to be more concerned with storage and release phenomena than with production.

Except for the often-appearing general statement that glycosyl transferases are tightly bound to membrane, little is known at present concerning the exact relationship. In particular, it would be helpful if synthesis, intracellular transport, and turnover of these enzymes could be measured in a similar fashion to the Rockefeller studies of the membrane components of liver cells (Omura et al., 1967; Widnell and Siekevitz, 1967; Bock and Siekevitz, 1970). This relationship will undoubtedly turn out to be important. Tetas et al. (1970) report that neutral detergents first activate and then labilize transferase activity and that the inactivation could be prevented with high nucleotide-sugar-substrate concentrations. Mookerjea and Chow (1970) have reported that added CDP-choline stimulated the GlcNAc transferase of endoplasmic reticulum fractions but not a similar transferase of Golgi fractions. Rosen and Rosen (1970) have reported the activation of frog erythrocyte-adenylcyclase particles (which have the characteristics of a sialoglycolipoprotein membrane subunit) by a *Clostridium perfringens* toxin that may well be a phospholipase (Kraemer, 1968). The role of lipid in complex carbohydrate biosynthesis is also underscored by the discovery of labile sugar-lipid structures (Caccam *et al.*, 1969; Zatz and Barondes, 1969; De Luca *et al.*, 1970; Tetas *et al.*, 1970) that turn over very rapidly and may be intermediates in glycoprotein synthesis. Tetas *et al.* (1970) have suggested that one possible function of these substances may be in transport of activated monosaccharides across the membranes so as to make them available to reactions *inside* the cisternae.

IV. Postbiosynthetic Processes

A. General Features

Following glycosylation, the newly created complex carbohydrate molecule has a large number of alternative or sequential fates. For one thing, further modification of the molecule may take place within the membranous passages and vesicles of the smooth endoplasmic reticulum and/or Golgi apparatus. Examples of these include *sulfation* of mucopolysaccharides (Godman and Lane, 1964) or certain glycoproteins (Pamer *et al.*, 1968), *complex formation* with lipid or other membrane components (Bosmann *et al.*, 1969), or possibly *specific scissions* of the molecule as perhaps occur for serum albumin in liver cells (Eylar, 1965) or hyaluronic acid in rooster comb (Swann, 1968) or heparan sulfate of the cell sap of Chinese hamster cells (Kraemer, 1971c).

Regardless of the relative importance of the Golgi apparatus in terminal glycosylation of complex carbohydrates, it is clear that these vesicles have a *storage* function in the sense that cells pulse-labeled with radioactive sugar precursors *continue* to show concentration of radioactivity over the Golgi zone for many hours [e.g., the HeLa cell study of Reith *et al.* (1969) confirming several studies of tissues by Leblond and co-workers (Rambourg *et al.*, 1969)]. Additional storage loci of complex carbohydrates may include the cell sap for immunoglobulins (Swenson and Kern, 1967) and thyroglobulin (R. G. Spiro and Spiro, 1966). In the latter instance, storage may also follow transfer from vesicles to colloid (Whur *et al.*, 1969).

Most workers now believe that complex carbohydrates are *incorporated into cellular membranes*, most prominently into plasma membranes (Rambourg and Leblond, 1967). There is a rapidly growing literature on complex carbohydrates of the plasma membrane (reviewed by Winzler, 1970; Kraemer, 1971a); in addition, it appears that other cellular membranes also have integral complex carbohydrate moieties [e.g., mitochondrial membranes (Bosmann and Martin, 1969), nuclear membrane (P. I. Marcus et al., 1965; Kashnig and Kasper, 1969), and endoplasmic reticular membranes (Wallach and Kamat, 1966; Bosmann et al., 1968c; Evans, 1970)]. The exact role that such complex carbohydrates play in membrane structure, in any general sense, is still somewhat conjectural but has been recently discussed in some detail by the author (Kraemer, 1971a).

Secretion from the cell of origin, of course, constitutes a well-recognized fate of complex carbohydrates, and the process of secretion may well be an important *raison d'être* of glycosylation (Eylar, 1965). It seems to be less well recognized, however, that passage of complex carbohydrate to the outside of the cell of origin includes a broad continuum of possibilities of which secretion, in the sense of complete liberation of a fairly stable exported species, is only one. For one thing, peripherally oriented complex carbohydrates of the cell periphery may have various binding relationships to the plasma membrane (Blumenfeld, 1968), may be stable or rapidly turning over in concert with other plasma membrane components (Warren and Glick, 1968), and may desquamate continuously from the cell surface (Kraemer, 1969). Extracellular complex carbohydrates are also known to be capable of complex interactions in the immediate environment of their cells of origin; such a process can result in formation of extracellular basement membranes that have vital implications for tissue formation and function (Kefalides and Winzler, 1966; Grobstein, 1968; R. G. Spiro, 1970). Finally, extracellular complex carbohydrates may be incorporated into the structure of the periphery of other cells (Sneath and Sneath, 1959; D. M. Marcus and Cass, 1969), which may have important implications for phenomena of cellular interactions (Rubin, 1967).

Complex carbohydrates can also be degraded in their cells of origin. Batteries of specific glycosidases capable of hydrolyzing most of the linkages of a wide variety of complex carbohydrates are apparently ubiquitous in animal cells (Ginsburg and Neufeld, 1969; R. G. Spiro, 1970). Most of these enzymes are confined to lysosomal particles; however, some appear to be free in the cell sap (Tulsiani and Carubelli, 1970), localized in plasma membranes (Fleischer and Fleischer, 1969), or secreted products (Makino et al., 1966; De Salegui and Pigman, 1967). Not only *can* complex carbohydrates be degraded but it appears that they *are* being continuously degraded in living cells: one of the interesting features of the studies of kinetics of labeling of various cellular organelles by the Montreal group was the finding that lysosomes became heavily labeled almost as soon as the biosynthetic membranous apparatus (Rambourg and Leblond, 1967; Rambourg et al., 1969). One can probably go even further and assert that many cellular complex carbohydrates *must* be continuously degraded on the basis that genetic absence of particular glycosidases is frequently associated with serious cytopathology [disorders of degradation have recently been reviewed by R. G. Spiro (1970)].

The sequential or alternative fate of a particular newly synthesized complex carbohydrate species is often complex. Woodward and Davidson (1968) cite an interesting case of alternative fates of chondroitin chains: sulfate is transferred to the 4-O position on galactosamine only when the chains are attached to protein, but sulfate is attached to the 6-O position after the apoprotein core is degraded. Shen et al. (1968) have correlated a particular glycosylation step (Fuc $1 \rightarrow 2$ Gal . . .) with whether A, B, and O blood-group specificities occur as both secreted glycoprotein and cell-bound glycolipid or only as glycolipid. Hayden et al. (1970) have detected a very early stimulation of glucosamine incorporation into phytohemagglutinin-stimulated lymphocytes which they postulate is related to enhanced glycosylation and secretion of preformed γ-globulin. Forstner (1970), in his recent report on the intracellular fates of glycoproteins of rat intestinal mucosa, found early transfer to various loci: surface plasma membrane, cell sap, and mitochondrial compartments. Later, labeled product appeared in the intestinal lumen, but this exodus apparently came from the cell sap compartment rather than surface plasma membrane. In a recent study of the heparan sulfate of Chinese hamster cells in culture (Kraemer, 1971c), four intracellular compartments were detected: rapidly turning-over compartments associated with both internal membranes and cell surface, of which the latter was much greater in mass; and intracellular membrane-bound and cell sap species, both of which turned over more slowly. The cell sap species appeared to be a partially degraded (in terms of peptide moiety) storage form.

B. *Control Mechanisms*

In the context of this review, control mechanisms must include those phenomena that assure the presence of complex carbohydrates, in the right forms, correct amounts, and proper cellular loci, for physiological cellular function. Most authors understandably admit that very little is known along these lines. Nevertheless, from fragmentary precedents and analogies with other systems, one can construct an impressive list of potentially controlling factors of complex carbohydrate metabolism, some of which might even turn out to be important.

Quantal (all or nothing) control, of course, is implied by the genetic component as it determines both glycosyl transferase availability and production of suitable acceptor aglycone species. It follows then that quantitative control might be exerted on this genetic component by the general mechanisms of control of protein synthesis. Along the same line, control of membrane biosynthesis in general would be expected to influence the membrane-associated glycosylation process as, in fact, Warren and Glick's data (1968) on surface membrane turnover indicate. Coordinate regulation of the glycosylation process and energy metabolism (Kalckar, 1965; Paul et al., 1966) or nucleotide metabolism (Vail, 1968) have also been suggested.

As suggested earlier, feedback inhibition of pathways leading to specific nucleotide precursors (S. Kornfeld et al., 1964; Neufeld and Hall, 1965; S. Kornfeld and Ginsburg, 1966) provides an obvious control point; it seems likely, however, that this control is mainly protective of general nucleotide pools rather than responsive to variable cellular requirements (however, see Mazlen et al., 1969). In addition, there is some evidence (e.g., Tetas et al., 1970) that nucleotide sugar substrates may stabilize glycosyl transferases; hence, the nucleotide-sugar concentration might affect the efficiency of glycosylation as acceptors pass through cisternae of the endoplasmic reticulum. Similarly, control might be exerted by other variables of the relationship of glycosyl transferases to membranes (Mookerjea and Chow, 1970).

There is also abundant evidence that complex carbohydrate levels can be influenced by vitamins (e.g., Kochhar et al., 1968; Schafer et al., 1968) and hormones (e.g., Ozzello and Bembry, 1964; Dukes, 1968; Carubelli and Griffin, 1970). Although exact mechanisms of these effects are not clear, it is apparent that the primary effects of such agents may be exerted on any combination of biosynthesis, modification, transfer, or degradation.

The existence of more mysterious control mechanisms is evident from the literature concerning cellular interactions (which is further considered in a later section), which imply control by subtle factors external to the cell. A simple model system for inferring such effects has been the study (Kraemer, 1966, 1967a) of the terminal sialic acid residues of the outwardly oriented glycoproteins of suspension-cultured cells. When cells were treated with neuraminidase, removing only those terminal sialic acid residues of the externally exposed sugar chains, no change in cellular growth or division occurred. Resuspension of cells in medium containing glucosamine-1-^{14}C resulted in a 50-fold increase in surface sialic acid specific activity by 4 hours, treated compared to untreated cells. By 12 hours, the mass of surface sialic acid of the treated cells had returned to "normal" values. These studies indicated that the biosynthetic and transfer-to-the-surface processes were responsive to changes in the outermost periphery of the cell. Both of these processes, for a brief period, were resistant to either actinomycin D or puromycin; hence, it was concluded that the "inductive" phenomena observed influenced, in part at least, the postribosomal processes of glycosylation and transport to the surface membrane.

V. Cellular Physiology and Complex Carbohydrates

A. Complex Carbohydrates and the Mitotic Cycle

Biochemical studies of cell cycle-dependent processes are motivated by at least two general areas of interest. First, of course, such studies might yield information concerning the mechanisms that cause an orderly cell cycle traverse culminating in the mitotic process itself; second, there are abundant (though not consistent) data linking cell cycle-dependent genetic expression with cytodifferentiation. Synchronized cell cultures have been a favorite tool for a wide variety of cell cycle studies (recent general reviews in Baserga, 1971), including a number of studies concerning complex carbohydrates.

Because of the well-known relationship of sialic acid residues of cell-surface complex carbohydrates and cellular electrophoretic mobility (Eylar et al., 1962), the latter parameter has been a popular technique in a variety of contexts (reviewed by Weiss, 1967). Mayhew and O'Grady (1965) and Mayhew (1966), using suspension-cultured RPMI No. 41 cells, synchronized by double-thymidine block or cold shock, reported a transient increase in electrophoretic mobility that coincided with the time of maximum mitosis. Neuraminidase treatment decreased electrophoretic mobility of the cells at all stages of the life cycle and, in addition, eliminated the transient increase associated with mitosis. Similar findings were also reported by Brent and Forrester (1967) using HeLa cells synchronized by mitotic selection. Mayhew (1966) and Weiss (1966) favored the hypothesis that increased mobility was due to a transient elevation of surface sialic acid density at the electrokinetic shear layer, perhaps by a burst of synthesis or transport to the cell surface in cells nearing mitosis. However, Kraemer (1967b), using thymidine-blockade synchronization of suspension-cultured Chinese hamster cells, reported that, in mass terms, surface sialic density remained constant throughout the cell cycle and suggested that the electrophoretic mobility shift might be due to a transient change in the orientation of surface sialoglycans to the electrokinetic shear plane. That such configuration shifts are possible has been shown experimentally by Ward and Ambrose (1969). In any case, the transient electrophoretic mobility shift itself could not be demonstrated with L5178Y murine lymphoma cells synchronized by a thymidine-colcemid technique (Buckhold and Burki, 1969).

Rapid cell cycle dependent changes in the expression of surface complex carbohydrate antigens have also been reported by Cikes (1970), Fox et al. (1971), and Kuhns and Bramson (1968). In the last study, expression of H blood-group antigen of HeLa cells synchronized by thymidine-blockade was pronounced only near mitosis. These authors (1970b) have further reported that the changing antigenic expression is due to a change of synthesis rate and transport to the surface, since H^+ cells incorporated twice as much fucose-^{14}C as H^- cells (Fuc $1 \rightarrow 2$ Gal . . . is the H determinant). However, Kuhns and Bramson (1970a) also reported that some H^- cells could be rendered H^+ by brief trypsin treatment, suggesting that another source of changing antigenic expression in their system might be due to a variable "masking" of the H determinants by other surface glycoprotein species. It is not yet completely clear which processes are cell-cycle-dependent and which are variable for other reasons, such as genetic variability within their HeLa cells.

Synchronized (with thymidine or thymidine plus colcemid) human lymphoid cell lines have been used to study the cycle dependence of immunoglobulin synthesis, storage, and secretion (Buell and Fahey, 1969; Takahashi et al., 1969). Taken together, these reports indicate that all three of these processes occur maximally during the late G_1 and S periods of the cell cycle; these findings suggest that actively growing cells of this type would be more active in expression of their differentiated function, while cells arrested in a G_0 or G_1 phase might be inactive. It seems likely that glycosylation and subsequent processes are closely linked to aglycone production in this system.

Bosmann and his co-workers, using a mouse lymphoma cell line (L5178Y), have made more general studies of processes involved with complex carbohydrate metabolism of cells synchronized by thymidine and colcemid block. In a study (Bosmann and Winston, 1970) of 30-minute pulse labeling with a variety of radioactive precursors (thymidine-^3H, leucine-^3H, glucosamine-^{14}C, fucose-^3H, and choline-

[14]C), these workers found a striking cycle-dependent difference between incorporation into proteins (including glycoproteins) and into lipids (including glycolipids). Incorporation of leucine-^{3}H, glucosamine-^{14}C, and fucose-^{3}H into cellular proteins and glycoproteins all roughly paralleled incorporation of thymidine-^{3}H into DNA (the G_1 and G_2 periods are very brief in their study). Incorporation of leucine-^{3}H, glucosamine-^{14}C, and fucose-^{3}H into secreted glycoproteins, however, yielded much more scattered data but suggested that secretion of these species occurred throughout the cycle. By contrast, incorporation of glucosamine-^{14}C, fucose-^{3}H, and choline-^{14}C into either cellular or secreted lipids and glycolipids was almost nonexistent except for a sharp burst during the G_2 and M periods of the cycle.

The same study (Bosmann and Winston, 1970) showed that L5178Y cells contained glycolipid and glycoprotein transferases (with endogenous acceptors) that could transfer sugars from GDP-fucose, GDP-mannose, UDP-glucose, UDP-galactose, UDP-xylose, UDP-arabinose, and UDP-GalNAc, but cell cycle dependence of these activities was not determined. However, in another study with L5178Y cells (Bosmann, 1970b), transfer activity of glucose and galactose from nucleotide sugars into endogenous cellular glycoprotein material appeared as fairly sharp peaks during the S period. Activity with exogenous collagen acceptors also appeared as sharp peaks during the S period; the significance of the collagen glycosyl transferases in L5178Y cells appears obscure, since Green et al. (1966) have shown that the ability to synthesize collagen is completely repressed in lymphocytes and lymphoma cells. In any event, the studies of Davies et al. (1968), using a variety of cells that produce collagen in culture, show that synthesis of the protein portion of collagen continued steadily throughout the cycle but was more rapid during logarithmic than stationary growth (Priest and Davies, 1969).

Bosmann's group has also studied glycosidase activities of L5178Y cells (Bosmann and Bernacki, 1970); these activities included β-N-acetyl glucosaminidase, β-galactosidase, and β-N-acetyl galactosaminidase whose specific activities decreased in that order in exponentially growing cells. In synchronized cells, the specific activites of each increased continuously throughout the cycle, reaching 3- to 30-fold values by the G_2 and M periods of the first generation after release from block. These findings were in distinct contrast to the results from glycosyl transferases, the incorporation rates of precursors into cellular complex carbohydrates, and the activities of membrane marker enzymes (Bosmann, 1970a). In the latter study, enzymes characteristic of smooth endoplasmic reticulum (UDPase), rough endoplasmic reticulum (esterase), and plasma membrane (5'-nucleotidase) all showed peak activities in the S period. As

Bosmann and Bernacki (1970) point out, however, the cellular content of glycosidases at various times in the life cycle may not be the most important physiological variable, since these values largely represent lysosomal content rather than actual usage for degradation. These authors detected glycosidase release from random cells and suggested that the release function might be for degradation of the cell's own plasma membrane complex carbohydrates. They have not yet studied whether the release of these enzymes is cycle-dependent; however, Allison and Mallucci (1964) reported the premitotic release of lysosome content into the cytoplasm of HcLa cells.

Results with KB cells synchronized by double-thymidine block (Gerner *et al.*, 1970) were somewhat different than the L5178Y cell results. Gerner *et al.* (1970) pulse-labeled synchronized cultures for 1 hour at various times after release using leucine-^{14}C, glucosamine-^{14}C, or choline-^{14}C. The cells were then fractionated into plasma membrane, residual cellular particulates, and soluble protein fractions. All three precursors were incorporated into all three cell fractions throughout the period studied. In addition, a transient period of enhanced incorporation of all three precursors into the plasma membrane fraction occurred immediately following the time of maximum cell division. This period of enhanced incorporation was minor or completely absent in the cell fractions (particulates) containing most of the other cellular membranes.

There are several problems related to the use of synchronized cell cultures that make clear interpretations difficult in many studies. For one thing, the quality of synchrony is often either poor, is not clearly measured, or suffers rapid decay (Anderson and Peterson, 1965). Second, synchronizing agents such as thymidine, FUDR, and colcemid may cause serious perturbations of cellular metabolism (Petersen *et al.*, 1969) so that the phenomena measured after release may partly or wholly represent recovery from this perturbation. Nevertheless, taken as a whole, these studies suggest that there may be two general classes of cell cycle dependence of complex carbohydrate metabolism: "fast reactions" mostly associated with mitosis, and more sluggish changes associated with the remainder of the cycle. The functional implications of these changes are relatively obscure, although shifts in polysaccharide ion binding continue to be implicated in the mitotic process (Heilbrunn, 1956; Lippman, 1968; Robbins and Pederson, 1970a,b). At the very least, these studies certainly do not support the notion that cell cycle traverse and the expression of differentiated functions are always mutually exclusive processes. The induction of cell cycle traverse in cells arrested in G_0 or G_1 by cellular processes that occur after the interaction of external materials with cell-surface complex carbohydrates is exemplified by the

lymphocyte-phytohemagglutinin response and will be considered further
in the next section.

B. Complex Carbohydrate Antigens and Receptors of the Cell Surface

The external cell surface is enormously enriched in terms of the total
cellular content of complex carbohydrates. This is evident from data
concerning the proportion of all of the cellular membranes which is
plasma membrane (Bosmann et al., 1968c; Weinstein et al., 1970), combined with data concerning the proportion of total cellular complex
carbohydrate which is resident on the cell surface (Shen and Ginsburg,
1968; Kraemer, 1969). In general, it seems that from one half to three
fourths of the total cellular complex carbohydrate is on the outside
surface of the plasma membrane, which itself constitutes only from 0.5
to 2% of the total cellular membrane system. It is also probable that
this coat of complex carbohydrate consists of a complex melange of
molecular entities, at least as far as can be determined from the characteristics of glycopeptides removed from the cell surface with trypsin
(Winzler et al., 1967; Kraemer, 1969). There is even some reason to
believe that all the externally oriented plasma membrane proteins are
glycoproteins (Kraemer, 1971a) and, in addition, that the periphery
includes carbohydrate chains of glycolipids and glycosaminoglycans.

In parallel to this information of the complexity and prominence of
the cell-surface complex carbohydrates are the accumulated data of the
complexity and variety of reactive groups at the cell surface. These
accumulated data concern numerous enzymatic activities; receptor sites
for vitamins, hormones, and infectious agents; sites related to transport,
cell-surface involvement in phagocytosis, attachment, or cell movement;
as well as numerous cell-surface antigenic specificities. Not surprisingly,
cell-surface complex carbohydrates have, from time to time, been implicated in many of these reactivities [reviewed by Franks (1968), Winzler
(1970), and Kraemer (1971a)]. In many cases, evidence for complex
carbohydrate involvement has consisted of a change in cellular behavior
following treatment of the cells with glycosidases or trypsin. Winzler
(1970) has recently reviewed the numerous instances of altered cell
behavior following neuraminidase treatment and discusses some of the
difficulties in interpretation of many reports of this type.

One source of difficulty in interpretation of data deserves special com-
ment, since evidence for *indirect* behavioral changes following enzymatic
treatment of intact cells has arisen in a wide variety of contexts. For
instance, Yunis and his co-workers (Kassulke et al., 1969), comparing

normal and leukemic leukocytes, have shown that the apparent deficiency of A and H blood-group specificities of the leukemic cells was due to increased, trypsin-removable, M- and N-active glycoprotein masking the A and H sites. Wallach and de Perez Esandi's study (1964) of three different types of tumor cells emphasizes the fact that various sialic acid residues of the cell surface contribute quite differently to the overall behavior of the cell. The study of Kuhns and Bramson (1970a), in which H antigen of some H⁻ cells could be exposed by brief trypsin treatment, has already been mentioned, and further instances of indirect effects will be discussed later. For the moment, the generality can be formulated that the numerous complex carbohydrate molecular entities of the cell surface *do not* function independently *in situ*. On the contrary, manipulation or degradation of some of these entities, more often than not, is followed by altered configuration, exposure, and probably metabolism of others, thus making cause-and-effect relationships difficult to prove.

With these reservations in mind, nevertheless, there remain a number of instances where cellular reactivity and cell-surface complex carbohydrate structure have been correlated in some detail. One of these is the M,N blood group-active cell-surface glycoprotein. This substance is probably the same molecular entity as the cell-surface myxovirus receptor (Springer, 1967) and can be demonstrated on human erythrocytes and other cells (Winzler, 1970) such as HeLa cells (Kelus *et al.*, 1959). The reactive carbohydrate chains appear to be tetrasaccharides (Adamany and Kathan, 1969; Thomas and Winzler, 1969) linked to serine or threonine residues of the polypeptide chain and having the following structure:

$$\text{NANA } 2 \rightarrow 3 \text{ Gal } 1 \rightarrow 3 \text{ GalNAc-}O\text{-Ser (Thr)}$$
$$\uparrow 6$$
$$| 2$$
$$\text{NANA}$$

In human erythrocytes, the native molecule *in situ* appears to be a multiple subunit structure of several hundred thousand molecular weight (Winzler, 1970). The carbohydrate content is very high (~60%), and a majority of the available serine and threonine residues are substituted with carbohydrate chains, although not necessarily of identical structure. A portion of the carboxyl end of the polypeptide chain is probably carbohydrate-poor, rich in hydrophobic amino acids, and embedded in the membrane itself (Winzler, 1970). The topographical distribution of the molecules on the surface of the membrane appears as discontinuous patches according to Howe *et al.* (1970), a distribution that has been noted by Boyse and Old (1969) with several types of surface antigens. Recent studies utilizing the technique of freeze-etch electron microscopy combined with other techniques (Marchesi, 1970; Pinto da Silva *et al.*, 1970; Tillack *et al.*, 1970) have shown that about 25% of the peptide chain is rich in hydrophobic amino acids, is resistant to proteolysis *in situ*, and is associated with globular particles of 85-Å diameter situated in the interior of the membrane. These studies [as well as the study of Howe *et al.* (1970)] also suggest that such multisubunit glycoproteins, anchored in discrete membrane subunits, simultaneously bear reactivities not only for myxovirus and M,N systems but also to other reactivities such as ABO, phytohemagglutinin, and wheat germ agglutinin.

This conclusion is of particular interest, since it is known that the carbohydrate determinants of these other reactivities are structurally quite different. The ABO determinants (including H and Lewis specificities) involve α-GalNAc → Gal, α-Gal → Gal, Fuc → Gal, and Fuc → GlcNAc determinants (Watkins, 1966; Lloyd and Kabat, 1968), none of which appears in the M,N system; conversely, sialic acid is not involved in the ABO (H) system. As in the case of M,N system reactivities, the ABO system reactivities are detectable in many cells in culture (Franks, 1968). When found as determinants of glycoproteins (these determinants also exist in glycolipid form), they belong to the same general "mucin" class as do the M,N glycoproteins (namely alkalilabile linkage to peptide as O-glycoside bonds between GalNAc and serine or threonine).

Marchesi's report (1970) that the above erythrocyte glycoprotein also carried phytohemagglutinin (PHA) receptors is of particular interest, since the latter structure belongs to the alkali-stable "plasma" variety of glycopeptides, and the presence of additional glycopeptides of this type has been noted by other workers studying the M,N-active glycoprotein. R. H. Kornfeld and Kornfeld (1970) have recently isolated PHA-inhibitory glycopeptides from human erythrocytes and propose the following structure for their most active fraction:

$$
\text{NANA } 2 \rightarrow 6\beta \text{ Gal } 1 \rightarrow 3,4 \text{ GlcNAc} \rightarrow (\text{Man})_2 \rightarrow \overset{\displaystyle |}{\text{GlcNAc-NH-Asp}}
$$
$$
\underset{\text{Gal} \rightarrow \text{GlcNAc}}{\uparrow} \qquad |
$$

The terminal sialic acid residue was irrelevant to biological activity; however, removal of both galactose residues resulted in almost complete loss of activity. Attempts to show PHA-inhibitory activity with other structurally similar glycopeptides (from fetuin and transferrin) were less than 10% as effective.

Determination of the PHA receptor structure is a very exciting piece of work since, in lymphocytes at least, reaction between PHA and the cell-surface receptor initiates a profound series of metabolic changes that include both cell cycle traverse and the derepression of differentiated functions that, in many respects, resemble the natural immunogenic events (see review by Ling, 1968). Although the exact mechanisms by which these processes are stimulated are still unknown, recent evidence indicates that some of the earliest events include stimulation of synthesis of surface membrane components (Fisher and Mueller, 1968; Hayden et al., 1970) and activation of the membrane K^+ transport system (Quastel and Kaplan, 1970).

In general, the expression of these blood-group and allied reactivities is quite variable in cells in culture (Franks, 1968), and their functional implications are obscure. Many people have speculated that their primary function relates to cell-surface cation control and exchange. In this connection, it is of interest that of all the known cellular or secreted blood-group-active substances, a physiological function has been assigned only to one: gastroferrin, which regulates iron transport in the intestine, appears to have all the chemical characteristics of a blood-group-active secreted mucin (Multani et al., 1970).

Unlike the cultured-cell blood-group substances, the major transplantation antigens are very stable cell-surface components (Franks, 1968). For instance, Gaugal et al. (1966) showed that, after more than 20 years in culture, mouse L cells continued to show all the H-2 specificities of the original C3H mouse from which the line derived. Similarly, the specificities of cultured murine somatic hybrid cells were consistent with perfect codominance in H-2 antigen expression (Spencer et al., 1964). Considerable progress has been made in the last few years on the chemical characterization of these materials (see reviews by Kahan and Reisfeld, 1969; Nathenson, 1970; Nathenson et al., 1970; Reisfeld and Kahan, 1970a,b), and the recent development of human lymphocyte cell lines by Papermaster and others appears to offer great promise for elucidating further detailed structural information. Nathenson and his co-workers have studied soluble fragments released from cellular membranes with papain, purified by ammonium sulfate fractional precipitation, Sephadex G-150 column chromatography, and other fractionation techniques including, in some cases, isolation of antigen-antibody complexes formed with alloantibody of known specificity. The purified fragments carried the appropriate reactivities such as the ability to induce accelerated allograft rejection as well as measurable activity in vitro (inhibition of immune cytolysis) that could be used for routine assays of various fractions. Both mouse (H-2) and human (HL-A) papain-solubilized preparations included fragments of two classes separable on Sephadex G-150, of which each size class carried an incomplete series of antigenic determinants. For instance, cells from DBA/2 mice of genotype H-2^d express specificities 3, 4, 6, 8, 10, 13, 14, 27, 28, 29, and 31; it is known that some of these specificities come from the "K" region of the genetic locus and others from the "D" region. Nathenson and co-workers found that specificity 31 (from the K region) was found uniquely on papain fragments estimated to be about 34,000 molecular weight by Sephadex G-150 column chromatography, while specificities 3, 4, and 13 from the D region were found uniquely on the larger class of fragments of molecular weight about 70,000. Analogous (although not yet complete) findings were reported for both H-2^b specificities and human HL-A specificities.

As isolated by Nathenson and co-workers, all these active papain fragments from either mouse or human material are glycoprotein fragments. Their recent studies (Muramatsu and Nathenson, 1970a,b), using isolated fragments from murine tumor cells cultured in the presence of various radioactive sugars, are compatible with the idea that each fragment bears two sugar chains, each containing 12–15 sugar residues attached to the polypeptide chain by GlcNAc-N-Asp linkage. The sugar chains, as implied by the linkage, appear to be typical plasma-type chains in that they contain N-acetyl glucosamine, mannose, galactose, fucose, and sialic acid but no N-acetyl galactosamine. Whether the antigenic specificities are determined by the carbohydrate side chains, as favored by some workers (e.g., Amos, 1970), or by the amino acid sequence of the polypeptide chain is not yet completely clear. Nathenson and his co-workers currently favor the latter notion, since (1) they were unable to find any gross carbohydrate differences between H-2^d and H-2^b fragments from Meth-A and EL-4 tumor cells, respectively, while (2) peptide mapping of cyanogen-bromide, trypsin treated material from analogous fragments of allogenic mouse spleen cells showed that 38 identifiable peptides were common to both, while there were a total of 7 differences. However, more detailed information is necessary concerning the sugar chains before they can be eliminated as participants.

Kahan and Reisfeld (1969) and Reisfeld and Kahan (1970a), working with guinea pig and human material, also favor the idea that the amino acid sequence is the antigenic determinant, since they found specific amino acid differences between highly purified transplantation antigens from individuals with defined genetic differences. Their work differed strikingly from that of Nathenson et al. in that (1) their active guinea pig material had a molecular weight of about 14,000, while their HL-A material was mostly about 35,000 molecular weight, and (2) they detected no carbohydrate at all. One possible explanation for these differences concerns differences in methods of antigen solubilization: Kahan and Reisfeld liberate their antigens from membrane material by very mild sonication (diaphragm-mediated, 9–10 kHz, 15.5 μ/cm², 5 minutes with 4×10^7 cells), and their yield is only about 15% of the total cellular antigen content. From experience with attempts to liberate membrane-bound enzymes by various amounts and types of sonic energy (e.g., Gregg, 1967), it seems doubtful whether the material isolated by Kahan and Reisfeld should be considered membrane-bound. Hence, an alternative view is that these workers have been isolating an intracellular, cisternally localized, nonglycosylated precursor of the final molecule which later becomes glycosylated and bound to the external face of the plasma membrane. If this is true, then the fact that the nonglycosylated precursor is antigenically active would confirm the importance of the polypeptide chain as the antigenic determinant.

While considerable information concerning myxovirus receptor site structure is now available, the chemical structure of most other receptor sites for infectious agents is still obscure (see review of Phillipson, 1963). Carbohydrate groups appear to be important for some (e.g., polyoma virus) and less important for others [e.g., enterovirus (McLaren *et al.*, 1968); however, see Phillipson *et al.*, 1964]. It is clear, however, that even closely related viruses do not necessarily have identical receptor sites (e.g., Zajac and Crowell, 1969). Some viruses have membrane envelopes (such as arboviruses, myxoviruses, herpes simplex, and the RNA tumor viruses) which are thought to represent viral controlled modifications of host cellular membrane biosynthesis (e.g., Duesberg *et al.*, 1970; Klenk and Choppin, 1970; Strauss *et al.*, 1970). In these instances, initial interaction of virus and cell has the characteristics of a membrane-membrane interaction; just exactly what happens and why is not clear, but the general area of viral-modified membranes is of great interest in terms of cell behavior (Keller *et al.*, 1970).

C. Cellular Interactions

Short-range cellular interactions are generally thought to be primary determinants of cytodifferentiation and embryogenesis. Tissue culture systems are being widely used in attempts to arrive at mechanistic explanations for these phenomena and, in particular, complex carbohydrates of the cell surface or in the immediate cellular microenvironment have become favored candidates for an important role. Mainly, this has come about for two reasons: First, recognition that the cell periphery is enormously enriched in outwardly facing carbohydrate moieties raises the understandable suspicion that these moieties would be the first to interact between approaching cells. Second, a large series of studies of isolated and reaggregating sponge cells (see reviews of Humphreys, 1967; Moscona, 1968a,b; Lilien, 1969) have shown that, in sponges at least, specific cell-surface glycoproteins operate as intercellular ligands that can account for initiation of species-specific multicellular grouping of cells.

A wide variety of studies points to *cellular adhesiveness,* in one guise or another, as being the premier determinant of cellular interactions. Studies with vertebrate cells have clearly shown that adhesiveness, as measured by cellular aggregation systems, is under a variety of metabolic controls including tissue-specific differences between cells (see review of Lilien, 1969). In fact, differential adhesiveness between various types of cells, combined with cellular motility, has been a popular starting point for explanations of the mechanism of tissue formation (Steinberg, 1970). Tissue-specific aggregation factors, presumably desquamated from the cell surface, have been reported (Lilien and Moscona, 1967; Kuroda, 1968; Lilien, 1968), and involvement of complex carbohydrates has been suggested (Moscona, 1962; Richmond et al., 1968; Oppenheimer et al., 1969; Kemp, 1970). However, as of this writing, chemical information on cellular adhesiveness is still minuscule, and most of the major theories of cellular adhesiveness are equally compatible with a variety of chemical mechanisms. Jones and co-workers (Gröschel-Stewart et al., 1970; Jones and Kemp, 1970) have suggested involvement of contractile, surface, actinomyosin-like proteins in cellular adhesiveness; the relation of their findings to other work is not yet clear.

The *property* of cell adhesiveness initiates phenomena that lead to the *state* of cell adherence (to substrates such as glass or plastic or to other cells), a state which is associated with a variety of changes in cell behavior. These changes may manifest themselves as renewed or halted division cycle traverse, expression of differentiated cell functions, specific changes in uptake of metabolites, cell fusion, immune lymphocyte cytotoxicity, altered cell movements, extension of axonlike processes or other parameters, depending on the properties of the cell system under study and the particular interests of the investigator. That adherence per se has important functional implications is evident from studies concerning attachment of cells to glass or plastic substrates. For instance, it is widely recognized that nonmalignant euploid cells cannot sustain division cycle traverse unless they are in the adherent state (Hayflick, 1967). On the other hand, in what might be called a hyperadherent state (density-dependent inhibition), division cycle traverse is again inhibited (Stoker and Rubin, 1967). Horwitz and Dorfman

(1970) have reported an interesting exception to this rule. They found that chick cartilage cells in culture, presumably euploid and nontumorigenic, would grow in soft agar or liquid suspension and that these cells continued to synthesize a chondromucoprotein capsule. One interpretation of this result is that these cells achieved an adherent state by interaction with their own secreted substrate. An analogous interpretation can be made for the Sanders and Smith study (1970) of the effect of added matrix materials on growth of cells in soft agar. The adherent state has also been reported to be correlated with Forssman antigen synthesis in cells in culture (Fogel and Sachs, 1964) and with the ability of cells from clonal lines of a neural tumor to extend axons *in vitro* (Seeds *et al.*, 1970; Schubert and Jacob, 1970). A similar interpretation can be applied to the inducibility of glutamine synthetase activity by hydrocortisone in *in vitro* studies of chick neural retina (Morris and Moscona, 1970) and to the induction of UDP-glucuronyltransferase in chick embryo liver cells in culture (Skea and Nemeth, 1969). Conceptually, these consequences of the adherent state should be distinguished from changes in adhesiveness per se. For instance, Ballard and Tomkins (1970) found that glucocorticoids induced increased adhesiveness and tyrosine aminotransferase activity in HTC hepatoma cells in culture but that they were separate phenomena in the sense that the enzyme induction proceeded in suspension culture; hence, it did not depend on achieving an adherent state.

Another consequence of the adherent state is formation of specialized junctions between cells that allow ion communication (see reviews of Furshpan and Potter, 1968; Loewenstein, 1967, 1968). These junctions may be functionally related to such *in vitro* phenomena as "metabolic cooperation" described by Subak-Sharpe *et al.* (1969) or *in vitro* complementation (Danes and Bearn, 1970). It appears probable that only low-molecular-weight materials are passaged through such junctions (Baker and Mintz, 1969; Cox *et al.*, 1970).

The role of the intercellular matrix complex carbohydrates in determining the functional consequences of the adherent state is still obscure. For instance, while it is well known that collagen matrix has a role in the *in vitro* muscle formation process (Hauschka and Konigsberg, 1966), it is not clear what the collagen effect is on each of the various parts of the total process [cell division, cell movement, cell fusion (Nameroff and Holtzer, 1969)], nor, for that matter, what the role is of the carbohydrate moieties of the collagen. In some cases, close packing of cells may merely encourage accumulation of intercellular materials that would otherwise be washed away (e.g., Amborski *et al.*, 1969) or enhance particular degradative phenomena by secreted glycosidases

(Bosmann and Bernacki, 1970). Interactions within such materials might modify important cell peripheral zones that would operate as molecular sieves and/or ion sinks. Precedent for such notions comes from information concerning the structure, biogenesis, and function of basement membranes (R. G. Spiro, 1970). In this connection, Nagata and Rasmussen (1970) have proposed that a fundamental and primitive mechanism for transmittal of a "message" from the cell surface to the interior involves a calcium shift mediated by an adenylcyclase-dependent cell membrane permeability change. A peripheral ion depository with physiologically variable withdrawal characteristics (based perhaps on conformation changes of surface carbohydrate chains) would be an important aspect of such a system.

D. Malignant Transformation

Malignancy can be considered as a special case of cellular interaction; in the context of the above discussion of cellular interactions, it may be regarded as an aberration of the adherent state. The recent florescence of *in vitro* studies concerning malignant transformation has strongly implicated alterations of cell-surface complex carbohydrates in this aberration and has, therefore, emphasized the role of these materials in the various functional manifestations of the adherent state in general.

As studied in culture, malignant cells commonly differ from their "normal counterparts" in one or more of the following parameters: pattern of growth in monolayers, cell density achieved at confluency, chromosomal changes, cytological changes (e.g., number of nucleoli, cytoplasmic basophilia), minimum serum concentration required for growth, ability to grow in suspension culture, ability to grow indefinitely in culture, changes in glycolytic controls, response to various mitosis-stimulating factors, changes in transport of metabolites, changes in the prevalence of tight junctions and/or ionic communication between cells, and changes in the apparent amount of intercellular matrix material stainable with colloidal iron or ruthenium red [examples and further references to these changes can be found in Hayflick (1967), Loewenstein (1968), Cunningham and Pardee (1969), Dulbecco (1970), Eagle *et al.* (1970), Hatanaka and Hanafusa (1970), Martinez-Palomo (1970), Sanford *et al.* (1970), and Temin (1970)]. It seems probable with a carefully selected model pair of cultures that all these differences might be demonstrable between the members of a single pair. However, when a larger series of cell types is examined, flagrant exceptions are generally found (e.g., Mora *et al.*, 1969; Eagle *et al.*, 1970). In particular, poor correlation between "transformation," which is an *in vitro* parameter, and tumorigenicity has frequently been noted. Therefore, the problem is not to find differences between malignant and nonmalignant cells per se; there are obviously ample differences between members of various pair combinations. Rather, the problem is to interpret the differences in the correct context. The question is, which changes are relevant in themselves to the peculiar behavior of malignant cells *in vivo;* which differences represent "epiphenomena" that reflect secondary changes due to indirect alterations in genetic expression? For instance, chick embryo fibroblasts transformed by Rous sarcoma virus often show strikingly increased synthesis of hyaluronic acid (Erichsen *et al.,* 1961; Ishimoto *et al.,* 1966) which might explain their altered behavior in other respects such as adherence to glass. However, other malignant cells may not be altered in this respect (Castor and Naylor, 1969) or may even show diminished synthesis of hyaluronic acid (Hamerman *et al.,* 1965). Hence, it seems likely that hyaluronic acid synthesis differences would not per se explain the malignant behavior of these cells *in vivo.* There are analogous reasons to be cautious in interpreting many other "pair differences" currently in the literature. It is necessary, therefore, to construct a more appropriate conceptual point of departure to explain, in biochemical terms, the malignant phenotype (this is a distinctly different quest than an attempt to explain the mechanism of formation of the malignant genotype as, for instance, the question of how does a virus cause this change).

As a conceptual point of departure, conceiving the malignant state as an aberration of the adherent state has a definite appeal. For one thing, as discussed above, the adherent state of cellular interaction has, in itself, been demonstrated to have a controlling, albeit complex, influence on a wide variety of behavioral characteristics. Second, the adherent state focuses attention on the biochemical properties of the cell surface and intercellular matrix, thus encouraging biochemically testable hypotheses. Third, many of the changes that have consistently been correlated at the cellular level with malignant transformation are readily visualizable as altered cell-surface properties.

An interesting example of the relationship of the adherent state to malignancy is the study of Silagi and Bruce (1970) on mouse melanoma cells in culture. These workers found that growth of the cells in medium containing 1–3 μg/ml of 5-bromodeoxyuridine (BUdR) resulted in (1) suppression of tumorigenicity, (2) suppression of the differentiated function (pigment production), and (3) altered adherence to plastic substrate. As mentioned above, Schubert and Jacob (1970) also found alteration in the adherent state following exposure to BUdR but, in that case, differentiation was enhanced. Holtzer *et al.* (1969) believe that BUdR specifically represses "luxury" molecules (i.e., differentiated functions) at the translational level. However, the above results suggest that a more generalized change in state (such as adherence) may be involved.

A variety of recent studies of cell-surface antigens and receptor sites have yielded an inkling of a biochemical explanation for the change in adherence state that accompanies malignant transformation. Considering one block of data, cells transformed by SV40 virus possess a large number of new cell-surface reactivities. These include (1) an SV40 tumor-specific transplantation antigen (Defendi, 1962), (2) a surface antigen detectable by immunofluorescence (Tevethia *et al.*, 1965), (3) a surface antigen detectable by mixed hemagglutination (Häyry and Defendi, 1970), (4) Forssman reactivity (Robertson and Black, 1969), (5) reactivity with an agglutinin from wheat germ (Burger, 1969), (6) reactivity with concanavalin A (Inbar and Sachs, 1969a,b), (7) specific reactivity with antimouse egg cytotoxic antibody (Baranska *et al.*, 1970), (8) specific aggregation with ornithine-leucine copolymers in the presence of serum (Duksin *et al.*, 1970), and (9) reactivity with soybean agglutinin (Sela *et al.*, 1970). In each case, studies of SV40-transformed cell lines and nontransformed counterpart lines indicated cell-surface reactivity only with the transformed cells. In several studies the association with SV40 transformation was made in several pairs of cells of different species of origin. Hence, at first glance, these reactivities might be assumed to represent direct alterations of cell-surface macromolecules by the SV40 viral genome-coded products, analogous to the alterations induced by herpesvirus as discussed by Roizman and Spring (1967). It might also be suspected that all these reactivities reflect a single, newly acquired, cell-surface reactive site. It turns out that neither of these explanations is correct.

The cellular sites for the various neoreactivities are clearly not all the same in chemical terms. This is indicated by the following: (1) the Forssman site includes structures similar to α-GalNAc $1 \rightarrow 3$ Gal $1 \rightarrow 4$ Gal $1 \rightarrow 4$ Glc-ceramide (Robertson and Black, 1969); (2) concanavalin A appears to be quite specific for α-mannosides and α-glucosides (Hassing and Goldstein, 1970); (3) the wheat germ agglutinin site appears to involve structures containing N-acetyl glucosamine (Burger and Goldberg, 1967) while the soybean agglutination is inhibited with N-acetyl galactosamine (Sela *et al.*, 1970); and (4) the specific polyornithine-leucine site is apparently different from any of the above, but the chemical groups involved in it, as well as in the other listed reactivities, are not yet known. It is already clear, however, that SV40 transformation is associated with the appearance of a variety of chemically distinct reactive sites.

That involvement of the viral genome is completely indirect is indicated by the fact that several of these reactivities associated with transformation can be demonstrated to exist in nontransformed cells in a masked form. These reactivities can be commonly demonstrated if the "normal" cells are first gently treated with proteolytic enzymes. This has been shown for the wheat germ agglutinin site (Burger, 1969), the concanavalin A site (Inbar and Sachs, 1969a), the surface antigen detected by mixed hemagglutination (Häyry and Defendi, 1970), and the soybean agglutinin site (Sela *et al.*, 1970). Furthermore, some of these sites, such as the three plant-lectin agglutinin sites, are exposed following viral or nonviral transformation other than with SV40 virus.

It still might be claimed that the various reactivities, expressed by transformed cells or exposed by proteolysis on the surface of normal cells, represent different portions of a single surface macromolecule. This seems unlikely since, when the proteolytic exposure of the three plant-lectin agglutinin sites were followed simultaneously, they became exposed at distinctly different rates (Sela *et al.*, 1970). It is also evident from this work that, in comparing different (by virus and by species of cell) transformed cells, exposure of the three sites is not always coordinate. Further evidence that the viral effect causing exposure of these sites is related to the *in vitro* transformed behavior of the cell comes from studies which showed correlated reversion of behavior and exposure (Inbar *et al.*, 1969; Ben-Bassat *et al.*, 1970), the failure of nontransforming mutants of virus to cause site exposure (Benjamin and Burger, 1970), the quantitative correlation of graded behavioral characteristics and site exposure (Pollack and Burger, 1968) and, finally, the reversion of behavior upon coverage of the exposed sites by reaction with exogenous agglutinin fragments (Burger and Noonan, 1970).

Taken as a whole, these studies definitely implicate the arrangement of cell-surface complex carbohydrates in the altered adherence states of transformed cells; by inference, therefore, cell-surface complex carbohydrates must be vital determinants of adherence states in general, which, in turn, suggests for them a vital role in such general processes as cytodifferentiation, cell movement, division cycle traverse, and embryogenesis. However, there is a certain ambiguity inherent in this construction. One wonders which class of cell-surface complex carbohydrates should be considered determinants of the malignant adherence state: those that react with reagents such as the plant-lectin agglutinins or those degraded with proteolytic enzymes, thereby exposing these sites? Burger and Noonan's data (1970) would seem to suggest that nonspecific (i.e., with concanavalin fragments rather than any cellular structure) coverage of the lectin-reactive sites is sufficient to return the cell to normal behavior, thus implicating the first alternative. However, the converse situation (i.e., absence of coverage causing malignant cellular behavior) is apparently not always true, since it is well known that agglutinin sites are also naturally exposed in some nonneoplastic cells (Liske and Franks, 1968; Gantt et al., 1969). In one such instance (normal human lymphocytes), reaction of exposed sites with concanavalin A is followed by blastogenic transformation similar but not identical to blastogenesis induced by phytohemagglutinin (Perlmann and Nilsson, 1970; Powell and Leon, 1970). However, Burger's group has found increased exposure of wheat germ agglutinin sites on murine lymphoma cells, compared to normal mouse lymphocytes (Jansons et al., 1970), while S. Kornfeld (1969) reports decreased phytohemagglutinin receptor sites in human chronic lymphatic leukemia cells. It seems likely that "masking" and "exposure" of cell-surface reactive sites will turn out to be very complex parameters but may have a physiological role in embryogenesis.

In parallel with these studies of complex carbohydrate reactive sites on cell surfaces, a number of biochemical comparisons have recently been made concerning the glycolipids and glycoproteins of "normal," transformed, and/or tumorigenic cell lines. In a series of papers by Hakomori and his co-workers, the glycolipids of a wide variety of cells in culture were analyzed (Hakomori and Murakami, 1968; Hakomori et al., 1968; Hakomori, 1970a,b; Siddiqui et al., 1970). The cells selected for study included mouse, hamster, rat, chicken, and human material: nonmalignant, spontaneously transformed and/or tumorigenic, and transformed by polyoma, SV_{40}, or Rous virus. In general, their results indicated that nonmalignant-nontransformed cells consistently had glycolipids of longer carbohydrate chain length that did various "counterpart" tumorigenic, and/or virally transformed, and/or established heteroploid cell lines. For instance, studies of three BHK hamster lines (one "normal," one "spontaneously transformed," and one transformed with polyoma virus) indicated inverse correlation of hematoside content (NANA $1 \rightarrow 3$ Gal $1 \rightarrow 4$ Glc-ceramide) and lactosyl ceramide (Gal $1 \rightarrow 4$ Glc-ceramide) content, which also correlated with the behavior of the cells in monolayer (Hakomori and Murakami, 1968). Hematoside content was highest in the "normal" BHK subline and lowest in the one transformed by polyoma. This report also included the finding that wheat germ agglutination paralleled the reduction of hematoside content and that the agglutination of the polyoma-transformed cells could be inhibited by a glycolipid fraction from the polyoma-transformed line but not when extracted from the "normal" cells unless the latter were first subjected to Smith degradation. The other reports of this group cited above have generally confirmed the idea of reduced glycolipid chain length in a wide variety of comparisons of cells in culture; in some cases, the preponderant chain length number included zeros as manifested by a striking increase in free ceramide (Siddiqui et al., 1970). Furthermore, these general results confirmed their comparative analyses of human adenocarcinoma and normal tissues, in which case, the normal ABH-active glycolipids were missing and in their place were found increased amounts of the truncated, fucose-rich, Le^a- and Le^b-active versions of the same molecule (Hakomori and Jeanloz, 1964; Hakomori and Andrews, 1970).

In principle, the work of Hakomori and his co-workers has been confirmed by a group at the National Institutes of Health. In studies of the gangliosides of various mouse cell lines, Mora et al. (1969) and Brady and Mora (1970) reported strikingly decreased amounts of the larger gangliosides (GM_1 and DG_{1a}) in cells transformed by DNA tumor virus. However, hematosides (GM_3) were the same or increased, and, furthermore, no clear correlations were evident when saturation density or tumorigenicity per se was evaluated. Comparisons between rat hepatocyte and hepatoma lines showed decreased disialotetra-hexoside ganglionsides (GD_{1a}) and (presumably) compensatory increases in hematoside (GM_3) and monosialotetrahexoside (CM_1) glycolipids. In considering the question of whether these changes primarily represented differences in degradation or biosynthesis, these workers have recently reported (Cumar et al., 1970) that the DNA-tumor-virus-transformed cell lines had a striking deficiency of a key glycosyl transferase. This transferase is responsible for the addition of N-acetyl galactosamine to hematoside as a prerequisite to formation of the higher gangliosides. In addition, no differences in ganglioside neuraminidase activity could be found. As in the case of the mass data, the transferase deficiency was found only in cell lines transformed by DNA tumor viruses and was normal or increased in other tumorigenic cell lines. However, Murray's group, comparing the established, heteroploid, tumorigenic, but nontransformed mouse L cell line with secondary mouse embryo fibroblasts, have recently reported that both neutral glycolipids and gangliosides have, in general, shorter chain length in the L cell (Yogeeswaran et al., 1970).

Studies complementary to the glycolipid studies have also been reported for glycoproteins (Meezan et al., 1969; Wu et al., 1969; Buck et al., 1970). Comparing the overall sugar content of 3T3 and SV_{40}-transformed 3T3 cells, Wu et al. (1969) reported that most of the major sugars of glycoproteins were decreased in the latter type of cells; this was especially true of terminal sugars such as sialic acid. Spontaneously transformed 3T3 cells yielded intermediate values. Molecular sieve chromatography of glycoproteins and pronase-generated glycopeptides from these cells indicated multiple differences in the size spectrum of materials isolated from all cell fractions (nuclei, mitochondrial, endoplasmic reticulum, surface membrane, soluble), but no clear model for the reduced sugar content could be distinguished (Meezan et al., 1969). The work of Warren and co-workers (Buck et al., 1970), comparing BHK cells and BHK cells transformed by Rous virus (an RNA virus), also indicated multiple size differences of both glycoproteins and pronase-generated glycopeptides; particularly prominent differences concerned fucose-rich material that eluted early.

A general interpretation of all these cell-surface reactivity and biochemical results might be that altered cell behavior, including malignant behavior, is the result of defective glycosylation of cell-surface complex carbohydrates. However, there are a number of reasons that ardor for this notion should remain restrained. For one thing, there remains the possibility that some of the glycolipids of cells in culture derive from serum of the medium (Sneath and Sneath, 1959; D. M. Marcus and Cass, 1969; Dawson and Sweeley, 1970), although it seems clear that de novo biosynthesis also occurs (Brady and Mora, 1970; Weinstein et al., 1970; Yogeeswaran et al., 1970). This possibility could especially complicate the interpretation of both mass and labeling data if such exchangeable molecules were subject to catabolism and/or further glycosylation by the cells.

Second, the data that derive from either mass analyses or prolonged labeling do not yield a dynamic concept of cellular processes. While Cumar et al. (1970) failed to find enhanced neuraminidase activities for the gangliosides of transformed cells and did find lowered hematoside-UDP-GalNAc transferase activity, Bosmann and co-workers have reported a generally different picture. Biosynthetically, transformation of 3T3 cells was associated with enhanced transferase activity for a variety of nucleotide sugars (Bosmann and Eylar, 1968b; Bosmann et al., 1968b) except for collagen-glycosyl transferases (Bosmann, 1969a). These studies and one other (Bosmann, 1969b) also indicated that degradative activities were higher in transformed cells. In any case, measurement of the enzyme activities of broken-cell preparations does not necessarily reflect the actual rates of synthetic and degradative processes in the living cell monolayer, since many other factors other than the total cellular enzyme concentration could be rate-limiting for each step.

Finally, there is now ample reason to suspect that many of the differences between "normal" and "transformed" cells, as described above,
can also be found at different stages of the monolayer development
of the "normal" cell. This has been recognized by many of these same
workers (e.g., Meezan et al., 1969; Buck et al., 1970; Fox et al., 1971),
as well as others (Cox and Gesner, 1968; Foster and Pardee, 1969; however, see also Barland and Scher, 1970). Hakomori (1970b) has recently
incorporated this phenomenon into his general theory of malignant cell
behavior by suggesting that normal cells, upon attaining close contact,
undergo an "extension response." That is, the glycosylation rate and
the tendency to extend the carbohydrate chain length are enhanced
at this time. By contrast, malignant and/or transformed cells can be
considered defective in this response. Therefore, such cells would be
unable to achieve a normal adherent state and could, therefore, exhibit
various functional aberrations that follow aberrant adherent states. The
major ambiguity of this, of course, is that the cause-and-effect relationship might equally well be reversed. In either event and to conclude
on a more optimistic note, the possibilities of artificially manipulating
the adherent state, such as exemplified by the studies of Burger and
Noonan (1970) and Silagi and Bruce (1970), combined with the biochemical expertise that has gradually developed among workers using
cell cultures, should yield an explanation for the miscreant behavior
of malignant cells within a few years.

REFERENCES

Adamany, A. M., and Kathan, R. H. (1969). Isolation of a tetrasaccharide common
    to MM, NN, and MN antigens. Biochem. Biophys. Res. Commun. 37, 171–178.
Aleo, J. J., Orbison, J. L., and Hawkins, W. B. (1967). Histochemical and biochemi-
    cal studies of strain L fibroblasts treated with a lathyrogen. Lab. Invest. 17,
    425–435.
Allison, A. C., and Mallucci, L. (1964). Lysosomes in dividing cells, with special
    reference to lymphocytes. Lancet 2, 1371–1373.
Amborski, L., Parker, L. T., and Amborski, G. F. (1969). Accumulation of an
    extracellular acid mucopolysaccharide in the aggregated state of the "L" strain
    mouse fibroblast. Exp. Cell Res. 53, 673–678.
Amos, D. B. (1970). Genetic aspects of human HL-A transplantation antigens.
    Fed. Proc., Fed. Amer. Soc. Exp. Biol. 29, 2018–2025.
Anderson, E. C., and Petersen, D. F. (1965). Synchronized mammalian cells: An
    experimental test of a model for synchrony decay. Exp. Cell Res. 36, 423–426.
Baker, W. W., and Mintz, B. (1969). Subunit structure and gene control of mouse
    NADP-malate dehydrogenase. Biochem. Genet. 2, 351–360.
Ballard, P. L., and Tomkins, G. M. (1970). Glucocorticoid-induced alteration of
    the surface membrane of cultured hepatoma cells. J. Cell Biol. 47, 222–234.

Baranska, W., Koldovsky, P., and Koprowski, H. (1970). Antigenic study of un-fertilized mouse eggs: Cross reactivity with SV₄₀-induced antigens. *Proc. Nat. Acad. Sci. U.S.* **67**, 193–199.

Barland, P., and Scher, I. (1970). Protein and glycoprotein components of cultured cell surface membranes. *J. Cell Biol.* **47**, 13a.

Bartholomew, B. A., and Jourdian, G. W. (1966). V. Colostrum sialyl-transferases. *Methods Enzymol.* **8**, 368–372.

Baserga, R. (1971). *In* "The Cell Cycle and Cancer" (R. Baserga, ed.). Dekker, New York.

Ben-Bassat, H., Inbar, M., and Sachs, L. (1970). Requirement for cell replication after SV₄₀ infection for a structural change of the cell surface membrane. *Virology* **40**, 854–859.

Benjamin, T. L., and Burger, M. M. (1970). Absence of a cell membrane alteration function in non-transforming mutants of polyoma virus. *Proc. Nat. Acad. Sci. U.S.* **67**, 929–934.

Blumenfeld, O. O. (1968). The proteins of the erythrocyte membrane obtained by solubilization with aqueous pyridine solution. *Biochem. Biophys. Res. Commun.* **30**, 200–205.

Bock, K. W., and Siekevitz, P. (1970). Turnover studies on membrane NAD glycohydrolase. *Fed. Proc., Fed. Amer. Soc. Exp. Biol.* **29**, 540.

Bosmann, H. B. (1968a). Collagen-galactosyl transferase: Subcellular localization and distribution in fibroblasts transformed by oncogenic viruses. *Life Sci.* **8**, 737–746.

Bosmann, H. B. (1969b). Glycoprotein degradation. Glycosidases in fibroblasts transformed by oncogenic viruses. *Exp. Cell Res.* **54**, 217–221.

Bosmann, H. B. (1969c). Glycolipid biosynthesis: Biosynthesis of mannose- and fucose-containing glycolipids by HeLa cells. *Biochim. Biophys. Acta* **187**, 122–132.

Bosmann, H. B. (1970a). Cellular membranes: Membrane marker enzyme activities in synchronized mouse leukemic cells L5178Y. *Biochim. Biophys. Acta* **203**, 256–260.

Bosmann, H. B. (1970b). Glycoprotein synthesis: Activity of collagen-galactosyl and collagen-glucosyl transferases in synchronized mouse leukemic cells L5178Y. *Exp. Cell Res.* **61**, 230–233.

Bosmann, H. B., and Bernacki, R. J. (1970). Glycosidase activity. Glycosidase activity in L5178Y mouse leukemic cells and the activity of acid phosphatase, β-galactosidase and β-N-acetylgalactosaminidase and β-N-acetylglucosaminidase in a synchronous L5178Y cell population. *Exp. Cell Res.* **61**, 379–386.

Bosmann, H. B., and Eylar, E. H. (1968a). Glycoprotein biosynthesis: The biosynthesis of the hydroxylysine-galactose linkage in collagen. *Biochem. Biophys. Res. Commun.* **33**, 340–346.

Bosmann, H. B., and Eylar, E. H. (1968b). Collagen-glucosyl transferase in fibroblasts transformed by oncogenic viruses. *Nature (London)* **218**, 582–583.

Bosmann, H. B., and Martin, S. S. (1969). Mitochondrial autonomy: Incorporation of monosaccharides into glycoprotein by isolated mitochondria. *Science* **164**, 190–192.

Bosmann, H. B., and Winston, R. A. (1970). Synthesis of glycoprotein, glycolipid, protein, and lipid in synchronized L5178Y cells. *J. Cell Biol.* **45**, 23–33.

Bosmann, H. B., Hagopian, A., and Eylar, E. H. (1969a). Glycoprotein biosynthesis: The characterization of two glycoprotein:fucosyl transferases in HeLa cells. *Arch. Biochem. Biophys.* **128**, 470–481.

Bosmann, H. B., Hagopian, A., and Eylar, E. H. (1968b). Membrane glycoprotein biosynthesis: Changes in levels of glycosyl transferases in fibroblasts transformed by oncogenic viruses. *J. Cell. Physiol.* **72**, 81–88.

Bosmann, H. B., Hagopian, A., and Eylar, E. H. (1968c). Cellular membranes: The isolation and characterization of the plasma and smooth membranes of HeLa cells. *Arch. Biochem. Biophys.* **128**, 51–69.

Bosmann, H. B., Hagopian, A., and Eylar, E. H. (1969). Cellular membranes: The biosynthesis of glycoprotein and glycolipid in HeLa cell membranes. *Arch. Biochem. Biophys.* **130**, 573–583.

Boyse, E. A., and Old, L. J. (1969). Some aspects of normal and abnormal cell surface genetics. *Annu. Rev. Genet.* **3**, 269–290.

Brady, R. O., and Mora, P. T. (1970). Alteration in ganglioside pattern and synthesis in SV$_{40}$ and polyoma virus-transformed mouse cell lines. *Biochim. Biophys. Acta* **218**, 308–319.

Brady, R. O., Borek, C., and Bradley, R. M. (1969). Composition and synthesis of gangliosides in rat hepatocyte and hepatoma cell lines. *J. Biol. Chem.* **244**, 6552–6554.

Brent, T. P., and Forrester, J. A. (1967). Changes in surface charge of HeLa cells during the cell cycle. *Nature (London)* **215**, 92–93.

Brew, K., Vanaman, T. C., and Hill, R. L. (1968). The role of lactalbumin and the A protein in lactose synthesis: A unique mechanism for the control of a biological reaction. *Proc. Nat. Acad. Sci. U.S.* **59**, 491–497.

Buck, C. A., Glick, M. C., and Warren, L. (1970). A comparative study of glycoproteins from the surface of control and Rous sarcoma virus transformed hamster cells. *Biochemistry* **9**, 4567–4576.

Buckhold, B., and Burki, H. J. (1969). Electrophoretic mobility of tissue-cultured mouse lymphoblasts through the cell cycle. *12th Annu. Meet., Biophys. Soc.* Abstract A-188.

Buell, D. N., and Fahey, J. L. (1969). Limited periods of gene expression in immunoglobulin-synthesizing cells. *Science* **164**, 1524–1524.

Burger, M. M (1969). A difference in the architecture of the surface membrane of normal and virally transformed cells. *Proc. Nat. Acad. Sci. U.S.* **62**, 994–1001.

Burger, M. M., and Goldberg, A. R. (1967). Identification of a tumor-specific determinant on neoplastic cell surfaces. *Proc. Nat. Acad. Sci. U.S.* **57**, 359–366.

Burger, M. M., and Noonan, K. D. (1970). Restoration of normal growth by covering of agglutinin sites on tumor cell surface. *Nature (London)* **228**, 512–515.

Caccam, J. F., Jackson, J. J., and Eylar, E. H. (1969). The biosynthesis of mannose-containing glycoproteins: A possible lipid intermediate. *Biochem. Biophys. Res. Commun.* **35**, 505–511.

Carubelli, R., and Griffin, M. J. (1970). Neuraminidase activity in HeLa cells: Effect of hydrocortisone. *Science* **170**, 1110–1111.

Castor, C. W., and Naylor, B. (1969). Characteristics of normal and malignant human mesothelial cells studied *in vitro*. *Lab. Invest.* **20**, 437–443.

Cifonelli, J. A. (1968). Structural features of acid mucopolysaccharides. *In* "The Chemical Physiology of Mucopolysaccharides" (G. Quintarelli, ed.), pp. 91–105. Little, Brown, Boston, Massachusetts.

Cikes, M. (1970). Relationship between growth rate, cell volume, cell cycle kinetics, and antigenic properties of cultured murine lymphoma cells. *J. Nat. Cancer Inst.* **45**, 979–988.

Clarris, B. J., and Fraser, J. R. E. (1968). On the pericellular zone of some mammalian cells *in vitro*. *Exp. Cell Res.* **49**, 181–193.

Claude, A. (1970). Growth and differentiation of cytoplasmic membranes in the course of lipoprotein granule synthesis in the hepatic cell. I. Elaboration of elements of the Golgi complex. *J. Cell Biol.* **47**, 745–766.

Cox, R. P., and Gesner, B. M. (1968). Studies on the effects of simple sugars on mammalian cells in culture and characterization of the inhibition of 3T3 fibroblasts by L-fucose. *Cancer Res.* **28**, 1162–1172.

Cox, R. P., Kraus, M. R., Balis, M. E., and Dancis, J. (1970). Evidence for transfer of enzyme product as the basis of metabolic cooperation between tissue culture fibroblasts of Lesch-Nyhan disease and normal cells. *Proc. Nat. Acad. Sci. U.S.* **67**, 1573–1579.

Cumar, F. A., Brady, R. O., Kolodny, E H., McFarland, V. W., and Mora, P. T. (1970) Enzymatic block in the synthesis of gangliosides in DNA virus-transformed tumorigenic mouse cell lines. *Proc. Nat. Acad. Sci. U.S.* **67**, 757–764.

Cunningham, D. D., and Pardee, A. B. (1969). Transport changes rapidly initiated by serum addition to "contact inhibited" 3T3 cells. *Proc. Nat. Acad. Sci. U.S.* **64**, 1049–1056.

Dallner, G., Siekevitz, P., and Palade, G. E. (1966). Biogenesis of endoplasmic reticulum membranes. I. Structural and chemical differentiation in developing rat hepatocyte. *J. Cell Biol.* **30**, 73–117.

Danes, B. S., and Bearn, A. G. (1970). Correction of cellular metachromasia in cultured fibroblasts in several inherited mucopolysaccharidoses. *Proc. Nat. Acad. Sci. U.S.* **67**, 357–364.

Davies, L. M., Priest, J. H., and Priest, R. E. (1968). Collagen synthesis by cells synchronously replicating DNA. *Science* **159**, 91–93.

Dawson, A., and Sweeley, C. C. (1970). *In vivo* studies on glycosphingolipid metabolism in porcine blood. *J. Biol. Chem.* **245**, 410–416.

Defendi, V. (1962). Effect of SV_{40} immunization on growth of transplantable SV_{40} and polyoma virus tumors in hamsters. *Proc. Soc. Exp. Biol. Med.* **113**, 12–16.

De Luca, L., Rosso, G., and Wolf, G. (1970). The biosynthesis of a mannolipid that contains a polar metabolite of $15\text{-}^{14}C$-retinol. *Biochem. Biophys. Res. Commun.* **41**, 615–620.

De Salegui, M., and Pigman, W. (1967). The existence of an acid-active hyaluronidase in serum. *Arch. Biochem. Biophys.* **120**, 60–67.

Dietrich, C. P., and De Oca, H. M. (1970). Production of heparin related mucopolysaccharides by mammalian cells in culture. *Proc. Soc. Exp. Biol. Med.* **134**, 955–962.

Duesberg, P. H., Martin, G. S., and Vogt, P. K. (1970). Glycoprotein components of avian and murine RNA tumor viruses. *Virology* **41**, 631–646.

Dukes, P. P. (1968). Erythroprotein-stimulated incorporation of $1\text{-}^{14}C$-glucosamine into glycolipids of bone marrow cells in culture. *Biochem. Biophys. Res. Commun.* **31**, 345–354.

Duksin, D., Katchalski, E., and Sachs, L. (1970). Specific aggregation of SV_{40}-transformed cells by ornithine, leucine copolymers. *Proc. Nat. Acad. Sci. U.S.* **67**, 185–192.

Dulbecco, R. (1970). Behavior of tissue culture cells infected with polyoma virus. *Proc. Nat. Acad. Sci. U.S.* **67**, 1214–1220.

Eagle, H., Foley, G. E., Koprowski, H., Lazarus, H., Levine, E. M., and Adams, R. A. (1970). Growth characteristics of virus-transformed cells. *J. Exp. Med.* **131**, 863–879.

Ebner, K. E., and Schanbacher, F. L. (1970). Galactosyl acceptor specificity of the A protein of lactose synthetase. *Fed. Proc., Fed. Amer. Soc. Exp. Biol.* **29**, 873.

Erichsen, S., Eng, J., and Morgan, H. R. (1961). Comparative studies in Rous sarcoma with virus, tumor cells, and chick embryo cells transformed *in vitro* by virus. *J. Exp. Med.* **114**, 435–439.

Evans, W. H. (1970). Glycoproteins of mouse liver smooth microsomal and plasma membrane fractions. *Biochim. Biophys. Acta* **211**, 578–581.

Eylar, E. H. (1965). On the biological role of glycoproteins. *J. Theor. Biol.* **10**, 89–113.

Eylar, E. H., Madoff, M. A., Brody, O. V., and Oncley, J. L. (1962). The contribution of sialic acid to the surface charge of the erythrocyte. *J. Biol. Chem.* **237**, 1992.

Fisher, D. B., and Mueller, G. C. (1968). An early alteration in the phospholipid metabolism of lymphocytes by phytohemagglutinin. *Proc. Nat. Acad. Sci. U.S.* **60**, 1396–1402.

Fleischer, B., and Fleischer, S. (1969). Glycosidase activity of bovine liver plasma membranes. *Biochim. Biophys. Acta* **183**, 265–275.

Fleischer, B., and Fleischer, S. (1970). Preparation and characterization of Golgi membranes from rat liver. *Biochim. Biophys. Acta* **219**, 301–319.

Fleischer, B., Fleischer, S., and Ozawa, H. (1969). Isolation and characterization of Golgi membranes from bovine liver. *J. Cell Biol.* **43**, 59–79.

Fogel, M., and Sachs, L. (1964). The induction of Forssman-antigen synthesis in hamster and mouse cells in tissue culture, as detected by the fluorescent-antibody technique. *Exp. Cell Res.* **34**, 448–462.

Forstner, G. G. (1970). [1-,^{14}C]Glucosamine incorporation by subcellular fractions of small intestinal mucosa. Identification by precursor labeling of three functionally distinct glycoprotein classes. *J. Biol. Chem.* **245**, 3584–3592.

Foster, D. O., and Pardee, A. B. (1969). Transport of amino acids by confluent and nonconfluent 3T3 and polyoma virus transformed 3T3 cells growing on glass coverslips. *J. Biol. Chem.* **244**, 2675–2681.

Fox, T. O., Sheppard, J. R., and Burger, M. M. (1971). Cyclic membrane changes in animal cells: Transformed cells permanently display a surface architecture detected in normal cells only during mitosis. *Proc. Nat. Acad. Sci. U.S.* **68**, 244–247.

Franks, D. (1968). Antigens as markers on cultured mammalian cells. *Biol. Rev.* **43**, 17–50.

Frot-Coutaz, J., Louisot, P., Got, R., and Colobert, L. (1968). Les sites subcellulaires d'incorporation du 1-,^{14}C-α-fucose dans les glycoprotéines des cellules normales et cancereuses en culture *in vitro*. *Experientia* **24**, 1206–1207.

Furshpan, E. J., and Potter, D. D. (1968). Low-resistance junctions between cells in embryos and tissue culture. *Curr. Top. Develop. Biol.* **3**, 95–127.

Gantt, R. R., Martin, J. R., and Evans, V. J. (1969). Agglutination of *in vitro* cultured neoplastic and non-neoplastic cell lines by a wheat germ agglutinin. *J. Nat. Cancer Inst.* **42**, 369–373.

Gaugal, S. G., Merchant, D. M., and Shreffler, D. C. (1966). Characterization of the H-2 antigens of L-M mouse cells grown in culture. *J. Nat. Cancer Inst.* **36**, 1151–1159.

Gerner, E. W., Glick, M. C., and Warren, L. (1970). Membranes of animal cells. V. Biosynthesis of the surface membrane during the cell cycle. *J. Cell. Physiol.* **75**, 275–280.

Ginsburg, V. (1964). Sugar nucleotides and the synthesis of carbohydrates. *Advan. Enzymol.* **26**, 35–88.

Ginsburg, V., and Neufeld, E. F. (1969). Complex heterosaccharides of animals. *Annu. Rev. Biochem.* **38**, 371–388.

Glaumann, H., and Ericsson, J. L. E. (1970). Evidence for the participation of the Golgi apparatus in the intracellular transport of nascent albumin in the liver cell. *J. Cell Biol.* **47**, 555–567.

Godman, G. C., and Lane, N. (1964). On the site of sulfation in the chondrocyte. *J. Cell Biol.* **21**, 353.

Got, R., Frot-Coutaz, J., Colobert, L., and Louisot, P. (1968). Biosynthèse des glycoprotéines dans les cellules de culture *in vitro. Biochim. Biophys. Acta* **157**, 599–606.

Gottschalk, A. (1969). Biosynthesis of glycoproteins and its relationship to heterogeneity. *Nature (London)* **222**, 452–454.

Green, H., and Goldberg, B. (1965). Synthesis of collagen by mammalian cell lines of fibroblastic and nonfibroblastic origin. *Proc. Nat. Acad. Sci. U.S.* **53**, 1360–1365.

Green, H., Goldberg, B., and Todaro, G. J. (1966). Differentiated cell types and the regulation of collagen synthesis. *Nature (London)* **212**, 631–633.

Gregg, C. T. (1967). Preparation and assay of phosphorylating submitochondrial particles: Particles from rat liver prepared by drastic sonication. *Methods Enzymol.* **10**, 181–185.

Grobstein, C. (1968). Developmental significance of interface materials in epitheliomesenchymal interaction. *In* "Epithelial-Mesenchymal Interactions" (R. Fleischmajer and R. E. Billingham, eds.), pp. 173–176. Williams & Wilkins, Baltimore, Maryland.

Gröschel-Steward, D., Jones, B. M., and Kemp, R. B. (1970). Detection of actomyosin-type protein at the surface of dissociated embryonic chick cells. *Nature (London)* **227**, 280.

Hagopian, A., and Eylar, E. H. (1969). Glycoprotein biosynthesis: The purification and characterization of a polypeptide:N-acetyl-galactosaminyl transferase from bovine submaxillary glands. *Arch. Biochem. Biophys.* **129**, 515–524.

Hagopian, A., Bosmann, H. B., and Eylar, E. H. (1968). Glycoprotein biosynthesis: The localization of the polypeptidyl:N-acetylgalactosaminyl, collagen:glucosyl, and glycoprotein:galactosyl transferases in HeLa cell membrane fractions. *Arch. Biochem. Biophys.* **128**, 387–396.

Hakomori, S. (1970a). Cell density-dependent changes of glycolipid concentrations in fibroblasts, and loss of this response in virus-transformed cells. *Proc. Nat. Acad. Sci. U.S.* **67**, 1741–1747.

Hakomori, S. (1970b). Physiological variation of glycolipids in cultured cells, and change of the pattern in malignant-transformed cells. *160th Meet. Amer. Chem. Soc.,* Abstract No. 33.

Hakomori, S., and Andrews, H. D. (1970). Sphingoglycolipids with Le^b activity, and the co-presence of Le^a-, Le^b-glycolipids in human tumor tissue. *Biochim. Biophys. Acta* **202**, 225–228.

Hakomori, S., and Jeanloz, R. W. (1964). Isolation of a glycolipid containing fucose, galactose, glucose, and glucosamine from human cancerous tissue. *J. Biol. Chem.* **239**, PC3606–PC3607.

Hakomori, S., and Murakami, W. T. (1968). Glycolipids of hamster fibroblasts and derived malignant-transformed cell lines. *Proc. Nat. Acad. Sci. U.S.* **59**, 254–261.

Hakomori, S., Teather, C., and Andrews, H. (1968). Organizational difference of cell surface "hematoside" in normal and virally transformed cells. *Biochem. Biophys. Res. Commun.* 33, 563–568.

Hamerman, D., Todaro, G. J., and Green, H. (1965). The production of hyaluronate by spontaneously established cell lines and viral transformed lines of fibroblastic origin. *Biochim. Biophys. Acta* 101, 343–351.

Hassing, G. S., and Goldstein, I. J. (1970). Ultraviolet difference spectral studies on concanavalin A. Carbohydrate interaction. *Eur. J. Biochem.* 16, 549–556.

Hatanaka, M., and Hanafusa, H. (1970). Analysis of a functional change in membrane in the process of cell transformation by Rous sarcoma virus, alteration in the characteristics of sugar transport. *Virology* 41, 647–652.

Hauschka, S. D., and Konigsberg, I. R. (1966). The influence of collagen on the development of muscle colonies. *Proc. Nat. Acad. Sci. U.S.* 55, 119–126.

Hayden, G. A., Crowley, G. M., and Jamieson, G. A. (1970). Studies of glycoproteins. V. Incorporation of glucosamine into membrane glycoproteins of phytohemagglutinin-stimulated lymphocytes. *J. Biol. Chem.* 245, 5827–5832.

Hayflick, L. (1967). Oncogenesis *in vitro*. *Nat. Cancer Inst., Mongr.* 26, 355–385.

Häyry, P., and Defendi, V. (1970). Surface antigen(s) of SV_{40}-transformed tumor cells. *Virology* 41, 22–29.

Heilbrunn, L. V. (1956). "The Dynamics of Living Protoplasm." Academic Press, New York.

Helgeland, L. (1965). Incorporation of radioactive glucosamine into submicrosomal fractions isolated from rat liver. *Biochim. Biophys. Acta* 101, 106–112.

Herscovics, A. (1970). Biosynthesis of thyroglobulin: Incorporation of [³H]-fucose into proteins by rat thyroids *in vitro*. *Biochem. J.* 117, 411–413.

Holtzer, H., Bischoff, R., and Chacko, S. (1969). Activities of the cell surface during myogenesis and chondrogenesis. *In* "Cellular Recognition" (R. T. Smith and R. A. Good, eds.), pp. 19–25. Appleton, New York.

Horwitz, A. L., and Dorfman, A. (1968). Subcellular sites for synthesis of chondromucoprotein of cartilage. *J. Cell Biol.* 38, 358–368.

Horwitz, A. L., and Dorfman, A. (1970). The growth of cartilage cells in soft agar and liquid suspension. *J. Cell Biol.* 45, 434–438.

Horwitz, A., Stoolmiller, A. C., and Dorfman, A. (1970). Purification of the chondromucoprotein (CM-P) synthetase complex. *Fed. Proc., Fed. Amer. Soc. Exp. Biol.* 29, 337.

Howe, C., Spiele, H., Minio, F., and Hsu, K. C. (1970). Electron microscopic study of erythrocyte receptors with labeled antisera to membrane components. *J. Immunol.* 104, 1406–1416.

Humphreys, T. (1967). The cell surface and specific cell aggregation. *In* "The Specificity of Cell Surfaces" (B. D. Davis and L. Warren, eds.), pp. 195–210. Prentice-Hall, Englewood Cliffs, New Jersey.

Inbar, M., and Sachs, L. (1969a). Structural difference in sites on the surface membrane of normal and transformed cells. *Nature (London)* 223, 710–712.

Inbar, M., and Sachs, L. (1969b). Interaction of the carbohydrate-binding protein concanavalin A with normal and transformed cells. *Proc. Nat. Acad. Sci. U.S.* 63, 1418–1425.

Inbar, M., Rabinowitz, Z., and Sachs, L. (1969). The formation of variants with a reversion of properties of transformed cells. III. Reversion of the structure of the cell surface membrane. *Int. J. Cancer* 4, 690–696.

Ishimoto, N., Temin, H. M., and Strominger, J. L. (1966). Studies of carcinogenesis by avian sarcoma viruses. II. Virus-induced increase in hyaluronic acid synthetase in chicken fibroblasts. *J. Biol. Chem.* **241**, 2052–2057.

Jamieson, J. D., and Palade, G. E. (1967). Intracellular transport of secretory proteins in the pancreatic exocrine cell. I. Role of the peripheral elements of the Golgi complex. *J. Cell Biol.* **34**, 577–596.

Jansons, V. K., Sakamoto, C. K., and Burger, M. M. (1970). Partial isolation of a receptor site exposed on a leukemia cell but not the parent lymphocyte. *Fed. Proc., Fed. Amer. Soc. Exp. Biol.* **29**, 410.

Jeanloz, R. W. (1970). Mucopolysaccharides of higher animals. *In* "The Carbohydrates" (W. Pigman and D. Horton, eds.), Vol. 2B, pp. 590–625. Academic Press, New York.

Jones, B. M., and Kemp, R. B. (1970). Inhibition of cell aggregation by antibodies directed against actomyosin. *Nature (London)* **226**, 261–262.

Kahan, B. D., and Reisfeld, R. A. (1969). Transplantation antigens. *Science* **164**, 514–521.

Kalckar, H. M. (1965). Galactose metabolism and cell "sociology." *Science* **150**, 350–313.

Kashnig, D. M., and Kasper, C. B. (1969). Isolation, morphology, and composition of the nuclear membrane from rat liver. *J. Biol. Chem.* **244**, 3786–3792.

Kassulke, J. T., Stutman, O., and Yunis, E. J. (1969). Effect of neuraminidase on the surface antigens of human normal and malignant cells. *Fed. Proc., Fed. Amer. Soc. Exp. Biol.* **28**, 364.

Kaufman, R. L., and Ginsburg, V. (1968). The metabolism of L-fucose by HeLa cells. *Exp. Cell Res.* **50**, 127–132.

Kefalides, N. A., and Winzler, R. J. (1966). The chemistry of glomerular basement membrane and its relation to collagen. *Biochemistry* **5**, 702–713.

Keller, J. M., Spear, P. G., and Roizman, B. (1970). Proteins specified by herpes simplex virus. III. Viruses differing in their effects on the social behavior of infected cells specify different membrane glycoproteins. *Proc. Nat. Acad. Sci. U.S.* **65**, 865–871.

Kelus, A., Gurner, B. W., and Coombs, R. R. A. (1959). Blood group antigens on HeLa cells shown by mixed agglutination. *Immunology* **2**, 262.

Kemp, R. B. (1970). The effect of neuraminidase (3:2:1:18) on the aggregation of cells dissociated from embryonic chick muscle tissue. *J. Cell Sci.* **6**, 751–766.

Klenk, H. D., and Choppin, P. W. (1970). Plasma membrane lipids and parainfluenza virus assembly. *Virology* **40**, 939–947.

Kochhar, D. M., Dingle, J. T., and Lucy, J. A. (1968). The effects of vitamin A (retinol) on cell growth and incorporation of labelled glucosamine and proline by mouse fibroblasts in culture. *Exp. Cell Res.* **52**, 591–601.

Kolodny, E. H., Brady, R. O., Quirk, J. M., and Kanfer, J. N. (1970). Preparation of radioactive Tay-Sachs ganglioside labeled in the sialic acid moiety. *J. Lipid Res.* **11**, 144–149.

Kornfeld, R. H., and Ginsburg, V. (1966). Control of synthesis of guanosine-5′ diphosphate D-mannose and guanosine-5′ diphosphate L-fucose in bacteria. *Biochim. Biophys. Acta* **117**, 79–87.

Kornfeld, R. H., and Kornfeld, S. (1970). The structure of a phytohemagglutinin receptor site from human erythrocytes. *J. Biol. Chem.* **245**, 2536–2545.

Kornfeld, S. (1969). Decreased phytohemagglutinin receptor sites in chronic lymphocytic leukemia. *Biochim. Biophys. Acta* **192**, 542–545.

Kornfeld, S., and Ginsburg, V. (1966). The metabolism of glucosamine by tissue culture cells. *Exp. Cell Res.* **41**, 592–600.

Kornfeld, S., and Gregory, W. (1968). The conversion of glucosamine to glucose by white blood cell cultures. *Biochim. Biophys. Acta* **158**, 468–470.

Kornfeld, S., Kornfeld, R., Neufeld, E. F., and O'Brien, P. J. (1964). The feedback control of sugar nucleotide biosynthesis in liver. *Proc. Nat. Acad. Sci. U.S.* **52**, 371–379.

Kraemer, P. M. (1966). Regeneration of sialic acid on the surface of Chinese hamster cells in culture. I. General characteristics of the replacement process. *J. Cell. Physiol.* **68**, 85–90.

Kraemer, P. M. (1967a). Regeneration of sialic acid on the surface of Chinese hamster cells in culture. II. Incorporation of radioactivity from glucosamine-l-^{14}C. *J. Cell. Physiol.* **69**, 199–207.

Kraemer, P. M. (1967b). Configuration change of surface sialic acid during mitosis. *J. Cell Biol.* **33**, 197–200.

Kraemer, P. M. (1968). Cytotoxic, hemolytic and phospholipase contaminants of commercial neuraminidases. *Biochim. Biophys. Acta* **167**, 205–208.

Kraemer, P. M. (1969). Glycopeptides from suspension-cultured Chinese hamster cells: Intracellular, cell surface, and "desquamating" species. *9th Ann. Meet., Amer. Soc. Cell Biol.*

Kraemer, P. M. (1971a). Complex carbohydrates of animal cells: Biochemistry and physiology of the cell periphery. *In* "Biomembranes" (L. Manson, ed.), Vol. I. Plenum Press, New York.

Kraemer, P. M. (1971b). Heparin sulfates of cultured cells. II. Acid-soluble and precipitable species of different cell lines. *Biochemistry* **10**, 1445–1451.

Kraemer, P. M. (1971c). Heparin sulfate of cultured cells. I. Membrane associated and cell-sap species in Chinese hamster cells. *Biochemistry* **10**, 1437–1445.

Kraemer, P. M., and Gurley, L. R. (1969). Unpublished observation.

Kuhns, W. J., and Bramson, S. (1968). Variable behaviour of blood group H on HeLa cell populations synchronized with thymidine. *Nature (London)* **219**, 938–939.

Kuhns, W. J., and Bramson, S. (1970a). Blood groups H on HeLa cells: A marker for cellular differentiation. *J. Cell Biol.* **47**, 113a.

Kuhns, W. J., and Bramson, S. (1970b). Cellular variants for blood group H in an established cell line. *Fed. Proc., Fed. Amer. Soc. Exp. Biol.* **29**, 505.

Kuroda, Y. (1968). Preparation of an aggregation-promoting supernant from embryonic chick liver cells. *Exp. Cell Res.* **49**, 626–637.

Leloir, L. F. (1964). Nucleoside diphosphate sugars and saccharide synthesis. *Biochem. J.* **91**, 1–8.

Lilien, J. E. (1968). Specific enhancement of cell aggregation *in vitro*. *Develop. Biol.* **17**, 657–678.

Lilien, J. E. (1969). Toward a molecular explanation for specific cell adhesion. *Curr. Top. Develop. Biol.* 169–195.

Lilien, J. E., and Moscona, A. A. (1967). Cell aggregation: Its enhancement by a supernant from cultures of homologous cells. *Science* **157**, 70–72.

Ling, N. R. (1968). "Lymphocyte Stimulation." North-Holland Publ., Amsterdam.

Lippman, M. (1968). Glycosaminoglycans and cell division. *In* "Epithelial-Mesenchymal Interactions" (R. Fleischmajer and R. E. Billingham, eds.), pp. 208–229. Williams & Wilkins, Baltimore, Maryland.

Liske, R., and Franks, D. (1968). Specificity of the agglutinin in extracts of wheat germ. *Nature (London)* **217**, 860–861.

Lloyd, K. O., and Kabat, E. A. (1968). Immunochemical studies on blood groups. XLI. Proposed structures for the carbohydrate portions of blood groups A, B, H, Lewis[a], and Lewis[b] substances. *Proc. Nat. Acad. Sci. U.S.* **61**, 1470–1477.

Loewenstein, W. R. (1967). On the genesis of cellular communication. *Develop. Biol.* **15**, 503–520.

Loewenstein, W. R. (1968). Some reflections on growth and differentiation. *Perspect. Biol. Med.* **12**, 260–272.

Louisot, P., Lebre, D., Pradal, M. B., Gresle, J., and Got, R. (1970). Biosynthèse des glycoprotéines. XIV. Role des sites cytoplasmiques dans les cellules embryonnaires et néoplastiques en culture en suspension. *Can. J. Biochem.* **48**, 1082–1086.

McCluer, R. H. (1968). Gangliosides and related glycolipids. In "Biochemistry of Glycoproteins and Related Substances" (E. Rossi and E. Stoll, eds.), pp. 203–224. Karger, Basel.

McKibbin, J. M. (1970). Glycolipids. In "The Carbohydrates" (W. Pigman and D. Horton, eds.), 2nd ed., Vol. 2B, pp. 711–738. Academic Press, New York.

McLaren, L. C., Scalletti, J. V., and James, C. G. (1968). Isolation and properties of enterovirus receptors. In "Biological Properties of the Mammalian Surface Membrane" (L. A. Manson, ed.), pp. 123–135. Wistar Inst. Press, Philadelphia, Pennsylvania.

Makino, N., Kojima, T., and Yamashina, I. (1966). Enzymatic cleavage of glycopeptides. *Biochem. Biophys. Res. Commun.* **24**, 961–966.

Marchesi, V. T. (1970). Some properties of the major glycoproteins of the human red cell membrane. *J. Cell Biol.* **47**, 129a.

Marcus, D. M., and Cass, L. E. (1969). Glycosphingolipids with Lewis blood group activity uptake by human erythrocytes. *Science* **164**, 553–554.

Marcus, P. I., Salb, J. M., and Schwartz, V. G. (1965). Nuclear surface N-acetyl neuraminic acid terminating receptors for myxovirus attachment. *Nature (London)* **208**, 1122–1124.

Martinez-Palomo, A. (1970). The surface coat of animal cells. *Int. Rev. Cytol.* **29**, 29–75.

Mayhew, E. (1966). Cellular electrophoretic mobility and the mitotic cycle. *J. Gen. Physiol.* **49**, 717–725.

Mayhew, E., and O'Grady, E. A. (1965). Electrophoretic mobilities of tissue culture cells in exponential and parasynchronous growth. *Nature (London)* **207**, 86–87.

Mazlen, R. G., Muellenberg, C. G., and O'Brien, P. J. (1969). The control of α-glutamine-D-fructose-6-phosphate amidotransferase in bovine retina. *Biochim. Biophys. Acta* **117**, 352–354.

Meezan, E., Wu, H. C., Black, P. H., and Robbins, P. W. (1969). Comparative studies on the carbohydrate-containing membrane components of normal and virus-transformed mouse fibroblasts. II. Separation of glycoproteins and glycopeptides by Sephadex chromatography. *Biochemistry* **8**, 2518–2524.

Melchers, F., and Knopf, P. M. (1967). Biosynthesis of the carbohydrate portion of immunoglobulin chains: Possible relation to secretion. *Cold Spring Harbor Symp. Quant. Biol.* **32**, 255–262.

Meyer, K. (1966). Introduction: Mucopolysaccharides. *Fed. Proc., Fed. Amer. Soc. Exp. Biol.* **25**, 1032–1052.

Montgomery, R. W. (1970). Glycoproteins. *In* "The Carbohydrates" (W. Pigman and D. Horton, eds.), 2nd ed., Vol. 2B, pp. 628–709. Academic Press, New York.

Mookerjea, S., and Chow, A. (1970). Stimulation of UDP-*N*-acetyl glucosamine:glycoprotein *N*-acetyl glucosaminyl transferase activity by cytidine 5'-diphosphocholine. *Biochem. Biophys. Res. Commun.* **39,** 486–493.

Mora, P. T., Brady, R. O., Bradley, R. M., and McFarland, V. W. (1969). Gangliosides in DNA virus-transformed and spontaneously transformed tumorigenic mouse cell line. *Proc. Nat. Acad. Sci. U.S.* **63,** 1290–1296.

Morre, D. J., Merlin, L. M., and Keenan, T. W. (1969). Localization of glycosyl transferase activities in a Golgi apparatus-rich fraction isolated from rat liver. *Biochem. Biophys. Res. Commun.* **37,** 813–819.

Morris, J. E., and Moscona, A. A. (1970). Induction of glutamine synthetase in embryonic retina: Its dependence on cell interactions. *Science* **167,** 1736–1738.

Moscona, A. A. (1962). Analysis of cell recombinations in experimental synthesis of tissues *in vitro*. *J. Cell. Comp. Physiol.* **60,** Suppl. 1, 65–80.

Moscona, A. A. (1968a). Cell aggregation: Properties of specific cell-ligands and their role in the formation of multicellular systems. *Develop. Biol.* **18,** 250–277.

Moscona, A. A. (1968b). Aggregation of sponge cells: Cell-linking macromolecules and their role in the formation of multicellular systems. *In Vitro* **3,** 13–21.

Multani, J. S., Cepurneek, C. P., Davis, P. S., and Saltman, P. (1970). Biochemical characterization of gastroferrin. *Biochemistry* **9,** 3970–3976.

Muramatsu, T., and Nathenson, S. G. (1970a). Isolation of a chromatographically unique glycopeptide from murine *histocompatibility-2* (*H-2*) membrane alloantigens labelled with H^3-fucose or H^3-glucosamine. *Biochem. Biophys. Res. Commun.* **38,** 1–8.

Muramatsu, T., and Nathenson, S. G. (1970b). Studies on the carbohydrate portion of membrane-located mouse H-2 alloantigens. *Biochemistry* **9,** 1875–4883.

Nagata, N., and Rasmussen, H. (1970). Parathyroid hormone, 3',5'-AMP, CA⁺⁺, and renal gluconeogenesis. *Proc. Nat. Acad. Sci. U.S.* **65,** 368–374.

Nameroff, M., and Holtzer, H. (1969). Contact mediated reversible suppression of myogenesis. *Develop. Biol.* **19,** 380–396.

Nathenson, S. G. (1970). Biochemical properties of histocompatibility antigens. *Annu. Rev. Genet.* **4,** 69–90.

Nathenson, S. G., Shimada, A., Yamane, K., Muramatsu, T., Cullen, S., Mann, D. L., Fahey, J. L., and Graff, R. (1970). Biochemical properties of papain-solubilized murine and human histocompatibility alloantigens. *Fed. Proc., Fed. Amer. Soc. Exp. Biol.* **29,** 2026–2033.

Neufeld, E., and Hall, C. W. (1965). Inhibition of UDP-D-glucose dehydrogenase by UDP-D-xylose: A possible regulatory mechanism. *Biochem. Biophys. Res. Commun.* **19,** 456–461.

Neutra, M., and Leblond, C. P. (1966). Radioautographic comparison of the uptake of galactose-³H and glucose-³H in the Golgi region of various cells secreting glycoproteins or mucopolysaccharides. *J. Cell Biol.* **30,** 137.

Omura, T., Siekevitz, P., and Palade, G. E. (1967). Turnover of constituents of the endoplasmic reticulum membranes of rat hepatocytes. *J. Biol. Chem.* **242,** 2389–2396.

Oppenheimer, S. B., Edidin, M., Orr, C. W., and Roseman, S. (1969). An L-glutamine requirement for intercellular adhesion. *Proc. Nat. Acad. Sci. U.S.* **63,** 1395–1402.

Ozzello, L., and Bembry, J. Y. (1964). Effects of 17-β-oestradiol on the production of acid mucopolysaccharides by cell cultures of human fibroblasts. *Nature (London)* **203**, 80–81.

Pamer, T., Glass, G. B. J., and Horowitz, M. (1968). Purification and characterization of sulfated glycoproteins and hyaluronidase-resistant mucopolysaccharides from dog gastric mucosa. *Biochemistry* **7**, 3821–3829.

Paul, J., Broadfoot, M. M., and Walker, P. (1966). Increased glycolytic capacity and associated enzyme changes in BHK21 cells transformed with polyoma virus. *Int. J. Cancer* **1**, 207–218.

Perlmann, P., and Nilsson, H. (1970). Inhibition of cytoxicity of lymphocytes by concanavalin A *in vitro*. *Science* **168**, 1112–1115.

Petersen, D. F., Tobey, R. A., and Anderson, E. C. (1969). Synchronously dividing mammalian cells. *Fed. Proc., Fed. Amer. Soc. Exp. Biol.* **28**, 1771–1779.

Phillipson, L. (1963). The early interaction of animal viruses and cells. *Progr. Med. Virol.* **5**, 43–78.

Phillipson, L., Bengtsson, S., Brishammer, S., Svennerholm, L., and Zetterquist, O. (1964). Purification and chemical analysis of the erythrocyte receptor for hemagglutinating enteroviruses. *Virology* **22**, 580–590.

Pinto da Silva, P., Douglas, S. D., and Branton, D. (1970). Location of A₁ antigens on the human erythrocyte membrane. *J. Cell Biol.* **47**, 159a.

Pollack, R. E., and Burger, M. M. (1968). Surface-specific characteristics of a contact-inhibited cell line containing the SV₄₀ viral genome. *Proc. Nat. Acad. Sci. U.S.* **62**, 1074–1076.

Powell, A. E., and Leon, M. A. (1970). Reversible interaction of human lymphocytes with the mitogen concanavalin A. *Exp. Cell Res.* **62**, 315–325.

Priest, R. E., and Davies, L. M. (1969). Cellular proliferation and synthesis of collagen. *Lab. Invest.* **21**, 138–142.

Quastel, M. R., and Kaplan, J. G. (1970). Lymphocyte stimulation: The effect of ouabain on nucleic acid and protein synthesis. *Exp. Cell Res.* **62**, 407–420.

Raisys, V. A., and Winzler, R. J. (1970). Metabolism of exogenous D-mannosamine. *J. Biol. Chem.* **245**, 3203–3208.

Rambourg, A., and Leblond, C. P. (1967). Electron microscope observations on the carbohydrate-rich cell coat present at the surface of cells in the rat. *J. Cell Biol.* **32**, 27–53.

Rambourg, A., Hernandez, W., and Leblond, C. P. (1969). Detection of complex carbohydrates in the Golgi apparatus of rat cells. *J. Cell Biol.* **40**, 395–414.

Redman, C. M., Siekevitz, P., and Palade, G. E. (1966). Synthesis and transfer of amylase in pigeon pancreatic microsomes. *J. Biol. Chem.* **241**, 1150–1158.

Reisfeld, R. A., and Kahan, B. D. (1970a). Biological and chemical characterization of human histocompatibility antigens. *Fed. Proc., Fed. Amer. Soc. Exp. Biol.* **29**, 2034–2040.

Reisfeld, R. A., and Kahan, B. D. (1970b). Transplantation antigens. *Advan. Immunol.* **12**, 117–200.

Reith, A., Oftebro, R., and Seljelid, R. (1969). Incorporation of ³H-glucosamine in HeLa cells as revealed by light and electron microscopic autoradiography. *Exp. Cell Res.* **59**, 167–170.

Richmond, J. E., Glaeser, R. M., and Todd, P. (1968). Protein synthesis and aggregation of embryonic cells. *Exp. Cell Res.* **52**, 43–58.

Robbins, E., and Pederson, T. (1970a). Iron: Its intracellular localization and possible role in cell division. *Proc. Nat. Acad. Sci. U.S.* **66**, 1244–1251.

Robbins, E., and Pederson, T. (1970b). Nucleolar polysaccharide and its possible role in mitosis. *J. Cell Biol.* **47**, 172a.

Robertson, H. T., and Black, P. H. (1969). Changes in surface antigens of SV_{40}-virus transformed cells. *Proc. Soc. Exp. Biol. Med.* **130**, 363–370.

Robinson, E. A., Kalckar, H. M., and Troedsson, H. (1966). Metabolic inhibition of mammalian uridine diphosphate galactose 4-empimerase in cell cultures and in tumor cells. *J. Biol. Chem.* **241**, 2737–2745.

Roizman, B., and Spring, S. B. (1967). Alteration in immunological specificity of cells infected with cytolytic viruses. *In* "Cross-Reacting Antigens and Neoantigens" (J. J. Trentin, ed.), pp. 85–96. Williams & Wilkins, Baltimore, Maryland.

Roseman, S. (1968). Biosynthesis of glycoproteins, gangliosides and related substances. *In* "Biochemistry of Glycoproteins and Related Substances" (E. Rossi and E. Stoll, eds.), pp. 244–269. Karger, Basel.

Roseman, S. (1970). Multiglycosyltransferase systems in intercellular adhesion. *160th Meet., Amer. Chem. Soc.*, Abstract No. 36.

Rosen, O. M., and Rosen, S. M. (1970). A bacterial activator of frog erythrocyte adenyl cyclase. *Arch. Biochem. Biophys.* **141**, 346–352.

Rubin, H. (1967). The behavior of normal and malignant cells in tissue culture. *In* "The Specificity of Cell Surfaces" (B. D. Davis and L. Warren, eds.), pp. 181–194. Prentice-Hall, Englewood Cliffs, New Jersey.

Sanders, F. K., and Smith, J. D. (1970). Effect of collagen and acid polysaccharides on the growth of BHK/21 cells in semi-solid media. *Nature (London)* **227**, 513–515.

Sanford, K. K., Barker, B. E., Parshard, R., Westfall, B. B. Woods, M. W., Jackson, J. L., King, D. R., and Peppers, E. V. (1970). Neoplastic conversion *in vitro* of mouse cells: Cytologic, chromosomal, enzymatic, glycolytic, and growth properties. *J. Nat. Cancer Inst.* **45**, 1071–1096.

Schachter, H., Jabbal, I., Hudgin, R. L., Pinteric, L., McGuire, E. J., and Roseman, S. (1970). Intracellular localization of liver sugar nucleotide glycoprotein glycosyltransferases in a Golgi-rich fraction. *J. Biol. Chem.* **245**, 1090–1100.

Schafer, I. A., Sullivan, J. C., svejcar, J., Kofoed, J., and Robertson, W. V. B. (1968). Study of the Hurler syndrome using cell culture: Definition of the biochemical phenotype and the effect of ascorbic acid on the mutant cell. *J. Clin. Invest.* **47**, 321–328.

Schenkein, I., and Uhr, J. W. (1970a). Immunoglobulin synthesis and secretion. I. Biosynthetic studies of the addition of the carbohydrate moieties. *J. Cell Biol.* **46**, 42–51.

Schenkein, I., and Uhr, J. W. (1970b). Glycosyl transferase for mouse IgG[1]. *J. Immunol.* **105**, 271–273.

Schubert, D. (1970). Immunoglobulin biosynthesis. IV. Carbohydrate attachment to immunoglobulin subunits. *J. Mol. Biol.* **51**, 287–301.

Schubert, D., and Jacob, F. (1970). 5-Bromodeoxyuridine-induced differentiation of a neuroblastoma. *Proc. Nat. Acad. Sci. U.S.* **67**, 247–254.

Seeds, N. W., Gilman, A. G., Amano, T., and Nirenberg, M. W. (1970). Regulation of axon formation by clonal lines of a neural tumor. *Proc. Nat. Acad. Sci. U.S.* **66**, 160–167.

Sela, B., Lis, H., Sharon, N., and Sachs, L. (1970). Different locations of carbohydrate-containing sites in the surface membrane of normal and transformed mammalian cells. *J. Membrane Biol.* **3**, 267–279.

Shen, L., and Ginsburg, V. (1968). Release of sugars from HeLa cells by trypsin.

In "Biological Properties of the Mammalian Surface Membrane" (L. A. Manson, ed.), pp. 67–71. Wistar Inst. Press, Philadelphia, Pennsylvania.

Shen, L., Grollman, E. F., and Ginsburg, V. (1968). An enzymatic basis for secretor status and blood group substance specificity in humans. *Proc. Nat. Acad. Sci. U.S.* **59**, 224–230.

Sherr, C. J., and Uhr, J. W. (1969). Immunoglobulin synthesis and secretion. III. Incorporation of glucosamine into immunoglobulin on polyribosomes. *Proc. Nat. Acad. Sci. U.S.* **64**, 381–387.

Sherr, C. J., and Uhr, J. W. (1970). Immunoglobulin synthesis and secretion. V. Incorporation of leucine and glucosamine into immunoglobulin on free and bound polyribosomes. *Proc. Nat. Acad. Sci. U.S.* **66**, 1183–1189.

Siddiqui, B., Hakomori, S., Vogt, R. K., and Saito, T. (1970). Deviation of glycolipid pattern in minimal deviation tumors and virus transformed cells. *Fed. Proc., Fed. Amer. Soc. Exp. Biol.* **29**, 928.

Silagi, S., and Bruce, S. A. (1970). Suppression of malignancy and differentiation in melanotic melanoma cells. *Proc. Nat. Acad. Sci. U.S.* **66**, 72–78.

Skea, B. R., and Nemeth, A. M. (1969). Factors influencing premature induction of UDP-glucuronyltransferase activity in cultured chick embryo liver cells. *Proc. Nat. Acad. Sci. U.S.* **64**, 795–802.

Sneath, J. S., and Sneath, P. H. A. (1959). Adsorption of blood-group substances from serum on to red cells. *Brit. Med. Bull.* **15**, 154–164.

Spencer, R. A., Hanschka, T. S., Amos, D. B., and Ephrussi, B. (1964). Co-dominance of isoantigens in somatic hybrids of murine cells grown *in vitro. J. Nat. Cancer Inst.* **33**, 893–903.

Spiro, M. J., and Spiro, R. G. (1968a). Glycoprotein biosynthesis: Studies on thyroglobulin. Thyroid sialyltransferase. *J. Biol. Chem.* **243**, 6520–6528.

Spiro, M. J., and Spiro, R. G. (1968b). Glycoprotein biosynthesis: Studies on thyroglobulin. Thyroid galactosyltransferase. *J. Biol. Chem.* **243**, 6529–6537.

Spiro, R. G. (1970). Glycoproteins. *Annu. Rev. Biochem.* **39**, 599–638.

Spiro, R. G., and Spiro, M. J. (1966). Glycoprotein biosynthesis: Studies on thyroglobulin. *J. Biol. Chem.* **241**, 1271–1282.

Spiro, R. G., and Spiro, M. J. (1968a). Enzymatic synthesis of the hydroxylysine-linked disaccharide of basement membranes and collagens. *Fed. Proc., Fed. Amer. Soc. Exp. Biol.* **27**, 345.

Springer, G. F. (1967). Human MN glycoproteins: Dependence of blood-group and anti-influenza virus activities on their molecular size. *Biochem. Biophys. Res. Commun.* **28**, 510–513.

Steinberg, M. S. (1970). Does differential adhesion govern self-assembly process in histogenesis? Equilibrium configurations and the emergence of a hierarchy among populations of embryonic cells. *J. Exp. Zool.* **173**, 395–434.

Stoker, M. G. P., and Rubin, H. (1967). Density dependent inhibition of cell growth in culture. *Nature (London)* **215**, 171–172.

Stoolmiller, A. C., and Dorfman, A. (1969). The metabolism of glycosaminoglycans. *Compr. Biochem.* **17**, 241–275.

Strauss, J. H., Jr., Burge, B. W., and Darnell, J. E. (1970). Carbohydrate content of the membrane protein of Sindbis virus. *J. Mol. Biol.* **47**, 437–448.

Subak-Sharpe, H., Bürk, R. R., and Pitts, J. D. (1969). Metabolic cooperation between biochemically marked mammalian cells in tissue culture. *J. Cell. Sci.* **4**, 353–367.

Svennerholm, L. (1970). Ganglioside metabolism. *Comp. Biochem.* **18**, 201–227.

Swann, D. A. (1968). Studies on hyaluronic acid. II. The protein component(s) of rooster comb hyaluronic acid. *Biochim. Biophys. Acta* **160**, 96–105.

Swenson, R. M., and Kern, M. (1967). Synthesis and secretion of γ globulin by lymph-node cells. II. The intracellular segregation of amino acid labeled and carbohydrate labeled γ globulin. *J. Biol. Chem.* **242**, 3242.

Takahashi, M., Yagi, Y., Moore, G. E., and Pressman, D. (1969). Immunoglobulin production in synchronized cultures of human hematopoietic cell lines. I. Variation of cellular immunoglobulin level with the generation cycle. *J. Immunol.* **103**, 834–843.

Temin, H. M. (1970). Malignant transformation of cells by viruses. *Perspect. Biol. Med.* **14**, 11–26.

Tetas, M., Chao, H., and Molnar, J. (1970). Incorporation of carbohydrates into endogenous acceptors of liver microsomal fractions. *Arch. Biochem. Biophys.* **138**, 135–146.

Tevethia, S. S., Katz, M., and Rapp, F. (1965). New surface antigen in cells transformed by simian papova virus SV$_{40}$. *Proc. Soc. Exp. Biol. Med.* **119**, 896–901.

Thomas, D. B., and Winzler, R. J. (1969). Structural studies on human erythrocyte glycoproteins. Alkalic labile oligosaccharides. *J. Biol. Chem.* **244**, 5943–5946.

Tillack, T. W., Scott, R. E., and Marchesi, V. T. (1970). Studies on the chemistry and function of the intramembranous particles observed by freeze-etching of red cell membranes. *J. Cell Biol.* **47**, 213a.

Tulsiani, D. R. P., and Carubelli, R. (1970). Studies on the soluble and lysosomal neuraminidases of rat liver. *J. Biol. Chem.* **245**, 1821–1827.

Uhr, J. W., and Schenkein, I. (1970). Immunoglobulin synthesis and secretion. IV. Sites of incorporation of sugars as determined by subcellular fractionation. *Proc. Nat. Acad. Sci. U.S.* **66**, 952–958.

Vail, J. M. (1968). Relation of energy metabolism to growth and differentiation in L-cell suspension cultures. *Diss. Abstr.*, 2687-B.

Wallach, D. F. H., and de Perez Esandi, M. V. (1964). Sialic acid and the electrophoretic mobility of three tumor cell types. *Biochim. Biophys. Acta* **83**, 363–366.

Wallach, D. F. H., and Kamat, V. B. (1966). The contribution of sialic acid to the surface charge of fragments of plasma membrane and endoplasmic reticulum. *J. Cell Biol.* **30**, 660–663.

Warburg, O. (1956). On the origin of cancer cells. *Science* **123**, 309–314; **124**, 269–270.

Ward, P. D., and Ambrose, E. J. (1969). Electrophoretic and chemical characterization of the charged groups at the surface of murine CL3 ascites leukaemia cells. *J. Cell Sci.* **4**, 289–298.

Warren, L. (1966). The biosynthesis and metabolism of amino sugars and amino sugar-containing compounds. *In* "Glycoproteins" (A. Gottschalk, ed.), 1st ed., Chapter 12, pp. 570–593. Elsevier, Amsterdam.

Warren, L. (1969). The biological significance of turnover of the surface membrane of animal cells. *Curr. Top. Develop. Biol.* **4**, 197–222.

Warren, L., and Glick, M. C. (1968). Membranes of animal cells. II. The metabolism and turnover of the surface membrane. *J. Cell Biol.* **37**, 729–745.

Watkins, W. M. (1966). Blood-group substances. *Science* **152**, 172–181.

Weinstein, D. B., Marsh, J. B., Glick, N. C., and Warren, L. (1970). Membranes of animal cells. VI. The glycolipids of the L cell and its surface membrane. *J. Biol. Chem.* **245**, 3928–3937.

Weiss, L. (1966). Effect of temperature on cellular electrophoretic mobility phenomena. *J. Nat. Cancer Inst.* **36**, 837–847.

Weiss, L. (1967). "The Cell Periphery, Metastasis, and Other Contact Phenomena." North-Holland Publ., Amsterdam.

Whur, P., Herscovics, A., and Leblond, C. P. (1969). Radioautographic visualization of the incorporation of galactose-^{3}H and mannose-^{3}H by rat thyroids *in vitro* in relation to the stages of thyroglobulin synthesis. *J. Cell Biol.* **43**, 289–311.

Widnell, C. C., and Siekevitz, P. (1967). The turnover of the constituents of various rat liver membranes. *J. Cell Biol.* **35**, 142A.

Winzler, R. J. (1970). Carbohydrates in cell surfaces. *Int. Rev. Cytol.* **29**, 77–125.

Winzler, R. J. (1972). *In* "Glycoproteins—Their Structure and Function" (A. Gottschalk, ed.), 2nd ed. Elsevier, Amsterdam (in press).

Winzler, R. J., Harris, E. D., Peaks, D. J., Johnson, C. A., and Weber, P. (1967). Studies on glycopeptides released by trypsin from intact human erythrocytes. *Biochemistry* **6**, 2195–2202.

Woodward, C., and Davidson, E. A. (1968). Structure-function relationships of protein polysaccharide complexes: Specific ion-binding properties. *Proc. Nat. Acad. Sci. U.S.* **60**, 201–205.

Wu, H. C., Meezan, E., Black, P. H., and Robbins, P. W. (1969). Comparative studies on the carbohydrate-containing membrane components of normal and virus-transformed mouse fibroblasts. I. Glucosamine-labeling patterns in 3T3, spontaneously transformed 3T3, and SV$_{40}$-transformed 3T3 cells. *Biochemistry* **8**, 2509–2517.

Yogeeswaran, G., Wherrett, J. R., Chatterjee, S., and Murray, R. K. (1970). Gangliosides of cultured mouse cells. Partial characterization and demonstration of C^{14}-glucosamine incorporation. *J. Biol. Chem.* **245**, 6718–6725.

Yuen, R., Nwokoro, N., and Schachter, H. (1970). Pathway of L-fucose metabolism in pork liver. *Fed. Proc., Fed. Amer. Soc. Exp. Biol.* **29**, 675.

Zagury, D., Uhr, J. W., Jamieson, J. D., and Palade, G. E. (1970). Immunoglobulin synthesis and secretion. II. Radioautographic studies of sites of addition of carbohydrate moieties and intracellular transport. *J. Cell Biol.* **46**, 52–63.

Zajac, I., and Crowell, R. L. (1969). Differential inhibition of attachment and eclipse activities of HeLa cells for enteroviruses. *J. Virol.* **3**, 422–428.

Zatz, M., and Barondes, S. H. (1969). Incorporation of mannose into mouse brain lipid. *Biochem. Biophys. Res. Commun.* **36**, 511–517.

AUTHOR INDEX

Numbers in italics refer to the pages on which the complete references are listed.

SUBJECT INDEX

A